AF786647

Mikael Nilsson

Tools of Hegemony

Military Technology and Swedish-American Security Relations 1945–1962

Santérus
Academic Press
Sweden

www.santerus.se

All rights reserved. No part of this publication may be reproduced, stored in a retrieval system, or transmitted, in any form or by any means, electronic, mechanical, photocopying, recording, or otherwise, without the prior written permission of the publisher, except in the case of brief quotations embodied in critical articles and reviews.

© 2006 Mikael Nilsson and Santérus Academic Press Sweden
ISBN: 978-91-7335-007-5
Layout: Mikael Nilsson
Cover art: *Complex Presentiment: Half-Figure in a Yellow Shirt* by Kazimir Malevich
Cover profile: Sven Bylander
Santérus Academic Press is an imprint of
Santérus Förlag, Stockholm, Sweden
academicpress@santerus.se
Printed by BOD, Germany 2012

To my parents whom I love very much

Contents

CHAPTER 3
A Challenge to U.S. Hegemony:
U.S. Technology, Guided Missile Development, and the SDU, 1945–1948

CHAPTER 4
The Consent to U.S. Hegemony:
Sweden and the ERP, COCOM, and the MDAA Agreement, 1948–1952

CHAPTER 5
The Turning Tide:
How Sweden gained Access to Guided Missiles, 1953–1958

Acknowledgments

...and so a long and arduous time has finally come to an end. For more than four years this thesis has been an integral part of my life. Even when I have not been writing or researching I have had it in the back of my mind, wondering what this day would feel like. So how does it feel? Well, everyone who has ever finished their Ph.D. will testify that it is extremely satisfying. My friends and family will no doubt also rejoice now that this day has arrived. No more will they have to listen to me going on and on about guided missiles and Swedish-American security relations during the Cold War. Although, unfortunately for them, I intend to continue researching this subject in the future. On the other hand, I will miss this thesis and the tug-of-war that I have had with it for so long. The finished book leaves behind it a vacuum which will have to be filled by something else. I will also miss being a doctoral student. There is a special feeling of brotherhood and sisterhood between doctoral students which for me is now no more. It is a strange feeling to be one of the 'senior' researchers, to cross over to the other side, so to speak. New students will take our place and look to us for guidance and advice... I wish them the best of luck!

The thesis before you is the result not only of my effort, but also of the efforts of many other people. There is no way that I could possibly mention them all by name so I apologize to anyone who might feel left out. Having said this, I will now mention the people and institutions that have provided help, support, and advice during the process of writing this thesis. The research for this book was paid for by the Bank of Sweden Tercentenary Foundation and I am thankful for its support. My studies have been undertaken at the Department of the History and Philosophy of Science and Technology at the Royal Institute of Technology in Stockholm. There are a number of people there who need to be acknowledged.

Among them is my advisor Hans Weinberger with whom I have had many discussions of draft versions of this book. I know I tired you with my stubborn insistence upon certain matters, and I thank you Hans for believing in me all this time. My co-advisor Niklas Stenlås has supported me and given me invaluable advice all the way. You saw potential in me already at the undergraduate level, and you were the one who introduced me to the Department and the NNET project. Without you Niklas, I would not be where I am today. You are a rock and I thank you for everything. I also want to thank my second co-advisor Nina Wormbs. You were not with me for as long as Hans and Niklas but your knowledge and integrity as a researcher have inspired me greatly. I am also especially grateful to Ola Tunander and Robert Dalsjö for having read, criticized, and commented on draft versions of the thesis. Robert, your views were of immense importance when I was finishing this work. Thomas Jonter definitely deserves all my gratitude. You agreed, at short notice, to comment on and criticize my text, and I enjoyed and benefited greatly from our discussions! Also, Bernard Vowles, thank you so very much for checking my English and making all the necessary corrections. Thank you also to Karl Molin for the valuable assistance at the end.

My thanks also go out to my fellow doctoral students in the NNET project Sara Collmar and Johan Gribbe. Sara, you have supported me when the straits looked dire and I have enjoyed our conversations a lot. Johan, I thank you for sharing some of your archival findings with me. You are a workhorse and I know that your thesis will turn out great. Stefan Lindström, who used to be a part of the NNET project has earned my gratitude as well. You never fail to astound me with your knowledge of Swedish political history. Your wife Lou is also in my good books for being so cheerful every time I see her.

I am much indebted to Professor Arne Kaijser at the Department. Your insightful comments on the draft versions of this book have meant a lot to me, and your vast knowledge in the field of the history of technology is a constant inspiration! Professor Thomas Kaiserfeld, you gave me extremely valuable and pertinent feedback on an early version of the text, and I will always owe you thanks for your advice! I also want to thank Isa Dussauge, Martin Emanuel, David Nilsson, and Julia Lindquist, who all, at one time or another shared an office with me, and who put up with my frenetically tapping my fingers and feet to imaginary music. Isa, you and Janne hold a special place in my heart and I have appreciated our talks so very much.

Other people at the Department (and elsewhere) whom I want to thank are: Dag Avango, Anna Storm, Professor Maths Isacson, Maja Fjaestad, Professor Sverker Sörlin, Per Lundin, Brita Lundström, May-Britt Öhman, Anders Houltz, Per Högselius, Pär Blomkvist, and Professor Marie Nisser. Dag, for your coolness, insight, and for our moments of subversive discussion at the coffee table. Let's continue the fight for a better tomorrow! Anna, your sense of humor, warmth, and shrewd comments at the seminars have been a blessing. I know you will go far. Maths, for always sharing your great knowledge of economic and socio-political history with us all. Maja, for being smart, studious, and funny. Give your husband my best because he is the nicest guy. I hope you will be very happy together. Sverker… you are one of a kind for sure. Highly intelligent, inexplicably productive, an active outdoors man, and extremely humorous! I feel truly blessed to be your colleague. Per L., for being an extremely capable and intellectual person. Your clever sense of humor saves lives, for real! Brita, for always being a very cool girl, and a very able scholar. May-Britt, for always being positive, and for having the guts to go your own way! Anders, for always offering perceptive remarks and observations. Per H., for being the intelligent and cool person that you are. Pär, for being there for me when I needed you the most. I would not be where I am today if it were not for you! Marie, for never giving up, and for your incredible knowledge! You set an example and inspire others with your energy!

Lars Ulfving, Petter Wulff, and Nils Bruzelius must have special mention here. Lars, you never cease to amaze me with your knowledge of military matters, big and small. Petter, for being an intelligent man and scholar of great integrity. I always enjoy talking to your kind and wonderful wife Maria Elena. Nils, for the dignity and respect with which you have always treated me, many years your junior and a civilian, in our discussions at seminars and dinner tables. Even though we do not always share the same view of things I have the greatest respect for you. I have always much enjoyed the dinner parties at your house, and the company of your lovely wife Birgitta and daughter Josefine.

I would like to thank Kent Zetterberg and Simon Moores for their diligence and scholarly aptitude. Kent, because you combine the mind of a great historian with warmth, hospitality, and a wonderful sense of humor. Simon, for your admirable skills as a historian and for always being nice to me. I wish to thank Professors Roy MacLeod of the University of Sydney, Thomas P. Hughes of the MIT, and Philip Scranton of the State University of New Jersey Rutgers. Roy, for sharing your endless wisdom with me, and for bearing with my less than adequate administrative skills. Philip, for the enlightening conversations regarding history of technology over dinner and in private. Thomas, for inspiration and for taking the time to be interviewed by me. Furthermore, I want to thank Thomas Zeller and Robert Friedel of the University of Maryland. Thomas, for taking such good care of me during the four months I spent in Washington, D.C. in the winter of 2005/2006. I enjoyed the Thanksgiving dinner with you and your family so very much! Robert, for being a wonderful person and for sharing your knowledge with me.

Other groups of people to which I owe this thesis are the archivists and librarians. The archivists and staff at the Military Archives and the

National Archives in Stockholm; the librarians at the Library of the Swedish Parliament, the University of Stockholm Library, the National Library of Sweden, the library at the Royal Institute of Technology. I want to give special thanks to the librarians at the Anna Lindh Library, and Chief Librarian Eva Hesselgren Mortensen, for always being of great help and for never being flustered by silly questions. I want to send a huge hug and thank you to all the archivists and staff of the National Archives and Records Administration in College Park, Maryland. Without you guys this thesis could not have been written! I especially want to thank Richard L. Boylan, Kenneth Schlessinger, and John Taylor. A big thanks also to Charles the security guard for the warm welcome and our many conversations. Moreover, thank you Jessica (sugar momma!), Kaeten, Nitsan (thanks for everything!), Jeremy, Eric and Becky, Mary Lee and Kona (big furry pants!), Scott (keep on rockin' in the free world!), Ben and Megan (and now baby Condoleezza), Bella, Adrienne, Sheldon, and everyone else at (or connected to) Mother Irving. Also, thank you all you friendly clerks at Safeway on Adams Morgan. You are all working-class heroes! I sincerely hope that I have not forgotten anyone.

I want to thank my dear sister Monica, and her lovely kids Thomas and Magnus. You all give me great joy when I visit Norrköping and I love you all very much! I also want to give a very special thanks to you Mojca. Without you I would probably not have finished this thesis on time, and even though we have parted ways you will always reside in a special place in my heart. Thank you so much for your love and support! I also want to thank my friends Jenny, Niklas, Felix and Julian. Your understanding and support has been greatly appreciated.

The reason why I have saved you, my parents, until last is not because you have contributed the least to this thesis. On the contrary, without you I would not be here today. This is true not only in the very literal sense that you gave birth to me, but also because if it were not for the love, support (emotional as well as financial), understanding, and patience that you have given me during my studies (and all other years) I would not be writing these words today. As the first academic in my family, I want to say thank you from the bottom of my heart, and I want you to know that I will always be proud of you and my working-class background!

Stockholm in August 2007

Mikael Nilsson

Abbreviations

AIM	*Air Intercept Missile:* American designation of air-to-air missiles.
ARGMA	U.S. *Army Rocket and Guided Missile Agency.*
BNA	*Office of British Commonwealth and Northern European Affairs:* Office within the American State Department which handled matters relating to Sweden after 1949.
CD	*Central Directorate for Reaction and Rocket Propulsion:* The name of the agency charged with the responsibility of overseeing and coordinating Swedish guided missile development, from 1945 to April 1948.
CEEC	*Committee for European Economic Cooperation.*
CIA	*Central Intelligence Agency:* U.S. foreign intelligence agency, established through the National Security Act of 1947.
CINCUSAFE	*Commander-in-Chief United States Air Force Europe.*
CNO	*Chief of Naval Operations:* The Chief of the U.S. Navy.
COCOM	*The Coordinating Committee:* Committee formally established on January 1, 1950 to coordinate the Western trade embargo against the Soviet Union and its satellites.
COMECON	*Council for Mutual Economic Assistance:* Communist organization for international economic cooperation formed in 1949 as a response to the Marshall Plan.
COMINFORM	Communist Information Bureau: Organization created in September 1947 as a response to the Marshall Plan, intended to coordinate the

activities of Communist parties in Eastern Europe, as well as France and Italy, under Soviet direction.

COMINT	*Commiunications Intelligence:* A sub-category of Signals Intelligence (SIGINT) that intercepts messages or voice information from foreign communication.
COS	*Chiefs of Staff:* Britain's supreme military command.
CW	*Continuous Wave:* Radar that transmits a continuous signal towards the target.
DCI	*Director of Central Intelligence:* The title of the Chief of the CIA.
DNS	*Danmark, Norge, Sverige:* Secret plans for military cooperation between the Scandinavian countries in wartime, which had their roots in the aftermath of the sdu negotiations in 1948/49.
DOD	*Department of Defense* (U.S.).
ECA	*Economic Cooperation Administration:* American agency in charge of administrating the ERP in Western Europe.
ECE	*Economic Commission for Europe:* The United Nations' organization for European economic development and cooperation.
EHP	*European Hawk Program:* NATO's Program for manufacturing the Hawk missile in Europe.
EPU	*European Payments Union:* Machinery within the ERP for handling payments deficits and surpluses between the Western European countries.
ERP	*European Recovery Program:* The official name of the Marshall Plan, 1948-1951.
EUCOM	*European Command:* U.S. military headquarters in Stuttgart, Germany.
EUR	*Office of European Affairs:* Office within the State Department handling matters related to Europe.
FB	*Försvarsbeslut:* Defense Decisions decided on by the parliament which stipulates the development of the Swedish Armed Forces for the following years.
FFI	*Forsvarets forskningsinstitutt:* Norway's defense research agency.

FMV	*Försvarets materielverk:* Agency created in 1968 that co-ordinates the supply of military equipment to the Swedish Defense Forces.
FOA	*Försvarets forskningsanstalt:* Sweden's defense research agency, established in 1945.
GAR	*Guided Air Rocket.*
GMB	*Armed Forces' Guided Missile Bureau:* In connection with an administrative re-organization in 1948 the the CD was abolished and replaced by the GMB.
GMC	*Armed Forces' Guided Missile Council:* The CDRRP was re-organized in 1948, and re-named GMC. Its task was still to facilitate cooperation between the three services of the Swedish Armed Forces in the guided missile field. The organization was officially disbanded in 1956.
GMIORG	*Guided Missile Interdepartmental Operational Requirements Group.*
GNP	*Gross National Product.*
HAC	*Hughes Aircraft Company:* American aircraft and guided missile manufacturer. Produced the Falcon missiles.
HM	*Hughes Model:* HAC's own designation for its missiles.
HPO	*Hawk Production Organization:* The body that co-ordinated and supervised the production of Hawk missiles in European NATO countries.
IB	The informal name of a covert branch of the Swedish military intelligence, which registered and spied on 'suspected' leftists etc. in collution with the Social Democratic Party from 1965 to about 1975.
IFF	*Identification, Friend or Foe:* A system for electronic identification of military aircraft.
IR	*Infrared.*
ISA	*Assistant Secretary of Defense, International Security Affairs:* The principal assistant and advisor to the U.S. Secretary of Defense for formulating international security and political-military policy.
JCS	*Joint Chiefs of Staff:* Committee consisting of the highest military chiefs of all U.S. defense services.

JIC	*Joint Intelligence Committee:* Organization with representatives from all the U.S. armed services evaluating collected intelligence.
JSSC	*Joint Strategic Survey Committee:* The Committee provided the JCS with advice on long-range planning, and acted as a counterbalance to the advice coming from the individual services.
MAAG	*Military Assistance Advisory Group:* Local agency responsible for the distribution of U.S. military aid, and the training of personnel in the recipient countries.
MAP	*Military Assistance Program:* Program for distributing grant military aid to U.S. allies around the world.
MAPAG	*Military Assitance Program Advisory Group:* Regional agency dealing with issues relating to the Military Assistance Program.
MC	*Military Committee:* Committee within the NATO organization.
MDAA	*Mutual Defense Assistance Act of 1949:* Bill that paved the way for the Military Assistance Program (MAP).
MDAP	*Mutual Defense Assistance Program:* Program for the development of common defense efforts between the U.S. and its allies, established subsequent to the MDAA.
MK	*Mark:* Term used to designate different versions of, for instance, a certain tank or missile; e.g. Bloodhound MK I, MK II etc.
MOD	*Ministry of Defense* (U.K.).
MOS	*Ministry of Supply:* Department of the British government from 1939 to 1959 in charge of co-ordinate the supplying of equipment to the British Armed Forces.
MOU	*Memorandum of Understanding.*
MSP	*Mutual Security Program:* U.S. economic and military assistance program established through the Mutual Security Act of 1951.
MWDP	*Mutual Weapons Development Program:* Program designed to facilitate common weapon development between the U.S. and its allies.
NAC	*North Atlantic Council:* This body has the effective political authority within NATO and

consists of permanent civilian representatives from all member countries.

NADGE — NATO *Air Defense Ground Environment:* NATO's early warning radar system.

NAMC — *North Atlantic Military Committee:* Consists of the military representatives to NATO; subordinate to the NAC.

NAMSA — NATO *Maintenance Supply Agency:* Organization handling the maintenance of NATO's common military infrastructure.

NASA — *National Aeronautics and Space Administration.*

NAT — *North Atlantic Treaty:* Treaty on collective defense signed by Belgium, Denmark, France, Iceland, Italy, Luxemburg, The Netherlands, Norway, Portugal, Britain, the United States, and Canada on April 4, 1949. Turkey and Greece joined in 1952, followed by the Germany Federal Republic in 1955.

NATO — *North Atlantic Treaty Organization:* The formal defense organization of the NAT countries, established in 1951.

NOE — *Division of Northern European Affairs:* Division within the American State Department wich handled matters relating to Sweden (i.e. in the Northern European office) until 1949.

NOFORN — *Not Releasable to Foreign Nationals:* U.S. information security classification.

NOHAB — *Nydqvist & Holm aktiebolag:* Company whose aircraft engine department was sold to Volvo and Bofors in 1941, and subsequently became Volvo Flygmotor AB.

NOTS — *Naval Ordnance Test Station:* Research and devlopment facility at China Lake, California. Sidewinder was developed and tested there.

NSC — *The National Security Council:* Council intended to advice the U.S. President on issues regarding national security, established through the National Security Act of 1947.

OEEC — *Organization for European Economic Cooperation:* Agency established in 1948 in connection to the ERP, intended to further closer economic cooperation between the

	Western European countries after the end of the ERP.
OMA	*Office of Military Assistance:* Central agency in charge of U.S. military assistance.
OSP	*Off-Shore Production Program:* Program under which production of military equipment for European and U.S. forces in Europe would be placed in Western European countries instead of in the United States.
PCK	*Provskjutningscentralen i Karlsborg:* Test firing area for guided missiles by Lake Vättern in south-central Sweden.
PPS	*Policy Planning Staff:* Division of the U.S. State Department formed in 1947 dealing with policy formulation; initially headed by George Kennan 1947-1950.
PRO	*Public Record Office:* Britain's national archive.
RA	*Office of European Regional Affairs:* Office within the American State Department in charge of matters relating to Europe. The RA took over this task from the NOE after 1949, and was the home of the BNA.
RAF	*Royal Air Force* (U.K.).
RB	*Robot:* Swedish abbreviation for guided missile.
R&D	*Research and Development:*
RFN	*Raketförsöksplats Norrland:* The Swedish Guided Missile Bureau's missile test firing range at Vidsel in northern Sweden.
RSA	*Royal Swedish Army.*
RSAB	*The Royal Swedish Army Board:* The administrative arm of the Swedish Army.
RSAF	*Royal Swedish Air Force.*
RASFB	*Royal Swedish Air Force Board:* The administrative arm of the Swedish Air Force.
RSN	*Royal Swedish Navy.*
RSNB	*Royal Swedish Navy Board:* The administrative arm of the Swedish Navy.
SAAB	*Svenska aeroplanaktiebolaget:* Swedish automobile, aircraft, and guided missile producer.
SAC	*Strategic Air Command:* USAF's strategic nuclear bomber straike force.
SACEUR	*Supreme Allied Commander Europe:* NATO's Supreme Commander; always an American.

SAPO	*Socialdemokratiska arbetsplatsombud:* Social Democratic Worker Representatives who sought to contain and control communist influence among the Swedish workforce.
S-DMICC	*State-Defense Military Information Control Committee:* Committee consisting of representatives of the U.S. State and Defense Departments, the Armed Services, and the CIA. Evaluates security procedures in countries receiving U.S. military technology.
SDU	*Scandinavian Defense Union:* Initiative to establish a formal defense collaboration between the Scandinavian countries 1948-1949.
SEB	*Stockholms Enskilda Bank:* The Wallenberg family's bank.
SEK	*Svensk enkrona:* Sweden's currency.
SETEL	Société Européenne de Téléguidage: French registered consortium consisting of the European contractors that manufactured Hawk in Europe.
SHAPE	*Supreme Headquarters Allied Powers Europe:* NATO's military headquarters in Europe.
SKF	*Svenska kullagerfabriken:* Multi-national Swedish ball bearing manufacturer based in Gothenburg on the Swedish west coast.
SNOWCAT	*Support of Nuclear Operations With Conventional Attacks:* SACEUR planning for from mid-1950s to mid-1970s wartime operations under which Norwegian fighter bombers would overfly Sweden to destroy Soviet radar defense, and saturate the Soviet command and control capacity so that the NATO Atomic Strike Force could get through to their targets.
SRA	*Svenska radioaktiebolaget.*
STANAG	*Standadization Agreement:* Standardization specifications for military equipment in the NATO countries.
SVENORDA	*Sverige, Norge, Danmark:* Designation for the cooperation between the Scandinavian countries in the area of aerial safety (search and rescue etc.), which had its roots in the SDU negotiations in 1948/49, but which was formalized in 1960.
SÖ	*Sveriges överenskommelser med främmande*

makter: Governmental series in which agreements with foreign powers are published.

UN	*United Nations.*
UP	*United Press.*
USAF	*United States Air Force.*
USAFE	*United States Air Force Europe:* The European branch of the American Air Force based in Wiesbaden, Germany.
USMC	*United States Marine Corps:* i.e. U.S. Marines.
USRO	*The United States Mission to the North Atlantic Treaty Organization and European Regional Organizations:* Based in Paris.
USSR	*Union of Soviet Socialist Republics.*
V1	*Vergeltungs-waffe 1:* German cruise missile developed during World War II. Provided the inspiration for the early Swedish guided missile development.
V2	*Vergeltungs-waffe 2:* German ballistic missile developed during World War II.
ÖB	*Överbefälhavaren:* The Commander-in-Chief of the Swedish Armed Forces.

Footnote Abbreviations

AAB	Air Attaché Branch.
AGMP	Records Relating to the Army Guided Missiles Program.
AIDF	Army-Intelligence Document File.
AIPDF	Army-Intelligence Project Decimal File.
AOR	Aircraft Operations Reports.
AR	Admiral Radford.
ARGMA	The U.S. Army Rocket & Guided Missile Agency.
BNA	Records of the Office of British Commonwealth and Northern European Affairs.
CDF	Central Decimal File.
CGR	Classified General Records.
ChF	Chairman's File.
CLR	Centrala ledningen för reaktions- och raket-drift.
COD	Collection Operations Division.
ConF	Confidential File.
CorrF	Correspondence Files.
CounF	Country Files.
DACRU	Office of the Director of Administration Communications and Records Unit.
DCSO	Deputy Chief of Staff, Operations.
DDCD	Dep Dir for Collection & Dissemination.
DF	Decimal File.
DI	Director of Intelligence.
EOHB	Executive Office Historical Branch.
EUR	Bureua of European Affairs.
EURONOE	Bureau of European Affairs Office of Northern European Affairs.
EUR/SCAN	Bureau of European Affairs Country Director for Scandinavia, Iceland and Finland.
ExSec	Executive Secretariat.

FKC	Försökscentralen.
FRB	Försvarets robotvapenbyrå.
FRR	Försvarets robotvapenråd.
FS/KH	Flygstaben kvalificerat hemlig.
Hgkv	Högkvarteret.
INR	Subject Files of the Bureau of Intelligence and Research.
ISA	International Security Affairs.
KATF	Kungliga arméstygförvaltningen.
KFF	Kungliga flygförvaltningen.
MAP/MSProF	Military Assistance Policy/Mutual Security Program Files.
MOFASSEA	Miscellaneous Office Files of the Assistant. Secretaries of State for European Affairs.
NARA	National Archives and Records Administration.
NATO	North Atlantic Treaty Organization.
NPCA	The Neutrality Policy Commission Archive.
OAPOMA	Office of Atlantic Political and Military Affairs.
OMA	Office of Military Assistance.
PA	Planeringsavdelningen.
PDF	Project Decimal File.
PG	Policy Guidelines.
PMNF	Political-Military Numeric Files.
RA	Records of the Office of European Regional Affairs.
RAJOCa	Records of Ambassador John M. Cabot.
Ravd	Robotavdelningen.
Ravdvid	Robotavdelningens försöksplats i Vidsel.
RBK	Robotkontoret.
RBN	Robotbyrån.
RecDir	Records of the Director.
ReEUR	Records of the Bureau of European Affairs.
RG	Record Group.
RRE	Records Relating to Sweden.
RRSWA	Records Relating to Swedish Affairs.
SF	Subject Files.
SFREDA	Subject Files Relating to European Defense Agreements.
SLE	Stockholm Legation and Embassy.
SRE	Special Representative in Europe.
StaGRes	Standing Group Records Section.
StF	Station File.
SUMCon	The Secretary's and Undersecretary's Memoranda of Conversation.
SWSE	Sweden; Stockholm Embassy.
SÖ	Sveriges överenskommelser med främmande makter.
TSGR	Top Secret General Records.
Vavd	Vapenavdelningen.

Chapter 1

Introduction:

Research Questions and Theoretical Definitions

Introduction

The Royal Swedish Air Force (RSAF) found itself in an odd position in the mid-1950s. On the one hand, it had built up considerable strength (in terms of number of aircraft it was the fourth largest Air Force in the world),[1] and was the hub in the defense against an attack upon Sweden. On the other hand, it had considerable weaknesses; of which perhaps the most important one was the great shortage of pilots,[2] and the complete lack of guided missiles. This dissertation will investigate the latter issue, or rather Sweden's efforts to develop guided missiles domestically, as well as purchase such weapons from the United States.

The subject of guided missiles has gone largely unnoticed by earlier research into Swedish Cold War history, with one notable exception.[3] The lack of interest in the issue seems strange considering the importance of the technology for Sweden's defense capability. It is of course

1 SOU 1994:11 *Had There Been a War… Preparations for the reception of military assistance 1949–1969, Translation of* SOU 1994:11 (Stockholm: Fritzes, 1994) [henceforth: *Had There Been a War…*], p. 253; Cole, Paul Marion, *Neutralité du jour: The conduct of Swedish Security policy since 1945.* (Baltimore, MD: unpublished Ph.D. diss., Johns Hopkins School of Advanced International Studies, 1990), p. 327; Dörfer, Ingemar, "System 37 Viggen – vapenteknisk forskning inom industrin" in *Daedalus* (Stockholm: Sveriges Tekniska Museums Årsbok, 1986), p. 150. For a detailed overview of the number of aircraft in the RSAF between 1942 and 1961, see: Wennerholm, Bertil, *Fjärde flygvapnet i världen? Doktrinutveckling i det svenska flygvapnet i försvarsbesluten 1942–1958. Underlag, beslut och genomförande i nationellt och internationellt perspektiv.* (Stockholm: Försvarshögskolan, 2006), pp. 155–159. Full translations of Swedish titles are provided in the bibliography.

2 Former Supreme Commander (Överbefälhavaren, ÖB) of the Swedish Armed Forces Bengt Gustafsson said in an interview that there was a great fear that the pilots would get murdered prior to a Soviet attack, see: Askelin, Jan-Ivar, "Mardrömsscenariot var att piloterna blev mördade" in *Framsyn*, 1:2004, p. 29.

3 Moores, S., *'Neutral on Our Side':* US*-Swedish Military and Security Relations during the Eisenhower Administration* (London: unpublished Ph.D. diss., The London School of Economics and Political Science, September, 2004).

interesting that during the height of the Cold War the RSAF was essentially impotent; why has this not generated more interest among Swedish Cold War historians? Well, one reason is probably that aircraft have a much more suggestive quality than air-to-air and surface-to-air missiles. As a builder of scale model aircraft in my teens I cannot remember ever paying much attention to the guided missiles—they were just additions, details that stuck to my fingers instead of to the fuselage.

It just so happens that guided missiles, particularly air-to-air missiles, and the process by which this technology was transferred to U.S. allies, has not spawned any major American study either,[4] while ballistic missiles have generated much more scholarly interest.[5] The simplest explanation for this lacuna is probably that the destructive force of nuclear payloads delivered over vast distances to great population centers evoke images of total devastation, and the end of the world as we know it, and that such potent technologies somehow 'deserve' preferential treatment. Several studies include references to surface-to-air missiles, such as for example Nike, often in connection with the deployment of these missiles within NATO. However, there is as yet no book dedicated to the transfer of this technology and what the considerations of the U.S. government were as it decided to release such missiles to the Western European allies.

By and large, historians of U.S. foreign relations have, with exception of some studies of the atomic and hydrogen bomb and the resulting diplomacy, not paid much attention to technology and its effects. This fact was pointed out by Walter LaFeber in his Presidential address to the Society for Historians of American Foreign Relations in 1999. In his speech he urged diplomatic historians to follow the example of historians of technology who view technology not as artifacts dispersed in the landscape, but as an open system integrated with its surroundings.[6] The situation is largely the same in the history

4 Ron Westrum's book about the American air-to-air missile Sidewinder is concerned with the development process, and does not address the transfer of this missile to other coountries. See: Westrum, Ron, *Sidewinder: Creative Guided Missile Development at China Lake* (Annapolis, MD: Naval Institute Press, 1999). The same is true for Elizabeth Babcock's short history, see: Babcock, Elizabeth, *Sidewinder: Invention and Early Years* (Annapolis, MD: Naval Institute Press, 1999).

5 Cf.: Hill, C. N., *A Vertical Empire: The History of the UK Rocket and Space Programme, 1950–1971* (London: Imperial College Press, 2001); MacKenzie, Donald, *Inventing Accuracy: A Historical Sociology of Nuclear Missile Guidance* (Cambridge, MA: The MIT Press, 1990).

6 Weinberger, Hans, "The Neutrality Flagpole: Swedish Neutrality Policy and Technological Alliances, 1945–1970" in Michael Thad Allen and Gabrielle Hecht (eds), *Technologies of Power: Essays in Honor of Thomas Parke Hughes and Agatha Chipley Hughes* (Cambridge, Mass. & London: MIT Press, 2001), p. 295.

of Sweden's foreign policy during the Cold War, with some notable exceptions.[7]

In the pages below I will provide an investigation of how the transfer of a new, vitally important, highly advanced, technology (i.e. guided missiles) affected, and was affected by, U.S. hegemonic policy towards a small Western European nation (one which was not an official ally of the United States).

Purpose and Research Questions

It is impossible to understand non-aligned Sweden's collaboration with the United States, or indeed the history of any country's foreign relations, without placing it in a larger context. In this case, that larger context is the Cold War and the economic, political, and cultural patterns created by this global conflict. The purpose of this study is to investigate how the issue of military technology, and guided missile technology in particular, fitted into Sweden's secret military collaboration[8] with the United States, and how the acquisition of the Swedish government's consent to American hegemony by the U.S. government interplayed with the deliveries of military technology to Sweden during the period 1945–1962.

The case of Sweden's guided missile purchases at the U.S. in the end of this period is used to highlight both the importance of this technology to the United States, and the significance of its final release to Sweden in the late 1950s and early 1960s. The case of guided missile technology serves, better than perhaps all other technologies, to illustrate just how deeply enmeshed Sweden had become in the Western defense system. In order to correctly understand the importance of American technology and assistance in this field, and with it the impact of the denial of such assistance for the defensive capability of Sweden, it is necessary to study the domestic development of guided missiles in Sweden. I will now formulate my research questions and briefly suggest why they are necessary to provide a satisfactory answer to the purpose of the study.

Firstly, it is assumed in this study that the United States had a hegemonic relationship with Western Europe during the Cold War.[9] American policymakers were deeply skeptical of Swedish foreign policy in the late 1940s; yet in the summer of 1952 the Swedish and U.S. governments exchanged notes confirming that Sweden could now

7 For these exceptions, see Chapter 2.

8 The term 'collaboration', apart from being simply synonymous with 'cooperation', is also used to designate the act of cooperating as a traitor with an enemy of one's own country. In this thesis I use the word 'collaborate' strictly in the more common way, i.e. as synonymous with cooperation.

9 That this is a reasonable assumption will be made clear by the discussion of the concept of hegemony and relevant literature provided below.

procure military technology from the United States on a reimbursable basis. Clearly something had changed, but what? Was it U.S. policy towards Sweden that had gone through a drastic transformation over just a few years, or had the Swedish government been bullied into abandoning the policy of neutrality? Or was it a mixture of the two? The main question that I will answer in this thesis is:

> *How did the United States act when confronted with Sweden's policy of neutrality, and how did Sweden handle* U.S. *preponderance during the period covered by this study. What was required for Sweden to gain access to advanced American military technology, such as guided missiles? Furthermore, what role did technology, i.e. the material objects, play in this process?*

Secondly, considering the fact that Sweden was made eligible to procure military equipment from the United States on a reimbursable basis in early 1952, it seems strange that guided missile technology was for a long time denied the Swedes. It is tempting to assume that this had to do with the fact that Sweden refused to align itself with the United States, and that the delay was thus a consequence of Sweden's dissent. But was this really the case? In consonance with this, another question I try to answer in this thesis is:

> *Why did Sweden not get access to American guided missile technology before the late 1950s? How can we explain the apparent delay in this field compared with other types of military technology?*

Thirdly, I aim to study the Swedish efforts to develop guided missiles domestically, and ask why this ambitious program generated so little end result in terms of operational, domestically developed and produced, missile systems. One could with justification ask why this question has been neglected for so long. Guided missiles were (and still are today) expensive weapons, and the procurement of such weapons consumed a relatively large portion of the Air Force's budget during the second half of the 1950s. Many books and articles have been written about the aircraft of the RSAF, but nothing has so far been said about the weapons with which they were armed, which would have rendered these planes effective in the disastrous event of a war. Hence, the third question that I want to investigate in this thesis is:

> *Why did so many of the early Swedish guided missile development projects not result in operational missile systems? What purpose did the domestic* R&D *efforts have? Which were the priorities behind this ambitious* R&D *effort?*

Fourthly, there can be little doubt that the policy of neutrality was a factor to be reckoned with in the machinations to establish the military connections with the West, but what exactly were its effects in the case of guided missile procurement? An understanding of how the policy of neutrality affected the collaboration with the Americans is essential to correctly ascertain Sweden's role during the Cold War. Thus, the fourth problem that I want to study, and assess, is:

> *What role, if any, did Sweden's policy of neutrality, i.e. the material practices that constituted it, play in the overall secret military cooperation with the United States during this period? How did it affect the material actions undertaken, and how did it constrain and/or enable the historical subjects (i.e. actors) in the* U.S. *and in Sweden?*

A clarification, or refinement, is in order here: when I, in this thesis, speak of the policy of neutrality, and whether it had any effect on the Swedish-American cooperation, I do not view this policy as having been generated *ex nihilo,* and then somehow put into the minds of historical actors in Sweden and the United States. Sweden's policy of neutrality, or the policy of 1812 as it has also been called, came about as a result of a particular historical experience, i.e. it was created *ex materia*. It had its origin in the fact that Sweden was geographically caught between Russia and Britain by the end of the Napoleonic Wars in 1812, being too weak economically and militarily to balance Russian power in the Baltic region.[10] The Policy of 1812 became the means with which consecutive Swedish governments managed to stay out of all armed conflicts until the end of the Cold War. Naturally, different material factors motivated the policy at different times, and it was certainly motivated by domestic concerns as well, but the at the core it remained the same; Sweden was thought to benefit materially from this policy.[11]

So what do I mean when I say that the policy of neutrality affected the Swedish-American military cooperation? Well, no idea could ever influence anything unless it was transformed into material practice. In this case this practice entailed making decisions such as if aircraft or guided missiles ought to be manufactured domestically, or if they should be purchased from abroad. Decisions regarding whether to enter a military alliance, or to remain non-aligned etc. Speeches held, and declarations

10 Talvitie, Heikki, "Sverige och 1812 års politik" in Tapani Suominen (ed.), *Sverige i fred. Statsmannakonst eller opportunism? En antologi om 1812 års politik.* (Stockholm: Atlantis, 2002), p. 16.

11 The same material circumstances could of course have lead to another response from the Swedish government (if other circumstances had been different), e.g. an effort to build an alliance with another powerful nation. This would, however, not change the point of my argument: that policies (ideas) are born when the human mind confronts her material environment.

proclaimed; all such actions made up the policy of neutrality. The result of these practices then, in turn, affected the military cooperation (which was itself a material practice in continuous transformation) by transforming the environment in which this cooperation took place. It is in this way that a spoken message can have considerable effect in human history—because it is a material practice that generates a material response. This is *not* the same as saying that ideas change the world. In fact, it is quite the opposite; i.e. it is to say that what changes the world is the material practices triggered by ideas; which themselves generate from the material circumstances and realities, as they are perceived and interpreted by historical actors. In otherwords: objective reality, confronted by the subjective human mind transformed into material practice.

I will tackle each of these four sets of questions in chapters 3–6, which are the empirical chapters in this thesis. I have chosen to arrange the chapters chronologically because it provides for the best historical narrative. Question-complex number four, concerning what effects Sweden's policy of neutrality had on the secret military collaboration regarding guided missiles, is dealt with and analyzed continuously in all the empirical chapters. Question-groups one and two, on the other hand, are more specific and narrowed in on particular chapters. Consequently, the first issue, regarding how the United States managed to gain the Swedish government's consent to American hegemony, is the subject of chapter 3. Chapter 5 provides an answer to the second set of questions, which concern why Sweden did not gain access to guided missiles until the late 1950s. Having made manifest my purpose in undertaking this study, and the research questions that will be answered by it, it is now time to move on to a presentation of the theoretical framework utilized to interpret the source-material of the thesis.

The History of the Concept of Hegemony

My approach will be somewhat eclectic, which means that I will borrow from a wide variety of sources in the theoretical spectrum dealing with the concept of hegemony, in order to construct what is—in my opinion—a comprehensive definition of my theoretical apparatus. I choose this approach to theory because I share the opinion of Joseph Nye and Robert Keohane that one model cannot explain all situations, but even so, theory is inescapable, because all empirical or practical analysis rests on it.[12] The term hegemony has a long history. Being derived from its Greek root, the term *hegemon* literally means leader, and traditionally it

12 Keohane, Robert O. and Nye, Joseph S., *Power and Interdependence* (third edition) (New York: Longman, (1977) 2001), p. 4. For the same, see: Gaddis, John Lewis, *The Landscape of History: How Historians Map the Past* (Oxford: Oxford University Press, 2002), pp. 62, 109.

has signified a combination of authority, leadership, and domination.[13] It was used by the early Russian Marxists Plekhanov (in the late nineteenth century) and Lenin (in the late nineteenth and early twentieth centuries). Lenin's and Plekhanov's usage referred to the role of the working class in leading an alliance of classes (proletarians and peasants) in overthrowing the feudal/capitalist Tsarist Empire.[14] Perhaps the most famous theorist connected to the concept of hegemony is the Italian Communist (and co-founder of the Italian Communist Party) Antonio Gramsci, who both elaborated and radically extended it. He was elected leader of the Italian Communist Party in 1924 and was imprisoned by Mussolini's fascist government in 1926. Gramsci remained in prison for the rest of his short life. Over the years Gramsci and the concept of hegemony coalesced and have become almost synonymous.[15]

The international literature utilizing the concepts of *hegemony* and *hegemonic power* is very dense, and span almost the entire scientific spectrum. It has been used by Marxists and Liberals alike.[16] The term is, however, often used rather offhandedly, and is seldom given a specific, scientific definition. In many studies of U.S. foreign policy in the post-World War II era, for example, hegemony is often used in the traditional way, as being synonymous with *domination* or *omnipotence*, or sometimes even with *empire*. I will return to these interpretations, and why I do not agree with them, below. Although Gramsci himself did not manage to provide one definite and unambiguous definition of his most central concepts, one of his greatest contributions to a theoretical understanding of the world was his emphasis of *the complex relation-*

13 Ives, Peter, *Language and Hegemony in Gramsci* (London: Pluto Press, 2004), p. 63.

14 Therborn, Göran, *What Does the Ruling Class Do When It Rules?* (London: Verso, 1980), p. 157; Ives, P., *Language and Hegemony in Gramsci*, pp. 63–64; Bocock, Robert, *Hegemony* (London: Tavistock Publications, 1986), p. 25.

15 Therborn, G., *What Does the Ruling Class Do When It Rules?*, p. 157; Joseph, Jonathan, *Hegemony: A Realist Analysis*, (London: Routledge, 2002), p. 19; Bocock, R., *Hegemony*, p. 21.

16 For a Liberal/Neo-Realist approach to hegemony, see: Keohane, Robert O., *After Hegemony: Cooperation and Discord in the World Political Economy* (Princeton: Princeton University Press, 1984); for a Marxist approach, see: Stephen Gill (ed.), *Gramsci, Historical Materialism and International Relations* (Cambridge: Cambridge University Press, 1993); in cultural studies, see: Williams, Raymond, "Gramsci's relevance for the study of race and ethnicity", *Journal of Communication Inquiry, 10* (1986), pp. 5–27, and Artz, Lee and Ortega Murphy, Bren, *Cultural hegemony in the United States* (London: Sage Publications, 2000); for an application of the concept in a study of the Swedish Social Democratic Party's usage of history, see: Linderborg, Åsa, *Socialdemokraterna skriver historia. Historieskrivning som ideologisk maktresurs, 1892–2000* (Stockholm: Atlas, 2001).

ships between coercion and consent in democratic capitalist societies.[17] Furthermore, Gramsci never occupied himself with international relations, but studied only the relations of forces *within* a nation (particularly in Italy). His concepts have nevertheless been expanded to cover the field of international relations, as well as other fields.[18] The concept of hegemony has been analyzed in great detail in numerous works.[19]

Hegemony Defined

Hegemony has been defined as a moral and philosophical leadership, which is attained through the active consent of major groups in society.[20] It is a relationship in which one consents to the leadership of another *because it is beneficial.*[21] The way in which hegemony works in the international system has been very well described by Peter M. Sánchez:

> A hegemonic system exists when one nation-state has achieved a preponderance of military and economic power and has also convinced subordinate states that it is in their best interest to accept the leadership of the dominant power, since the hegemon's ideology will most likely promote the national and collective interest of the subordinate states.[22]

17 Ives, P., *Language and Hegemony in Gramsci*, pp. 64–65. The main reason for the vagueness of his definitions is that Gramsci wrote a large part of his works in fascist prison, and writing obliquely and guardedly was a way to avoid the prison censor.

18 Cox, Robert W., "Gramsci, Hegemony and International Relations: An Essay in Method" in Stephen Gill (ed.), *Gramsci, Historical Materialism and International Relations* (Cambridge: Cambridge University Press, 1993), p. 49.

19 See for example: Buci-Glucksmann, Christine, *Gramsci and the State* (London: Lawrence and Wishart, 1980), p. 63; Bocock, R., *Hegemony*; Joseph, J., *Hegemony: A Realist Analysis*; Femia, Joseph V., *Gramsci's Political Thought. Hegemony, Consciousness and Revolutionary Process* (Oxford: Clarendon, 1987); for a so-called post-Marxist analysis, see: Laclau, Ernesto and Mouffe, Chantal, *Hegemony and Socialist Strategy. Towards a Radical Democratic Politics* (London: Verso Press, 1985).

20 Bocock, R., *Hegemony*, p. 11.

21 Artz, L. and Ortega Murhpy, B., *Cultural hegemony in the United States*, quote on p. viii, and: pp. 1–70; for similar definitions, see: Cox, Robert W., "Gramsci, Hegemony and International Relations: An Essay in Method" in S. Gill (ed.), *Gramsci…*, p. 50; Gill, Stephen and Law, David, "Global Hegemony and Structural Power of Capital" in S. Gill (ed.), *Gramsci…*, p. 93; and: Dobson, Alan P. and Marsh, Steve, US *Foreign Policy Since 1945*, (London: Routledge, 2001), p. 65.

22 Sánchez, Peter M., *Panama Lost?* U.S. *Hegemony, Democracy, and the Canal* (Gainesville, FL: University Press of Florida, 2007), p. 8.

The subordinate states consent, and submit, to the hegemonic power because the hegemon provides them with a sense of security, sovereignty, as well as prosperity. International anarchy is replaced by cooperation under a unipolar regime.[23] Robert Keohane has called this particular form of mutually beneficial relationship *hegemonic cooperation*.[24] Hegemonic cooperation does not, however, imply that the relationship is harmonious at all times (i.e. free from conflict). Indeed, as Keohane firmly underlines, cooperation is itself a result of a percieved real, *or potential*, conflict. If there was harmony, there would not really be any need for cooperation.[25] This follows quite naturally from the unequal balance of power between the cooperating parties. Indeed it is the asymmetries in dependence in the hegemonic relationship that provide the less dependent part (i.e. the hegemonic state) with its sources of power and influence over events.[26]

Moreover, hegemony is not based solely on consent, but also entails a certain amount of coercion, although the consensual aspect is the key to understanding how hegemony is established and upheld.[27] Faced with challenges to its domination the hegemon can choose to respond in one of two ways. Either it can accept the challenge, i.e. if it makes the judgement that the opposition does not threaten the basis for its hegemony, and take no immediate action; or it can take action with the aim of upholding its hegemony. If the latter path is chosen, it can, according to Peter M. Sánchez, opt for a soft approach (the 'carrot') and provide incentives and concessions, or a harsh approach (the 'stick') and employ threats and force, i.e. coersion.[28]

The exact relation between the forces of coercion and consent, however, is a matter of empirical investigation, and one cannot make *a priori* predictions about their relationship. There is also the issue of how you define coercion. Basically, there are two kinds of coercion readily available to the hegemonic power, 1) political/economic pressure tactics (which are farely common), and 2) outright military force and threats of such force (which is less common or even abnormal). Robert Cox has stated that the latter form of coercion "is always latent but is only applied in marginal, deviant cases."[29] It is also reasonable to assume that

23 Ibid., p. 9.

24 Keohane, R. O., *After Hegemony...*, p. 55.

25 Ibid., pp. 53–54. Keohane also stresses that it may be just as erroneous to infer conflicting interests from discord, pointing to the fact that actors might fail to cooperate even if their interests are identical (ibid., p. 65).

26 Keohane, R. O. and Nye, J. S., *Power and Interdependence*, p. 9.

27 Cox, R. W., "Gramsci, Hegemony and ..." in S. Gill (ed.), *Gramsci,...*, p. 52.

28 Sánchez, P. M., *Panama Lost?...*, pp. 15–16.

29 Cox, R. W., "Gramsci, Hegemony and ..." in S. Gill (ed.), *Gramsci,...*, p. 52. Apart from the odd case of armed force being applied to keep subordinated

the former type of coercion is more prevalent when a hegemonic relationship is being established; i.e. when the nation that covets hegemony seeks to persuade others to conform to its vision of world order, and make these other nations defer to its leadership, and even more so when it is seriously contested, e.g. in times of hegemonic crisis.

Writes Peter M. Sánchez:

> [In a hegemonic] system, the hegemon is mostly successful in determining the rules of the game in the international system. The hegemon also assumes the role of enforcing the rules when one or more of the states in the system attempt to challenge the hegemon's leadership or ally with an extrasystemic power. In these situations, the hegemon will often forcefully employ its military and economic instruments of power.[30]

As will become evident in Chapter 3, this description fits perfectly with the way in which the United States treated Sweden during the period 1945–1952. True, the U.S. government never employed military means in order to gain Sweden's consent, but American military power played an important role through the fact that Sweden was dependent upon the United States for arms deliveries and military support in case of war. We must remember that American political and economic leadership in Western Europe did not exist in isolation from NATO, and even if economic issues were rarely explicitly tied to military ones it "remained an important factor in the background [...]."[31] It is thus very important that we do not let the relative importance of the consensual aspect make us forget about the role of coercion in a hegemonic relation.

The strength of the concepts of hegemony and hegemonic cooperation is that they enable us to make a difference between *domination due to force/threat of force*, i.e. non-hegemonic power, and *domination*

---

nations in check, military preponderance is necessary due to the hegemonic power's need to be able to resist incursions in its sphere of influence from outside powers. According to Keohane: "[A] hegemonic state must possess enough military power to be able to protect the international political economy that it dominates [...]. [...] The military conditions for [...] hegemony are met if the economically preponderant country has sufficient military capabilities to prevent incursions by others that would deny it access to major areas of its economic activity." Yet this does not mean that the hegemonic power needs to be military dominant worldwide. No hegemonic nation could be expected to possess such omnipotence, and neither British nor American power ever extended so far (Keohane, R. O., *After Hegemony...*, p. 39).

30 Sánchez, P. M., *Panama Lost?...*, p. 8.

31 Keohane, R. O., *After Hegemony...*, p. 137. For a similar argument, see: Calleo, David P., *Beyond American Hegemony: The Future of the Western Alliance* (New York: Basic Books, Inc., Publishers, 1987), pp. 8–9.

due to consent, i.e. power and influence based on hegemony. It is true, however, that the focus on the force/consent dichotomy risks becoming one-sided, in its extreme form defining hegemony as exclusively representing the actions and plans of social agents, i.e. groups and individuals. This tends to lead to a conspiratorial view of hegemony, portraying it as being simply the result of conscious and cunning agents.[32]

This view ignores the structural aspect of hegemony, on which both the hegemon and the subordinate nations partly depend. In contrast to this, Charles S. Maier has acknowledged that the huge American predominance in the immediate postwar period, almost inevitably had to give rise to American hegemony. "Hegemony", he wrote, "was in the cards [...]."[33] There is a large consensus in the scholarly community today that the United States was hegemonic vis-à-vis Western Europe during the period dealt with in this thesis, i.e. 1945–1962.[34] United States hegemony involved an effort to rally the Western European nations around a common set of values and ideas, and to isolate the governments that did not share the same ideology. In doing so, the U.S. government resorted to a great deal of bargaining, and used the appeal of a liberal ideology to establish a centrist political mindset among the European populace. A large part of this effort was of course to establish a new economic order in Western Europe, what Maier has called *the politics of productivity*.[35]

32 Joseph, J., *Hegemony: A Realist Analysis*, p. 1.

33 Maier, Charles S., "Hegemony and autonomy within the Western Alliance" in Melvyn P. Leffler and David S. Painter (eds.), *Origins of the Cold War: An International History* (London: Routledge, 1994), p. 155.

34 See for instance: Maier, C. S., "Hegemony and autonomy..." in M. P. Leffler and D. S. Painter (eds.), *Origins of the Cold War:...*, pp. 154–174; McCormick, Thomas, J., *America's Half-Century: United States Foreign Policy in the Cold War and After* (second edition), (Baltimore, MD: The Johns Hopkins University Press, 1995); Lundestad, Geir, *The American "Empire" and Other Studies of* US *Foreign Policy in a Comparative Perspective* (Oxford: Oxford University Press, 1990); Keohane, R. O., *After Hegemony:...*; Leffler, M. P., *A Preponderance of Power: National Security, the Truman Administration, and the Cold War* (Stanford: Stanford University Press, 1992), p. 164. Although Leffler do not use the term hegemony to any considerable extent, his usage of the term *preponderance* largely corresponds with the concept of hegemony. See e.g. his own conflation of the two in his concluding arguments (ibid., pp. 499–500, 504, 516, 518). In general, all scholars using the term hegemony more or less agree that the floating of the dollar in 1971 and the oil crisis in 1973 mark the decline, but not the end, of the United States as a hegemonic power.

35 Maier, C. S., "Hegemony and Autonomy..." in M. P. Leffler and D. S. Painter (eds.), *Origins of the Cold War:...*, pp. 155–157. See also: Maier, C. S., "The Politics of Productivity: Foundations of American International Economic Policy after World War II" in Peter J. Katzenstein (ed.), *Between Power and Plenty: The Foreign Economic Policies of Advanced Industrial States* (Madison, WI: University of Wisconsin Press, 1978), pp. 23–49.

In order to pinpoint and explain that consent to the hegemony of another does not at all preclude strategies that to a certain degree contradict this consent, Göran Therborn has developed the term *accommodating opposition* to account for the defiant acts on the part of the consenting parties in a hegemonic relationship.[36] Even the closest of friends may want to retain a sense of autonomy and independence vis-à-vis the other. This does not mean that they cease to be friends. Likewise, in a hegemonic relationship, such acts do not mean that hegemony ceases to affect the relationship. Hegemony is a dynamic relation, and the concept of accommodating opposition is valuable to have in mind when we interpret Sweden's behavior in her relations with the United States during the Cold War.

It is possible to discern two conditions that the hegemonic power has to live up to, if it is to be able to elicit consent from the subordinated nations. Firstly, concrete material benefits have to be forthcoming if continued support by the subordinated nations is to be maintained. Secondly, the hegemonic state has, in some way, to transcend its own narrow self-interest in the areas of politics and the economy. This means (in an international relations context) that the consent of the majority of nations can only be secured politically through agreements and/or alliances with these nations;[37] this is often done through what have been called *hegemonic apparatuses*.[38] All of this also meant that the price that the United States paid in order to sustain its leading role was quite high.[39]

Hegemonic apparatuses are similar to international regimes, i.e. they are formal, or informal, institutions (consisting of rules, norms, and procedures) that are relatively autonomous and stable (over time), and that are established in order to coordinate expectations and organize behavior between states with the intention of minimizing real (or potential) conflict.[40] A crucial difference between hegemonic apparatuses and international regimes, however, is the power asymmetry built into the former. While international regimes can be established essentially by any country and provide equal influence to all its members, a hegemonic

36 Therborn, Göran, *The Ideology of Power and the Power of Ideology* (London: Verso, 1980), p. 95.

37 Artz, L. and Ortega Murphy, B., *Cultural hegemony in the United States*, pp. 24–27.

38 Ibid., pp. 39–45.

39 Calleo, D. P., *Beyond American Hegemony:…*, p. 14.

40 Keohane, R. O., *After Hegemony…*, pp. 49–64; Pempel, T. J., "Restructuring Social Coalitions: State, Society, and Regime" in Rolf Torstendahl (ed.), *State Theory and State History* (London: Sage Publications, 1992), pp. 118–148; Ruggie, John Gerard, *Constructing the World Polity: Essays on International Institutionalization* (London: Routledge, 1998), pp. 85–101.

apparatus cannot. A hegemonic apparatus is created by the hegemon (sometimes in concert with others, but always on its terms), with the purpose of administrating the hegemon's relationship with the consenting nations. This means that hegemonic apparatuses are 'rigged' to the hegemon advantage. That does not mean that they are vehicles for outright domination, however. The United States sought to provide special benefits to its partners, but also ensure deference to its policies, through the creation of hegemonic apparatuses.[41]

Hegemonic apparatuses, and this cannot be stressed heavily enough, should be thought of as an open concept that allow a variety of contradicting practices to take place within its boundaries. This is indeed necessary in order to elicit the consent of subordinate nations. Nevertheless, they serve to establish order within the international community. Thus, the concept of hegemony is not rigid, i.e. in the sense that it does not permit quite a considerable degree of freedom of action for the subordinated nations. Maier has written that the degree to which the U.S. hegemony allowed scope for European autonomy is an intriguing question. "The relationship worked out between Washington and the European centers during the formative Truman years", he says, "provided cohesive political purpose but simultaneously allowed significant national independence."[42]

Hence, the fact that Sweden consented to the American hegemony did not prevent a domestic production of guided missiles, but it opened a window of opportunity to acquire American guided missiles that would (I will argue in this thesis) otherwise have been closed. What has been said about hegemonic apparatuses in a cultural context is also applicable to the case of international relations, namely that:

> Hegemonic apparatuses do not prevent challenges to hegemony, but they do provide an important mechanism for stemming those challenges.[43] Hegemony is never completely secured.[44]

Furthermore, it is important to point out that hegemonic apparatuses are something other than vehicles for outright domination, or

41 Keohane, R. O., *After Hegemony...*, p. 137; Keohane, R.; Nye, Joseph S.; and Hoffman, Stanley, *After the Cold War: International Institutions and State Strategies in Europe, 1989–1991* (Cambridge, MA: Harvard University Press, 1993), p. 105. Keohane does not apply the term hegemonic apparatus, but uses international regime instead. However, I find that the distinction is an important one, and that it serves to mark the structural difference between the two terms.

42 Maier, C. S., "Hegemony and autonomy..." in M. P. Leffler and D. S. Painter (eds.), *Origins of the Cold War:...*, pp. 155–156.

43 Artz, L. and Ortega Murphy, B., *Cultural hegemony in the United States*, p. 44.

44 Ibid., p. 45.

manipulation, of the subordinated nations. Just as reality is a complex and contradictory process, so we must allow our historical account of past events to reflect this complexity (within limits, of course). Hence, flexibility is an important quality in, or indeed the very definition of, scientific concepts.

Critique of the Concept of Hegemony

Critics of the concept of hegemony, such as James C. Scott, has called it "a theory of false consciousness", since, according to him, the proponents of this concept claim a superior knowledge of what the subordinated groupings (in this case nations) really want, since the term subordinated implies a wish not to be dominated.[45] Scott argues that proponents of the concept of hegemony present the dominant ideology as being omnipotent, "as impenetrable".[46] Further, he says, the concept "assumes that the subordinate group is [...] relatively quiescent, [...] relatively disadvantaged, and that it is not directly coerced."[47] He also underlines that there is no warrant for supposing that the "acceptance of a broad, idealized version of the reigning ideology prevents conflict—even violent conflict [...]."[48] Even though Scott's critique is aimed at the traditional intra-national and class-based version of hegemony (with which Gramsci was concerned), and not at the international version of the concept as it has been developed and used by Cox et al., the basis of the criticism leveled also applies to the international usage of the concept, and the defense presented here is mainly intended to defend the latter.

Then there is the fact that subordinated groups, or in this case nations, actually manage to put up resistance despite the overwhelming dominance of the hegemonic power.[49] This is an anomaly in the theory of hegemony, Scott argues, and points out that even extreme efforts at ensuring subordination, even under total domination, often does not succeed, and takes the process of extracting confessions from prisoners in the prisoner-of-war camps in North Korea during the Korean War as an example. Despite the draconian efforts, he concludes, "the captors [were only able] to exercise a temporary hegemony"[50] over their prisoners. Nicholas Abercrombie has put forward a similar critique.[51]

45 Scott, James C., *Domination and the Arts of Resistance: Hidden Transcripts* (London: Yale University Press, 1990), p. 72; see further: pp. 70–76.

46 Scott, J. C., *Domination and the Arts of Resistance:...*, p. 78.

47 Ibid., p. 71.

48 Ibid., p. 74.

49 Scott does not touch upon the concept of hegemony in an international relations context. However, I find his criticisms as valid for this field as for any other.

50 Scott, J. C., *Domination and the Arts of Resistance:...*, p. 84.

51 Abercrombie, Nicholas, et al., *The Dominant Ideology Thesis* (London: George Allen & Unwin, 1980), pp. 59–94.

I will now show that Scott's critique is built on a misunderstanding of the concept he has criticized, and thus do not present a serious threat to the concept as it is laid out in this thesis. Firstly, in international relations hegemony is not a theory of false consciousness because the theory aims to explain a power relation between a hegemonic state and its consenting allies. Governments consent to their subordinate status because they perceive this position to be beneficial to them. Secondly, the hegemonic ideology can certainly be dense and opaque, but it could never prevent the expression of dissenting views. If it did, then it would be a totalitarian order, and there could thus be no hegemony, since there would be no need for the dominant state to seek the consent of other nations. The point is instead another: the dominant ideology can permit dissenting views to be expressed, precisely because it is preponderant. In fact, the possibility for the subordinated countries to voice their opposition, coupled with the will of the hegemon to accommodate its allies, actually strengthens hegemony because it appears more benevolent.

Thirdly, the reference to efforts to persuade prisoners of war is not relevant, since these efforts are not built on, and in fact have no need for, the consent of the prisoners. Indeed that is probably why the persuasions failed in the first place. As we have seen, a nation consents to another's hegemony because it derives benefits from it, if it did not there would be no hegemony. The concept of hegemony and hegemonic cooperation in no way precludes accommodating opposition. Instead, such opposition is a necessary part of any hegemonic relationship, and to assume that subordinated groups or nations would be absolutely unable to imagine the reversal of the present power relations would be very unrealistic. This is indeed a point made most forcefully by Scott himself.[52] The critique of the concept of hegemony often misses the target completely due to a misunderstanding of what is criticized.

Hegemony or Empire?

It is not only Scott and Abercrombie who interpret the concept of hegemony as essentially the same as domination without consent. This is indeed very common.[53] This practice manifests itself most clearly in the way many writers deal with U.S. armed interventions in, for instance, South East Asia and the Western Hemisphere (i.e. North, Central, and South America). Alan P. Dobson and Steve Marsh has written an entire chapter where the focus is on U.S. relations in the Western Hemisphere, which appear to them as having "been truly hegemonic since 1945."[54]

52 Scott, J. C., *Domination and the Arts of Resistance:...*, e.g. p. 80.

53 Cf. McCormick, T., J., *America's Half-Century:...*; Dobson, A. P. and Marsh, S., US *Foreign Policy Since 1945*, pp. 65–75.

54 Dobson, A. P. and Marsh, S., US *Foreign Policy Since 1945*, p. 65. For the Latin American case, see also: Sánchez, P. M., *Panama Lost?...*, pp. 11–21.

In these accounts the fact that the United States has been striving to dominate an area is thought of as an expression of its hegemonic power. I would not agree with that conclusion. Perhaps the United States have been striving for an enlargement of its hegemonic power, but not succeeded (i.e. only achieved domination by force/threat of force, or sometimes suffered outright defeat), or only succeeded in a few cases. In these areas, where U.S. domination has mostly been attained through dictatorial puppet regimes,[55] there has been no need for consent in any meaningful sense of the word. In these cases it is better to talk of U.S. actions and domination in terms of empire.

In contrast, it is the relations between the United States and its (mostly) Western European allies that particularly well fit the description of hegemony and hegemonic cooperation. It is in its interaction with the democratic capitalist nations, such as Sweden, that the United States has been able to (maybe even forced to) develop a hegemonic cooperation. Here, the consent of these nations has been crucial for the hegemony of the United States and, hence, these countries were given large material incentives in order to accept their subordinate position. But even here, of course, there is contradiction and non-linearity.

Geir Lundestad does something different when he states that *empire* and *hegemony* are essentially the same thing.[56] He argues that the European nations invited the United States (which would be totally consistent with hegemony), but he uses the term *empire by invitation* to describe the result (and empire is not consistent with hegemony). What Lundestad does, really, is use the term empire when speaking about hegemony. If one necessarily wants to make a point out of the obvious fact that the Europeans wanted the protection of the United States, then perhaps '*hegemony* by invitation' would be a more proper designation in this particular case. John Krige has embraced the term coproduction, i.e. that the subordinated nation participates in its own subordination, to describe the same phenomenon.[57]

55 The body of literature on the United States foreign policy in Latin and South America and the Caribbean is immense. For just a few examples, see: Horowitz, David, *The Free World Colossus: A critique of American Foreign Policy in the Cold War* (New York: Hill and Wang, 1965); Hahn, Peter L. and Heiss, Mary Ann, *Empire and Revolution: The United States and the Third World since 1945* (Columbus: Ohio State University Press, 2001); Prados, John, *Presidents' Secret Wars: CIA and Pentagon Covert Operations from World War II Through the Persian Gulf* (Chicago: Ivan R. Dee, Inc., 1996); Blum, William, *Killing Hope: U.S. Military and CIA Interventions since World War II* (Monroe, Me: Common Courage Press, 1995); Dobson, A. P. and Marsh, S., US *Foreign Policy Since 1945*.

56 Lundestad, G., *The American "Empire"*.... For the same, see: Calleo, D. P., *Beyond American Hegemony:...*, p. 14.

57 Krige, John, *American Hegemony and the Postwar Reconstruction of Science in Europe* (Cambridge, Massachusetts: The MIT Press, 2006), pp. 4–9.

However, such a formulation appears redundant, since the concept of hegemony already includes this feature implicitly in its definition. Despite the fact that there are almost as many interpretations of the meaning of empire as there are scholars researching the subject,[58] British historian Stephen Howe has provided a description which I think is general enough to please all but the most extreme positions:

> Empires always involve a mixture of direct and indirect rule. The central power has ultimate sovereignty, and exercises some direct control, especially over military force and money-raising powers, in all parts of its domain.[59]

This definition does not fit well with the relations which the United States had towards the Western European countries during the Cold War. Empire should be reserved for situations where a power center enlarges its territory by force, occupies the same territory and forces its laws and institutions upon the conquered inhabitants.[60]

Furthermore, what Lundestad's conceptualization implies, although he does not say it explicitly, is that, basically, the United States became hegemonic in Western Europe only because it was invited by the Western European nations. Underlying such an argument is the assumption that the state of the world is due only to the intentions of the actors involved, i.e. that United States hegemony had no material basis. An interpretation of the postwar transatlantic relationship has to acknowledge that there were in fact material factors in play, which provided for the opportunity of American hegemony.

The United States could never have attained hegemony, and would never have been allowed to do so by the Western European governments, if the economic and political devastation of war-torn Europe had not been so complete. Pre-war Britain certainly had no interest in 'inviting' the United States to take over its responsibilities on the international arena. Conversely, it could not have taken on this monumental task if the North American continent had not been untouched by the destruction of war. John Krige and Charles S. Maier differs from Lundestad in that they, take the material basis for hegemony into consideration. The position

58 The literature dealing with the concept of *empire* is immense; and the meaning of empire and its implications for the subordinate territories has been very fiercely contested. A very good overview of the historiography of the meaning of *formal* empire is provided in: Howe Stephen, *Empire: A Very Short Introduction* (Oxford: Oxford University Press, 2002); see also: Porter, Andrew, *European Imperialism, 1860–1914* (Basingstoke: Macmillan, 1994).

59 Howe, S., *Empire: A Very Short History*, p. 15.

60 A useful discussion regarding the distinctions between empire and hegemony is presented for a Scandinavian audience in: Knutsen, Torbjörn, "Amerikas fire ansikter" in *Internasjonal politikk*, Vol. 62, No. 3, 2004, pp. 303–325.

of the United States after World War II was so totally preponderant that hegemony, as well as empire (only in other parts of the world), was probably more or less inevitable.[61] They nevertheless share Lundestad's interpretation that hegemony is basically the same as empire, or informal empire.[62] My definition of hegemony thus differ from the description provided above, by disconnecting the empire and hegemony from one another, and making them into two distinct concepts designating separate realities.

It might be correct to state, as Scott does, that hegemony has been presented as impenetrable and omnipotent, at least in some works. However, I do not think that my definition purports such a stand. Scott, however, is right on the mark when he says that hegemonic cooperation does not necessarily prevent conflict. There is certainly room for conflict within a hegemonic relationship, i.e. what has been called accommodating opposition.[63] In such a situation a subordinated ally might resist the adoption of a certain policy if, for instance, it feels that the hegemonic state is being too inattentive to its needs—e.g. if it is not providing enough material reward to secure its consent, or if political sentiment prevents it from accepting a certain line of action. This forces either the hegemonic nation to change its strategy, or it might result in a modified strategy being proposed by the subordinate nation.

In the case of violent conflict, however, I am not sure that I would agree with Scott's conclusion. Violence, I would say, is symptomatic of non-hegemonic relations—i.e. the balance of power has to be settled by explicit force (there is no consent from the subordinated nations). Furthermore, it is hard to see how hegemonic cooperation (or any cooperation at all) could still go on if there is an armed conflict between the two nations in question. But I would not take this for granted either, however; a careful empirical study of each historic event is of course needed before any definite conclusion can be drawn. In general, though, I would say that the explicit use of military force does not fit the concepts of hegemony and hegemonic cooperation. The concept of hegemony, as described above, will be used in my analysis of the Swedish efforts to provide its armed forces with air-to-air, surface-to-air, and air-to-surface guided missiles during the period 1945–1967, and the collaboration between Sweden and the United States in this field.

61 Krige, J., *American Hegemony…*, p. 2; Maier, C. S., "Hegemony and Autonomy…" in M. P. Leffler and D. S. Painter (eds.), *Origins of the Cold War:…*, p. 155.

62 Krige, J., *American Hegemony…*, pp. 2, 255, 260.

63 Therborn, G., *The Ideology of Power and the Power of Ideology*, p. 95.

Chapter 2

Prior Research & General Background:

Swedish Security Policy, and American Hegemony-Building

In this chapter I aim to do three things: Firstly, I will start with providing the reader a historiography of prior research into Swedish security policy during the Cold War period. I do not pretend to cover all the works produced on this subject, but mainly the ones that are relevant for the study and period concerned. Secondly, I will move on to a description of the history of Sweden's security policy, the policy of neutrality. The intention of this section is to offer the reader a deeper understanding of the rationale behind this policy. Finally, I intend to provide a account of American hegemony-building from the immediate postwar years up to about 1950. I believe that such an account is necessary in order to better understand one of the central arguments in this thesis, namely Sweden's acceptance of U.S. hegemony and the importance that this consent had to the U.S. administration.

Three Schools of Revisionist History

In an article in *Internationella Studier* [International Studies] in 1999, Kjell Engelbrekt identified three versions of revisionist history, e.g. reinterpretations, which had developed in Swedish Cold War research since the early 1990s.[1] In his article, however, Engelbrekt did not propose any designations for these schools. Thus I suggest the terms *moderate*,

1 Engelbrekt, Kjell, "Den sjuttonde alliansmedlemmen?" in *Internationella Studier*, No. 4, 1999, pp. 61–72.

critical, and *radical* revisionism.[2] The different schools will be briefly, and admittedly very crudely, described below.

The *critical school* argues that the government's non-alignment policy was double-faced. These scholars have described Sweden's neutrality policy as being a lie used to deceive the Swedish populace, or as the policy of a nation 'free-riding' on Western security arrangements—i.e. it has a moral dimension.[3] The *radical school* argues that Sweden became so integrated into the Western Power's defense system that Sweden's ability to stay out of a future war became seriously circumscribed. Some proponents of this school even go so far as to say that it is justified to talk of Sweden as 'an informal, seventeenth, member of NATO', but it generally lacks the moral indignation so common to the critical school.[4]

2 The division into three revisionisms is Engelbrekt's, but the terms *moderate*, *critical*, and *radical* are my own. Just like all such classifications it is of course crude and not very precise, but it still serves its purpose. I think that these terms are more useful than the ones used to denote the re-writing of general Cold War history; i.e. *traditionalist, revisionist,* and *post-revisionist* (the latest position is a post post-revisionism referred to as the *new Cold War history*, which has been the result of the end of the Cold War and the opening of Russian archives (see: Lundestad, G., "How (Not) to Study the Origins of the Cold War" in Odd Arne Westad (ed.), *Reviewing the Cold War: Approaches, Interpretations, Theory* (London: Frank Cass, 2000), pp. 66–69). While these could probably be used even to describe Swedish Cold War historiography, these terms are strongly connected to different interpretations of the *origins* of the Cold War. This is not the case in Swedish historical debate on the Cold War. Moreover, in Sweden *all* three schools are revisionist to some degree. However, Mikael af Malmborg has suggested that the international terminology should also be used in the Swedish case (see: af Malmborg, M., "Sweden – NATO's Neutral 'Ally'? A Post-Revisionist Account" in Gustav Schmidt (ed.), *A History of* NATO—*The First Fifty Years*, Volume III (Hampshire: Palgrave, 2001), p. 295).

3 Agrell, Wilhelm, *Den stora lögnen. Ett säkerhetspolitiskt dubbelspel i alltför många akter.* (Stockholm: Ordfront, 1991); Cole, P. M., *Neutralité du jour:* ... For more examples of critical revisionism, see: Goldmann, Kjell, "Neutralitetspolitiken som dubbelpolitik", *Internationella Studier*, No. 4, 1998:, pp. 73–80; Steen-Johansson, Cecilia, *Ett Folkbedrägeri.* DC 3:*an och svensk säkerhetspolitik. Sverige och* NATO *1952–92*. (Stockholm: Carlssons Bokförlag, 1992); Johan Kristoffer Hell has, in a less than well supported study, termed the Swedish policy during the Cold War 'strategic deception' (Hell, Johan Kristoffer, *Strategic Vulnerability: Understanding Sweden's National Security Policies During the Cold War* (Münster: Waxmann, 1996).

4 Tunander, Ola, "The Uneasy Imbrication of Nation-State and NATO: The Case of Sweden" in *Cooperation and Conflict*, Vol. 34, No. 2, 1999, pp. 169–203; Weinberger, H., "The Neutrality Flagpole:..." in M. Thad Allen and G. Hecht (eds), *Technologies of Power:...*, pp. 295–331; Weinberger, H., ***Technology : and the de/construction of Swedish neutrality during the cold war*** (Stockholm: Department of History of Science and Technology, Royal Institute of Technology TRITA-HST, 2002/2); Weinberger H., "A Small/Medium-sized Great Power in the North: Swedish Neutrality During the Cold War Through the Geostrategic Lens" in K. L. Kleve (ed.), *The Cold War, Military Power...*, pp.

The radical revisionist school also contained a partly new approach to the study of Sweden's policies during the Cold War; i.e. the study of technology, the transfer of technology, and its consequences for Swedish neutrality policy and non-alignment. Questions were raised regarding the technological tie-in with the Western Powers created, or was preconditioned by, dependencies and obligations that were perhaps prejudicial to Sweden's freedom of action.[5] This point of departure was new, even if a number of prior studies of Sweden's technological and economical dependence upon the United States and other nations have stressed the inherent risks of interdependence with regard to freedom of action in foreign and security policy and future neutrality.[6]

45–64; Dahl, Ann-Sofie, "The Myth of Swedish Neutrality" in Cyril Buffet and Beatrice Heuser (eds.), *Haunted by History: Myths in International Relations* (Providence and Oxford: Berghahn Books, 1998), pp. 28–40; Holmström, Mikael, "Svenskt dubbelspel under kalla kriget" in *Svenska Dagbladet*, October 11, 2003; Haglund, Magnus, "Jovisst var vi hemliga NATO-medlemmar!" in *Kungl krigsveten-skapsakademiens handlingar och tidskrift*, No. 5, 2002, pp. 125–128.

5 Gribbe, Johan and Nilsson, Mikael, "The Foreign Domestic: Hard Artefacts and Soft Politics in Sweden During the First Half of the Cold War, 1945–1967" in *icon: Journal of the International Committee for the History of Technology*, VOL. 11, 2005, pp. 51–62; Weinberger, H., "The Neutrality Flagpole:..." in M. Thad Allen and G. Hecht (eds), *Technologies of Power:...*, pp. 295–331; Weinberger, H., *Technology: and the de/construction of Swedish neutrality...*; Weinberger, H., "A Small/Medium-sized Great Power in the North..." in K. L. Kleve (ed.), *The Cold War, Military Power...*, pp. 45–64; Weinberger, H., "'På sidan om de stora kraftlinjerna'? Teknik, vetenskap och det kalla kriget" in Sven Widmalm (ed.), *Vetenskapsbärarna. Naturvetenskapen i det svenska samhället, 1889–1950.* (Hedemora: Gidlunds Förlag, 1999), pp. 352–368; Weinberger, H., "'Det rörde sig väl om ofarliga ting' Svensk neutralitetspolitik och synen på teknik" in *Arbetarhistoria. Meddelande från Arbetarrörelsens Arkiv och Bibliotek*, NO. 92, 4/1999, pp. 31–36.

6 Mörth, Ulrika and Sundelius, Bengt, *Interdependens, konflikt och säkerhetspolitik. Sverige och den amerikanska teknikexportkontrollen.* (Stockholm: Nerenius & Santérus Förlag, 1998); Holmström, Mikael and von Sivers, Tom, *Tekniken som vapen.* (Stockholm: Ingenjörsförlaget, 1985). Both of these studies deal with the 1970s and 1980s, and thus their results are not directly applicable to my study. For a study of the American export control during the Cold War, and Sweden's participation in COCOM and the Western economic warfare against the Eastern Bloc, and its consequences for Swedish freedom of action in the field of foreign policy, see: Adler-Karlsson, Gunnar, *Western Economic Warfare 1947–1967: A Case Study in Foreign Economic Policy* (Stockholm: Ph.D thesis, Almqvist & Wiksell, 1968); Karlsson, Birgit, *Handelspolitik eller politisk handling. Sveriges handel med öststaterna 1946–1952* (Göteborg: Meddelanden från ekonomisk-historiska institutionen vid Göteborgs universitet 66, 1992); Karlsson, Birgit, "Sveriges relationer till Östeuropa 1945–1950" in Charles Silva and Thomas Jonter (eds.) *Sverige inför en ny världsordning, 1945–1950. Formativa år för svensk utrikespolitik?* (Stockholm: Utrikespolitiska institutet, 1995), pp. 43–51; Neuhold, Hanspeter, "Challenges To Neutrality in an Interdependent World"

The *moderate school* contests the claim that the government's actions circumscribed its freedom of action, and stresses with fervor that there are as yet no available documents that support the conclusion that Sweden was a hidden member of NATO. Swedish neutrality policy is seen as having been successful since it kept the country out of war (although Sweden hardly can take credit for there not being a war between the blocs), and the secret collaboration with the West was a necessary insurance policy in case the government's neutrality policy would have failed.[7] My study would fall within the radical school, and it argues that Sweden's technological (as well as its ideological, political, and economic) tie-in with the United States and the other Western Powers, circumscribed her ability to defend her neutrality in case of war between the two Power Blocs.

Prior Research: Sweden and Neutrality

Prior to his visit to Sweden in October 1981, the American Secretary of Defense Caspar Weinberger boldly told a group of Swedish journalists that "'I do not think we have a neutral Sweden.'"[8] At the time, the

---

in Michael H. Haltzel and Joseph Kruzel (eds.), *Between the Blocs: Problems and Prospects for Europe's Neutral and Nonaligned States* (Cambridge, MA: Woodrow Wilson International Center for Scholars and Cambridge University Press, 1989), p. 89–93; Sundelius, Bengt, "National Security Dilemmas and Strategies" in M. H. Haltzel and J. Kruzel (eds.), *Between the Blocs:...*, p. 108–110; Sundelius, Bengt, "Coping with Structural Security Threats" in Otmar Höll (ed.), *Small States in Europe and Dependence* (Wien: Wilhelm Braumüller, Universitäts- Verlagsbuchhandlung, 1983), pp. 281–305; Davidsson, Sune, in cooperation with Hans Bergendorff and Bo Ljung, *Sveriges utlandsberoende* (Stockholm: Centralförbundet Folk & Försvar, 1976); Dohlman, Ebba, *National Welfare and Economic Interdependence: The case of Sweden's Foreign Trade Policy* (Oxford: Clarendon Press, 1989), pp. 39–91.

7 Kronvall, Olof; Petersson, Magnus; Silva, C.; and Skogrand, Kjetil, "Comments on Ola Tunander's article 'The Uneasy Imbrication of Nation-State and NATO: The Case of Sweden'", in *Cooperation and Conflict*, Vol. 35, No. 4, 2000, pp. 417–429; Kronvall, O. and Petersson M., *Svensk säkerhetspolitik i supermakternas skugga* 1945–1991 (Stockholm: Santérus, 2005); Petersson, M., "'She would not *fight* unless attacked'. Några kommentarer rörande NATO:s syn på Sverige 1949–1965." in *Kungl krigsvetenskapsakademiens handlingar och tidskrift*, No. 3, 2002, pp. 73–88; Engelbrekt, K., "Den sjuttonde alliansmedlemmen?" in *Internationella Studier*, No. 4, 1999; Petersson, M., "Sverige och Väst. Det säkerhetspolitiska samarbetet mellan Sverige, Norge och Västmakterna 1949–1969" in Lars Wedin and Gunnar Åselius (eds.), *Mellan byråkrati och krigskonst. Svenska strategier för det kalla kriget.* (Stockholm: Försvarshögskolan, 1999), pp. 165–234; af Malmborg, Mikael, "Sweden – NATO's Neutral 'Ally'?..." in G. Schmidt (ed.), *A History of* NATO..., pp. 295–314, for explicit refutation, see: p. 295.

8 SOU 2002:108. *Fred och säkerhet. Svensk säkerhetspolitik 1969–89* (Stockholm: Fritzes, 2002), p. 342. This is a report from the Inquiry on Security Policy (Säkerhetspolitiska utredningen, SPU) appointed by the government in 2000. I

statement did not have any major effects, mainly because it seemed so utterly uninformed. Knowledge of Sweden's cooperation with the Western Powers during the Cold War was virtually non-existent, a fact which made Weinberger's words incomprehensible to both the Swedish public and the scholarly community. The utterance actually pointed to a significant discrepancy between the view of Swedish neutrality policy that the Swedish government officially wanted to convey to the United States, and the view that the latter actually held. The great increase in interest in Sweden's relations with the Western Powers during the Cold War did not gain real momentum until the beginning of the 1990s. Since then it has been growing rapidly.[9] Quite a few of these works have been published in English.[10] Even so, the number of monographs dealing

have quoted the Swedish version here because the English translation does not include Weinberger's comment. For the United States' view of Swedish neutrality over the years and its attitude towards neutrality in general, see: Schlesinger, Thomas O., *The United States and the European Neutrals* (Wien: Wilhelm Braumüller Universitäts-Verlagsbuchhandlung, 1991); for an interesting study of Swedish-American relations during a diplomatic crisis (although it is outside the time frame of my thesis), see: Leifland, Leif, *Frostens år. Om usa:s diplomatiska utfrysning av Sverige* (Stockholm: Nerenius & Santérus, 1997).

9 In her 1997 overview of the prior research in this field Ann-Marie Ekengren found that 48 titles had been produced between 1991–1995, with a total of 225 works had been written in the period 1945–1995 (Ekengren, Ann-Marie, *Sverige under kalla kriget 1945–1969. En forskningsöversikt.* (Stockholm: Forskningsprogrammet *Sverige under kalla kriget Arbetsrapport nr 1*, 1997), p.37.) For a broad overview in English, see: Bjereld, Ulf and Ekengren, Ann-Marie, "Cold War Historiography in Sweden" in Thorsten B. Olesen, *The Cold War – and the Nordic Countries: Historiography at a Crossroads* (Odense: University Press of Southern Denmark, 2004), pp. 143–175.

10 See: Aunesluoma, Juhana, *Britain, Sweden and the Cold War 1945–1954*, (Oxford: St Anthony's College, 1998); Aunesluoma, J., *Britain, Sweden and the Cold War 1945–1954: Understanding Neutrality* (Basingstoke: Palgrave Macmillan, 2003); Cole, Paul M., *Neutralité du jour:...*; Lindström, Stefan, "The open society and its stewards", in K. L. Kleve (ed.), *The Cold War, Military Power...*, pp. 127–150; Luders, Emma, *Sweden and Western Intelligence Collaboration: A Regime Study* (Stockholm: Stockholm International Studies 91:3, 1992); Molin, Karl, "The Central Issues of Swedish Neutrality Policy", in Michael Gehler and Rolf Steininger (eds.), *The Neutrals and the European Integration* (Wien, Köln, Weimar: Böhlau Verlag, 2000), pp. 261–275; Moores, Simon, "'Neutral on our side': us Policy towards Sweden during the Eisenhower Administration", in *Cold War History*, Vol. 2, No. 3, April 01, 2002, pp. 29–62; Moores, S., *'Neutral on Our Side':...*; Petersson, Magnus, "The theory and practice of the 'Nordic Balance' during the 1950s and 60s, with special emphasis on Swedish-Norwegian relations", in K. L. Kleve (ed.), *The Cold War, Military Power...*, pp. 25–34; Kronvall, Olof, "Finland in Swedish Security Policy, 1948–1962", in K. L. Kleve (ed.), *The Cold War, Military Power...*, pp. 35–44; Silva, C., *Keep Them Strong, Keep Them Friendly: Swedish-American Relations and the Pax americana, 1948–1952*, (Stockholm: Ph.D. diss., University of Stockholm, 1999); Tunander, O., "The Uneasy Imbrication of Nation-State..."

exclusively with Sweden's security relations with the Western Powers published in English is very limited. This means that knowledge of this aspect of the United States' foreign policy is relatively unfamiliar. This work is partly intended to remedy this situation, and to shed further light upon Sweden's role in the Cold War on NATO's northern flank. I also hope that my theoretical framework will provide for a better understanding of Swedish foreign, as well as domestic, policy during these years.

Research has shown that the precise content of the neutrality policy was debated all through the period from 1945 to 1962. Karl Molin has written about the experts' critique of the government's policy, and has found that the bureaucratic and military elite did their best to express their dissenting views—even contradicting official statements in their conversations with foreign officials and military officers.[11] Jacob Stridsman has studied more closely (although solely using Swedish sources) how one of these professional critics (Erik Boheman, Ambassador to Washington, 1948–1958) set about influencing the government's policies. A conclusion that he drew was that Boheman tried to influence the government by repeatedly emphasizing that the Soviet Union did not find Sweden's policy of neutrality credible, and that it only served to alienate Sweden from the Western Powers. Boheman also,

---

in *Cooperation and Conflict*, Vol. 34, No. 2, 1999, pp. 169–203; and the answer by Kronvall, O.; Petersson, M.; Silva C.; and Skogrand, K., "Comments on Ola Tunander's article..."" in *Cooperation and Conflict*, Vol. 35, No. 4, 2000; Weinberger, H, "The Neutrality Flagpole:..." in M. Thad Allen and G. Hecht (eds), *Technologies of Power:...*, pp. 295–331; Weinberger, H., *Technology: and the de/construction...*; Weinberger H., "A Small/Medium-sized Great Power..." in K. L. Kleve (ed.), *The Cold War, Military Power...*, pp. 45–64; Dahl, A.-S., "The Myth of Swedish Neutrality" in C. Buffet and B. Heuser (eds.), *Haunted by History:...*, pp. 28–40; af Malmborg, Mikael, *Neutrality and State-building in Sweden* (Hampshire: Palgrave, 2001); *Had There Been a War...*; Bengt Sundelius (ed.), *The Committed Neutral: Sweden's Foreign Policy* (Boulder: Westview Press, 1989); Swedish-American relations 1945–1962 are dealt with briefly in: Hanhimäki, Jussi M., "Security and Identity: the Nordic Countries and the United States since 1945" in G. Lundestad (ed.), *No End to Alliance: The United States and Western Europe: Past, Present and Future* (New York: St. Martin's Press, 1998), pp. 81–91; see also: Hanhimäki, J. M., *Scandinavia and the United States: An Insecure Friendship* (New York: Twayne Publishers, 1997); Aalders, Gerard, "The Failure of the Scandinavian Defence Union, 1948–1949" in *Scandinavian Journal of History*, Vol. 15, No. 1, 1990, pp. 125–153; Aalders, G., *Swedish Neutrality and the Cold War, 1945–1949* (Amsterdam: Ph.D. diss., University of Nijmegen, 1989); *SOU 2002:108 Peace and Security: Swedish Security Policy 1969–89. Abridged Version and Translation of SOU 2002:108* (Stockholm: Swedish Government Official Reports, 2004); Dalsjö, Robert, *Life-Line Lost: The Rise and Fall of 'Neutral' Sweden's Secret Reserve Option of Wartime Help from the West* (Stockholm: Santérus Academic Press, 2006).

11 Molin, Karl, *Omstridd neutralitet. Experternas kritik av svensk utrikespolitik 1948–1950* (Stockholm: Tiden, 1991).

due to the fact that he did not have share the views of his government, tried to portray the Swedish policy as less than definite to the British and the Americans.[12]

The Dutch historian Gerard Aalders has written about Sweden's neutrality policy during the immediate postwar period, as well as the failure to establish a formal Scandinavian Defense Union (SDU). His conclusions are that the Cold War did not in any notable way affect the policy of neutrality, and that Finland's situation was of little or no significance in the formulation of Swedish foreign policy, despite the occasional public hint to the contrary. He also concludes that Swedish officials saw Norway as the party guilty of dooming the SDU by purposefully seeking to align itself with the Western Powers and NATO.[13]

The American attitude toward Sweden's neutrality policy can be found in Norwegian historian Geir Lundestad's classic study *America, Scandinavia, and the Cold War, 1945–1949*. He showed that American policymakers were very critical of Sweden and its choice of security

12 Stridsman, Jacob, "Erik Boheman – västvänlig svensk diplomat under det kalla kriget" in Kent Zetterberg (ed.), *Hett och kallt – Tre studier i svensk säkerhetspolitik och strategi gentemot västmakterna i början av det kalla kriget. Hot and Cold—Three Studies in Swedish Security Politics and Strategy towards the Western Powers in the beginning of the Cold War* (Stockholm: Försvarshögskolan, 1999), pp. 99–141.

13 Aalders, G., *Swedish Neutrality and the Cold War, 1945–1949*; Aalders, G., "The Failure of the Scandinavian Defence Union, 1948–1949" in *Scandinavian Journal of History*, Vol. 15, No. 1, 1990, pp. 125–153. A lot has been written on the Scandinavian Defense Union, cf.: Wahlbäck, Krister, "Norden och blockuppdelningen 1949–49" in *Internationella Studier, 1973:b* (Stockholm: The Swedish Institute of International Affairs, 1973); Molin, K., *Omstridd neutralitet…*, pp. 46–82; Ekman, Jan G., "Förhandlingarna om Skandinaviskt Försvarsförbund, 1948–49 (Paper presented at the Department for Political Science at Stockholm University, spring 1967); Haskel, Barbara G., *The Scandinavian Option: Opportunities and Opportunity Costs in Postwar Scandinavian Foreign Relations* (Oslo: Universitetsforlaget, 1976), pp. 39–87; Skodvin, Magne, "Nordic or North Atlantic Alliance?: The Postwar Scandinavian Security Debate" in *Forsvarsstudier*, No. 3, 1990 (Oslo: Institutt for forsvarsstudier, 1990); Zetterberg, Kent, "Skandinavisk samverkan? 'N-D-S-1945–1949.' En studie i militärpolitiskt samarbete, strategisk analys och den säkerhetspolitiska prövningen av ett skandinaviskt försvarsförbund mellan krigsslutet och Atlantpaktens bildande." in K. Zetterberg (ed.), *Hotet från öster. Svensk säkerhetspolitik, krigsplanering och strategi, 1945–1958* (Stockholm: Försvarshögskolan, 1997), pp. 17–74. For Soviet perspectives and actions during the period of the Scandinavian Defense Union negotiations, see: Dörfer, I., "Soviet Policy in the North: Finland, Sweden, Norway, 1948–1949" (Paper prepared at Harvard University, 1964); see also: Holtsmark, Sven G., "Enemy springboard or benevolent buffer?: Soviet attitudes to Nordic cooperation, 1920–1955" in *Forsvarsstudier*, No. 6, 1992 (Oslo: Institutt for forsvarsstudier, 1992), pp. 59–65.

policy during these years. The book points out that this judgment was shared by basically all government branches, and that Sweden took a less prominent place than Norway in American strategic considerations.[14]

Sten Ottosson has studied the use of moral arguments in the foreign policy debate by politicians and opinion makers, from the end of the war until the early 1950s. The actors included in his study shared a common aversion to Communism, even though their views upon Swedish neutrality policy differed. None of the actors argued for a Swedish membership in a Western Alliance, except the newspaper *Dagens Nyheter* and its editor Herbert Tingsten. Ottosson concluded that the debate can be viewed as a struggle over the implications and content of the policy of neutrality.[15] The domestic inter-party debate on Sweden's neutrality policy has been analyzed by Fredrika Björklund. She points to the fact that even though the political parties held differing political views, there was a consensus regarding neutrality as the only viable policy open to Sweden—a phenomenon she calls 'unity under discord'.[16] Per Ahlin has studied the way in which Swedish governments have made use of international law in the design and implementation of Swedish security policy after 1945. One conclusion he reached was that the interpretations of the norms have shifted according to national self-interest.[17]

In his dissertation *Brobygge eller blockbildning?* Erik Noreen investigated the security policy perceptions of the Norwegian and Swedish governments. His focus was on the policy considerations leading to the failure of the Scandinavian Defense Union negotiations in 1948.[18] The bridge-building policy was, argues Noreen, only a short intermezzo in Swedish foreign policy, and ended with the intensification of the Cold War in early 1948. After this point, states Noreen, the government (and Undén in particular) reverted to a stricter interpretation of the neutrality

14 Lundestad, G., *America, Scandinavia, and the Cold War, 1945–1949* (New York: Columbia University Press, 1980). For another study of the same period, but from a different perspective, see: Sevón, Cay, *Visionen om Europa. Svensk neutralitet och europeisk återuppbyggnad 1945–1948* (Helsinki: Suomen Historiallinen Seura, 1995).

15 Ottosson, Sten, *Den (o)moraliska neutraliteten. Tre politikers och tre tidningars moraliska värdering av svensk utrikespolitik 1945–1952* (Stockholm: Santérus Förlag, 2000).

16 Björklund, Fredrika, *Samförstånd under oenighet. Svensk säkerhetspolitisk debatt under det kalla kriget* (Stockholm: Almqvist & Wiksell International, 1992).

17 Ahlin, Per, *Folkrätten i svensk säkerhetspolitik* (Stockholm: Juristförlaget, 1993).

18 Noreen, Erik, *Brobygge eller blockbildning? De norska och svenska utrikesledningarnas säkerhetspolitiska föreställningar 1945–1948* (Stockholm: Carlsson Bokförlag, 1994).

policy.[19] Ottosson in a sense continued where Noreen left off, and produced a similar study of how eight Swedish Foreign Ministers viewed Sweden from 1950 until the end of the Cold War. His main conclusion was that the Swedish self-image became more confident in the early 1960s, and that this led to a shift towards a more active foreign policy.[20] Mikael af Malmborg has written what he calls a post-revisionist account of Sweden's relations with the NATO powers.[21] He has suggested that rather than discussing whether Sweden was 'neutral' or not, "the focus should be on how neutrality policy was and is transformed."[22] This is a very wise attitude that I intend to abide by in my study.

The inter-scholarly debate, on the other hand, has centered on what consequences the Swedish government's secret preparations for collaboration with the West had for the officially endorsed view of Sweden as non-aligned[23] in the Cold War.

Prior Research: Sweden and the Western Powers During the Cold War

The first scholars to question the official declarations regarding Sweden's behavior during the Cold War belonged to the so-called critical revisionist school. This group, of which Paul Marion Cole and Wilhelm Agrell were the most prominent, made Caspar Weinberger's utterance regarding Sweden's neutrality understandable by putting it into a historical context. In 1990, American historian Paul Cole argued in his dissertation *Neutralié du jour:...* (and in a couple of articles in Swedish newspapers in 1990)[24] that the Swedish policy of neutrality had been nothing but a chimera. The Swedes, said Cole, had all along collaborated with the Americans, expecting to get all the benefits of being an ally without making any sacrifices themselves.[25] Cole's results were, however, made

19 Noreen, E., *Brobygge eller blockpolitik?...*, pp. 225–226.

20 Ottosson, Sten, *Svensk självbild under det kalla kriget. En studie av stats- och utrikesministrarnas bild av Sverige 1950–1989*. (Stockholm: Utrikespolitiska institutet, 2003).

21 af Malmborg, M., "Sweden – NATO's Neutral 'Ally'?..." in G. Schmidt (ed.), *A History of* NATO..., pp. 295–314.

22 Ibid., p. 313.

23 The term "non-alignment" will be used in this thesis since it has been frequently used to describe the Swedish position in the Cold War. Traditionally, however, the concept has been used to designate the policies of (most often) countries such as India and Yugoslavia, and Third World countries, which are the nations attributed with having invented this policy. For this, see: Vukadinovic, Radovan, "The Various Concepts of European Neutrality" in M. H. Haltzel and J. Kruzel (eds.), *Between the Blocs:...*, pp. 40–44.

24 See: Cole, P. M., "USA-samarbetet hemlighölls" in *Dagens Nyheter*, October 21, 1990 and "Neutraliteten gavs upp" in *Dagens Nyheter*, October, 31, 1990.

25 Cole, P. M., *Neutralité du jour....*

problematic by his clearly expressed moral indignation over Swedish behavior and the way this indignation biased his reading of his sources. Nevertheless, it was an important study because it came so early, and it remains so today due to its rich source material.

The first Swedish scholar to grapple with the subject was military historian Wilhelm Agrell. In 1991 he published the book *Den stora lögnen* which attracted much attention in the media. Agrell argued approximately along the same lines as Cole, i.e. that Sweden's policy of neutrality and non-alignment was essentially a sham. Agrell showed that Sweden had collaborated extensively with the Western Powers during the Cold War, primarily through its contacts with NATO members Norway and Denmark. He did not, however, express the same indignation regarding the Swedish 'free riding' that Cole displayed in his thesis, although he too was certainly normative in his judgment.[26]

These, as well as intermittent media revelations, created a political stir, and repeated calls for an investigation into this matter caused the Conservative government to appoint a commission in 1992 that was supposed to 'get to the bottom' of the issue. Neutralitetspolitikkommissionen [Henceforth: the Neutrality Policy Commission][27] presented its report *Om kriget kommit…* in 1994, which became the first major contribution to the moderate revisionist school. The report managed to do two things at the same time: re-write Swedish post-War history, as well as set the official record straight and confirm the *status quo*. Interestingly, it confirmed the critical revisionists' findings, but in addition uncovered even more details about Sweden's collaboration with the West and its preparations for military assistance in event of war. Despite this, the Commission concluded that none of this had infringed upon Swedish sovereignty or freedom of action. The neutrality and non-alignment policy were still credible because no agreement had been signed that would force Sweden in on any side at the outbreak of hostilities. According to the Commission, all was calm; nothing had changed.[28] And yet everything was different.

The Commission's report generated harsh criticism from the scholarly community. The critique concerned the limited time span available to

26 Agrell, W., *Den stora lögnen…*. The issue was also investigated in the early 1990s in a series of broadcast programs on Swedish Radio's Program 1 by journalist Christer Larsson.

27 The Commission was to look into the Swedish preparations for receiving military assistance from the West 1949–1969. It presented its report SOU *1994:11 Om kriget kommit… Förberedelser för mottagande av militärt bistånd 1949–1969* (Stockholm: Fritzes, 1994) in 1994. For an English version, see: *Had There Been a War…*. The latter version will be used for quotations throughout this thesis.

28 *Had There Been a War…*, p. 264–268.

the Commission, its investigative methods, as well as the conclusions drawn from the available sources.[29] Perhaps because of its controversial results, it spurred a research frenzy regarding Sweden's behavior during the Cold War. That frenzy has yet to die out, and this thesis is but one small result of it.[30] Magnus Petersson examined Sweden's cooperation with Norway, 1949-1969, in his dissertation *"Brödrafolkens väl"*. He showed how, through bilateral cooperation with Sweden, Norway was made effectively into a proxy for NATO in the north. Norway became a conduit that made possible exchange of information, personal contacts, and military planning between Sweden, Norway and the Western Powers. Petersson concluded that this arrangement could work because the leading actors in this process, i.e. politicians, diplomats, and military officials, had similar views on strategic issues.[31]

Olof Kronvall has written about Finland in Swedish security policy from the late 1940s to the early 1960s in his dissertation *Den bräckliga barriären*.[32] Kronvall has found that 'the Finland argument', i.e. that a closer Swedish connection to the West would lead to Soviet inclusion of Finland, was less important to the Swedish government than to the West-friendly opposition and diplomats—in a sense re-stating Aalders'

29 Nilsson, Mikael, "'Gud vare tack och lov så talade de inte om allt de gjorde' : en källkritisk granskning av Neutralitetspolitikkommissionens betänkande SOU 1994:11.", (Stockholm: TRITA-HST Working paper, 2003/2; Blomberg, Mats, "Efter Neutralitetspolitik-kommissionen: 50-talets verkliga säkerhetspolitik är bara delvis tecknad" in *Kungl Krigsvetenskapsakademiens handlingar och tidskrift*, No. 1, 1995, pp. 97-118; Dörfer, I., "Neutralitetsbygget – ett spökslott för inbördes beundran.", *Kungl Krigsvetenskapsakademiens handlingar och tidskrift*, No. 1, 1995, pp. 17-19; Weinberger, H., "'Det rörde sig väl om ofarliga ting'..." in *Arbetarhistoria...*, No. 92, 4/1999, pp. 31-36. For a critique of the Commission's report, and a defense of the Social Democratic government, see: Björk, Kaj, "Anteckningar om neutralitetspolitikkommissionen" in *Tiden*, 1994, 86:7, pp. 31-34.

30 Two research programs were established. The first was the large interdisciplinary *Sverige under kalla kriget* (SUKK) [Sweden During the Cold War], a collaboration between several universities in Stockholm, Gothenburg, and Uppsala. The second was the much smaller *Nätverk, neutralitet och teknik* (NNET) [Networks, Neutrality, and Technology], which was based at Kungliga teknisk högskolan (KTH) [the Royal Institute of Technology] (my work is the first dissertation that has been produced within the latter project).

31 Petersson, M., *"Brödrafolkens väl". Svensk-norska säkerhetsrelationer 1949–1969.* (Stockholm: Santérus Förlag, 2003); see also: Petersson, M., "Sverige och Väst. Det säkerhetspolitiska samarbetet mellan Sverige, Norge och Västmakterna 1949–1969", in Lars Wedin and Gunnar Åselius (eds.), *Mellan byråkrati och krigskonst. Svenska strategier för det kalla kriget.* (Stockholm: Försvarshögskolan, 1999), p. 165–234.

32 Kronvall, Olof, *Den bräckliga barriären. Finland i svensk säkerhetspolitik 1948–1962.* (Stockholm: Försvarshögskolan, 2003).

conclusions on the issue. Even though the government sometimes found this argument handy in conversations with Western diplomats and statesmen, they very seldom used it publicly. He related this fact to the government's fear of officially tying a reassessment of non-alignment to a Soviet annexation of Finland—the possibility of remaining non-aligned even in the face of such a move was of prime importance, according to Kronvall.[33]

Of central importance to the history of Swedish-American relations from 1948–1952 is Charles Silvas' dissertation *Keep Them Strong, Keep Them Friendly*. In 1999, Silva was the first scholar to study this subject in depth. He was also the first to attempt to apply an interesting theoretical framework, the concept of 'informal empire', to explain Swedish foreign policy behavior during this period. But this theoretical perspective, as is often the case in historical works, was noticeably absent in the empirical study and also curiously missing in the conclusions and analysis.[34] In fact, in the concluding chapter the term Pax Americana, which is presented as an important theoretical concept in the opening chapter, is mentioned only once, and then in passing.[35] The book looks at Swedish-American security relations from the perspective of both governments, and Silva's conclusion is that from being at loggerheads in 1948 the Swedish and American governments, after a shift in American attitude in early 1950, eventually found a working relationship by 1952. This gradual development was clearly discernable in the areas of economy, trade, and security. He also found that the Swedish-American relationship verified the aspects of 'informal empire'. This was done by formal as well as informal agreement, with the United States expanding its influence, until Sweden had become a consenting subject to Washington's hegemony—the Pax Americana.[36] On the surface, Silva's conclusions and mine seem to be approximately the same, but there are important differences, all of which will be revealed in this study.

Thomas Jonter argued on the same lines as Silva in his dissertation *Socialiseringen som kom av sig*, i.e. that Sweden was integrated in a new world order during the late 1940s, and that the Swedish government was essentially pressured to accept this order by the United States. However, his study only covered the period 1945–1949.[37] My study has a similar

33 Ibid., pp. 101–112, 192–200, 324–336, 359.

34 Silva, C., *Keep Them Strong, Keep Them Friendly:...*, pp. 8–10. Silva has chosen to use this concept instead of hegemony, although he refers to the latter in passing.

35 Ibid, pp. 8–13, 356.

36 Ibid., p. 349.

37 Jonter, T., *Socialiseringen som kom av sig. Sverige, oljan och usas planer på en ny världsordning 1945–1949* (Stockholm: Carlssons Bokförlag, 1995).

argument, but I will show that it was more complicated than that, since the Swedish government wanted Washington's presence and support. Silva and Jonter have both argued that Sweden's foreign and security policy underwent a profound change from 1945 and up to 1950; a process during which the Swedish government adopted a foreign policy position more acceptable to the policymakers in Washington than prescriptive neutrality.[38] The main reason for this, writes Silva, was the fear that Swedish neutrality policy would otherwise, ironically, lose its credibility in the eyes of the Americans.[39]

In his dissertation, *Hemlig utrikespolitik*, Stefan Ekecrantz investigated the parliamentary opposition's ability to influence foreign policy through the Advisory Council on Foreign Affairs (Utrikesnämnden), and the government's choices when provided the choice of informing or not informing the opposition in the Council. The conclusion Ekecrantz drew was that the government acted differently in each of the cases studied. This was due to perceived demands forced on the government from external or domestic situation, and the government sometimes chose to provide the Council with satisfactory information, sometimes informed it after the fact, and sometimes did not inform it at all, keeping the matter completely secret.[40]

The most recent study of Swedish-American relations is Robet Dalsjö's *Life-Line Lost*.[41] In his well-researched study Dalsjö analyzes how Sweden's foreign policy declarations during four decades were connected to what he calls the decline of the Western security option. He concludes that "changes in declaratory doctrine […] made the maintenance of preparations too sensitive. Over time", he adds, "rhetoric changed reality."[42] In contrast to Dalsjö, Hans Lödén concluded in his dissertation that the opportunity for making the changes in Swedish foreign policy was provided by the *détente* between the superpowers.[43]

38 Jonter, T., "En tänkbar tolkningsram för svensk-amerikanska studier under kalla kriget" in T. Jonter and C. Silva, *Sverige inför en ny världsordning 1945–50…*, pp. 29–42; Silva, C., "Att omvärdera svensk utrikespolitik 1945–1950" in T. Jonter and C. Silva (eds.), *Sverige inför en ny världsordning 1945–50…*, pp. 9–28.

39 Silva, C., "Att omvärdera svensk utrikespolitik 1945–1950", p. 13.

40 Ekecrantz, Stefan, *Hemlig utrikespolitik. Kalla kriget, utrikesnämnden och regeringen 1946–1959* (Stockholm: Santérus Förlag, 2003).

41 Dalsjö, R., *Life-Line Lost:…*

42 Ibid., p. 279.

43 Lödén, H., *För säkerhets skull…*, p. 405.

Although Dalsjö provides a very brief argument that such an explanation is not plausible,[44] he does not explicitly address Lödén's results.

Prior Research: Guided Missiles

Although a lot has been written on Swedish security and neutrality policy in general, it is also correct, as Simon Moores has pointed out, that comparatively little attention has been given to the military and security cooperation between the United States and Sweden during the Eisenhower administrations, 1952–1960.[45] This thesis is a part of the continuing effort to fill this knowledge gap, and to complement the results obtained from other studies made on other subjects. Research on development, and technology transfer, of guided missile technology has indeed not been abundant, not even internationally. Ron Westrum's book *Sidewinder*, which deals with the development of the famous U.S. Navy heat-seeking guided missile, is only partly an exception. The book deals extensively with the *development* aspect but does not concern itself at all with the dispersion of this technology to the United States' allies.[46] As I mentioned in Chapter 1, research regarding guided missiles, also in the history of technology field, has concentrated on nuclear missiles.[47]

A few scant references were made to Swedish guided missiles in Ingemar Dörfer's *System 37 Viggen* published in 1973.[48] Dörfer's book was for a long time the only Swedish scholarly work that mentioned guided missiles at all. In the non-scholarly catalogue, a short history of surface-to-air missiles could be found in *Luftvärnets historia*.[49] As with so many other publications, though, it was totally void of theoretical perspective and historical analysis. Another example is the official eulogy *Robot 50 år*, written for the 50th anniversary of the guided missile development at the Swedish automobile and aircraft manufacturer SAAB,[50] which for a long time was the most complete account of Sweden's guided missile development. Thomas Jonter briefly addressed the Swedish

44 Dalsjö, R., *Life-Line Lost:...*, pp. 27–28.

45 Moores, S., *'Neutral on Our Side':...*, p. 9. Moore's dissertation has initially the same title as an article he wrote for *Cold War History*; I have quoted only the dissertation in my thesis.

46 Westrum, R., *Sidewinder:...*. See also: Babcock, E., *Sidewinder: Invention and Early Years...*.

47 Cf. MacKenzie, D., *Inventing Accuracy:...*.

48 Dörfer, I., *System 37 Viggen. Arms, Technology and the Domestication of Glory* (Oslo: Universitetsforlag, 1973).

49 Carl Herlitz (ed.), *Luftvärnets historia* (Stockholm: Militärhistoriska Förlaget, 1980).

50 Cervin-Ellqvist, Gunnar and König, Bertil, *Robot 50 år* (Linköping: SAAB, 1997).

missile purchases, as well as the concomitant security agreement, in his article “Det amerikanska spåret”, which was a part of the governmental Intelligence Committee’s report.[51]

Of central importance to my work has been Simon Moores’ *‘Neutral on Our Side’:...*. One of the issues studied by Moores was Sweden’s efforts to get technical information regarding guided missiles in the 1950s, and the eventual purchase of American guided missile systems by Sweden at the end of the decade.[52] Moores covered the period of the two Eisenhower administrations and, when dealing with the guided missile question, his focus was on the machinations on the part of Swedish and American diplomats and military personnel to try and convince the Pentagon and the Department of Defense (DOD) to release guided missile technology to Sweden. The work performed by Moores has served as an excellent point of reference for my investigation. Our source material differs and while Moores has concentrated on American archives he has not touched upon any of the Swedish archival material that I have made use of. This is probably because Moores has not been interested in the Swedish guided missile development but only in certain aspects of how the Swedes finally gained access to American guided missiles. Researching the guided missile issue was but one of four objectives which Moores formulated in his thesis; the other three concerned the overall military and security relations between the United States and Sweden during Eisenhower’s tenure as President.[53] This probably also contributed to the neglect of the domestic guided missile development.

What Moores did do in his thesis, however, (besides providing a penetrating history of Swedish-American security relations generally) was to point out that guided missiles were important in bolstering Sweden’s defense capacity.[54] Guided missiles were regarded “as being an intrinsic component of the Armed Forces’ military capabilities.”[55] Moores’ results further underline the pertinent nature of the subject of this thesis. This is quite natural if one pauses to think about it. What good would it do to have the fourth largest Air Force in the world if one had no weapons to arm them with? The effectiveness of the defense of Sweden’s ‘non-aligned’ air space in the 1960s and onwards hung in the balance.

51 Jonter, T., “Det amerikanska spåret. En undersökning av ib:s bildande och eventuella kopplingar till USA.” in SOU 2002:95 *Forskarrapporter till Säkerhetstjänstk ommissionen* (Stockholm: Fritzes offentliga publikationer, 2002), pp. 135–191.

52 Moores, S., *‘Neutral on Our Side’:...*

53 Ibid., p. 10.

54 Ibid., pp. 64, 124–125.

55 Ibid., p. 129.

It is now time to identify the blind spots in the history of Swedish foreign and security policy during the Cold War, and to state what I will do in this dissertation that no one has done before me. As I stated above, my contribution to the bibliography of Swedish-American security relations during the Cold War is twofold, i.e. on both a theoretical and an empirical level. It is clear from the summary of prior research made above that there is no scholarly account of the Swedish guided missile development program, and that my study will break new ground in this area. Although Moores has written about Sweden's efforts to procure guided missiles from the United States, I have made use of American documents not utilized by Moores, and I have also re-interpreted many of the documents used by him. Furthermore, no prior study has put the Swedish experience into an international context, i.e. no one has so far explored the similarities and differences between Sweden and other nations with regard to military cooperation the United States.

There has been a tendency to view the Swedish experience as unique, due to the nature of the government's foreign policy. However, as this thesis will show this view actually obscures more than it reveals, and Swedish Cold War history would benefit from more comparisons with other Western democracies. Such an approach would not only make it possible to isolate what was truly distinctive about the Swedish experience, but it would also reveal the more or less universal traits in the U.S. government's relations with the European countries. For example, there is good reason to assume that U.S. policy towards Switzerland followed the same lines as towards Sweden.[56] Moreover, such comparative histories are necessary if we want to know how Sweden was treated compared to the NATO countries. The hegemony perspective has provided me with an excellent opportunity to fill at least a part of this lacuna in Swedish Cold War history. The fact that there has been almost nothing written on the use of technology for achieving hegemonic consent, even internationally, also sets my thesis in a truly unique position.

My hope is that I have been able to make the best of this opportunity and that it might spawn similar histories elsewhere in the future. As Nikolaj Petersen has remarked, distinguished Cold War historian John Lewis Gaddis' claim that 'we now know' what there is to know about the Cold War "was pretentious because of its underlying idea of 'the end of history' […]." Historical knowledge is always dependent upon what questions we ask, and "we will never arrive at a definitive answer:

56 Cf. Ammann, Klaus, "Swiss Trade with the East in the Early Cold War" in Jari Eloranta and Jari Ojala (eds.), *East-West Trade and the Cold War* (Jyväskylä: University of Jyväskylä, 2005), pp. 113–131.

Now We Know!"[57] We can, however, come much closer to the truth than at present.

A Note on Sources

When writing this dissertation I have, with very few exceptions, made use of publicly available documents from archives in Sweden and the United States (and to some degree in Belgium[58]). The exception to this rule is the report *saabs robot- och elektronikhistoria* [SAAB's Guided Missile and Electronics History] from SAAB Dynamics in Linköping, which provided the background material for the book *Robot 50 år*. This report has not been published but can be accessed by contacting the author of the present study.

I have also made use of documents readily accessible on the internet. The internet presents some delicate problems from the point of view of source criticism, and it can sometimes be hard to know whether or not the information found on a particular website is accurate. However, I have made use of articles found on academic websites, or documents on official websites, such as nato.int. It has not been possible to go through all the relevant public and private archives when writing this book. I have made use of secondary sources that are, at least partly, based on documents from archives I have not been able to study. The majority of the documents used have been found in the Swedish Military Archives (Krigsarkivet), the National Archives of Sweden (Riksarkivet), both in Stockholm, Sweden, and the National Archives and Records Administration (NARA) in College Park, Maryland, in the United States. I have also made limited use of the Neutrality Policy Commission Archive at the Swedish Government Offices Records Center (Regeringskansliets arkiv- och dokumentcenter) in Stockholm. A detailed list of the records used in the thesis can be found in the bibliography. It is important to note that the archival material does not tell the whole story. Sweden's secret military cooperation with the United States was so sensitive that a lot was never put down on paper in the first place, and some of the things that were put into print or written down, were often not entered

57 Petersen, Nikolaj, "'We Now Know'. The Nordic Countries and the Cold War: An Assessment", in T. B Olesen *The Cold War – and the Nordic Countries:...*, p. 177. See also: Gaddis, John Lewis, *We Now Know: Rethinking Cold War History* (Oxford: Clarendon Press, 1997).

58 I have not myself visited the NATO Archives, but I have made use of material gathered there by my advisor Hans Weinberger. The same is true for the Eisenhower Library in Abilene, Kansas. I thank him for letting me use the material gathered by him during his research trips. I am also indebted to my colleague Johan Gribbe, who has been kind enough to share some documents from the Swedish Armed Forces Headquarters (Högkvarteret, Hgkv).

in ledgers, thus creating the illusion that they never existed.[59] There has also been a tendency, within the Swedish military establishment, to burn sensitive records. Much is also still classified, especially records from the 1960s onwards.

The American archives thus contain a lot of information that is not available in the Swedish archives, such as the American version of the reasoning of Swedish diplomats and military officers when engaged in candid conversations with their American counterparts. Many American documents relating to Sweden are still classified as well, however, and I have not been able to get access to any previously classified U.S. documents, despite several Freedom of Information Act (FOIA) requests (these are all pending a decision from U.S. authorities). Even so, the American documents have been extremely helpful when trying to re-construct Swedish-American security cooperation. Most of the information found during my four months at NARA has not been included in this dissertation.

Other sources that have been used are published memoirs and diaries. These also present a special kind of problem for historians. While they can provide valuable insights into a central person's reasons for acting in a certain way etc, they can also be misleading if their content is *a priori* assumed to be the true version of events. Often, people might have had considerable incentive not to be completely honest about some sensitive issue, or they may remember the incident only vaguely. Sometimes, the historian is presented with a striking dissonance between the archival documents and the diary or memoir. It is most often the case that the version that is less flattering for the person, or country, in question is closest to what actually happened. One must also make room for the possibility that the person writing did not have all the facts clear to him or her, and hence that they are stating what they believed to be correct at the time. Most of these problems can be quite successfully managed if one compares several different types of sources.

Newspaper articles have been used in this study. While journalistic material provides a poor basis for conclusions regarding major events in Swedish-American security relations, such articles can be favorably used, e.g. when trying to discern what the military wanted the public to know concerning the guided missile program. I have also used published primary source-material from the Swedish parliament (Riksdagstrycket), the Riksdag, as well as the semi-official publications (statens offentliga utredningar) from various publicly appointed commissions on Swedish foreign and security policy, such as *Had There Been A War…* and *Peace*

59 Dalsjö, R., *Life-Line Lost:…*, p. 35; Lindström, S., "The open society and its stewards", in K. L. Kleve (ed.), *The Cold War, Military Power…*, pp. 127–130.

and Security:.... As stated above, the result of these investigations have been heavily criticized, but they nonetheless provide the historian with a valuable minimum level for what went on during the Cold War.

Another kind of problematic sources are the eulogies published by, for example, companies and/or private enthusiasts or former military employees, i.e. historical actors writing their own history. Such publications often provide the 'history' of aircraft or missiles (as in the case of *Robot 50 år* or FOA *informerar om Robotvapen*) with calendaric precision, but contain little analysis or critical thinking of value to the historian. Even though *Robot 50 år* is a celebration of SAAB's missile activity since 1945, I have, often due to scant archival records, made use of it when giving the background of Swedish guided missile development activities. The same is true with regard to *saabs robot- och elektronikhistoria*, even if this report could perhaps be assumed to be less problematic from a source-critical perspective than the book it resulted in, since it was not intended for publication in unedited form. However, I have found no major discrepancies when comparing the published and unpublished parts of the report. I have made an effort to remain critical of its content. The unpublished report contained a lot more on the different missile projects developed and worked on by SAAB than did either the book or the documents at KRA, and in this way it turned out to be a valuable source. The book *Robot 50 år*, and other works just like it, occupy a peculiar place, since they have not been subjected to scholarly review and criticism. Moreover, the report on which this book was based is not a primary source, but has been put together probably using primary sources from SAAB's company archive. I have tried to remain critical of the content of these sources.

A fact that could be regarded as a drawback for this study is the fact that I have not done any interviews in the course of writing it. This omission has been a conscious decision on my part. Most of the people involved during the late 1940s and early 1950s are no longer alive, and I have been able to answer my research questions in a satisfactory way even without interviews. Hence, it is entirely based upon written document, except where the secondary literature may contain information obtained through oral history interviews. An exception is an interview from the Neutrality Policy Commission Archive. This archive contains a lot of transcripts of interviews and conversations held with influential Swedish political officials and military officers. Oral history interviews are valuable to the historian, but can sometimes be treacherous as well.[60]

60 For a short, but good, discussion of the dilemmas presented to the historian by oral history, see: Hoddeson, Lillian, "The conflict of memories and documents: Dilemmas and pragmatics of oral history" in Ronald E. Doel and Thomas

For example, people may have only scant recollection of the period in question, and they may have conflated their own memories with things they have read or with information obtained after conversations with friends or colleagues. Interviews have, perhaps, their greatest value when there is a scarcity of written sources, or when there is a need to complement the written documents. This has not been a major problem in writing this thesis. I have tried, where possible, to corroborate the information obtained from these transcripts with primary or secondary sources.

Many of the sources are written in Swedish, which means that their content has been translated into English. I have done the translation myself, but since English is not my native language there is a good chance that they are flawed. In order to guard against accusations of misrepresenting the sources, I have provided the original Swedish text in the footnotes so that the translations can be checked by the reader. I have also chosen to translate the names of Swedish military institutions, departments, organizations, etc. In those cases where I have used an abbreviation to designate such an entity, I have abbreviated the English translation.[61] The names of Swedish companies have not been altered. I wish to point out that even though the text has been reviewed by a professional translator, I am solely responsible for any errors in spelling or syntax.

Finally, I would like to address some methodological problems which have to do with how we, as historians, interpret and evaluate the information contained in the sources. How can we determine the importance of the information related in a document? Which actors carried a lot of influence? Is the word of an actor from higher up in the government hierarchy worth more than the word of someone subordinate to him? All these issues pertain to our understanding of Swedish-American relations during the Cold War. Can we for example assume that U.S. policy was conducted in a certain way just because the American Ambassador, or some other lower official at the State Department, said so? While we cannot make any firm *a priori* statements as to how the sources should be read in every instance, since each document has to be evaluated in their context, Geir Lundestad has provided us with a good starting point.

Söderqvist (eds.), *The Historiography of Contemporary Science, Technology, and Medicine: Writing Science* (London: Routledge, 2006), pp. 187–198.

61 There is one exception to this rule: when referring to the National Defense Research Establishment (Försvarets forskningsanstalt), I have used the Swedish abbreviation FOA. The reason is that this acronym is so well known that it has virtually replaced the full version of the name in everyday use. In that way it resembles the name of civilian companies which are also better known by their acronyms, e.g. IBM, SAAB, or AEG.

According to him, the Scandinavian policy was handled almost exclusively by the regional experts in the State Department and by the personnel at the local Embassies.[62] Thus the lower level officials had considerable influence when it came to policy formulation and execution through their everyday handling of the cooperation, and through the information which they purveyed to senior officials such as the Secretary of State and the President. They, and other briefing officers, were the ones informing their superiors prior to meetings with, e.g. the Swedish Ambassador. Indeed, the President and the Secretary of State had only remote knowledge of U.S. everyday dealings with Sweden, and were dependent upon the briefing officers to provide them with pertinent knowledge. Nor could it had been any other way, for there were too many, and far more important, issues to deal with, and the President also had to concern himself with domestic issues. The Embassy in Stockholm, and the officials in the regional divisions of the State Department were the ones that, on the micro level, controlled and handled American hegemonic policy.

Most of the time these lower level officials had a lot of latitude when working to implement the policies decided upon in the higher echelons of the civilian and military bureaucracy, i.e. State Department, the Pentagon, and the White House. The latter entities, in their turn, worked within a framework laid down by the U.S. Congress. This means that the top officials knew much less about the details of Swedish-American cooperation than did, e.g. the American Ambassador in Stockholm. The same was probably true for the military, where it can be expected that the Joint Chiefs of Staff (henceforth: JCS) knew considerably less than the officers that met regularly with their Swedish counterparts.

Sweden: "Neutral as far as Possible, and Even A Bit Further"[63]

The word neutral is derived from the Latin roots *ne uter*, or 'neither of the two',[64] and as defined by international law (paragraphs V and XIII of the Hague Convention of 1907) neutrality refers to non-participation in open hostilities between other states, i.e. neutrality can only exist in times of war. Unlike Switzerland (whose neutrality was guaranteed at the Congress of Vienna, 1814–1815) and Austria (whose neutrality is written into its constitution), Sweden was never permanently neutral,

62 Lundestad, G., *America, Scandinavia, and…*, p. 48.

63 Stated by the Swedish King Oscar II in 1890, cited in: Kux, Stephan, "Europe's Neutral States: Partners or Profiteers in Western Security?" in *European Security Studies*, No. 4, 1986, p. 18.

64 Haltzel, Michael H., "The Complex Position of the European Neutrals" in Michael H. Haltzel and Joseph Kruzel (eds.), *Between the Blocs:…*, p. xvii.

and thus never had any special rights or obligations in peacetime.[65] Sweden's particular form of neutrality has been called *de facto* permanent neutrality, due to the fact that it has been the guiding principle for its foreign policy for so long.[66]

The guideline for Sweden's foreign policy during the Cold War was summarized in the axiom "'non-participation in alliances in peacetime, aiming at neutrality in war.'" This eventually came to be known as 'the Swedish neutrality policy',[67] even though the policy was actually not formulated in this way until October 1956.[68] Since 1815 Sweden has experienced a period of uninterrupted peace, but the interval up to World War I was not a period of uninterrupted neutrality. Rather, Sweden pursued something that could be called neutrality *ad hoc*, departing from neutrality in the Prussian-Danish war of 1848, and contemplating an alliance with Britain and France during the Crimean War 1853–1856. This makes it hard to put a definite date to the beginning of a consistent Swedish neutrality policy.[69]

65 *Had There Been a War...*, p. 49. Aunesluoma, J., *Britain, Sweden and...*, p. xvii. For a splendid comparison between the different types of European neutrality, see: Kux, S., "Europe's Neutral States: ..." in *European Security Studies*, No. 4, 1986. How Sweden has made use of, and interpreted, international law in its design and utilization of its security policy has been studied in: Ahlin, P., *Folkrätten i svensk säkerhetspolitik*.

66 Lindholm, Rolf H., *Sveriges neutralitet. Swedish Neutrality* (Lund: Juridiska Föreningen i Lund, 1987), p. 38. For arguments for calling Sweden a permanent neutral even though it is not bounded by law, see: Vukadinovic, R., "The Various Concepts of European Neutrality" in M. H. Haltzel and J. Kruzel (eds.), *Between the Blocs:...*, p. 31.

67 *Had There Been a War...*, pp. 50, 53, 261.

68 Ottosson, Sten, *Svensk självbild under det kalla kriget...*, p. 6, note 15.

69 Norman, Torbjörn, "Stages in Swedish Neutrality" in Jukka Nevakivi (ed.), *Neutrality in history: La neutralité dans l'histoire : proceedings of the Conference on the History of Neutrality organized in Helsinki 9–12 September 1992 under the auspices of the Commission of History of International Relations* (Helsinki: Suomen historiallinen seura, 1993), p. 304. For short histories of the evolution of Swedish neutrality in English, see also: Andrén, Nils, "Swedish Defense: Traditions, Perceptions, and Policies" in M. H. Haltzel and J. Kruzel (eds.), *Between the Blocs:...*, pp. 177–181; Carlgren, Wilhelm M., "The Emergence of Sweden's Policy of Neutrality" in "Neutrality and Defence: The Swedish Experience" in *Revue Internationale d'Historie Militaire*, No. 57 (Stockholm: 1984), pp. 11–33; Johansson, Alf W. and Norman, Torbjörn, "The Swedish Policy of Neutrality in a Historical Perspective" in "Neutrality and Defence:..." in *Revue Internationale d'Historie Militaire*, No. 57, pp. 69–94; Logue, John, "Sweden" in S. Victor Papacosma and Mark R. Rubin (eds.), *Europe's Neutral and Nonaligned States: Between* NATO *and the Warsaw Pact* (Wilmington, DE: SR Books, 1989), pp. 71–102. The most thorough account of the history of Swedish neutrality is: af Malmborg, M., *Neutrality and State-Building in Sweden.* Official documents relating to Swedish neutrality after 1945 can be found in:

The Swedish policy during these years has been described as "'a free-forming neutral policy, opportunist and interest-laden, with priority of policy before law'."[70] One aspect that has been said to be more or less constant since 1812 is the tendency of Swedish foreign policy not to be directed *against* Russia. Rather, it has been argued, Sweden's governments have constantly tried to placate the colossus in the east. According to this reasoning Sweden has taken this position due to the presence of great powers in the Nordic area, in essence, a kind of small state realism.[71] This conclusion, i.e. the continuity of the policy of 1812, has not been uncritically accepted, however.[72] By World War I, with the triumph of popular democracy, the policy of neutrality had become somewhat of an axiom for Sweden, the purpose of which was to keep Sweden out of war.[73] Despite the frequent deviations from the neutrality line in the past, writes historian Bo Stråth:

> [...] a line of continuity was established in historical and social sciences. A modern Swedish foundation myth emerged around the concept of neutrality, mediating feelings of Swedishness and historical continuity.[74]

In reality, however, the Swedish government acknowledged from the start that the membership of the United Nations (henceforth: U.N.) constituted a break with the past, and that it would make strict neutrality impossible.[75] Erik Noreen has argued that Swedish membership in the U.N. was made possible by a change of principle in Sweden's security

---

Lindholm, Rolf H., *Sverige under kalla kriget* (Stockholm: Santérus Förlag, 2000).

70 Luntinen, Pertti, "Neutrality in Northern Europe before the First World War" in J. Nevakivi (ed.), *Neutrality in History...*, p. 111.

71 For this, see: Talvitie, H., "Sverige och 1812 års politik" in T. Suominen (ed.), *Sverige i fred...*, pp. 15–74.

72 For this, see: Wahlbäck, K., "Sveriges naturliga ovilja att ta risker" in T. Suominen (ed.), *Sverige i fred...*, pp. 75–86. The question whether Sweden's secret cooperation with the United States during the Cold War was complementary, or an anomaly, to the policy of 1812 is a matter that could be debated at length. I thank Robert Dalsjö for drawing my attention to this book.

73 Andrén, N., "Swedish Defense: Traditions..." in M. H. Haltzel and J. Kruzel (eds.), *Between the Blocs:...*, p. 177; Logue, J., "Sweden" in S. V. Papacosma and M. R. Rubin (eds.), *Europe's Neutral and...*, p. 73.

74 Stråth, Bo, "The Swedish Image of Europe as the Other" in B. Stråth (ed.), *Europe and the Other and Europe as the Other* (Brussels: P.I.E.-Peter Lang, 2000), pp. 370–371.

75 *Had There Been A War...*, p. 61.

and foreign policy when Östen Undén became Foreign Minister in the summer of 1945; a change from passive neutrality to active bridge-building between East and West.[76] It was designed not only to keep the country out of war, but also to relieve tensions between the two Power Blocs and provide for regional stability in northern Europe.[77]

From the above follows that Sweden's unilaterally declared intention to remain neutral in case of war was purely a statement of will. This being the case, restrictions upon Sweden's possible foreign policy choices were largely self-imposed.[78] The late Social Democratic Prime Minister Olof Palme put it succinctly when he said: "'Our neutrality policy is for us to define.'"[79] Moreover, even a permanently neutral state is allowed, according to international law, to plan for contingencies that could arise if the nation was attacked. That includes preparations for military support from another nation or alliance of nations. Not even staff talks with other nations or with alliances can be considered a breach of neutrality, concludes the Neutrality Policy Commission, as long as they do not involve binding arrangements.[80] However, the Swedish Prime Minister Tage Erlander explicitly excluded staff talks in a statement in 1950.[81] Consequently, it is very hard to argue that the Swedish government acted in contradiction of the policy of neutrality. Per Ahlin has written that peacetime neutrality policy can never be denounced as 'phony', as long as the option to assume a neutral position in war is still available.[82]

76 Noreen, E., *Brobygge eller blockpolitik?...*, pp. 97–98; see also: Noreen, E., "Från neutralitet till brobygge. Svensk utrikespolitik och säkerhetspolitik under de första efterkrigsåren" in T. Jonter and C. Silva (eds.), *Sverige inför en ny världsordning, 1945–50...*, pp. 53–55.

77 Andrén, N., *Maktbalans och alliansfrihet. Svensk utrikespolitik under 1900-talet* (Stockholm: Norstedts juridik, 1996), p. 89; Andrén, N., "Swedish Defense: Traditions, Perceptions, and Policies" in M. H. Haltzel and J. Kruzel (eds.), *Between the Blocs:...*, pp. 181–182.

78 *Had There Been a War...*, p. 50; Cramér, Per, *Neutralitetsbegreppet* (Stockholm: Norstedts Förlag, 1989), pp. 97, 100. Interestingly, suggestions to bind Swedish neutrality by international law have been turned down on the grounds that permanent neutrality would circumscribe the government's freedom of action. (Cramér, P., *Neutralitetsbegreppet*, p. 96)

79 Andrén, N., *Maktbalans och alliansfrihet.*, p. 95. "Vår neutralitetspolitik bestämmer vi själva."

80 *Had There Been a War...*, pp. 49–50.

81 *Had There Been A War...*, p. 52; Gyldén, Nils, *Sweden's Security and Defence Policy—Through the Cold War and Towards the Turn of the Century* (Stockholm: Ministry of Defence, 1994), p. 20.

82 Ahlin, P., *Folkrätten i svensk säkerhetspolitik*, p. 5. A problem is of course that it is not always up to the nation itself to decide if that option is still available to it. This decision will often be made elsewhere, i.e. in the capitals of the warring parties.

But the above said is only true from a strictly legalistic point of view, and there is more to a neutrality policy than that.[83] Writers on international law and neutrality policy have pointed to the fact that if other nations cannot be reassured that the neutral will remain a non-participant, i.e. if the neutrality policy is not credible, the policy will fail.[84] "Credibility and respectability are the main criteria for evaluating the success of peacetime neutral diplomacy" writes Harto Hakovirta, and continues to state that the "neutrality of a given state is more *credible* the more likely it is that it will—intends to and is able to—maintain its neutral position and policy in future times of war and peace."[85] [Italics in original.] This meant that *de jure* permanent neutrality indirectly set limits to Sweden's *de facto* during this period, since the government could not contradict the peacetime obligations following from international law if it wished the policy of neutrality to remain credible.[86]

Accordingly, credibility was essential to the Swedish government, and it was deemed central to establish international confidence in the country's ability and will to defend its territory if the neutrality policy was to have a chance to succeed.[87] This also means that other nations' confidence in Sweden's determination to remain neutral can be used to judge whether the neutrality policy was successful or not. A strong defense was considered to be the backbone of a credible neutrality policy, and the government acted accordingly. A substantial military force, largely equipped by domestic suppliers in war has been one of the building blocks used to make Swedish neutrality credible.[88]

83 This point has been made elsewhere, cf.: Lindström, S., "The open society and its stewards", in K. L. Kleve (ed.), *The Cold War, Military Power...*, pp. 135–136.

84 Karsh, Efraim, *Neutrality and Small States* (London: Routledge, 1988), pp. 35–38; Lindholm, R. H., *Sveriges neutralitet. Swedish Neutrality*, p. 66.

85 Hakovirta, Harto, *East-West Conflict and European Neutrality* (Oxford: Clarendon Press, 1988), pp. 26–38; quotes on p. 26 and p. 27 respectively.

86 Cramér, P., *Neutralitetsbegreppet*, p. 100. For example, Sweden could not enter an alliance and still expect other governments to believe that it would remain neutral if the alliance was attacked. This was the case even though the Swedish government was not formally bound by international law to refrain from joining formal alliances. It is obvious that this inability to formally join an alliance stemmed from the fact that the perception of Sweden's *de facto* neutrality was tainted by the limitations imposed on a nation by *de jure* neutrality. Otherwise the Swedish government should have been able to do whatever it wished to do without fear of having its policies interpreted the wrong way as long as it issued statements saying that it would remain neutral in case of war.

87 Åström, Sverker, "Swedish Neutrality: Credibility Through Commitment and Consistency" in Bengt Sundelius (ed.), *The Committed Neutral*, pp. 15–16.

88 *Had There Been A War...*, p. 261; Logue, J., "Sweden" in S. V. Papacosma and M. R. Rubin (eds.), *Europe's Neutral and...*, p. 93. This was ironic, since the

But maintaining a strong defense was not the only reassuring strategy used by the Swedish government during the period of this investigation. In the early 1950s the government emphasized Sweden's long tradition of peace, as a sign of the credibility of its intentions. Another longstanding theme was that Sweden was described as a nation which, because of its neutral position, could be counted on to perform important missions within the United Nations framework.[89] It was said that any change in the Swedish policy would endanger the stability of the Nordic region;[90] a hypothesis later designated the 'Nordic Balance Theory'.[91]

During the first half of the 1960s there was a visible shift in how the Swedish government chose to portray Sweden's role on the international arena. A bolstered confidence in the neutral stance was displayed, and a more active foreign policy was embraced. Torsten Nilsson's ascendancy to the post of Foreign Minister in 1962 (Östen Undén had held this post since 1945) marked this shift in a definite way.[92] Accordingly, I agree with Donald Odom, Jr.'s statement regarding the "importance of neutrality to an understanding of Swedish foreign policy behavior [...]."[93] Apparently,

---

Social Democrats had criticized the Conservatives in the 1920s when the latter promoted neutrality, guaranteed by a strong defense, and the Social Democrats argued for international cooperation and disarmament (Stråth, B., "Poverty, Neutrality and Welfare: Three Key Concepts in the Modern Foundation Myth of Sweden" in B. Stråth (ed.), *Myth and Memory in the Construction of Community: Historical Patterns in Europe and Beyond* (Brussels: P.I.E.-Peter Lang, 2000), p. 393).

89 Ottosson, S., *Svensk självbild under kalla kriget*, p. 20.

90 Ibid., p. 21.

91 af Malmborg, M., *Neutrality and State-Building in Sweden*, pp. 155–156. This situation (where the Finnish-Soviet relationship was weighed up by Norway and Denmark's preconditioned membership of NATO, with non-aligned Sweden as a hub in the middle) was developed into a theory in the mid-1960s by Arne Olav Brundtland at the Norwegian Institute of International Affairs (NUPI). This theory has been criticized for lacking empirical ground, but has also gained support, cf.: Dalsjö, Robert, "Tungt vägande kritik? En granskning av kritik mot teorin om nordisk balans" in *Militärhistorisk tidskrift*, 1987, pp. 131–185; Petersson, M., "The theory and practice of the 'Nordic Balance' during the 1950s and 60s, with special emphasis on Swedish-Norwegian relations", in K. L. Kleve (ed.), *The Cold War, Military Power...*, pp. 25–34.

92 Ottosson, S., *Svensk självbild under kalla kriget*, p. 25. For a thorough account of the evolution of Sweden's active foreign policy, see: Lödén, H., *"För säkerhets skull."*.... This new activism has been given the pejorative label 'The Moral Super Power", see: Nilsson (now Dahl), Ann-Sofie, *Den moraliska stormakten. En studie av socialdemokratins internationella aktivism* (Stockholm: Timbro, 1991).

93 Odom, Jr., Donald, *Swedish Foreign Policy Behavior: An Event Data Analysis* (Stockholm: Stockholm University, International Graduate School, 1990) p. 51; Swedish historian Stefan Ekecrantz has also pointed to this fact, see: Ekecrantz,

it cannot be pointed out too many times that the neutrality policy was itself rooted in a perceived material need for such a stance. A materiality connected with foreign policy, Sweden's geostrategic position, as well as domestic factors. This is a leading assumption in this study.[94]

The concept of neutrality took on a life of its own in a very literal sense, and was a very important part of what constituted the identity of the Swedish people. Bo Stråth has called the Swedish neutrality concept "a myth and [...] a doctrine",[95] and he also states that the neutrality policy "gave rise to a kind of 'third-way identity'" which "radiated a feeling of standing outside big-power ties."[96] The concept became very closely knit to the Social Democratic Party, and hence, since the Social Democrats were in power uninterruptedly from 1945–1962, neutrality, guarded and guaranteed by them, totally dominated Sweden's security and foreign policy during the period covered by this thesis.[97]

This also meant that neutrality was never really questioned or scrutinized critically, except by certain opinion-makers in the Swedish media. Such criticism, however, never challenged the Social Democratic view of what Swedish foreign policy should be during the period of my study. The Social Democrats had acquired the prerogative of interpreting the meaning of the neutrality policy. Writes Bo Stråth:

> If the government argued that something was incompatible with the neutrality policy, that issue was, as a rule, removed from the political agenda and the action was justified by the collective memory of historical continuity.[98]

The image of Sweden as the neutral bastion in the North combined

Stefan, "'Neutralitet', 'neutralitetspolitik' och 'alliansfrihet' – en begreppshistorisk propå till forskning och debatt" in Ragnar Björk and Alf W. Johansson (eds.), *Samtidshistoria och politik. Vänbok till Karl Molin* (Stockholm: Hjalmarsson & Högberg, 2004), p. 53.

94 It is important here to avoid the idealist fallacy of assuming that there were no material underpinnings to the policy of neutrality. The government had invested heavily in the concept of neutrality and stood to lose considerably if the secret collaboration with the Americans was made public. Likewise, it was the need for security and deliveries of American military technology that necessitated the cooperation in the first place.

95 Stråth, B., "The Swedish Image..." in B. Stråth (ed.), *Europe and the Other and...*, p. 374.

96 Ibid., p. 371.

97 Stråth, B., "Poverty, Neutrality and Welfare:..." in B. Stråth (ed.), *Myth and Memory...*, p. 394.

98 Ibid., p. 394. "This memory", writes Stråth, "completely ignored the situation in the 1920s when it was the Conservatives who had stood for Swedish neutrality." (ibid., p. 394).

with the welfare discourse into a consistent Swedish self-image. This process, of which the perceived progressive organization of Swedish society was an important part, did not only take hold within Sweden, but also became the dominating view of Sweden internationally, for example, it emerged in political debates in Germany and Britain.[99]

However, the Social Democratic monopoly of the definition of the neutrality policy did not result in a fundamentalist approach to foreign policymaking. Rather, it was very pragmatic and nebulous. The central feature of this policy was adjustment to the given situation, and the requirements put before the government at a certain time (once again, the policy of neutrality was a response to material factors as they appeared to the Social Democratic leadership). The goal of the policy was to keep the country out of a future military conflict between the major powers.[100] This gave the Government a large degree of latitude in pursuing its foreign policies.

The Social Democratic dominance in Swedish political history is unequaled in any Western democracy. As Bo Rothstein has remarked, the party participated in every government from 1932 to 1976, either by itself or in coalition with other parties. During this time it established an ever closer relationship with the trade union movement, which organized about 85 per cent of the country's workforce.[101] But even though this obviously meant that the party came to have a large influence on the nation's policy, it did not give it total control of the bureaucratic apparatuses of the state.[102]

Rearmament and Non-Alignment

The build-up of Sweden's defense industry began during World War II. It is possible to discern four phases in this development: 1938 (the Anschluss crisis), 1939 (the German invasion of Poland), April 1940 (the German invasion of Denmark and Norway) and 1942.[103] The establishment of a Swedish aircraft industry, however, had already been a part of the rearmament decision of 1936.[104] In terms of military spending

99 Ibid., p. 395.

100 Ibid., p. 401.; Stråth, B., "The Swedish Image..." in B. Stråth (ed.), *Europe and the Other and...*, p. 371.

101 Rothstien, Bo, *The Social Democratic State: The Swedish Model and the Bureaucratic Problem of Social Reforms* (Pittsburgh: University of Pittsburgh Press, 1996), p. 3.

102 Ibid., p. 182.

103 Olsson, Ulf, *The Creation of a Modern Arms Industry: Sweden 1939–1974* (Göteborg: Ekonomisk-historiska institutionen, Göteborgs universitet, 1977), p. 11.

104 Dörfer, I., *System 37 Viggen. ...*, p. 53.

Sweden was a midget compared to the major allies, even relatively, and Sweden's defense expenditures during the war years (as a percentage of the GNP)[105] roughly equaled that of the United States and Great Britain *before* they entered the war.[106]

This early rearmament process was not so much a consequence of the efforts of the domestic industry, though. Initially, in order to achieve a rapid build-up, most of it consisted of imports.[107] During World War II production of aircraft suffered considerable delays primarily because of the dependence on foreign designs and technology.[108] The so-called Framework Agreement [Ramavtalet], of the fall of 1940, was conceived in order to solve this problem, creating a de facto monopoly on the production of aircraft for the companies involved: SAAB (airframes) and NOHAB Flygmotor (engines)[109].

This strategy was rather successful. By 1948, SAAB had produced about 1,000 aircraft, of which only 204 had been produced under license. In May 1950, the Swedish government approved a long-time Framework Agreement with SAAB (actually covering the period July 1948–June 1960),[110] and in May the following year Sweden was the first European country to introduce a swept-wing jet fighter aircraft (the J29 Tunnan).[111] Over the next decade "the buildup of the RSAF was guaranteed by the production of SAAB aircraft under this contract."[112] As stated above, at the beginning of the 1950s (from 1949 to be exact) the RSAF was the fourth largest in the world—only the United States, the Soviet Union, and Great Britain presided over a larger number of aircraft.[113] Most of the production of military equipment in Sweden (both during the War

105 Gross National Product. The Swedish equivalent to this would be BNP, or Bruttonationalprodukt.

106 Olsson, U., *The Creation of a Modern Arms Industry*..., p. 27.

107 Ibid., pp. 29, 53.

108 Holmström, Per and Olsson, U., "Sweden" in Nicole Ball and Milton Leitenberg (eds.), *The Structure of the Defense Industry* (London: Croom Helm, 1983), p. 140.

109 Olsson, U., *Furthering a fortune. Marcus Wallenberg. Swedish banker and industrialist.* (Stockholm: Ekerlid, 2001), p. 292. In 1941 NOHAB, then owned by Volvo and Bofors, changed its name to Svenska Flygmotor AB, or SFA (Dörfer, I., *System 37 Viggen*..., p. 54).

110 Dörfer, I., *System 37 Viggen*..., p. 67.

111 Ibid., p. 67. Tunnan means 'Barrel in Swedish, and pretty accurately described the shape of the aircraft.

112 Ibid., p. 67.

113 *Had There Been A War*..., p. 253; Cole, Paul M., *Neutralité du jour:*..., p. 327; Dörfer, I., "System 37 Viggen – vapenteknisk...", p. 150.

and in the post-War period) took place in private companies.[114] Only about 10 percent of the total purchases of military equipment were from state-owned corporations.[115] In this respect, the Swedish war material production sector bore great resemblance to the corresponding sector in the United States.[116] All of this, of course, had implications for the ownership structure in the Swedish economy, since the leading arms producers also dominated the Swedish economy as a whole (as producers for the civilian domestic and foreign markets). These corporations were tied to one another by family relations and by commercial banks, almost entirely dominated by the Wallenberg family financial empire.[117] The power concentration was even more emphasized on the field of guided missile development, since the number of companies operating there was even fewer. This situation should not, however, blind us to the fact that the Swedish aircraft industry was comparatively small; most of its European counterparts were five times its size.[118]

The patterns established during World War II permeated into the Cold War period and turned into hardy, tenacious structures—i.e. the same companies that produced weapons materials 1939–1945 (e.g. SAAB, Bofors, Volvo [as the owner of Svenska Flygmotor AB (henceforth: SFA)], LM Ericsson [and its subsidiary Svenska Radioaktiebolaget, SRA]), were also at the core of Cold War rearmament.[119] This invariance could be found in the United States too, where essentially the same corporations performed production of military hardware since the end of World War II.[120]

Defense spending soared, and doubled in constant value from the mid-1950s to the end of the 1960s. However, as a percentage of the GNP it actually decreased by one fifth, from about 5 percent to about 4 percent. In the mid-1950s about half of the defense grants were used for procurement, but this share also decreased towards the end of the 1960s,

114 Holmström, P. and Olsson, U., "Sweden" in N. Ball and M. Leitenberg (eds.), *The Structure of the Defense Industry*, p. 162 and p. 145; see also: Olsson, U., *The Creation of a Modern Arms Industry...*, p. 191.

115 Holmström, P. and Olsson, U., "Sweden" in N. Ball and M. Leitenberg (eds.), *The Structure of the Defense Industry*, p. 141.

116 Reppy, Judith, "The United States" in N. Ball and M. Leitenberg (eds.), *The Structure of the Defense Industry*, p. 23.

117 Ibid., p. 162.

118 Ibid., p. 155; see also: Dörfer, I., *System 37 Viggen...*, p. 58.

119 Holmström, P. and Olsson, U., "Sweden" in N. Ball and M. Leitenberg (eds.), *The Structure of the Defense Industry*, p. 141; see also: Olsson, U., *The Creation of a Modern Arms Industry...*, pp. 189-190.

120 Reppy, J., "The United States" in N. Ball and M. Leitenberg (eds.), *The Structure of the Defense Industry*, p. 27. Reppy writes that variability was typically achieved through corporate mergers, rather than new entry (ibid., p. 27).

at which time it was roughly 36 percent.[121] Sweden's military capabilities during these years have led Thomas O. Schlesinger to state that it would be somewhat misleading to characterize Sweden as a 'small state'.[122] Similarly, Hans Weinberger has pointed out that Sweden in this period was *militarized*, i.e. a society where neutrality, defense, the looming world war, and national security consumed a considerable amount of resources with great effects upon the Swedish social order in general.[123] The large, and sophisticated, defence forces have been motivated by the fact that Sweden's policy of neutrality would not have seemed credible to other nations if Sweden was not capable of defending its territory. In the words of Terry Carlbom, the "primary aim of this expensive defence system [was] to raise the threshold cost of an invasion to what [were] presumed to be unacceptable levels for any potential invader."[124]

It is now time to turn to the international stage and delve a little deeper into American foreign policy during the early Cold War period. It is important for the subsequent study to illustrate what animated American policymakers in the late 1940s and early 1950s, and to provide an account of the build-up of U.S. hegemony, and the thinking that underpinned the crucial policy decisions of this time such as the Truman Doctrine, the Marshall Plan, the trade embargo vis-à-vis the Eastern Bloc, and the formation of NATO – or the creation of East and West in European and world politics. An understanding of these events is pertinent if one wants to correctly comprehend the serious challenges facing the Swedish government at this time.

The Truman Administration and the Origins of the Cold War

President Franklin Delano Roosevelt's death on April 12, 1945, and Harry S. Truman's ascendancy to the American presidency, brought with it an about-face in U.S. foreign policy.[125] While Roosevelt had thought it important for the United States to pay close heed to Stalin's 'security objectives, Truman was of a different view entirely.[126] Roosevelt's death

121 Holmström, P. and Olsson, U., "Sweden" in N. Ball and M. Leitenberg (eds.), *The Structure of the Defense Industry*, pp. 142–144.

122 Schlesinger, T. O., *The United States and the European Neutrals*, p. 94.

123 Weinberger, H., "'På sidan om de stora kraftlinjerna'? Teknik, vetenskap och det kalla kriget" in S. Widmalm (ed.), *Vetenskapsbärarna...*, pp. 358–362.

124 Carlbom, Terry, *Rethinking Swedish Neutrality* (Lampeter: Saint David's University College, 1989), p. 17.

125 Leffler, M. P., *A Preponderance of Power:...*, p. 25; Yergin, Daniel, *Shattered Peace: The Origins of the Cold War and the National Security State* (Boston: Houghton Mifflin Company, 1977), p. 68.

126 Leffler, M. P., *A Preponderance of Power:...*, p. 57. For Truman's character and

cleared a path for a revision of strategy that elements within government circles had long sought. The U.S. Ambassador in Moscow Averell Harriman had tried to stiffen the former President's position, but to little avail.[127] The State Department housed the most ardent critics of Roosevelt's efforts to accommodate the Kremlin. Men like Charles E. Bohlen and George F. Kennan had been posted in Moscow during the trials and purges of the late 1930s and were appalled by the oppressive rule of Stalin and his henchmen.[128]

Faced with the opportunity that an untutored Chief Executive in the White House provided, the old structures redoubled their efforts to 'educate' the new President.[129] Truman took the advice that was given him, and proceeded to speak very bluntly to the Soviet Foreign Minister Vyacheslav Molotov when they met for the first time in the evening of April 23, 1945.[130] Bohlen, who acted as interpreter during

U.S. policies during his administration, cf.: McCullough, David, *Truman* (New York: Simon & Schuster, 1992); Haynes, Richard F., *The Awesome Power: Harry S. Truman as Commander in Chief* (Baton Rouge: Louisiana State University Press, 1973); Donovan, Robert J., *Conflict and Crisis: The Presidency of Harry S. Truman, 1945-1948* (New York: Norton & Company, Inc., 1977); Theoharis, Athan G., *The Truman Presidency: The Origins of the Imperial Presidency and the National Security State* (Stanfordville, NY: Earl M. Coleman Enterprises, Inc., Publishers, 1979). For a different view, see: Maddox, Robert, James, *From War to Cold War: The Education of Harry S. Truman* (Boulder: Westview Press, 1988).

127 Gaddis, J. L., *Strategies of Containment: A Critical Appraisal of Postwar American National Security Policy* (New York: Oxford University Press, 1982), p. 15.

128 Leffler, M., P., *A Preponderance of Power:...*, p. 31. Kennan was the author of 'the long telegram', a 8,000 word document sent to Washington on February 22, 1946 (For this, see: Kennan, George F., *Memoirs, 1925-1950* (London: Huntchinson of London, 1967), pp. 547-559; Barton J. Bernstein and Allen J. Matusow (eds.), *The Truman Administration: A Documentary History* (New York: Harper & Row Publishers, 1966) pp. 198-212). This document had a huge impact in the American capital, and subsequently became the basis for the United States' strategy towards the Soviet Union for the rest of the Cold War (the so-called 'containment' policy). It basically said that Soviet behavior had nothing to do with the behavior of the United States, and that Moscow would not change until it encountered a sufficiently long string of failures to convince some future Kremlin leader to change his attitude. It turned out that Stalin read 'the long telegram' himself, and immediately proceeded to have *his* Ambassador in Washington prepare a corresponding document on the United States (Gaddis, J. L., *The Cold War: A New History* (New York: The Penguin Press, 2005), pp. 29-30). Anti-Bolshevism had been "institutionalized" in the U.S. State Department as early as 1924 (Leffler, M. P., *The Specter of Communism: The United States and the Origins of the Cold War, 1917-1953* (New York: Hill and Wang, 1994), p. 19).

129 Gaddis, J. L., *Strategies of Containment:...*, p. 15.

130 Maddox, R J., *From War to Cold War:...*, p. 53.

the meeting, later said that "'he had never heard a top official get such a scolding'."[131] The meeting between Truman and Molotov signified an important change in the relations between the United States and the Soviet Union, and although it would be too much to say that this was the moment when the Cold War hostilities were resumed, it is no doubt true, as Daniel Yergin observes, that it was "a change that the Russians, engaged in their own calculations, could not miss."[132] Although there was not yet a coherent foreign policy in Washington, it was clear that U.S. hegemonic strategy was slowly taking form. There were "no longer any geographical limitations on the exercise of American power."[133]

By July, 1945, Truman had removed Secretary of State Edward Stettinius from his post, and the President's trusted foreign policy advisor James F. Byrnes, a man of vast experience in domestic affairs but almost none in diplomacy, was given Stettinius position (where he stayed until January 1947).[134] Since Truman was inherently uninterested in foreign policy details, he quickly surrendered the nation's diplomacy to the likewise inexperienced Byrnes, who was very soon replaced, too, by Army General George C. Marshall (who was, in turn, replaced by Dean Acheson in 1949). The frequent change at the top of the State Department took

131 Leffler, Melvyn, P., *A Preponderance of Power:...*, pp. 31-32; Yergin, D., *Shattered Peace:...*, p. 73; Horowitz, David, *The Free World Colossus: A Critique of American Foreign Policy in the Cold War* (New York: Hill and Wang, 1965), pp. 31-32; Bohlen, Charles E., *Witness to History 1929-1969* (London: Weidenfeld and Nicolson, 1973), pp. 212-213; Truman, H. S., *Memoirs*, Volume I, *Years of Decisions* (Garden City, NY: Doubleday & Company, Inc., 1955), pp. 79-82. Truman wrote in his memoirs that Molotov then stated that "'I have never been talked to like that in my life'", whereupon Truman allegedly quipped, "'Carry out your agreements and you won't get talked to like that'" (Ibid., p. 82). This apparently was only a figment of Truman's imagination for neither the official American record, Bohlen in his memoirs, nor the Soviet record can confirm that this exchange actually took place (Bohlen, C. E., *Witness to History...*, p. 213; Watson, Derek, *Molotov: A Biography* (Hampshire: Palgrave MacMillan, 2005), pp. 217-218). Bohlen simply states that "Molotov turned a little ashy", and then adds "[h]ow I enjoyed translating Truman's sentences! They were probably the first sharp words uttered during the war by an American President to a high Soviet official." (Bohlen, C. E., *Witness to History...*, pp. 213). In an interview some years later, Bohlen flatly denied that the words quoted by Truman were ever uttered (Maddox, R. J., *From War to Cold War:...*, p. 55). Truman's biographer David McCullough notes the divergent recollections, but does not offer any further insights. He simply states: "But whichever way the meeting ended, there had been no mistaking Truman's tone." (McCullough, D., *Truman*, p. 376).

132 Yergin, D., *Shattered Peace:...*, p. 73.

133 Leffler, Melvyn, P., *A Preponderance of Power:...*, p. 41.

134 Jenkins, Roy, *Truman* (London: William Collins Sons & Co. Ltd, 1986), pp. 69-70; Gaddis, J. L., *Strategies of Containment:...*, p. 16.

its toll on the patience of the military establishment. The administration dragged its feet on important decisions, and no integrated war plan had been approved by 1948. Being unable to establish what interests were vital for the security of the United States from the White House and the State Department, the officials in the Pentagon resorted to drawing up their own objectives and policies.[135] In an effort to remedy the lack of coordination in national security policy, the National Security Act of 1947 was signed into law by Truman on July 26 that same year. The Act created, among other things, the National Security Council (henceforth: NSC), the JCS, and the Central Intelligence Agency (henceforth: CIA).[136]

The Truman Doctrine—the Ideology of Intervention

February 21, 1947, was a fateful date in British foreign policy for on this day the British government informed Washington that due to their fiscal woes, they had to terminate their military and economic assistance to Greece and Turkey. They also let it be known that they intended to pull their troops out of both countries.[137] The days when Britain could wield any significant power outside of the Commonwealth were over, and the United States' Under Secretary of State for Economic Affairs William L. Clayton sensed that what was at stake was who should be the one to take up the "'reins of world leadership'" after Britain—the only two contestants being the United States and Russia.[138]

Three weeks later, on March 12, Truman delivered a speech to a joint session of the Congress where he recommended immediate action and urged that Greece and Turkey be granted $400 million worth of supplies and military equipment for the period ending June 30, 1948.[139] This address soon became referred to as the Truman Doctrine, and it

135 Leffler, Melvyn, P., *A Preponderance of Power:...*, pp. 38-39, 41; Tamnes, Rolf, *The United States and the Cold War in the High North* (Oslo: Ad Notam forlag AS, 1991), p. 44.

136 Hoxie, G. R., *Command Decision and the Presidency:...*, pp. 138-139; Haynes, R. F., *The Awesome Power:...*, pp. 105-107; Millis, Walter, "The National Security Act" in Raymond G. O'Connor (ed.), *American Defense Policy in Perspective: From Colonial Times to the Present* (New York: John Wiley & Sons, Inc., 1965) pp. 307-311). Under President Eisenhower the NSC became extremely institutionalized with firmly fixed procedures for planning and coordination of overall national security policy. A common criticism was "that the formalization resulted only in a requirement for a presidential 'yes' or 'no' with no other alternative being immediately available." (Bletz, Donald F., *The Role of the Military Professional in U.S. Foreign Policy* (New York: Praeger Publishers, 1972), p. 85).

137 Leffler, Melvyn, P., *A Preponderance of Power:...*, p. 142.

138 Yergin, D., *Shattered Peace:...*, p. 308; Leffler, Melvyn, P., *A Preponderance of Power:...*, p. 143.

139 Truman, H. S., *Memoirs*, Volume II, *Years of Trial and Hope* (Garden City, NY:

symbolized a definite "turning point in America's foreign policy, which now declared that wherever aggression, *direct* or *indirect*, threatened the peace, the security of the United States was involved."[140] [Italics added.]

In his speech, Truman had said that "'it must be the policy of the United States to support free peoples who are resisting attempted subjugation by armed minorities or by outside pressures."[141] By this he implied that the United States would support 'free peoples' everywhere; an unprecedented move by an American President.[142] Truman thus recognized that the Truman Doctrine extended American national security on a global scale, and it was indeed, as Walter LaFeber has remarked, "a milestone in American history".[143] It was also the formalization of the American strategy of 'containment'.[144] The real aim of the assistance awarded to Greece and Turkey was in fact another—namely to stabilize the precarious situation in Western Europe.[145] In the Kremlin, the Doctrine was predictably seen as an example of brazen-faced American imperialism, and was denounced as an imperialist "'smokescreen for expansion.'"[146]

Doubleday & Company, Inc., 1956), p. 105; Byrnes, James F., *Speaking Frankly* (New York: Harper & Brothers Publishers, 1947), p. 301

140 Truman, H. S., *Memoirs*, Volume II..., p. 106. Acheson had helped convince Arthur Vandenberg and other Senators to support the President by presenting them with an early version of the domino theory. Acheson had said: "Like apples in a barrel infected by one rotten one, the corruption of Greece would infect Iran and all to the east. It would also carry infection to Africa through Asia Minor and Egypt, and to Europe through Italy and France [...]." (Acheson, Dean, *Present at the Creation: My Years at the State Department* (London: Hamish Hamilton, 1969), p. 219). Indeed, it must have been hard for the Senators to miss the parallel to the Black Death of the 14th century that eventually killed a third of Europe's population.

141 Mee, Jr., Charles L., *The Marshall Plan: The Launching of the Pax Americana* (New York: Simon and Schuster, 1984), p. 50.

142 Ibid, p. 50.

143 LaFeber, Walter, *America, Russia, and the Cold War 1945-2002* (updated ninth edition) (New York: McGraw-Hill, 2004), p. 62.

144 The term 'containment' was coined by George Kennan in an article in *Foreign Affairs* in July 1947, which he wrote under the pseudonym 'x' (Ulam, Adam B., *The Rivals: America and Russia Since World War ii* (New York: The Viking Press, 1971), p. 121). Melvyn Leffler has pointed out that the spirit of 'containment' was U.S. state policy even before Kennan sent 'the long telegram' in February 1946 (Leffler, M. P., *A Preponderance of Power:...*, p. 61).

145 Ireland, Timothy P., *Creating the Entangling Alliance: The Origins of the North Atlantic Treaty Organization* (Westport, Connecticut: Greenwood Press, 1981), p. 27; Kolko, Gabriel and Kolko, Joyce, *The Limits of Power: The World and United States Foreign Policy, 1945-1954* (New York: Harper & Row, Publishers, 1972), p. 332.

146 Rubinstein, Alvin Z., *Soviet Foreign Policy Since World War ii: Imperial and Global* (Boston: Little, Brown and Company, 1981), p. 51.

Some American officials did not entirely approve of the picture that Truman painted in his speech. George Kennan, now the Director of the State Department's Policy Planning Staff (henceforth: PPS), for example, thought that military supplies for Turkey were "unfortunate" since the Soviet threat was political and not military.[147] Even Bohlen and Marshall, who were in Moscow preparing for the Council of Foreign Ministers conference, thought that there was too much "flamboyant anti-Communism in the speech", and that Truman was using "too much rhetoric."[148] Not sharing Truman's view of the speech as the definite turning point in U.S. foreign policy, Kennan also took exception to "the sweeping nature of the commitments which it [the Truman Doctrine] implied."[149] Kennan eventually made himself impossible because of his vision of the role of postwar America. At the same time as the United States' hegemony was being built, Kennan took the view that the United States was not fitted "to be an imperial power in the grand manner, and particularly not one holding the great peoples of Western Europe indefinitely *in some sort of paternal tutelage.*"[150] Kennan also modified his position towards a partition of Germany and the European continent, which he had initially advocated, making him an opponent of this move as well.[151] The PPS could not be headed by a man like that, and in late September 1949 he resigned.[152] This was perhaps an ironic ending for the main author of the strategy of containment.

As John Gimbel showed, in his seminal study on the origins of the Marshall Plan, there were plenty of signs of Russia's willingness to

147 Kennan, G. F., *Memoirs...*, p. 317.

148 Bohlen, C. E., *Witness to History...*, p. 261.

149 Ibid., pp. 319–320.

150 Ibid., p. 464.

151 Eisenberg, Carolyn, "Rethinking the Division of Germany" in Allen Hunter (ed.), *Rethinking the Cold War* (Philadelphia: Temple University Press, 1998), p. 56; Gaddis, J. L., *The Long Peace: Inquiries Into the History of the Cold War* (Oxford: Oxford University Press, 1987), p. 70.

152 Leffler, M. P., *A Preponderance of power:...*, p. 313; Kennan, G. F., *Memoirs...*, p. 466. Kennan had grown continuously more and more disillusioned with American policy and conflicts arose between his and the State Department's general view of the Soviet problem. The basic issue was whether the Soviet Union really posed a threat to Western Europe; Kennan thought it did not which put him in a serious minority position. (Kennan, G. F., *Memoirs...*, pp. 464-465). Kennan left the government service and the State Department, and later bitterly and dejectedly concluded that United States foreign policy was condemned to rely "almost exclusively on what we might call 'diplomacy by dilettantism.'" (Kennan, G. F., *American Diplomacy, 1900-1950* (London: Secker & Warburg, 1952), p. 94).

abide by the Potsdam Agreement and to unify Germany.[153] Contrary to what is often assumed, it was not the Russians, but the French, that were wreaking havoc on four-power agreement in Germany.[154] But the Americans had already decided upon dividing Germany "without Stalin, and if necessary, against his will."[155] The chronology of events also shows that studies concerning a European aid program had already been finished by the time Marshall came back from Moscow on April 28.[156]

The policymaking officials in Washington feared that the Soviet Union would try to exploit the economic weakness of Western Europe and extend its political control into this area, but most historians today agree that this was never actually part of the Kremlin leader's plans and that Stalin's reactions were mostly defensive.[157] For example, he did not interfere with the British and American campaign against the Communist guerrillas in Greece during the Greek Civil War.[158] At the time, American officials nevertheless saw increased Russian influence in Western Europe as a distinct possibility, and took steps that they thought would best prevent such a scenario.

The Soviet policy of non-intervention in Greece was largely the result of the Tolstoy Conference which had been held in the Kremlin in early October 1944, between British Prime Minister Winston Churchill, British Foreign Minister Anthony Eden, Stalin and Soviet Foreign Minister Molotov. What was concluded was nothing else than a spheres-of-influence agreement where Britain was guaranteed first say in Greece, and the Russians should have predominance in Romania, Bulgaria, and Hungary; the two powers would share responsibility in Yugoslavia.[159]

153 Gimbel, John, *The Origins of the Marshall Plan* (Stanford, CA: Stanford University Press, 1976), p. 127.

154 Ibid., pp. 4, 16, 35–49, 128–129

155 Pleshakov, Constantine and Zubok, Vladislav, *Inside the Kremlin's Cold War: From Stalin to Khrushchev* (Cambridge, MA: Harvard University Press, 1996), p. 50.

156 Eisenberg, C., *Drawing the Line:...*, p. 319; Acheson, D., *Present at the Creation:...*, pp. 227–228.

157 Leffler, M. P., *A Preponderance of Power:...*, p. 186.

158 Yergin, D., *Shattered Peace:...*, pp. 310f; see also: Leffler, M. P., *A Preponderance of Power:...*, on Stalin's character, p. 102, on Greece, pp. 73–75; on American and British support to fascist forces in Greece, see: Kolko, G. and Kolko, J., *The Limits of Power:...*, pp. 218–235; Fleming, D. F., *The Cold War and its Origins 1917–1960*, Volume 1, 1917–1950 (Garden City, NY: Doubleday & Company, Inc., 1961), pp. 174–187.

159 Gardner, Lloyd C., *Spheres of Influence: The Great Powers Partition Europe, from Munich to Yalta* (Chicago: Elephant Paperbacks, 1993), pp. 184–206; , Byrnes, J. F., *Speaking Frankly*, p. 53; Bohlen, C. E., *Witness to History*, pp. 161–163; Yergin, D., *Shattered Peace:...*, pp. 59–60; the Tolstoy Conference is

Robert Service has commented that both leaders no doubt "sensed that they were deciding something of historic importance."[160] With regard to American opinion, Churchill told Stalin, it was better to put the agreement in diplomatic terms, and to refrain from using the term 'spheres of influence' because "'the Americans might be shocked. *But as long as he and Marshal Stalin understood each other he could explain matters to the President.*'"[161] [Italics added.] The Soviets really went out of their way to ensure that the agreement they had reached with the British was also acceptable to the United States.[162] Despite the aversion that Roosevelt felt toward bilateral agreements, he "was ready to support arrangements in Moscow that promoted immediate Allied cooperation", and he received reports regarding the talks via both Churchill and Harriman.[163] There was hence no doubt that Roosevelt assented to this arrangement.[164]

The final percentages were hammered out by Eden and Molotov during discussions that were not always amiable. Molotov drove a hard

---

covered in detail in: Gilbert, Martin, *Winston S. Churchill*, Volume VII, *Road to Victory, 1941-1945* (William Heinemann Ltd, 1986), pp. 989-1033. See also: Francis L. Loewenheim, Harold D. Langley and Manfred Jonas (eds.), *Roosevelt and Churchill: Their Secret Wartime Correspondence* (London: Barrie & Jenkins, 1975); Churchill, Winston S., *The Second World War*, Volume VI, *Triumph and Tragedy* (London: Casell, 1954), pp. 197-204; Lundestad, G., *The American Non-Policy Towards Eastern Europe, 1943-1947: Universalism in an Area Not of Essential Interest to the United States* (Tromsø: Universitetsforlaget, 1978); Kimball, Warren F. (ed.), *Churchill & Roosevelt: The Complete Correspondence*, Volume III, *Alliance Declining: February 1944-April 1945* (New Jersey: Princeton University Press, 1984);

160 Service, Robert, *Stalin: A Biography* (London: Pan Books, 2005), p. 464.

161 Gilbert, M., *Winston S. Churchill*, Volume VII..., pp. 991-992. Churchill later tried to downplay the importance of what he had done in Moscow, by stating that he wrote a letter to Stalin where he explained the British "interpretation of the percentages". The only problem was that he never sent it (Churchill, W. S., *The Second World War*, Volume VI..., p. 201). This means that Stalin never knew the 'true' British intentions behind the deal. However, Molotov had carefully explained the Soviet interpretation to Eden without any protestations (Watson, D., *Molotov: A Biography*, p. 215). The importance Churchill attached to Stalin sticking to his word in not interfering in Greece points to the fact that he indeed thought of the agreement in terms of spheres of influence even long after hostilities had ended.

162 Gilbert, M., *Winston S. Churchill*, Volume VII..., p. 1001.

163 Dallek, R., *Franklin D. Roosevelt and American...*, p. 479; Churchill, W. S., *The Second World War*, Volume VI..., p. 201. See also: Kolko, Gabriel, *The Politics of War: Allied Diplomacy and the World Crisis of 1943-1945* (London: Weidenfield and Nicolson, 1968), p. 146; Dallek, R., *Franklin D. Roosevelt and American...*, p. 479; Kimball, W. F. (ed.), *Churchill & Roosevelt:...*, p. 350.

164 Dallek, R., *Franklin D. Roosevelt and American...*, p. 479.

bargain, but in the end the result was acceptable to both parties.[165] It has been argued that Molotov and Eden did not really know what their bosses intended with the percentage agreement, and also that the discussion was largely academic since the Red Army had effective control over, and would be the sole military occupiers of Hungary, Bulgaria, and Rumania, and that the deal struck was important for one country only—Greece.[166]The effects were not necessarily confined to Greece, however, since, as Vojtech Mastny has suggested, Churchill's initiative could have had the very serious effect of signaling to the Soviets a willingness on the part of the West to accept more Soviet influence than Moscow had actually so far been aiming at.[167]However, by the autumn of 1946, whatever had been decided at the Tolstoy Conference had long since been suppressed in the Truman administration's collective memory, and there was now a virtual consensus in Washington that the Russians were an ideological enemy, and that it had no legitimate claims or fears.[168]

The Marshall Plan—the Forging of the Western European Bloc

On June 5, 1947, Marshall delivered his famous speech at Harvard University in Cambridge, Massachusetts, in which was outlined what to the afterworld has become known as the Marshall Plan, or the European Recovery Program (henceforth: ERP).[169] At the core of the European crisis in 1947 was the dollar gap;[170] a crisis about which Thomas J. McCormick has said that it was "born of paradox and steeped in irony."[171] The reason for this apt formulation was the fact that the United State'

165 Watson, D., *Molotov: A Biography*, pp. 215–216.

166 Roberts, Geoffrey, *Stalin's Wars: From World War to Cold War, 1939-1953* (New Haven: Yale University Press, 2006), pp. 217–222.

167 Mastny, Vojtech, *The Cold War and Soviet Insecurity: The Stalin Years* (Oxford: Oxford University Press, 1996), p. 20. Despite Churchill's indignation over Russian policies behind the Iron Curtain expressed at Fulton in 1946, he apparently thought the agreement in Moscow was still valid when the British were fighting Communist rebels in Greece in 1947. Reflecting back upon this episode he later said: "Stalin always kept his word with me. I remember particularly saying to him when I visited Moscow in 1944, 'You keep Roumania [sic] and Bulgaria in your sphere of influence, but let me have Greece.' To this bargain he scrupulously adhered during months of fighting with the Greek Communists." (Gilbert, M., *Winston S. Churchill*, Volume VIII, *'Never Despair', 1945–1965* (William Heinemann Ltd, 1988), p. 1192.)

168 Leffler, M. P., *A Preponderance of Power:...*, p. 140.

169 Hogan, Michael J., *The Marshall Plan: America, Britain, and the Reconstruction of Western Europe, 1947–1952* (Cambridge: Cambridge University Press, 1987), p. 43.

170 Leffler, M. P., *A Preponderance of Power...*, p. 164.

171 McCormick, T., J., *America's Half-Century:...*, p. 73.

economic preponderance, upon which its hegemony was to rest, was so great that the resulting commercial imbalance threatened the whole system. American dominance almost undermined the very foundation for its own hegemony.[172] The Marshall Plan aimed simultaneously to prevent local Communist groupings from capitalizing on the crisis, and to stave off a defeat of the American design to include Western Europe in a multilateral international economy.[173]

Marshall was deliberately vague regarding the magnitude and duration of the program and stated that the Europeans must themselves come up with a comprehensive plan; only then could they count on American help.[174] The reason for this was that it had been concluded that it would be "neither fitting nor efficacious" for the administration to draw up the plan unilaterally. Kennan's staff no doubt realized that such an approach smacked of imperialist tendencies, and thus stressed that "[t]he *formal* initiative must come from Europe".[175] [Italics added.] Dean Acheson was even more candid about American hegemonic strategy when he wrote in his memoirs that "[s]urely the plan should be a European plan and come—or, at any rate, appear to come—from Europe. *But the United States must run the show.*"[176] [Emphasis in original.]

The question of Soviet participation presented American policymakers with a two pronged dilemma—one foreign and one domestic. From a foreign policy perspective, it was necessary to insulate the United States from accusations that it was seeking to create a Western Bloc. It was thus recommended that the U.S. government should include, or at least

172 Ibid., p. 73.

173 McCormick, T., J., *America's Half-Century:...*, pp. 74, 78; Leffler, Melvyn P., *A Preponderance of Power:...*, p. 160. To establish a new economic order is maybe the most classic example of hegemonic strategy. This is not to say that it was entirely successful. The intricacies of hegemony, the many-layered contradictions contained within it, and the interdependence of the international system, place real limits on the influence of the hegemonic nations. A good example is American policy versus Britain where the United States had to abandon its initial vision of multilateralism in favor of a kind of regionalism (see: Dobson, Alan P., "The USA, Britain, and the Question of Hegemony" in G. Lundestad (ed.), *No End to Alliance:...*, pp. 134–163). But as Geir Lundestad remarks: "the overall American objectives were rather flexible and [...] on the whole achieved [...]." (Lundestad, Geir, "American-European Cooperation and Conflict: Past, Present, and Future" in G. Lundestad (ed.), *No End to Alliance:...*, p. 250).

174 Leffler, Melvyn, P., *A Preponderance of Power:...*, p. 159.

175 PPS 1: *Policy with Respect to American Aid to Western Europe*, May 23, 1947, in Thomas H. Etzhold and John Lewis Gaddis (eds.), *Containment: Documents on American Policy and Strategy, 1945-1950* (New York: Columbia University Press, 1978), p. 104; for Marshall's repetition, see: Acheson, D., *Present at the Creation:...*, p. 234.

176 Acheson, D., *Present at the Creation:...*, p. 232.

purport to include, all the European countries in the offer,[177] "because any American plan that appeared to exclude the Soviet Union would have very little chance of being accepted in the world."[178] It was deemed imperative to free the United States from the responsibility of dividing Europe.[179] Thus Marshall, in his address, "left the door open to Soviet participation".[180] As it turned out, to paraphrase Hogan's analogy, the Americans left the door open, but were careful enough to furnish the room in such a way as to ensure that the Russians would not want to enter it.

The policymakers in Washington had all along been planning how to exclude Eastern Europe from the Marshall Plan. At the end of April, 1947, the JCS had completed a report, entitled JCS 1769/1, where the issue of foreign assistance was dealt with.[181] The JCS recommended that objective urgency of need should not be allowed to determine whether a country should get assistance. What mattered was the importance of that country to the national security of the United States.[182] The document read like a recipe for how to achieve hegemony, stating that "it is evident that, if we spread our available resources for aid over too large an area, no country is likely to receive assistance sufficient to be of major importance in the resurgence of its economy and military potential."[183] The most important area in the world, from a national security viewpoint, was of course Western Europe.[184] Hence, both the JCS and Kennan's PPS recommended that the USSR and every country under her influence, were to be specifically excluded from assistance, but it had to be done in such a way that those countries would exclude themselves, because if anyone was to be held responsible for dividing Europe it would have to be the Russians with their answer, and not the Americans with their offer.[185]

177 PPS 1:..., in T. H. Etzhold and J. L. Gaddis (eds.), *Containment:...*, p. 106.

178 Bohlen, C. E., *Witness to History...*, p. 264.

179 Acheson, D., *Present at the Creation:...*, p. 232. Acheson wrote: "I pointed out that Russian obstruction in developing a European plan could be overcome by not requiring her agreement [...]." (p. 232).

180 Hogan, M. J., *The Marshall Plan:...*, p. 44.

181 JCS 1769/1: *United States Assistance to Other Countries from the Standpoint of National Security*, April 29, 1947, in T. H. Etzhold and J. L. Gaddis (eds.), *Containment: Documents on...*, pp. 71–83.

182 JCS 1769/1: *United States Assistance...*, in T. H. Etzhold and J. L. Gaddis (eds.), *Containment:...*, p. 71.

183 Ibid., p. 72.

184 Ibid., p. 72.

185 For the exclusion of Eastern Europe: JCS 1769/1: *United States Assistance...*, in T. H. Etzhold and J. L. Gaddis (eds.), *Containment:...*, p. 72; for the 'voluntary' exclusion of the Eastern European nations: PPS 1:..., in T. H. Etzhold and J.

The PPS' first suggestion was that this be achieved in a seemingly egalitarian way, by advancing the idea in the Economic Commission for Europe (henceforth: ECE).[186] But the ECE was a U.N. body where all European countries were represented, and which offered no veto right to its members. This meant that the Soviet Union, via the eastern European satellites, could command six of the seventeen votes. Since the Americans could not control events by the use of the veto, they risked a congressional veto on the Marshall Plan.[187] The Chairman of the ECE Gunnar Myrdal had, moreover, but then as the Minister of Commerce in the Swedish government, negotiated the credit agreement, much despised by the Americans, between Sweden and the Soviet Union in 1946.[188] This proposal, writes Gaddis, thus "caused no little nervousness in Washington."[189] This gamble was not palatable to American government officials who did not heed the advice of the PPS in this instance, and consequently, bypassed the ECE. Instead the State Department lobbied for, and got, a new agency, under American control, to administer the ERP—the Economic Cooperation Administration (henceforth: ECA).[190]

Even so, Myrdal set out for Moscow in mid-June, 1947. He stayed there between June 16 and June 23, and met with both Molotov and the Vice Chairman of the Council of Ministers Anastas Mikoyan. According to Cay Sevón, Myrdal argued that the ECE was the best and indeed the only forum for discussions regarding the Marshall Plan, and he added that the lack of a positive Soviet response gave a green light to those who favored an *ad hoc* arrangement for handling the recovery program. Molotov apparently said that the Soviet Union had the intention of supporting the organization, but on the whole, Myrdal was struck by the evasive answers that the Soviet Foreign Minister provided. It became clear to him, writes Sevón, that there was no clear policy worked out in Moscow, and that the Russians did not know how to respond to the Marshall Plan.[191]

L. Gaddis (eds.), *Containment:...*, p. 106; for the responsibility for dividing Europe: Kennan, G. F., *Memoirs...*, p. 343.

186 PPS 1: *Policy with Respect to...*, in T. H. Etzhold and J. Lewis Gaddis (eds.), *Containment: Documents on...*, p. 106

187 Horowitz, David, *The Free World Colossus:...*, p. 75; Deighton, A., *The Impossible Peace:...*, p. 176.

188 Dohlman, E., *National Welfare and Economic Interdependence:...*, p. 74. For more on this, see Chapter 4, pp. 181–182.

189 Gaddis, J. L., "Dividing Adversaries: The United States and International Communism, 1945–1958" in J. L. Gaddis, *The Long Peace:...*, p. 156.

190 Hogan, M. J., *The Marshall Plan:...*, 103.

191 Sevón, C., *Visionen om Europa...*, pp. 141-145. Sevón bases his narrative on Molotov's papers in the Foreign Ministry's Archive in Moscow.

Even though the ECE had been outmaneuvered, there was still a risk that the Russians might accept aid on the American terms, and this created a domestic problem of wrestling the program through Congress. A positive response from the Soviets would have spelled the end of the Marshall Plan, since the representatives of the American people would not be very keen on appropriating aid to a country that was now seen as the single greatest threat to American national security.[192] "There must, therefore," writes Adam B. Ulam, "have been some private groans in Washington" when the Russians accepted the American invitation to attend the exploratory talks in Paris.[193] Myrdal was at the airport in Moscow when he received the news that Stalin had decided that the USSR would participate in the discussions in Paris, something which Myrdal interpreted as a consequence of his conversation with the Russian Foreign Minister.[194] However, Molotov soon angrily left the Paris negotiations after having received "a Soviet intelligence report describing the recent secret Anglo-American conversations in London, where the Marshall Plan was presented as a defense of Western Europe against a Soviet Threat."[195] This piece of information only confirmed to the Soviets their earlier suspicions, i.e. that the aid program would interfere in the internal affairs of the client nations.[196] The Soviet Union now began its own propaganda campaign, saying that American assistance posed a threat to the sovereignty of the recipient nations, as well as being a scheme to prevent industrialization of Eastern Europe.[197]

Inferring from the Russian experiences during the two recent World Wars, George Kennan concluded that the Marshall Plan "was forcing Moscow to consolidate its hold on Eastern Europe", and he stated prophetically (the Soviet consolidation of power in Prague was still three months in the future) that it would "probably have to

192 Kennan, G. F., *Memoirs…*, p. 52.

193 Ulam, A. B., *Expansion and Coexistence: Soviet Foreign Policy, 1917–73* (Second Edition) (New York: Holt, Reinhart and Winston, Inc., 1974), p. 433.

194 Sevón, C., *Visionen om Europa…*, p. 157.

195 Mastny, V., *The Cold War and Soviet Insecurity:…*, quote on p. 28; for a contemporary Western view, see: Acheson, D., *Present at the Creation:…*, pp. 234–235.

196 Kaplan, Lawrence S., *A Community of Interests: NATO and the Military Assistance Program, 1948–1951* (Washington D. C.: Office of the Secretary of Defense, Historical Office, 1980), p. 10.

197 Day, Richard, B., *Cold War Capitalism: The View from Moscow, 1945–1975* (Armonk, NY: M. E. Sharpe, 1995), p. 49. The British later implicitly supported the Soviet claim that sovereignty would be circumscribed, and voiced opposition to the Americans linking the Marshall Plan aid to trade restriction with Eastern Europe (ibid., p. 76).

clamp down completely on Czechoslovakia."[198] Czechoslovakia was the Soviet Union's most important source of uranium, and this fact obviously provided a strong reason for Stalin to keep the Czechs loyally and safely within the Soviet sphere of influence.[199] After the takeover in February 1948 Kennan accordingly stated that it "represented a defensive reaction—and one foreseen by ourselves—to the success of the Marshall Plan initiative."[200]

The Marshall Plan would, U.S. officials assumed, benefit all "and most certainly would foster American influence and power. It would [...] help the United States establish linkages with European elites, bind European economies to the American economy, and gain access to overseas markets and raw materials."[201] All these things were vital ingredients in establishing the United States as a hegemonic power. Economic interests often reinforced the United States' geostrategic considerations and its ideological predilections, mainly because they defined power in terms of control over or access to resources, and Truman later called the Doctrine and the ERP "two halves of the same walnut."[202] The Kremlin's answer was first the creation of the Council for Mutual Economic Assistance (COMECON), whose aim was to boost the economies of the Eastern Bloc so that the Soviet economy could derive profits from them. Furthermore, there was a partial reestablishment of the COMINTERN, the Communist International, now designated the Communist Information Bureau (COMINFORM).[203]

198 PPS 13: *Resumé of World Situation*, November 6, 1947, in T. H. Etzhold and J. L. Gaddis (eds.), *Containment: Documents on...*, p. 90; see also: Kennan, G. F., *Memoirs...*, p. 378. Melvyn P. Leffler writes that Kennan thought that "[a]lthough this was regrettable, it was not unacceptable." (Leffler, M. P., *A Preponderance of Power:...*, p. 186).

199 Yergin, D., *Shattered Peace:...*, p. 344.

200 Kennan, G. F., *Memoirs...*, p. 379; see also: Yergin, D., *Shattered Peace:...*, p. 350. The CIA analysis confirmed this view, stating that the psychological reaction to the Czech coup was out of all proportion to its actual significance. (Eisenberg, C., *Drawing the Line:...*, p. 369, note 26). The view of Soviet expansion into Eastern Europe had during the war been seen as essentially a defensive one. It was thought, then, that if the Soviet Union could be convinced of the West's peaceful intentions, the reason for seeking to expand its frontiers would disappear (Gaddis, J. L., *The Long Peace:...*, p. 51).

201 Leffler, Melvyn P., *A Preponderance of Power:...*, p. 163.

202 Leffler, M. P., *A Preponderance of Power...*, p. 14. There is, however, a tradition of interchangeability between 'bureaucrats' and 'businessmen' in American history (see: Kolko, G. and Kolko, J., *The Limits of Power...*, p. 381; Kolko, G., *The Triumph of Conservatism: A Reinterpretation of American History* (New York: The Free Press, 1963). For Truman's quote, see: Deighton, A., *The Impossible Peace:...*, p. 186; LaFeber, W., *America, Russia, and...*, p. 68.

203 LaFeber, W., *America, Russia, and...*, p. 75; Dmytryshyn, Basil, USSR: *A Concise*

American officials in government, military, and intelligence circles all recognized that the Soviet Union posed no military threat to Western Europe in the early postwar period (i.e. it was acknowledged that the Soviets did not want a war against the West and was not going to start one), and American fears of 'losing' Western Europe rested squarely on "appraisals of economic and political conditions".[204] International events merged with the American fear of a threat to domestic core values, a process which created the specific blend of policies that amounted to the U.S. administration's focus on national security.[205]

From the outset the Americans had recognized that for the Marshall Plan to fulfill its purpose, it was necessary to include the industrial western parts of Germany. The Americans, with the able assistance of the British, thus went ahead with a plan to permanently split Europe by dividing Germany.[206] On June 18, the Russian occupation authorities were given a forty-eight hour notice that the Americans and the British were introducing a new currency in their zone of occupation. The Russians responded less than a week later, on June 24, by stopping land traffic into and out of the western zones in Berlin.[207] The Americans were able to organize an airlift of civilian supplies, due to the fact that the Soviets did not block the air corridors, which were the only access routes to Berlin that were guaranteed in writing.[208] In the end, the Americans

---

History (New York: Charles Scribner's Sons, 1965), p. 258.

204 Leffler, M. P., "National Security and US Foreign Policy" in M. P. Leffler and D. S. Painter (eds.), *Origins of the Cold War:...*, pp. 15–52; quote on p. 28.

205 Leffler, M. P., "National Security" in Michael J. Hogan and Thomas G. Paterson (eds.), *Explaining the History of American Foreign Relations* (Cambridge: Cambridge University Press, 1991), pp. 202–213.

206 This decision has carefully studied in: Eisenberg, C., *Drawing the Line:...*; Deighton, A., *The Impossible Peace:...*; Eisenberg, C., "Rethinking the Division of Germany" in A. Hunter (ed.), *Rethinking the Cold War*; Backer, John, H., *The Decision to Divide Germany: American Foreign Policy in Transition* (Durham, N.C.: Duke University Press, 1978).

207 Eisenberg, C., *Drawing the Line:...*, pp. 409–410. This move had been contemplated for at least six months, as the Americans and the British had had freshly printed marks in vaults in Frankfurt since January (ibid., p. 409). William Stivers had shown that the blockade was less than complete, and that the Soviets never attempted to seal off West Berlin completely. No efforts were made to close West Berlin off from either East Berlin or the surrounding countryside. The airlift was a supplement to other sources of supply for the households of West Berlin, but the vast majority of the needs of the population and industry were still met through East-West trade; i.e. West Berlin was kept alive by supplies from the Soviet Zone (Stivers, William, "The Incomplete Blockade: Soviet Zone Supply of West Berlin, 1948–49" in *Diplomatic History*, Vol. 21, No. 4, Fall 1997, pp. 569–602).

208 Yergin, D., *Shattered Peace:...*, p. 377.

had been able, in keeping with Kennan's recommendations, to "place upon the Russians the onus of bringing about the final split of Germany, *just as we had done in the case of the Marshall Plan.*"[209] [Italics added.] The Marshall Plan was an enormous financial undertaking, and the United States could pull it off only because it had the power to truly shape the future. Thus, U.S. generosity rested upon an awareness of its hegemonic potential.[210]

The Brussels Pact and NATO—Materialized Military-Political Consensus

On January 22, 1948, British Foreign Secretary Ernest Bevin gave a speech in the House of Commons where he spoke of the need for a Western Union. This came as no surprise to the Americans, because the British Ambassador in Washington had already presented Secretary of State Marshall with a summary of Bevin's thinking already on January 13. The Ambassador had at that time made clear that this would be the groundwork for a broader Atlantic agreement directed specifically against the Soviet Union.[211] Hence, the concept of a Western defense pact had been dreamed up before the Communist takeover in Prague, but even so, this event did increase the sense of insecurity among the leaders of Western Europe. On March 17, 1948, the Heads of State from Britain, France, Belgium, Luxemburg, and the Netherlands signed the Brussels Treaty.[212] But the foe expressly identified in the Brussels Treaty was not the Soviet Union. Instead, it reflected French fears of a revived and belligerent Germany.[213] The focus upon Germany was something that the Americans, who considered German revival and participation indispensable to European recovery and security, intended to correct before assenting to any association with the Brussels Pact. The United States "wanted to be sure", writes Yergin, "that that new alliance was directed against a Soviet, and not a German, threat."[214]

209 Kennan, G. F., *Memoirs...*, p. 425.

210 Ibid., p. 425.

211 Ireland, T. P., *Creating the Entangling Alliance:...*, pp. 63, 66.

212 Yergin, D., *Shattered Peace:...*, p. 363; Ireland, T. P., *Creating the Entangling Alliance:...*, p. 69; for the similarity with the American approach to the Marshall Plan, see also: Schwabe, Klaus, "Atlantic Partnership and European Integration: American-European Policies and the German Problem, 1947–1969" in G. Lundestad (ed.), *No End to Alliance:...*, p. 41).

213 Germany was explicitly mentioned in two places in the Treaty; once in the Preamble, and once in Article VII ("Appendix 2: Extracts from the Brussels Treaty, 17 March, 1948" in NATO: *Facts About the North Atlantic Treaty Organization* (Paris: NATO Information Service, 1965), pp. 207–208); for French fears, see: Kolko, G. and Kolko. J., *The Limits of Power:...*, p. 486.

214 Yergin, D., *Shattered Peace:...*, p. 363.

Between March 22 and April 1, American, Canadian, and British officials met secretly in the Pentagon to discuss the negotiation of a transatlantic pact of mutual assistance.[215] This was the prelude to the formation of NATO.[216] The British wanted a firm military commitment from the United States, something that the American officials could not give. On the American side it was stressed that "the U.S. government had to retain the right to determine *when* aggression occurred and what type of military action it would take."[217] [Italics added.]

The crux of the matter was Article 5 of the North Atlantic Treaty (NAT), and what obligations it imposed on the signatory nations. The United States got its way, and the final clause remained true to the administrations assurances to Congress "that the proposed North Atlantic Treaty involved no obligation to go to war without its [congressional] consent."[218] The result was thus not really what the European nations had hoped for. The critical part of Article 5 read:

> The Parties agree that an armed attack against one or more of them in Europe or North America shall be considered an attack against them all and consequently they agree that, if such an armed attack occurs, each of them, in exercise of the right of individual or collective self-defense recognized by Article 51 of the Charter of the United Nations, will assist the Party or Parties so attacked by taking forthwith, individually and in concert with the other Parties, *such actions as it deems necessary*, including the use of armed force, to restore and maintain the security of the North Atlantic area.[219] [Italics added.]

Thus, as Dan Smith points out, a member state "could make an angry

215 Leffler, M. P., *A Preponderance of Power:...*, p. 211; Yergin, D., *Shattered Peace:...*, p. 363.

216 For a thorough account of the creation of NATO, see: Ireland, T. P., *Creating the Entangling Alliance:...*. The American delegation wanted to include an article that made provision for consultation "in the event of 'indirect aggression'." This was identified as being "'an internal *coup d'état* or political change favourable to an aggressor'" The clause was not approved by the British who foresaw problems with the French, who had rejected a similar article in the Brussels Treaty (Smith, Dan, *Pressure: How America Runs* NATO (London: Bloomsbury, 1989), pp. 73–74). This also corresponded perfectly to the essence of the Truman Doctrine; Truman described it in those terms himself in his memoirs. In a sense, therefore, one could say that this equaled an effort to extend the Truman Doctrine to Europe in writing (see above, p. 31).

217 Leffler, M. P., *A Preponderance of Power:...*, p. 211.

218 Gaddis, J. L., *The Long Peace:...*, p. 63.

219 "Appendix 4, 'The North Atlantic Treaty', Washington D. C., 4 April, 1949" in NATO: *Facts About...*, p. 211.

phone call if that were all it deemed necessary."[220] This meant that the Brussels Treaty nations failed to tie the United States to Europe in such an unambiguous way as they originally had intended; but even so, as so many writers have pointed out, for the United States the NAT evolved into an 'entangling' alliance that made sure that Europe would remain in the central focus for the Americans during the next fifty years.[221] Paraphrasing Truman's words, Thomas D. Lairson has called the Truman Doctrine, the Marshall Plan, and NATO "three 'halves' of the same walnut", viewing them as "the first major acts of U.S. hegemony."[222]

The Americans also insisted on broadening the alliance to include most European countries. "Scoffing at the possibility of neutrality", says Leffler, "and desiring a comprehensive anti-Soviet bloc" they even called for the partial inclusion of Sweden and Switzerland.[223] At the exploratory talks that began in Washington on July 6, 1948, a three-tiered membership was suggested, with Sweden and Italy being placed in the third tier recommending that they could be associated with the Atlantic alliance "'under such special arrangements as might be necessary owing to their geographical position or to their international obligations [...]."[224] Essentially the same view, with regard to Sweden, was repeated

220 Smith, D., *Pressure:...*, p. 73. Thus, contrary to what one might believe, there is no automatic procedure within NATO, and every country (except Germany whose government actually accepted an automatic procedure) decides for itself whether it will mobilize and submit troops to the inter-allied unified command. This means that "should Greece and Turkey be attacked, Norway and Great Britain, for instance, could refuse to place their forces under allied command." (Beaufre, André, (Translated from French by Joseph Green and R. H. Barry), *NATO and Europe* (London: Faber and Faber, 1967), pp. 44-45; quote on p. 45; see also: Knorr, Klaus, "The Strained Alliance" in Klaus Knorr (ed.), *NATO and American Security* (Princeton, New Jersey: Princeton University Press, 1959), pp. 3–4).

221 This term has been used in a series of works: Osgood, Robert Endicott, *NATO: The Entangling Alliance* (Chicago: University of Chicago Press, 1962); Ireland, T. P., *Creating the Entangling Alliance:...*; Kaplan, Lawrence S., *The Long Entanglement: NATO's First Fifty Years* (Westport, Connecticut: Praeger, 1999).

222 Lairson, Thomas D., "Revising Postrevisionism: Credibility and Hegemony in the Early Cold War" in A. Hunter (ed.), *Rethinking the Cold War*, p. 72.

223 Leffler, M. P., *A Preponderance of Power:...*, p. 211; Ireland, T. P., *Creating the Entangling Alliance:...*, pp. 83, 104; Paper from the Policy Planning Staff: "The Position of the United States with respect to Support for Western Union and Other Related Free Countries", dated June 24, 1948, p. 2, in: RG 59 General Records of the Department of State [henceforth: RG 59]; Records of the Office of European Regional Affairs [henceforth: RA]; Records Relating to the North Atlantic Treaty Organization, 1947–1953; Lot File 57D271; Box 8 of 10; NARA II, College Park, MD.

224 Lundestad, G., *America, Scandinavia, and...*, pp. 274-276, quote on p. 276.

in the policy paper NSC 28/1 in September 1948.[225]

On June 11, 1948, Congress approved the so-called Vandenberg Resolution authorizing the U.S. government to associate itself with "regional and other collective arrangements".[226] The Atlantic Pact was thus, as Senator Connally accurately put it in the Senate debate, "'the logical extension of the principle of the Monroe Doctrine.'"[227] The Vandenberg Resolution paved the way for the NAT, established on April 4, 1949 (which became NATO in 1951), and in October Truman signed the Mutual Defense Assistance Act of 1949, providing for the Mutual Defense Assistance Program (henceforth: MDAP).[228]

Expressing a profound understanding of the United States' role as hegemonic power in world affairs, Park Armstrong, Jr. at the State Department's Bureau of Intelligence and Research (INR) stated in late 1950 that:

> *The United States is the great conservative power in the world today.* This country stands for gradual evolutionary improvement as against proposals from elsewhere for quick revolutionary improvement. This situation is strange to us because at least until 1918 we were accustomed to regard ourselves, and to be regarded, as the radical young nation which gave active sympathy to movements against the old order in all parts of the world. *In many ways our role now is that of making conservatism attractive.* The American concept of freedom has many connotations which do not carry over to the rest of the world. The historical tradition with which we are unconsciously wrapped up [...] is now less meaningful in the most troublesome areas in the world. [...] [T]hese peoples have their own definitions

225 Ibid., pp. 276–277.

226 "Appendix 3, 'The Vandenberg Resolution', Senate Resolution 239, 80th Congress, 2nd Session, 11 June, 1948" in NATO: *Facts About...*, p. 209; see also: Kaplan, L. S., *The Long Entanglement:...*, p. 13.

227 LaFeber, W., *America, Russia, and...*, p. 91. Acheson also made this analogy at the public hearings on the Treaty on April 27, 1949 (Ireland, T. P., *Creating the Entangling Alliance:...*, p. 141). The Monroe Doctrine was issued by President James Monroe in 1823 under which the United States had 'pledged' to "oppose any outside intrusion into the Western Hemisphere [...]." (Cook, Don, *Forging the Alliance:* NATO, *1945-1950* (London: Secker & Warburg, 1989), p. 113; for more on the Monroe Doctrine, see: Perkins, Bradford, *The Cambridge History of American Foreign Relations*, Volume I, *The Creation of a Republican Empire, 1776-1865* (Cambridge: Cambridge University Press, 1993), pp. 163-169). The Monroe Doctrine of 1823, which ruled out foreign interference in Latin America, progressed into the Monroe Doctrine of 1912, which justified U.S. intervention in Latin American internal affairs (LaFeber, Walter, *The Cambridge History of American Foreign Relations,* Volume II, *The American Search for Opportunity, 1865-1913* (Cambridge: Cambridge University Press, 1993), pp. 60–82).

228 For the history of the MDAP, see: Kaplan, L. S., *A Community of Interests:* NATO *and...*

> of freedom which include *freedom from democratic colonial control, freedom from various forms of economic and social discrimination – all of which fail to fit in with the old political slogans.*[229] [Italics added.]

Thus when the United States set out to rebuild the European continent it also took it upon itself, implicitly, to restore as much as possible of the old regimes; under a democratic banner, where considered possible and desirable. It was towards this new United States, the great conservative power in the world, that Sweden turned for salvation in the late 1940s and early 1950s. Assistance would come eventually, but the road would be long and arduous, and entail a lot of giving in order to receive. The United States was looking to gain Sweden's consent to its hegemony in Western Europe; the Swedes were looking for ways in which to offer this consent, but at the same time preserve as much independence as possible in order not to clash with the official policy of neutrality. Moreover, the Swedish government had its own domestic political and economic agenda, which it did not want to be affected by the budding cooperation with the United States. The result will be, albeit partly, described in the following pages.

229 Memorandum by Armstrong, Jr. to Charles E. Marshall, dated November 24, 1950, p. 1, in: RG 59 General Records of the Department of State; Subject Files of the Bureau of Intelligence and Research (INR) [henceforth: INR], 1945-1960; Lot 58D776; Box No. 13; (Folder: Sweden); NARA II, College Park, MD.

Chapter 3

A Challenge to U.S. Hegemony:

U.S. Technology, Swedish Guided Missile Development, and the SDU, 1945–1948

The question of shipments of a number of P-51 aircraft to the Swedish Government is related to the foregoing inasmuch as the Department has taken the position that *these aircraft cannot be sent to Sweden unless Sweden goes along with us on the Argentine arms export policy*. I am informed that this was done on the express instructions of Dean Acheson, who felt that, [...] we should insist upon Sweden's cooperation. [Italics added.]

(Braden to Clayton, September 19, 1946.)[1]

Introduction

In this chapter I intend to show that the United States used technology as a means of influencing Swedish security policy from almost immediately after the end of World War II. This will give us a good basis for analyzing the effects that the policy of neutrality had on the set-up of secret military cooperation. Furthermore, I study the early formation of guided missile related research and development (henceforth: R&D) in Sweden. An understanding of the problems that Sweden encountered in this field, and their significance for future events, is pertinent for a correct interpretation of Sweden's subsequent efforts to procure guided missiles from the United States. Thus, it is crucial to grasp this early development to be able to answer the fourth set of questions as to why so few of these projects resulted in operational missiles. It is also necessary to include an account of the Scandinavian Defense Union (henceforth: SDU) negotiations here, because this issue is central to an understanding of U.S. policy towards Sweden in the years that followed, and thus to the overall purpose of this study. Swedish neutrality policy in general, and the SDU initiative in particular, challenged the United States policy

1 Letter from Braden to Clayton, dated September 19, 1946, pp. 1–2, in: RG 59; Records of the Office of British Commonwealth and Northern European Affairs [henceforth: BNA], 1941–1953; Subject Files [henceforth: SF], 1941–1953; Country (Sweden, External Political Affairs: U.S.S.R.-Sweden, Wood Pulp); Box No. 19; (Folder: Proposed Purchase of P-51); NARA II, College Park, MD.

objective of securing Norwegian and Danish participation in NATO, and hence the effectiveness of the Atlantic Alliance as a whole.

The P-51 Mustang Affair: An Early Attempt to Withhold Military Technology

The Germans had barely surrendered before the first example of U.S. pressure diplomacy, using technology as a lever, was levied upon the Swedes. It is hard to view this episode as anything other than the Americans wishing to teach the Swedish government a lesson, showing them who was the new political master in the Western World.

On April 16, 1946, Sweden and the United States signed a contract for Sweden's purchase of ninety P-51 Mustang aircraft from the Americans.[2] These planes came out of the surplus cornucopia that had opened up after the war, and fifty such aircraft had already been sold and delivered to Sweden in July 1945.[3] However, the second delivery was soon halted, and the reason for this abrogation of the contract was apparently twofold. One of the reasons was related in a letter from Secretary of War Patterson to Secretary of State Byrnes, according to which the contract conflicted with the current State Department policy prohibiting the sale of combat aircraft to any foreign government that had not been specifically designated for deliveries under previously approved sales programs. Examples of approved nations were the American Republics, Nationalist China, and the Philippine Islands.[4]

According to Under Secretary of State Dean Acheson, however, it was not so much a State Department policy as the policy of the U.S. Congress, i.e. United States law. At least that was what he had stated to the Swedish Minister in Washington Herman Eriksson in a conversation on June 1, 1946. Acheson said that the contract had been abrogated "not because of any lack of desire on our [the U.S.] part to see Sweden have these planes", but because the Mead Committee had published a bitterly critical report on U.S. transfers of military surplus materiel, and Acheson himself had made a most categorical statement before the Committee regarding to which countries such deliveries would be

2 See contract for the sale of 90 P-51s to Sweden, dated April 16, 1946, in: RG 59; BNA, 1941–1953; SF, 1941–1953; Country (Sweden, External Political Affairs: U.S.S.R.-Sweden, Wood Pulp); Box No. 19; (Folder: Proposed Purchase of P-51); NARA II, College Park, MD.

3 Letter from Secretary of War Robert P. Patterson to Secretary of State Byrnes, dated July 24, 1946, p. 1, in: RG 59; BNA, 1941–1953; SF, 1941–1953; Country (Sweden, External Political Affairs: U.S.S.R.-Sweden, Wood Pulp); Box No. 19; (Folder: Proposed Purchase of P-51); NARA II, College Park, MD.

4 Ibid., p. 1. See also: Weinberger, H., "The Neutrality Flagpole:..." in M. Thad Allen and G. Hecht (eds), *Technologies of Power:...*, p. 316.

made and why. Unfortunately, Sweden was not among those nations, said Acheson. Legislation had then been introduced in both Houses of Congress denying the administration the right to transfer such materiel, and even providing for call back of such deliveries that had already been made. "Under these circumstances", said Acheson, he "could not make any change in the decision already reached." He added that he could not say whether or not these considerations were permanent, but that he was "bound by statements made to the Congress."[5]

As we shall see, there is reason to believe that Acheson was not telling the whole truth to Eriksson. Overall, the Under Secretary's statements were somewhat cryptic; e.g. if there was actually a law prohibiting the government from exporting the P-51s to Sweden then why would he say that he was bound by statements made to Congress? If what he said was true, then he was bound by law, introduced and enacted by Congress, and could in fact do nothing at all to change the situation; the decision would thus be very permanent indeed. The only thing that could reasonably change the state of things would have been a reversal of the law by Congress. As it turned out, this was not what happened.

The other reason, probably the dominant one, given in internal State Department correspondence, concerned Swedish arms sales to Argentina. This matter went back to a sale of arms contracted during the war, in 1944, between the Argentine government and the Swedish arms manufacturer Bofors. As soon as the Americans heard of this contract they had leaned on the Swedish government to delay the deliveries, which they had agreed to do. Argentina had been blacklisted by the United States for trading with the enemy during the war, and the Americans had similar "informal agreements" with Britain and Canada to withhold arms shipments. The American attitude was a natural consequence of the Monroe Doctrine. The Swedes now felt that they had to commence deliveries in October for the reason that Bofors was after all bound by contract, but more importantly the Swedes felt that the so-called Truman Bill and the program for military standardization throughout the Western Hemisphere would shut them out of the Argentine market. Despite the diplomatic pressure, however, the Americans had as late as September been unable to persuade the Swedes to delay the shipments any longer. Braden at the State Department thus suggested to William Clayton that he should emphasize to the Swedish Minister "again what has already been said, namely:"

5 Memorandum of Conversation between Acheson and the Swedish Minister, dated June 1, 1946, pp. 1–2, in: RG 59; BNA, 1941–1953; SF, 1941–1953; South Africa, Union of, Policy – Sweden, External Political Affairs: U.S.; Box No. 18; NARA II, College Park, MD. I thank Hans Weinberger for pointing this document out to me.

> […] because of the possibility that the Swedish action might affect similar agreements which we have with the UK and Canada. […] Our policy in respect of shipments of arms to Argentina is independent of the plan for military standardization throughout the hemisphere. In any event there is no possibility that the so-called Truman Bill will be passed before January 1947 when Congress next meets. This means that the Swedes could delay shipments until the end of the year without prejudice to their position. While we have heretofore urged an indefinite postponement, perhaps, in view of the foregoing considerations, an agreement could be reached on the January deadline. […] If Sweden will go along on this, we will of course inform them in advance of any change in Argentine-US relations. […] If you agree that a fresh approach should be made along the lines suggested, it will be advisable to ask our Embassy in Stockholm to take the matter up again with the Swedish Foreign Office. The question of shipments of a number of P-51 aircraft to the Swedish Government is related to the foregoing inasmuch as the Department has taken the position that *these aircraft cannot be sent to Sweden unless Sweden goes along with us on the Argentine arms export policy*. I am informed that this was done on the express instructions of Dean Acheson, who felt that, […] we should insist upon Sweden's cooperation.[6] [Italics added.]

The American refusal to honor the P-51 contract with Sweden was thus directly linked to the efforts to coax the Swedes into withholding their deliveries to Argentina, and it was a policy initiated at the highest echelons of the United States government and the State Department; in fact, by Dean Acheson himself. This was in total contrast to what Acheson had told the Swedish Minister on June 1, and Braden did not mention anything regarding any laws preventing the government from transferring the planes to Sweden. He even said that this decision had been made on the instructions of Acheson. Instead of having his hands tied by Congress, or any statement made to the Mead Committee, it seemed that Acheson had wished to take a hard line with the Swedes in order to teach them a lesson.

However, this hardline approach toward Sweden happened to conflict with an objective of the U.S. War Department policy at this time; namely, the aerial photography project code-named *Casey Jones*. Project *Casey Jones* was an effort to photograph the European continent from the air, in order to make large-scale military maps. The War

6 Letter from Braden to Clayton, dated September 19, 1946, pp. 1–2, in: RG 59; BNA, 1941–1953; SF, 1941–1953; Country (Sweden, External Political Affairs: U.S.S.R.-Sweden, Wood Pulp); Box No. 19; (Folder: Proposed Purchase of P-51); NARA II, College Park, MD.

Department, as well as representatives of U.S. Air Force Europe (henceforth: USAFE) in Wiesbaden, Germany, already in July expressed concern that the cancellation of P-51 deliveries was "having a detrimental effect" upon the *Casey Jones* mapping project. Furthermore, Secretary of War Patterson wrote to Byrnes that the War Department felt that:

> [...] the Swedish Government could hardly be expected to collaborate on a mapping operation of this nature without some cooperation on our part in a matter which is of primary interest to Sweden. It is also significant to point out that the War Department is strongly supporting the "Casey Jones" project *and that the particular importance of this strategic area warrants further consideration of this sale of p-51 aircraft*, as it is felt such sale would materially facilitate negotiations concerning the War Department mapping interests in Sweden. [...] In the view of the importance the Swedish Government attaches to the purchase of additional P-51 aircraft and the importance which the War Department attaches to the "Casey Jones" project, *it is recommended that an exception be made in this instance to the State Department policy* governing sales of combat aircraft to foreign governments. Such action would reaffirm the friendly attitude of this government toward the government of Sweden and would serve to continue the hitherto harmonious relations between the two countries.[7] [Italics added.]

Although, Patterson was not referring to the Argentine question, perhaps because he did not know about it, his message to Byrnes was very clear; in order for the Swedish government to have an amicable attitude towards the War Department's mapping program the State Department would have to back away. Patterson did not mention anything about a congressional law preventing the sale either. He did, however, refer to a policy within the State Department. In fact, this was not the first time the War Department had tied the sale of aircraft to Sweden to the aerial mapping effort. In mid-July it was reported that the War Department wished to sell six F-9 type aircraft [modified B-17s] to Sweden in exchange for allowing the Americans to undertake their photographic flights. This deal was also stated to be of the utmost importance to the project, because "the War Department will be

7 Letter from Secretary of War Robert P. Patterson to Secretary of State Byrnes, dated July 24, 1946, pp. 1–2, in: RG 59; BNA, 1941–1953; SF, 1941–1953; Country (Sweden, External Political Affairs: U.S.S.R.-Sweden, Wood Pulp); Box No. 19; (Folder: Proposed Purchase of P-51); NARA II, College Park, MD. Hans Weinberger has written about both the P-51 affair and the *Casey Jones* mapping project, although he did not see just how explicit the connection between the two issues was (Weinberger, H., "The Neutrality Flagpole:..." in M. Thad Allen and G. Hecht (eds), *Technologies of Power:...*, pp. 315–316.

unable to carry out its aerial mapping plan in the event that the [State] Department does not permit the sale of these planes to Sweden."[8]

The War Department had approached the Swedes about the *Casey Jones* project in May 1946, but the informal request involving photographic missions made by U.S. personnel and planes had been refused by the Swedish Foreign Ministry; perhaps as a direct result of the withholding of the P-51 Mustangs. The Americans had then changed tactics, suggesting that the United States could provide the equipment for the Swedes to perform the survey themselves, but on July 11 the Swedish government refused to go along with this scheme as well. The reason was that the Swedes were worried lest the presence in Sweden of foreign personnel and planes "would preclude secrecy."

When the matter had not been resolved by August it was already too late to perform the photographic missions in 1946, but the Commanding General in Wiesbaden advised the War Department that the fall and winter should be used to conclude an agreement with the Swedes so that full advantage could be taken of the spring and summer photographic season (the reason being that there was no point in taking pictures of the ground if it was covered by snow). The project was considered so urgent and important that the USAF was prepared to train one Swedish officer to direct the operation and nine qualified Swedish personnel in the United States. The training of an officer would take six to nine months, and the other personnel could be made ready within 22–29 weeks; training these men in Europe would be preferable if the State Department were to agree to the idea. According to the USAF, training flights could be made in Sweden during the summer of 1947, with the actual photography starting in the summer of 1948. The Acting Chief of the Division of Northern European Affairs (NOE) J. H. Morgan did not find this idea feasible, however, and stated:

> In the opinion of NOE, the project will be acceptable to the Swedes only if no US personnel and no US planes* [Hand written footnote signifier] enter Sweden in connection with it. Furthermore, the training of Swedish personnel should take place in the U.S.A., as any attempt at such training in the US zone of Germany would be too obvious to foreign observers. In order to avoid embarrassment with other powers the Swedes will be inclined to approve the project only if its features give them assurance that secrecy can be maintained. In this connection, it would appear inadvisable to extend the project over two years, with

8 Letter from Bell to McGhee, dated July 19, 1946, in: RG 59; BNA, 1941–1953; SF, 1941–1953; South Africa, Union of, Policy – Sweden, External Political Affairs: U.S.; Box No. 18; NARA II, College Park, MD. See also: Weinberger, H., "The Neutrality Flagpole:..." in M. Thad Allen and G. Hecht (eds), *Technologies of Power:...*, p. 316.

the actual mapping to take place only in 1948. If Swedish approval is forthcoming, the Army should endeavor to hasten arrangements and complete the project as quickly as possible.
* [Handwritten footnote:] Suitable planes may be declared surplus and "sold" to Swedes.[9]

It seems as if the polices of the War and State Departments clashed on this issue, and that the State Department was not being entirely honest with its military counterparts. On October 2, Robert J. G. McClurkin, Director of the State Department's Aircraft Division, wrote to U.S. Army Air Force Brigadier General John A. Samford on the matter of the sale of P-51 and AT-6 aircraft to Sweden. In his letter McClurkin did not mention the issue of Swedish arms deliveries to Argentina when explaining to Samford why the aircraft had been withheld. Instead, only the question of U.S. policy of selling such aircraft merely to nations already included in a program was mentioned. McClurkin even stated that the delay in deliveries "reflects in no way upon the Government of Sweden." He also said that the decision to deny the aircraft to Sweden had been made by Acting Secretary of State Acheson himself, and that Acheson alone could reverse it. McClurkin added that a decision by Acheson was expected momentarily.[10]

Yet, a decision to reverse the decision had already been made the day before, at a meeting of the Secretary of State's Staff Committee, with explicit reference to the Swedish arms sales to Argentina. The memorandum from the meeting stated:

> [The meeting] APPROVED the transfer of ninety P-51 airplanes to Sweden and AGREED that the Swedish Government should be informed that *approval is given reluctantly in view of a Swedish contract for the sale of certain arms to Argentina* and that it is hoped that the Swedish

9 Letter from Morgan to Labouisse, dated August 15, 1946, pp. 1–3, in: RG 59; BNA, 1941–1953; SF, 1941–1953; Country (Sweden, External Political Affairs: U.S.S.R.-Sweden, Wood Pulp); Box No. 19; (Folder: Proposed Purchase of P-51); NARA II, College Park, MD.

10 Letter from McClurkin to Samford, dated October 2, 1946, pp. 1–2, quote on p. 1, in: RG 341 (Hq US Air Force) [henceforth: RG 341]; Deputy Chief of Staff, Operations [henceforth: DCSO]; Director of Intelligence [henceforth: DI]; Dep Dir for Collection & Dissemination [henceforth: DDCD]; Collection Operations Division [henceforth: COD]; Air Attaché Branch [henceforth: AAB]; Station File [henceforth: STF], 1948–49; Spain to Sweden; Box No. 67; (Folder: Sweden-Aircraft); NARA II, College Park, MD. For Acheson's decisions, see also: Weinberger, H., "The Neutrality Flagpole:..." in M. Thad Allen and G. Hecht (eds), *Technologies of Power:...*, p. 316.

Government or firms will not enter into similar contracts with Argentina in the future.[11] [Italics added.]

It thus seems clear that McClurkin was not playing it straight with Samford; or perhaps the real reason for denying Sweden the P-51s was a well-kept secret even within the State Department. The fact that the decision was reversed by Acheson, and reluctantly so if one is to believe the quotation above, indicates that the War Department's argument regarding the importance of the *Casey Jones* mapping project might have swung the Under Secretary of State around. This certainly seems to prove that Acheson's words to Eriksson in June were not entirely true; the Under Secretary did not have the power to override a law enacted by Congress.

The Swedish Chargé d'Affaires at the Swedish Embassy in Washington Alexis Aminoff was informed of the Staff Committee's decision by NOE Chief Morgan on October 7. Morgan also told Aminoff that there was a strategic angle to the choice of revising the decision. This "military question [...] did not appear in the official directive [but] the strategic consideration involved had been nevertheless of importance." Morgan had then "emphasized that what [he] had just said formed no part of the official decision and suggested that if he [Aminoff] reported this part of [Morgan's] remarks he should do so separately." Aminoff apparently stated that he appreciated the point and expressed his deepest thanks for the information.[12] Morgan apparently followed the Staff Committee's recommendation to the letter, and emphasized the reluctance with which the decision had been made in view of the Swedish arms sales to Argentina.[13] Although Morgan did not specify what the strategic angle to the American decision was, it is reasonable to assume that the NOE Chief was referring to the War Department's mapping program *Casey Jones*. It seems likely that this program was what had persuaded Acheson and the State Department to eventually approve the delivery of the ninety P-51 airplanes to Sweden, even though this could not be spelled out in the official motivation for the decision.

11 Memorandum from the Secretary's Staff Committee Meeting, dated October 1, 1946, in: RG 59; BNA, 1941–1953; SF, 1941–1953; Country (Sweden, External Political Affairs: U.S.S.R.-Sweden, Wood Pulp); Box No. 19; (Folder: Proposed Purchase of P-51); NARA II, College Park, MD.

12 Letter from Morgan to Hickerson, dated October 7, 1946, in: RG 59; BNA, 1941–1953; SF, 1941–1953; South Africa, Union of, Policy – Sweden, External Political Affairs: U.S.; Box No. 18; NARA II, College Park, MD. I thank Hans Weinberger for pointing this document out to me.

13 Letter from Aminoff to Sven Grafström, dated October 20, 1946, p. 3, in: HP 24 VI/U.S.A; "Tillstånd till in- och utförsel av vapen, ammunition och krigsmateriel"; 1946, okt.– aug. 1950; Folder 20; Box No. 1353; År: 1946–1952; Riksarkivet (henceforth: RIA), Stockholm.

Since Acheson had reversed his decision regarding the P-51 airplanes, the Americans now thought that the new scheme might be acceptable to the Swedes. Morgan at NOE, referring to the F-9 airplanes which had force-landed in Sweden during the war and then been left behind, felt that since there were already nine American B-17s in Sweden "the presence in Sweden of six additional planes of this type need not attract attention."[14]

Apparently, even though the P-51s were released to them, the Swedes turned the new proposal down, once again with reference to the fact that the project could not be expected to remain secret, and would risk subjecting Sweden "to embarrassing requests of a similar nature from other powers."[15] The power referred to was obviously the Soviet Union. The State Department, though, thought that this could be gotten around by portraying the matter as an exchange program; thereby any other nation could be invited to participate on a reciprocal basis, which of course the Soviets would never agree to. In addition, the State Department had agreed to authorize the U.S. Military Attaché in Stockholm to offer Sweden either one set of the best available topographic maps of any area in the United States that roughly corresponded to the area being photographed in Sweden, five sets of 1:1,000,000 scale aeronautical charts of North America and Europe, or one complete set of maps listed in the U.S. Army Map Service catalog.[16]

I have unfortunately been unable to find any documents that might show whether this barter trade actually took place or not. It is clear, however, that the War Department considered the *Casey Jones* project to be strategically important, for it was one of several steps preparing for possible air warfare against the Soviet Union. Sweden's geo-strategic position came into play here, since the American Strategic Air Command (henceforth: SAC), which had been established on March 21, 1946, had prepared war plans that involved flying through Swedish airspace on its way to the USSR.[17] This option was of great importance for the United States, because the state of the U.S. Armed Forces looked

14 Letter from Morgan to Hickerson, dated October 9, 1946, in: RG 59; BNA, 1941–1953; SF, 1941–1953; South Africa, Union of, Policy – Sweden, External Political Affairs: U.S.; Box No. 18; NARA II, College Park, MD. I thank Hans Weinberger for pointing this document out to me.

15 Letter from J. A. Michela at the War Department to the U.S. Military Attaché in Stockholm, dated November 19, 1946, p. 1, in: RG 59; BNA, 1941–1953; SF, 1941–1953; South Africa, Union of, Policy – Sweden, External Political Affairs: U.S.; Box No. 18; NARA II, College Park, MD. I thank Hans Weinberger for pointing this document out to me.

16 Ibid.

17 Weinberger, H., "The Neutrality Flagpole:..." in M. Thad Allen and G. Hecht (eds), *Technologies of Power:...*, pp. 315, 317; Ross, Steven T., *American War Plans 1945–1950* (London: Frank Cass, 1996), p. 12.

dismal to American planners. By the end of 1945 it was estimated that the USSR, despite rapid demobilization, could overrun Western Europe (excluding the British Isles) and Scandinavia at any time between 1945 and 1948. The American possession of the atomic bomb was intended to compensate the fact that U.S. conventional forces were greatly outnumbered by Russia and her allies in Europe. It was thus expected that the United States would have to rely on strategic and tactical air power, with long-range B-36 bombers attacking the the Soviet Union from U.S. bases. That was the basic assumption in the American war plan Bushwacker of March 8, 1948.[18] Under such circumstances it is easy to imagine why ready access to Sweden's airspace was important for the United States. As we shall see later on, overflying Swedish territory remained a central feature in the Western planning for its strategic air warfare.

The Heritage of Penemünde

The fact that Sweden was to be one of the first nations to attempt domestic guided missile development can be accredited to the Third Reich. In this case Sweden was no different from other nations that began R&D work on guided missiles in the immediate postwar period, including the United States, Britain and the Soviet Union. The most conspicuous difference was of course that Sweden was not a great power. 'The big three' quickly enrolled German scientists who had been involved in the German guided missile development, and Sweden too got some early assistance from individuals who had been involved in the German program.

In June 1943 a German V2-missile hit the ground just north of Kalmar,[19] a small town on the Swedish east coast. In November, that same year, a V1-missile crashed near Karlskrona (also on the Swedish east coast), and two weeks later yet another V1-missile was found not far from Ystad in the south of Sweden.[20] These missiles all came from the German missile base at Penemünde (all of them were unarmed test missiles which had gone off course).[21] All in all, pieces from six V1-mis-

18 Ross, S. T., *American War Plans 1945–1950*, pp. 5, 12, 79–89.

19 Weinberger, Hans, *Sievert: enhet och mångfald. En biografi över den svenska radiofysikens, radiobiologins och strålskyddets grundare Rolf Sievert* (Stockholm: Stockholm Papers in History and Philosophy of Technology, TRITA-HOT 2021, 1990), p. 77. it has been designated the 'Bäckebo Bomb', Bäckebobomben in Swedish, after the crash site in the village Bäckebo.

20 Cervin-Ellqvist, G. and König, B., *Robot 50 år*, p. 8.

21 The V1 was called "air torpedo" (lufttorped) by the Swedes. It flew at a low altitude and was fired from a slanted launching pad by a booster rocket and then propelled in its trajectory by a pulse jet engine. It was the world's first operational cruise missile. The V2, on the other hand, was a ballistic guided missile , i.e. its trajectory was arch-like, and it flew at a very high altitude (outside of the atmosphere most of the time). It was fired from an upright position and

siles were found in Sweden between 1943 and 1944, and the interest in this new weapon was great.[22] Almost immediately, the Swedes tried to establish international contacts, and in August 1944 one of the original V1-missiles was traded for British radar stations.[23] Extended range and increased accuracy were strong incentives for the different military services to study the different applications for this new weapon – and the possibilities it offered the Swedish Armed Forces. The National Defense Research Establishment (Försvarets forskningsanstalt; henceforth: FOA), stated in an information bulletin in 1968 that it was the tactical possibilities offered by this new weapon that took the center stage as the interest in guided missile development progressed, and that the guided missiles were seen as a complement to the conventional weapons arsenal.[24]

Sweden's dependence on foreign technology has been known for a long time.[25] What is not known, however, is how this dependency affected Swedish security policy during the Cold War. This dependence also had very concrete effects on the Swedish effort to develop guided missile, but up until know it has not been known what the effects were and what resulted from them. Centrala ledningen för reaktions- och raketdrift (the Central Directorate for Reaction and Rocket Propulsion, henceforth: CD) held its first meeting on February 8, 1945.[26] This organization became the central body of the guided missile development

had a considerably larger warhead than the V1. Both of these missiles were frequently deployed against Great Britain during World War II. The 'V' stood for Vergeltungs Waffe, or Vengeance Weapon. The reason for designating them "air torpedoes" was that there was no Swedish word for this weapon in those early days. The inequities of the language hence made this homemade denomination necessary. The projectiles looked rather kind of like underwater torpedoes, with the crucial difference that they traveled through air instead of water.

22 *FOA orienterar om Robotvapen*, nummer 9, juni, 1968, p. 10.

23 Dörfer, I., *System 37 Viggen…*, p. 96. Flight Director Henry Kjellson received the Order of the British Empire in 1944 for his reconstruction of a German V2-guided missile that had landed in Sweden (see: http://www.ne.se/jsp/search/article.jsp?i_art_id=225789&i_word=Kjellson, 2005-10-19, 15:22).

24 *FOA orienterar om Robotvapen*, p. 10.

25 For information regarding Sweden's dependence on foreign military technology during this period: Aunesluoma, J., *Britain, Sweden and the Cold War…*; *Had There Been A War…*; Silva, C., *Keep Them Strong, Keep Them Friendly:…*; Weinberger, Hans, "'Det rörde sig väl om ofarliga ting'. Svensk neutralitetspolitik …" in *Arbetarhistoria…*, NO. 92, 4/1999; Weinberger, Hans, "The Neutrality Flagpole…. Regarding the period after 1970, cf. Holmström, M. and von Sivers, T., *USAS exportkontroll. Tekniken som vapen*; Mörth, U. and Sundelius, B., *Interdependens, konflikt och säkerhetspolitik:…*.

26 "Protokoll över sammanträde med Centrala ledningen för samarbetet betr reaktions- och raketdrift, hållet den 8 februari 1945" in Kungliga flygförvaltningen [henceforth: KFF], Centrala ledningen för reaktions- och raketdrift (CLR) [henceforth: CLR], Serie: A II, "Protokoll från sammanträden", År: 1945–1948, VOL. 1, KrA.

effort over the following three years. It was created on an Executive Order from the Supreme Commander of the Swedish Armed Forces, with the explicit purpose of co-ordinating the incipient work in the field undertaken by the three Armed Services, the Royal Swedish Army (henceforth: RSA), the RSAF, and the Royal Swedish Navy (henceforth: RSN). One important task was to facilitate collective utilization of available technical resources.[27]

As the above-stated implies, missile-related R&D had already begun in all of the Armed Services, and parts of them were to be included in the guided missile research organization. It is important to remember that all initial organization and development work was done before the United States had established itself as a hegemonic power and was primarily concerned with finding a viable strategy to rebuild war-torn Western Europe. But U.S. officials recognized the connection between economic and political power and "were ready to assume Britain's former role as financial hegemon."[28] Hegemonic collaboration, with any nation, was thus impossible at this stage—Sweden was on its own in a volatile universe trying to enact relations with Britain and the United States which could act as conduits for military technology.

The problems associated with the capital-intensive and time-consuming guided missile research were acknowledged right from the start. The research institutions that had been contacted by the CD were unwilling, due to an already heavy workload, to accept the responsibility for a new project that would demand great contributions in terms of time and manpower.[29] The opinion of the CD was that this would be "[...] a matter well suited for the planned National Defense Research Establishment, which [would] be able to appoint a responsible institution or person, and allot funding to it."[30] From very early on it was decided that the CD would cooperate with tactical personnel from the

27 *foa orienterar om Robotvapen*, p. 10.

28 Leffler, M. P., *A Preponderance of Power:...*, p. 16.

29 "Protokoll över sammanträde med Centrala ledningen för samarbetet betr reaktions- och raketdrift, hållet den 8 februari 1945" in KFF, CLR, Serie: A II, "Protokoll från sammanträden", År: 1945–1948, VOL. 1, KrA, pp. 1–2.

30 "Protokoll över sammanträde med Centrala ledningen för samarbetet betr reaktions- och raketdrift, hållet den 8 februari 1945" in KFF, CLR, Serie: A II, "Protokoll från sammanträden", År: 1945–1948, VOL. 1, KrA, p. 2. "[...] ett typiskt ärende för den blivande Försvarets forskningsanstalt, som [skulle] kunna utse ansvarig institution eller person samt anvisa medel till denna." The Swedish Institute of Military Physics merged with the Defense Chemicals Establishment and a telecommunications branch in April 1, 1945, forming FOA, and it then became a central part of the guided missile research organization. Note that FOA was not formed as an appendix to Swedish guided missile development, but it was considered to be an important part in this development. It had an important role in enabling the guided missile research community to perform its task.

Defense Staff in order to guide the development work in a direction that corresponded with the tactical demands of the overall strategy.[31]

A plethora of guided missile projects were initiated during the ten-year period following the end of World War II and most of them were abandoned rather quickly. In the archives of Försvarets robotvapenråd (the Armed Forces' Guided Missile Council, GMC) from this period we find for example: air-to-air missile RB 321; air-to-surface missiles RB 300, RB 301, RB 302, RB 303 and RB 304; the ship-based[32] surface-to-surface missiles RB 310, RB 311, RB 312, RB 313, RB 315, RB 316 and the surface-to-air missile RB 322.[33] However, many of these were only testing and development projects, and most of them were probably never intended to go any further than that. Some missile projects where extremely short-lived, e.g. the air-to-surface missile RB 303, canceled in the spring of 1950 (efforts were instead directed towards the air-to-surface missile RB 304), and the RB 301 test guided missile.[34]

31 "NR 6/1945. "Protokoll över ordinarie sammanträde med CLR, hållet den 8 maj 1945." in KFF, CLR, Serie: A II, "Protokoll från sammanträden", År: 1945–1948, VOL. I, KrA, p. 1. The Defense Staff was the Swedish JCS; i.e. the highest decision-making military organization.

32 The Swedish term for this category is *sjörobotar*, which roughly translates as Sea Missiles.

33 In: KFF, Robotvapenbyrån Försvarets robotvapenråd (FRR) [henceforth: FRR], Serie: A I, "Protokoll från sammanträden 1948–1957", År: 1948–1957, VOL. I, KrA. Ingemar Dörfer has written that "[n]o less than 85 different prototypes were developed and abandoned at this stage." (Dörfer, I., *System 37 Viggen...*, p. 96.) This is not true. Dörfer probably came up with this number due to his misunderstanding of a caption to a picture of test missile RB 310 contained in *foa orienterar om Robotvapen*. The caption said that 85 missiles where manufactured and 190 test firings were made in the years 1946–1958 (*foa orienterar om Robotvapen*, p. 11). However, this number probably referred to the same kind of missile (i.e. RB 310), and not to 85 different types of missiles. Dörfer used *foa orienterar om Robotvapen* as a source in his book (Dörfer, Ingemar, *System 37 Viggen...*, p. 241).

34 "CLR INK 21/47", "Protokoll fört vid CLR sammanträde den 27/2 1947", in: KFF, CLR, Serie: A II, "Inkomna och avgångna handlingar med diarium", År: 1945–1948, VOL. I, KrA, p. 3. "CLR INK 22/47", "Protokoll från CLR sammanträde den 24/4 1947", in: KFF, CLR, Serie: A II, "Inkomna och avgångna handlingar med diarium", År: 1945–1948, VOL. I, KrA, p. 4. "NR 2/1950. Protokoll från FRR sammanträde den 16. mars 1950.", in KFF, FRR, Serie: A I, "Protokoll från sammanträden 1948–1957", År: 1948–1957, VOL. I, KrA, p. 1. Among the reasons for canceling RB 303 were, at least according to the GMC, problems with finding a suitable homing device for it. This was because RB 303 was intended to strike at ships from underneath, i.e. it was supposed to dive into the water prior to hitting the target. The RB 304 was also to be used against convoy ships and disembarking vessels, but would strike exclusively above water, a fact that made it less problematic. Another decisive factor was that air-to-surface missile RB 304 weighed significantly less (only 650 kilos), which meant that two RB 304s,

On October 9, 1945, the National Committee for Invention (Statens uppfinningsnämnd) ordered five empty hulls (intended for the future RB 310) from aircraft manufacturer SAAB.[35] The Committee acted on behalf of the Royal Swedish Navy Board (RSNB), represented by Captain Johan Gabriel Oxenstierna.[36] It is a fact that although the RSAF later became the dominant party in guided missile procurement and the bulk of all research was geared towards RSAF projects, it was the RSN (and its Torpedo Bureau where Oxenstierna was Chief) and the RSA that pioneered this field.[37]

At the time, there were no engines available that could be mounted onto these hulls, but work on prototypes was being carried out by industrial contractors STAL and AB Ljungström Ångturbiner. The contract eventually went to STAL, and on October 26, 1946, the first prototype was launched (unsuccessfully, it is true) from the Test Firing Center at Karlsborg (Provskjutningscentralen i Karlsborg, PCK) on the north-west shore of Lake Vättern.[38]

At this point the Swedes were heavily dependent upon the work done by the Germans during the war and this was sometimes made extremely obvious, e.g. when the RSNB requested that SAAB duplicate the autopilot, i.e. the steering mechanism, found in the German V1-guided missile. The German original was used as a model in a very literal sense, since Oxenstierna actually brought it with him to SAAB to be taken apart by the company's engineers.[39]

During this time the build-up phase of the RSAF had started to gain momentum. The Royal Swedish Air Force Board (RSAFB) had ordered empty de Havilland Vampire fighter jet fuselages for the RSAF from the British de Havilland Aircraft Company. Paralleling this they had

---

instead of just one RB 303, could be mounted on the Swedish fighter-bomber aircraft A 32 Lansen.

35 Cervin-Ellqvist, G. and König, B., *Robot 50 år*, p. 9.

36 Ibid., p. 12; see also: "CLR INK 17/46", "Protokoll fört vid CLR sammanträde den 12 mars 1946", in: KFF, CLR, Serie: A II, "Inkomna och avgångna handlingar med diarium", År: 1945–1948, VOL. 1, KrA, p. 4.

37 *SAABs robot- och elektronikhistoria, del 1*, "Så började det – 40-talet", 1:1:6.

38 Cervin-Ellqvist, G. and König, B., *Robot 50 år*, p. 9, 12. There were two test sites. The one at Karlsborg (RFK) covering most of the northern part of Lake Vättern, and one in Norrland (RFN),an area 65 kilometers long and about 35 kilometers wide close to the town Vidsel (*FOA orienterar om Robotvapen*, p. 65); see also: "CLR INK 17/46", "Protokoll fört vid CLR sammanträde den 12 mars 1946", in: KFF, CLR, Serie: A II, "Inkomna och avgångna handlingar med diarium", År: 1945–1948, VOL. 1, KrA, p. 4.

39 "CLR INK 17/46", "Protokoll fört vid CLR sammanträde den 12 mars 1946", in: KFF, CLR, Serie: A II, "Inkomna och avgångna handlingar med diarium", År: 1945–1948, VOL. 1, KrA, p. 5. The steering device was called 'styrautomat' in Swedish; see also: Cervin-Ellqvist, G. and König, B., *Robot 50 år*, pp. 11–13.

also ordered British Ghost and Goblin jet engines, and manufacturing licenses for these engines, from the de Havilland Engine Company. The contracts had been signed in January 1946 at which time the British government also gave the Swedes permission to manufacture the Goblin engine. Permission to manufacture the Ghost engine was given a year later in March, 1947. The intention was to mount the Swedish-made British engines in the empty Vampire fuselages.[40] By doing this Sweden became the third country in the world to introduce jet fighters into active service in its Air Force. The first de Havilland Vampire aircraft (designated J28A) arrived to the F13 Air Force Base in Norrköping in June, 1946.[41] Thus, while STAL and AB Ljungström Ångturbin were having a hard time trying to develop workable jet engines for future Swedish guided missiles, the RSAFB bought and manufactured British jet engines for its fighter squadrons. Unbeknownst to the Swedes at that time; a pattern had been set for the future.

The first test-firing was soon followed by several more, and the people from SAAB and the CD worked together to try to solve the problem of the stalling engines.[42] All in all, twelve test firings of the as yet unnamed missile were performed during the initial test phase, only three of which were designated a success. Test phase number two began with test number 13 of this particular kind of missile (which had by then been given the name RB 310) in November 1948.[43] In November 1947 only one of the original five missiles ordered from SAAB in October 1945 by the RSNB remained, and ten new missiles (which were to be designated RB 311) were ordered in September, due for delivery in February 1948.[44]

40 "CLR INK 17/46", "Protokoll fört vid CLR sammanträde den 12 mars 1946", in: KFF, CLR, Serie: A II, "Inkomna och avgångna handlingar med diarium", År: 1945–1948, VOL. 1, KrA, p. 7; see also: FO115/4367, "Supply to Sweden of Vampire Aircraft", September 30, 1948, p. 1, VOL. 16 in: The Neutrality Policy Commission Archive [henceforth: NPCA], Regeringens arkiv- och dokumentcenter.

41 Andersson, Lennart and Hellström Leif, *Bortom horisonten. Svensk flygspaning mot Sovjetunionen 1946–1952* (Stockholm: Stenbom, 2002), p. 56.

42 Cervin-Ellqvist, G. and König, B., *Robot 50 år*, p. 13; see also: "CLR INK 22/47", "Protokoll från CLR sammanträde den 24/4 1947", in: KFF, CLR, Serie: A II, "Inkomna och avgångna handlingar med diarium", År: 1945–1948, VOL. 1, KrA, p. 3.

43 Cervin-Ellqvist, G. and König, B., *Robot 50 år*, p. 9.

44 Ibid., pp. 9–10; "CLR INK 22/47", "Protokoll från CLR sammanträde den 24/4 1947", in: KFF, CLR, Serie: A II, "Inkomna och avgångna handlingar med diarium", År: 1945–1948, VOL. 1, KrA, p. 2. The RB 310 guided missile was a pretty bulky piece of work. Almost 5 meters long, it weighed in at 265 kilos and had a maximum velocity of 170 meters per second. It had a wingspan of 2.5 meters and its maximum range was 17 kilometers. But the RB 310 guided missile was overshadowed by the RB 311 that was over 6.5 meters long and weighed nearly

These two missiles were both replicas of the German V1-guided missile. They looked the same, and were launched the same way, i.e. from an inclined launch pad with the aid of a boost engine. The boost engine shut down after a few seconds and the guided missile was then propelled in its trajectory by a pulse jet engine.[45] The limited use of these guided missiles was apparently clear to the people involved because in the summer of 1948 it was decided that no more RB 311s would be produced. Instead plans were developed for making this guided missile into a target drone, because a drone was considered to be absolutely indispensable for the successful development and testing of a future surface-to-air missile (the RB 322) and the air-to-air missile RB 321.[46]

At about the same time, i.e. in early 1948, discussions had begun inside the CD regarding a reorganization of Sweden's guided missile development work; the disparate development did not yield any useful results and something had to be done about the situation. The importance of real and active cooperation between the three services was emphasized, because of the similarity of many of the problems encountered. Already, intra-service rivalry was evident in the struggle for control over the development of this new weapon. The omnipresent Captain Oxenstierna of the RSNB presented the idea of creating an entirely new organization to deal with this task at a meeting of the CD on February 3, 1948.[47] Oxenstierna probably thought that this could enhance the role of the RSN and cement its influence over the guided missile development. However, the RSAF beat him to it.

Prior to the next meeting a proposal for a letter to the Supreme Commander, Helge Jung, had been produced and sent to the members by the representatives of the RSAF. Accidentally or not, the letter contained a recommendation identical to Oxenstierna's (i.e. that a Bureau for the development of guided missiles should be created), but with the crucial difference that the new Bureau was to be under the command of the

---

one metric ton, or 900 kilos. The wing span measured 4.2 meters and it had a maximum velocity of 220 meters per second. The extra weight resulted in a shorter range, though: only 7.5 kilometers (Cervin-Ellqvist, G. and König, B., *Robot 50 år*, p. 11).

45 Cervin-Ellqvist, G. and König, B., *Robot 50 år*, p. 8. A pulse jet engine is (simplified) a jet engine with sub-sonic speed.

46 "NR 2/1948", "Protokoll från FRR sammanträde den 11/6 1948 (militära ledamöterna).", in KFF, FRR, Serie: A I, "Protokoll från sammanträden 1948–1957", År: 1948–1957, VOL. 1, KrA, p. 2; Annex NO 1 to "NR 3/1948", "Protokoll från FRR sammanträde den 4 aug 1948" entitled "Anteckningar vid sammanträde med MR sektionschefer den 2/7-48." in KFF, FRR, Serie: A I, "Protokoll från sammanträden 1948–1957", År: 1948–1957, VOL. 1, KrA, p. 2.

47 "CLR INK 28A/48", "Protokoll från CLR sammanträde den 3/2 1948", in: KFF, CLR, Serie: A II, "Inkomna och avgångna handlingar med diarium", År: 1945–1948, VOL. 1, KrA, p. 2.

RSAFB. It would merge the groups currently working on guided missile related issues in the three services into one.[48] The RSAF was maneuvering to gain control over the domestic development effort.

Accordingly, on April 22, 1948, the CD was officially re-organized and given the name Försvarets robotvapenråd (Armed Forces' Guided Missile Council; henceforth: GMC) in compliance with order NO 49:15 from the Military Headquarters.[49] Paralleling this development was the formation of Försvarets robotvapenbyrå (Armed Forces' Guided Missile Bureau, GMB). Captain Oxenstierna had failed to place the new organization under the aegis of RSNB's Torpedo Bureau, but was compensated by being made Chief of the new organization. This compromise was most likely made to strike a balance between the RSN and the RSAF, the two greatest rivals in the guided missile development area. The GMC was to consist of one or two persons representing the Supreme Commander as well as representatives from the three services, FOA, and the Chief of the GMB. One of the top priorities for the GMC was "to follow the overall technological, tactical, and organizational development in Sweden as well as abroad".[50]

The GMB, on the other hand, was organized into one Military Division and one Technical Division. It was the job of the Military Division to bring forth the guidelines that were to guide the work of the Technical Division.[51] Due to the multitude of guided missiles that was under way so far, as well as the lack of foreign aid, all projects were still at the experimental stage in November 1948. Serial production of any system was not realistic at this time, nor was it deemed possible to make a prediction about when such production might commence.[52]

It was not only the military services that suffered in this environment; the civilian company SAAB experienced problems too. Despite having created a Guided Missiles Division in 1948, SAAB was falling behind on

48 "CLR INK 29/48", "Protokoll från CLR sammanträde den 3/3 1948.", in: KFF, CLR, Serie: A II, "Inkomna och avgångna handlingar med diarium", År: 1945–1948, VOL. 1, KrA, pp. 1–3.

49 "NR 1/1948. Protokoll från FRR sammanträde den 19/5 1948" in KFF, FRR, Serie: A I, "Protokoll från sammanträden 1948–1957", År: 1948–1957, VOL. 1, KrA.

50 Annex NO 1 to "NR 1/1948. Protokoll från FRR sammanträde den 19/5 1948" entitled "Instruktion för Försvarets R-vapenråd (FRR)", in KFF, FRR, Serie: A I, "Protokoll från sammanträden 1948–1957", År: 1948–1957, VOL. 1, KrA, p. 1. "att följa den tekniska, taktiska och organisatoriska utvecklingen i stort såväl i Sverige som i utlandet".

51 "NR 1/1948. Protokoll från FRR sammanträde den 19/5 1948" in KFF, FRR, Serie: A I, "Protokoll från sammanträden 1948–1957", År: 1948–1957, VOL. 1, KrA, pp. 1–5.

52 Annex NO 1 to "NR 5/1948. Protokoll från FRR sammanträde den 4 nov 1948" entitled "PM angående förslag till krigsorganisation för MR." in KFF, FRR, Serie: A I, "Protokoll från sammanträden 1948–1957", År: 1948–1957, VOL. 1, KrA, p. 1.

deliveries as early as in the summer of 1949. The RB 310 guided missile lived on despite these problems, however, and despite the cancellation of RB 311. The GMC decided to support the suggested procurement of fifty RB 310s in the early summer of 1949.[53] Even though the RSNB was the most conspicuous spender thus far, the RSAF had almost twice as much funding available for activities related to guided missiles as the RSA, RSN and FOA, respectively, for the fiscal year 1949–1950. While the latter three hovered around SEK 1.4 million each, the RSAF could spend almost SEK 2.5 million out of a guided missile development budget of just over SEK 6.6 million.[54] These sums were of course dwarfed by the American outlays on their guided missile programs. For example, the Falcon project, which had been initiated with the Hughes Aircraft Corporation (henceforth: HAC) in 1947, had cost $15.1 million through FY 1951.[55] Test firings of Swedish missiles continued at a steady pace, and several more test guided missiles had been ordered, but it was not until early 1950 that the first Officer's Training Course started at the anti-aircraft regiment LV 3 in Karlsborg. Two more such courses had been held by early 1953.[56]

International Contacts and Influences on the Early Swedish Development

Generally, the access to information from foreign sources was very limited during this time, i.e. the late 1940s and the early 1950s. This was especially true with regard to input from the Americans. During this initial period it was the British that constituted Sweden's international

53 "NR 5/1949", "Protokoll från FRR sammanträde den 11 maj 1949.", in KFF, FRR, Serie: A 1, "Protokoll från sammanträden 1948–1957", År: 1948–1957, VOL. 1, KrA, p. 1.

54 "NR 7/1949", "Protokoll från FRR sammanträde den 9. september 1949.", in KFF, FRR, Serie: A 1, "Protokoll från sammanträden 1948–1957", År: 1948–1957, VOL. 1, KrA, p. 2.

55 "GM 206/2, Summary of Falcon Program Presentation Given to Guided Missile Committee, RDB 7 December 1950 (29th Meeting)", pp. 1–2, in: RG 156 Records of the Office of the Chief of Ordnance [henceforth: RG 156]; Executive Office Historical Branch [henceforth: EOHB]; Records Relating to the Army Guided Missiles Program [henceforth: AGMP], 1940–62; GM 80/Project Falcon MX-904 through GM 210/Snark; Box No. 13; NARA II, College Park, MD.

56 "NR 10/1949", "Protokoll från FRR sammanträde den 7. december 1949.", in KFF, FRR, Serie: A 1, "Protokoll från sammanträden 1948–1957", År: 1948–1957, VOL. 1, KrA, p. 1. "NR 2/1950", "Protokoll från FRR sammanträde den 16. mars 1950.", in KFF, FRR, Serie: A 1, "Protokoll från sammanträden 1948–1957", År: 1948–1957, VOL. 1, KrA, p. 2. There were five officers from the Army, five from the Navy, three from the Coastal Artillery, and four from the Air Force. Annex 1 to "NR 2/1953", "Protokoll från FRR sammanträde den 2 februari 1953.", entitled: "PM angående högre teknisk utbildning i robotvapentjänst för officerare och ingenjörer.", in KFF, FRR, Serie: A 1, "Protokoll från sammanträden 1948–1957", År: 1948–1957, VOL. 1, KrA, p. 2.

framework of contacts – and not just in the field of guided missiles.[57] The Swedes needed to obtain information from abroad, and this led to a series of journeys to England and the United States with the intention of studying guided missile development in these countries. Aircraft Director (flygdirektör) Henry Kjellson of the RSAFB visited England in the spring of 1945, but could only bring back detailed information about German guided missiles, since information concerning British ones ”[…] for obvious reasons had been hard to get access to.”[58] The CD put together a special plan for visits to foreign countries which was to contain participating personnel, objects of study, and countries of interest; [59] although England and the United States were the only nations considered at the time.[60]

The Swedish Defense Attachés in England and the United States then received letters inquiring about their attitudes towards the suggested visits to these countries.[61] In April, 1946, the RSA's suggested visits to England had been given a green light by the British; but the Attaché in the United States had thought it wise to wait and not ask the American authorities at this time since, in his judgment, it was very possible that they might turn the idea down.[62] At this point the Americans doubted that Sweden would consent to its demands because Sweden was in the midst of negotiating its large trade deal with the Soviets. In April the deal was mentioned in the largest Swedish daily newspaper, *Dagens Nyheter*,[63] so the Americans probably knew about the plans for an agreement. Moreover, and perhaps more importantly, during this period certain Swedish economic assets in the United States were confiscated

57 See: Aunesluoma, J., *Britain, Sweden and the Cold War…*; *Had There Been A War…*; Collmar, Sara, *forthcoming*.

58 ”Protokoll över sammanträde med CLR, hållet den 6 mars 1945.” in KFF, CLR, Serie: A II, ”Protokoll från sammanträden”, År: 1945–1948, VOL. 1, KrA, p. 1. ”[…] av naturliga skäl varit svåra att få taga del av.”

59 ”CLR INK 13/45”, ”Protokoll fört vid CLR sammanträde den 13 november 1945”, in: KFF, CLR, Serie: A II, ”Inkomna och avgångna handlingar med diarium”, År: 1945–1948, VOL. 1, KrA, pp. 1–2.

60 Annex 1 to ”CLR INK 15/46”, ”Protokoll fört vid CLR sammanträde den 15 januari 1946”, entitled “Förslag till studieresor i utlandet betr materiel inom CLR verksamhetsområde” in: KFF, CLR, Serie: A II, ”Inkomna och avgångna handlingar med diarium”, År: 1945–1948, VOL. 1, KrA, p. 1.

61 ”CLR INK 17/46”, ”Protokoll fört vid CLR sammanträde den 12 mars 1946”, in: KFF, CLR, Serie: A II, ”Inkomna och avgångna handlingar med diarium”, År: 1945–1948, VOL. 1, KrA, p. 1. Kungl. Maj:t is an abbreviation for Kunglig Majestät, i.e. His Royal Majesty.

62 ”CLR INK 18/46”, ”Protokoll fört vid CLR sammanträde den 9 april 1946.”, in: KFF, CLR, Serie: A II, ”Inkomna och avgångna handlingar med diarium”, År: 1945–1948, VOL. 1, KrA, p. 4.

63 Karlsson, B., *Handelspolitik eller politisk handling…*, p. 62.

due to front operations for German companies during World War II. Some Swedish companies came very close to being blacklisted by the Americans.[64] This could very well be part of the reason why the Attaché in Washington thought that there might be a risk that the Americans would turn down a request for a Swedish visit.

From very early on, then, the members of the CD realized the importance of sending people abroad for study purposes, and at the same time they acknowledged that it was necessary to concentrate development efforts on a few promising objects due to scarcity of fiscal and personnel resources. The latter point continued to be a matter of high priority and, in March 1948, it became one of the first tasks assigned to Captain Oxenstierna as Chief of the newly established GMB.[65]

The Swedes also had some contacts with their Western neighbor Norway. In this pre-hegemonic environment, where no hegemonic apparatuses had yet been erected to support information and technology exchange, sporadic collaboration such as the barter trade with the Norwegians served as a supporting infrastructure for Swedish guided missile development. For example, in 1947 the Norwegians provided the Swedes with the German PC 1400 X, also known as the FRITZ X-bomb.[66] This was part of a kind of barter trade between Sweden and Norway. In April the same year the Swedish Defense Staff's secret organization for unofficial collection of foreign intelligence, the T-Office [T-kontoret], smuggled a German missile over the Norwegian border, to let the Norwegian National Defense Research Establishment (Forsvarets forskningsinstitutt) have a look at it.[67] At the same time,

64 Aalders, G. and Wiebes, Cees, *The Art of Cloaking Ownership: The Secret Collaboration and Protection of the German War Industry by the Neutrals: The Case of Sweden* (Amsterdam: Amsterdam University Press, 1996), pp. 139–142.

65 "CLR INK 31/48", "Protokoll från CLR sammanträde den 31/3 1948.", in: KFF, CLR, Serie: A II, "Inkomna och avgångna handlingar med diarium", År: 1945–1948, VOL. 1, KrA, p. 4. Oxenstierna had been suddenly removed from his post as Naval Attaché to London in late 1943 because he was assumed, falsely, by the British to have leaked sensitive and secret information from his London office, via Stockholm, to Berlin (Denham, Henry, *Inside the Nazi Ring: A Naval Attaché in Sweden, 1940–1945* (London: John Murray, 1984), pp. 45–47, 68–71).

66 "CLR INK 18/47, 'Data för fallbomb PC 1400 X (FRITZ X).'", in: KFF, CLR, Serie: F I, "Inkomna och avgångna handlingar med diarium", År: 1947, VOL. 3, KrA. This guided missile engine was not propelled by an engine; instead it was dropped like an ordinary bomb from a delivering aircraft and then guided in its steep descent by the bombardier by radio remote control (Schollars, Todd J., "German wonder weapons: degraded production and effectiveness" in *Air Force Journal of Logistics*, Fall, 2003).

67 Petersson, Magnus, *"Brödrafolkens väl"* ..., p. 252. The Swedish-Norwegian collaboration is described thoroughly in this book. (The 'T' in the name of the Swedish intelligence organization stood for the first name of its director, Thede Palm.)

Norway provided Sweden with a few liquid-propellant rocket motors from the German Henschel HS 293 A1 air-to-surface guided missile.[68]

The Americans tried to keep themselves abreast of all developments in the guided missile field in Sweden through a number of channels. In July 1947 the Office of the Air Attaché reported on a conversation held with Captain Axel Wickström, Assistant to the Artillery Director at the Swedish munitions company Bofors. The intelligence report stated that no work pertaining to guided missiles was currently going on at Bofors, and neither was it planned in the near future. The reason was simply that the:

> [...] Bofors Company does not have the technical background and necessary electric engineers to develop the guidance and control equipment required. While Bofors has had considerable experience with solid propellant rockets they have had no experience with liquid fuel rockets or other types of propulsion units presently considered suitable for medium and long range guided missiles. The only known guided missiles being developed in Sweden at the present time consists of two air to surface missiles known as "301" and "302".[69]

The report concluded by promising more information regarding this subject.[70] It should be noted that the information given was somewhat flawed in that a lot more than two guided missile projects were being worked on in Sweden in 1947. This was remedied, however, in two reports submitted in November the same year. By that time, the American Air Attaché Office had gotten their hands on more accurate intelligence. Judging from the level of detail of the intelligence reported back to Washington, it is safe to assume that the source of the information was someone deeply involved in the Swedish development program.

The first report contained information regarding a pulse-jet guided missile called 250 LT [lufttorped] that was being developed for the RSN. It had the character of a questionnaire, and stated that no test had as yet been made so there was no way of telling how accurate the missile would be. Detailed information was provided regarding its type of

68 "CLR INK 22/47", "Protokoll från CLR sammanträde den 24/4 1947", in: KFF, CLR, Serie: A II, "Inkomna och avgångna handlingar med diarium", År: 1945–1948, VOL. 1, KrA, p. 5. This guided missile had been used with some success against Allied ships in the Mediterranean theater beginning in mid-1943 (http://www.nasm.si.edu/research/dsh/artifacts/RM-Hs293.htm.).

69 "ID 382341" from Air Attaché in Stockholm to Washington, dated July 7, 1947, p. 2, in: RG 319 Army-Intelligence Document File [henceforth: AIDF]; From ID 382201 To ID 382390; Box No. 2486; NARA II, College Park, MD.

70 Ibid., p. 2.

propellant and design, even stating that it would be launched from an aircraft catapult device. The company responsible for the development work was, of course, SAAB, and the Americans wished to know what the abilities of the personnel engaged in the development were:

> SAAB is, with the exception of electrical components, physically capable of carrying on experimental work in guided missile fields, and is further capable of limited production. Although some of the personnel engaged in the development are well qualified, they are still in the process of gaining experience. They are below U.S. or British standards.[71]

At the same time it was stated that "another unconfirmed source indicates that development studies on this missile may terminate in the near future."[72] The information is interesting since it so obviously flaunts the inadequacies of the development work performed at SAAB. Indeed, the statement that the company was physically capable of carrying on experimental work with the exception of electronic components was equal to saying the exact opposite, i.e. that it was not physically capable of doing such work at all. Electronic components were, after all, vital for the company to be able to function effectively. Providing such intelligence to the Americans was probably the best way to let them know about the areas in which Sweden needed help in development efforts. The Air Attaché's reports could thus serve as a conduit for relaying Swedish pleas for assistance in the guided missile field to the United States.

A couple of days later, yet another report was sent. This time it contained detailed information regarding most of the major projects under development at that time. The guided missiles covered RB 301, RB 303, RB 311, RB 321, RB 250 LT; including a drawing of anti-ship missile RB 303.[73] The RB 301, it was said, was a RSAF project adopted from an earlier version developed by the RSN. An interesting detail was the fuel and propulsion specifications. Apparently it had not been decided whether the missile would be propelled by a solid, or a liquid, propellant. The liquid propellant consisted of hydrogen peroxide and permanganate, and had been adopted from German 509 Walther missiles

71 "ID 416572" from Air Attaché in Stockholm to Washington, dated November 7, 1947, p. 2, in: RG 319 AIDF; From ID 416521 To ID 416640; Box No. 2685; NARA II, College Park, MD.

72 Ibid., p. 2.

73 "ID 416573" from Air Attaché in Stockholm to Washington, dated November 12, 1947, pp. 2–6, in: RG 319 AIDF; From ID 416521 To ID 416640; Box No. 2685; NARA II, College Park, MD.

that the Swedes had procured from Norway. It was, however, deemed uncertain whether these engines would actually be used.[74]

In his comments, Lieutenant Colonel Malcolm D. Seashore confirmed the intelligence provided in July by saying:

> Guided missile development in Sweden is in its infancy. The biggest bottlenecks are the lack of electrical equipment and mechanical and technical "know-how" for production of this type of equipment. Some of the personnel engaged in the development work are well qualified, but lack experience and can be considered below U.S. or British standards.[75]

The report was concluded by stating that "[a]nother visit to SAAB at Linköping is planned in the near future",[76] suggesting that the information contained in the above report had been gathered on site.

The Swedes also purveyed intelligence pertaining to Russian guided missile experiments to the Americans. During a visit by U.S. Navy Commander E. R. Dare from the Joint Research Board, the Swedes disclosed that they had received long wave radio signals that could "be attributed only to a system which probably is an electronic guidance system."[77] On the day following Dare's visit, U.S. Navy Lieutenant Commander Eric Pollard confirmed that what the Swedes had picked up was indeed signals from a guidance system. Pollard had made several radar search flights over Scandinavia and had received and identified similar electronic signals with a frequency of about 41.5 megacycles which could come "from only one location: Russian-occupied Germany in the vicinity of Rügen."[78] The intelligence gathered indicated, according to the Naval Attaché, that the Russians were experimenting with a radio-controlled aircraft or V1-type of missile.[79]

At about the same time the CD decided to send an engineer to Paris to meet with a Mr. Yardeney, an American who was working on problems related to remote control guidance, before Yardeney went back to the United States, to see if such matters were of interest to Sweden.[80] This

74 Ibid., p. 2.

75 Ibid., p. 5.

76 Ibid., p. 5.

77 "ID 394195" from Naval Attaché to Washington, dated July 31, 1947, p. 1, in: RG 319 AIDF; From ID 394901 To ID 395100; Box No. 2559; NARA II, College Park, MD.

78 Ibid., p. 2.

79 Ibid., p. 2.

80 "CLR INK 31/48", "Protokoll från CLR sammanträde den 31/3 1948.", in: KFF, CLR, Serie: A II, "Inkomna och avgångna handlingar med diarium", År: 1945–1948, VOL. 1, KrA, p. 3.

episode suggests that the relationship between Sweden and the United States had progressed somewhat and become more amicable by the end of March 1948. Note that this discussion took place only two months before Sweden signed up for participation in the Marshall Plan. Sweden was about to take its first major step towards consenting to American hegemony.

A major reason for the desire to go westward and study guided missile development in England and the United States was the series of serious problems that had plagued Swedish guided missile development from the very beginning. Some of these problems were reported by the American Attachés; as revealed above. The shortage of telecommunication components was just one of these problems, and purchases from the United States had take place with varying intensity at least since 1948.[81] But there was no certainty in the relationship with the Americans at this stage, and each deal was still made on an ad hoc basis.

This had the unfortunate effect that expectations concerning the Americans' desire to help the Swedes deviated widely among the different services and establishments. For example, the chief of the GMB, Captain Oxenstierna, seems to have been more confident than Martin Fehrm, Chief of FOA's Radar and Electronics Department (FOA 3), said that "Sweden can hardly count on more help from America on these issues *than can Russia*." [Emphasis added.][82] This is a statement that seems somewhat odd, especially since it had been Fehrm who recommended that a Swedish engineer be sent to France to meet with Mr. Yardeney just six months earlier. But perhaps Fehrm, too, believed that the American's would indeed assist in the Swedish development effort and that a corollary to this would be a drastic decline in prestige for FOA 3, which he himself managed. Should the Swedish development work be based on American research then the role of the Establishment would shrink correspondingly.

Challenging U.S. Hegemony in the North

In early January 1948 Marcus Wallenberg hosted a dinner at his house. Among the guests were the editor of Sweden's largest daily newspaper *Dagens Nyheter*, Herbert Tingsten, and several high military officers. At the party Tingsten, who was the most vociferous and energetic critic of the government's policy of neutrality, and who shared U.S. Ambassador

81 "FRR INK 20/48,'Protokoll från sammanträde vid FOA –3 den 9 sept 1948 för uppföljning av de FOAarbeten som direkt berör KFF/MR.'", in KFF, FRR, Serie: A I, "Protokoll från sammanträden 1948–1957", År: 1948–1957, VOL. 1, KrA.

82 "NR 4/48. Protokoll från FRR sammanträde den 15/9 1948.", in KFF, FRR, Serie: A I, "Protokoll från sammanträden 1948–1957", År: 1948–1957, VOL. 1, KrA, p. 3. "Sverige knappast kan påräkna större hjälp från Amerika i dessa frågor än Ryssland." There was three departments at FOA: FOA 1 (physics), FOA 2 (chemistry), and FOA 3.

Freeman H. Matthews' moral indignation toward the government's policy, urged Supreme Commander Helge Jung to argue publicly for Swedish adherence to the West. In his diary, Jung took the view that Tingsten was crazy to ask such a thing, for if he did he would surely lose his job without further delay.[83] Jung might have thought that Tingsten was crazy for suggesting this, but he no doubt also saw that the editor could be useful when trying to influence the government's policy away from strict neutrality. In late February Ambassador Matthews informed Secretary of State Marshall that he had been told "in strict confidence" by the Chief of the RSAF General Nordenskiöld, that he and other military officers supplied 'ammunition' which they possessed in order for Tingsten to conduct as efficient a campaign as possible for a closer association between Sweden and the West.[84] It thus seems as if the military elite, led by Jung, Swedlund, and Nordenskiöld, decided to do their share by supplying the editor with information and letting him mount the crusade for them. This way they could always plead ignorance, and work towards establishing secret ties with the West without risking their necks. The propaganda waged by Tingsten was very much to the liking of Matthews, who referred to it as a "splendid campaign".[85]

The Swedish business community, as well as the royal family and the royal court, provided willing support for the American hegemonic policy.[86] In early March, 1948, Marcus Wallenberg visited Washington, and Ambassador Matthews wrote to Under Secretary of State Robert A. Lovett from Stockholm that Wallenberg could need some 'education' regarding the realities of the Cold War, and insisted that Lovett make room in his busy schedule to see their "mutual friend".[87] Wallenberg would stop over shortly in Washington on semi-official business and Matthews' suggestion was made with the conviction that the message would be promptly carried back to the Social Democratic government in Sweden, since "the Wallenbergs are so powerful in this country [Sweden] and have such influence [...] even with a Social-Democratic government, that their advice carries considerable weight." However,

83 Molin, K., *Omstridd neutralitet...*, p. 38; see also: Weinberger, H., "The Neutrality Flagpole:..." in M. Thad Allen and G. Hecht (eds), *Technologies of Power:...*, p. 306.

84 Airgram from Matthews to the Secretary of State, dated February 21, 1948, in: RG 84; SLE; CONF 1948; 020-844; Box No. 6; (Folder: 800); NARA II, College Park, MD.

85 Letter from Matthews to Lovett, dated March 23, 1948, p. 3, in: RG 84; SWSE; TSGR; 1943–1949; Box No. 1; (Folder: 1948), NARA II, College Park, MD.

86 Molin, K., *Omstridd neutralitet...*, p. 31.

87 Airgram from Matthews to Lovett, dated February 17, 1948, p. 1, in: RG 84; Stockholm Legation and Embassy [henceforth: SLE]; Confidential File [henceforth: CONF] 1948; 020-844; Box No. 6; (Folder: 800); NARA II, College Park, MD.

even Wallenberg suffered from what Matthews referred to as "a strong 'neutrality' complex".[88] He added that since Sweden had considerable influence on its Scandinavian neighbors, due to its relative size, Lovett therefore ought to devote a few minutes to giving "him some straight-from-the-shoulder talk on the facts of life."[89]

Lovett agreed to Matthews' suggestion[90] and subsequently gave Wallenberg a lecture on the American apprehensions about the Swedish government's position. Lovett "endeavored to advance the 'educational process'" which Matthews had suggested to him, and he had a long and frank discussion with Wallenberg. For instance, although Wallenberg could not foresee any change in Sweden's official position in the near future, he was absolutely positive that the Swedes would fight if they were attacked. Wallenberg had added to this that he knew that:

> [the Swedish military] have set up their plans on the basis of a holding operation and have designed certain of their airfields, etc., to accommodate outside aid.[91]

According to Lovett, Wallenberg expressed admiration for Matthews' educational efforts in Stockholm, but said that it would take time to reorient the Swedish government; "they were stubborn", he said, and "they did not respond to 'prodding'"[92] His efforts even included trying to convince "certain extreme Left members of the Social Democrats" that the Soviet system did not have anything to offer them.

Perhaps because he was himself worried, but more likely because government officials in Sweden had asked him to, Wallenberg finally asked the American if the United States had written off Sweden as being unreliable, to which Lovett answered that they had not; adding that they "did not expect to have to do so." Pleased to hear this, the Swede offered his assurances "that Sweden was 'on our [the U.S.] side of the fence' and that he was confident it would stay that way fundamentally."[93]

In his answer to Lovett, Matthews intimated that he agreed fully with Wallenberg's words about the dangers of prodding the Swedish government. "My tactics", wrote the Ambassador, "are adjusted, consequently, to the indirect approach."[94] Note that this was six months

88 Ibid., p. 1.

89 Ibid., p. 2.

90 Note from Lovett, dated February 21, 1948, in: RG 84; SLE; CONF 1948; 020-844; Box No. 6; (Folder: 800); NARA II, College Park, MD.

91 Letter from Lovett to Matthews, dated March 5, 1948, p. 1, in: RG 84; SWSE; TSGR; 1943–1949; Box No. 1; (Folder: 1948), NARA II, College Park, MD.

92 Ibid., p. 2.

93 Ibid., p. 3.

94 Letter from Matthews to Lovett, dated March 23, 1948, p. 1, in: RG 84; SWSE;

before the President approved NSC 28/1, and it suggests that Matthews had realized what the best way to affect the Swedish policy was. At this time even the usually sullen Matthews could see some movement away from neutrality, albeit "very slowly". He wrote:

> Things have really begun to move at last in Sweden (though at a speed which in these days makes a tortoise seem like a derby winner).[95]

Even Wallenberg was surprised at this movement, and at "how far opinion had moved" in Sweden. In subsequent conversations with the Director of Svenska Handelsbanken (a representative of a rival sphere to the Wallenberg family) later the same month, it was no doubt confirmed to Matthews that his effort to co-opt prominent members of the Swedish business community had paid off. The Director assured him that business interests in Sweden were about to undertake "a quiet campaign" to "wean the government away from neutrality".[96] Matthews, while careful not to say too much, offered his views:

> While from this distance I could not say how much we would welcome any Swedish-American staff talks or how much or what type of equipment we might be able to supply, I felt certain that the closer Sweden associated itself with the West now, the more favorable would be the answers.[97]

Even though Undén's worries about the direction of Norwegian foreign policy went back further than to Bevin's Western Defense Union speech in March 1948, he apparently felt that the establishment of the Brussels Pact demanded action on his part. Undén was worried that Norway's drift to, and possible alignment with, the West would upset the precarious balance on the Scandinavian Peninsula, and Prime Minister Erlander felt that it would increase the risk of war.[98]

Undén's counter-proposal was to suggest the creation of an SDU that would remain non-aligned between the East and the West. The proposal actually had considerable implications for Swedish foreign

---

TSGR; 1943–1949; Box No. 1; (Folder: 1948), NARA II, College Park, MD.

95 Ibid., p. 1.

96 Ibid., pp. 1–2.

97 Ibid., p. 2.

98 Aalders, G., "The Failure of the Scandinavian Defence Union, 1948–1949" in *Scandinavian Journal of History*, Vol. 15, No. 1, 1990, p. 133; Cole, P. M., *Neutralité du jour:...*, pp. 186–187. For a thorough study on the motivations behind the Norwegian Foreign Ministry's attitude to nato and the SDU, see: Skodvin, Magne, *Norden eller* NATO*? Utenriksdepartementet og alliansespørsmålet 1947–1949* (Oslo: Universitetsforlaget, 1971).

policy, for even though the union itself would be neutral, Sweden pledged to assist its neighbors if they were attacked by the Soviets. This was equal to giving up the long cherished non-alignment policy, at least in a Scandinavian context.[99] However, the United States did not share this view in 1948. The Swedish initiative for a SDU was interpreted as a serious challenge to U.S. hegemony, and the Americans had no intention of sitting idly by and watching as Norway and Denmark got sucked into a neutral defense system.

The idea of an SDU was an old one. Already in the fall of 1946 the Swedish Defense Staff, in collaboration with the Foreign Ministry, had looked into the idea of a joint rearmament of the Scandinavian countries. Foreign Minister Undén took a positive view of this venture at the time, even if, being disinterested in military matters, according to Agrell, the full strategic implications of such an endeavor might not have been totally apparent to him. Later, in 1947, the British concluded a study where it was suggested that Soviet aggression could be countered by encouraging the Scandinavian countries to establish secret military ties, including standardization of material, and covert staff talks. Due to this, concludes Agrell, one can with good reason question the novelty of this proposal in mid-1948.[100]

Undén had declared the joint Scandinavian defense initiative pointless in early April 1948, but because of the dramatic international events in the spring of 1948, the SDU rested uneasy in the soft earth where it had been buried, and was almost immediately resuscitated and presented as new. Karl Molin has suggested that the reason for this was a combination of domestic and international factors. The idea, he says, was brought to the fore by the international events in the spring of 1948 (the communist takeover in Czechoslovakia and the Finnish-Soviet mutual 'Pact of Friendship'), but it was domestic reasons (the Social Democrats wanted to preempt any attempt by the opposition to gather political points before the elections in September by suggesting it first) that forced the government to act. The SDU had much support in important factions of the press, and the government found it opportune to make this issue their own.[101]

99 Agrell, W., *Alliansfrihet och atombomber. Kontinuitet och förändring i den svenska försvarsdoktrinen 1945–1982* (Stockholm: Liber Förlag, 1985), p. 279. For a penetrating analysis of the Swedish government's actions prior to, and during, the SDU negotiations, and the role played by the Advisory Council on Foreign Affairs in this process, see: Ekecrantz, S., *Hemlig utrikespolitik…*, pp. 61–99.

100 Agrell, W., *Den stora lögnen…*, pp. 60–63.

101 Molin, K., *Omstridd neutralitet…*, pp. 52–56, 59; Molin, Karl and Olesen, Thorsten B., "Security policy and Domestic Politics in Scandinavia 1948–49" in Thorsten B. Olesen (ed.), *Interdependence Versus Integration: Denmark, Scandinavia and Western Europe, 1945–1960* (Odense: Odense University Press, 1995), pp. 62, 65–70; for Undén's attitude towards the SDU before May,

Swedish Ambassador to London Erik Boheman also cited the election as a mitigating reason[102] when he tried to explain the Swedish initiative to Bevin, whose patience had been tested by the proposal of a neutral SDU, and who was irritated by what he construed as Sweden's pressure politics upon Norway and Denmark. Boheman apparently told Bevin that not too much should be made out of these talks, and gave his assurance that no pressure would be put on Sweden's neighbors.[103] It is perhaps important to note that even though Boheman was no friend of Undén's neutrality policy, and although he wished to see a more westward leaning Swedish foreign policy, he did not recommend that Sweden give up its non-alignment. He simply wanted the Swedish government to take a less 'doctrinaire' approach to matters relating to foreign policy, or as he himself called it, a more realistic one.[104]

---

see also: Aalders, G., *Swedish Neutrality and the Cold War, 1945–1949*, p. 82. In *Omstridd neutralitet* from 1991 Molin tried to position himself in opposition to earlier writers on the subject and specifically criticized Gerard Aalders for not having seen the domestic motive for the Swedish SDU initiative. He cites Aalders as talking rather vaguely about "'all those events of early 1948'" (Molin, K., *Omstridd neutralitet...*, p. 60). However, in 1995 Molin had swung around to a position much closer to Aalders. Now he had more or less taken over Aalders' view, stating that initiative was due to "the dramatic events during the spring of 1948" (Molin, K. and Olesen, T. B., "Security policy and Domestic Politics ..." in T. B. Olesen (ed.), *Inter-dependence Versus Integration:...*, p. 65).

102 Boheman had been appointed Swedish Minister to Paris after World War II and left Paris for London in June 1947 where he became Swedish Ambassador. He was then transferred to Washington a year later, in June 1948 (Stridsman, J., "Erik Boheman..." in K. Zetterberg (ed.), *...Hot and Cold...*, p. 99. Boheman, who was confessedly to the right politically, and who had aspirations to become Foreign Minister in a new rightist government, might seem an odd choice by a Social Democratic government for such an important position. It seems, however, as if there were domestic strategy behind the decision to appoint Boheman. When faced with the choice of sending Boheman to Washington, or keeping him in Sweden, Erlander opted for the former. The reason was that Erlander felt that Boheman was too dangerous to have as a critic of the government's policies in Sweden; better then to co-opt him by making him a servant of the government (*Tage Erlanders dagböcker, 1945–1949*, Published by Sven Erlander (ed.) (Hedemora: Gidlunds Förlag, 2001), p. 228). Erlander wrote in his diary: "I told Undén, that I had never been particularly fond of Boheman but that he was so skilled that it was rather unpleasant to have him as the government's opponent and critic. If we could get Hägglöf to take London I would approve of it [sending Boheman to Washington]. It seemed as if Undén for a while actually thought of following my advice." (ibid., p. 228.) "Jag sa Undén, att jag visserligen aldrig varit särskilt förtjust i Boheman men at than var så skicklig att det var ganska otrevligt att ha honom som regeringens motståndare och kritiker. Kunde man få Hägglöf att ta London tillstyrkte jag det därför. Det verkade som Undén faktiskt funderade en smula på att följa mitt råd."

103 Aunesluoma, J., *Britain, Sweden and the Cold War...*, pp. 62–64.

104 Stridsman, J., "Erik Boheman..." in K. Zetterberg (ed.), *...Hot and Cold...*, pp. 101–102, 104–107. See also: Oredsson, Sverker, "Erik Boheman" in Gunnar

Olof Kronvall has suggested that yet another issue hovered in the collective mind of the Swedish government at this time. This other issue was Finland. According to Kronvall, Finland's security was of certain importance to the government when suggesting negotiations on a Scandinavian Defense Pact.[105] It thus seems fair to conclude that there were multiple reasons for why the government chose to act when it did.

Whatever the immediate reason, even though he had previously condemned it, and even though he was very uncertain as to the chances of successfully establishing such a union, Undén nevertheless presented the idea to Norway and Denmark on May 3.[106] But this did not mean he had fundamentally altered his position on the issue; he was apparently surprised when he received a positive response from the Advisory Council on Foreign Affairs to his question if Sweden ought to probe the Norwegian and Danish governments on their attitude to the SDU.[107]

A vivid illustration of American sentiments, and of the strategy that was used to influence the Swedes, is the advice Hugh S. Cumming, Jr., Counselor of the American Embassy in Stockholm, gave to the PPS on May 4, 1948. Cumming reported the Stockholm Embassy's great efforts to 'educate' the Swedes toward a more realistic viewpoint of the Soviet Union, and Sweden's position relative to it. To this end it was making use, as we have seen, of all available contacts "among government, Army, Navy, Labor, business, banking and other circles."[108] The Embassy recommended that the U.S. government should withhold military assistance to Sweden pending an indication of the Swedes being willing to abandon their position. At the same time, said Cumming, the United States should provide assistance "as freely as possible" to

Artéus and Leif Leifland, *Svenska diplomatprofiler under 1900-talet* (Stockholm: Probus Förlag, 2001), pp. 196–203.

105 Kronvall, O., *Den bräckliga barriären…*, p. 97.

106 Aalders, G., "The Failure of the Scandinavian Defence Union, 1948–1949" in *Scandinavian Journal of History*, Vol. 15, No. 1, 1990, pp. 132–133; Aalders, G., *Swedish Neutrality and the Cold War, 1945–1949*, pp. 80–84.

107 Molin, K., *Omstridd neutralitet…*, p. 57. Stefan Ekecrantz has shown that the government did present this matter before representatives of the parliamentary opposition in the Advisory Council on Foreign Affairs, but that the Council "was not fully informed about the alliance initiative". Just prior to the top secret talks in Oslo, the government had yet another chance to inform the opposition leaders "about the precise nature of the talks without doing so" (Ekecrantz, S., *Hemlig utrikespolitik…*, p. 243). Furthermore, he points out that it is far from clear what Undén actually said, and that the participants of the meeting had divergent views regarding the content of his speech (ibid., pp. 64–74, 243).

108 "Summary of Remarks by Hugh S. Cumming, Jr. before the Policy Planning Staff of the State Department", dated May 4, 1948, p. 3, in: RG 84; SLE; CONF 1948; 020-844; Box No. 6; (Folder: 800); NARA II, College Park, MD.

Norway and Denmark to make Sweden "see that we are in earnest." It was the Counselor's view that such action would strengthen the hand of elements in Sweden desiring to associate themselves with the United States.[109]

The three Scandinavian countries agreed in May to form a joint Committee to investigate further the possible arrangements for the SDU. The Norwegians preferred an open-ended inquiry that did not preclude various forms of associations with the West. Deliberations were troubled by differing opinions, and it was not until August 1948 that an agreement was reached on how to publicly express the mandate to the Committee. At a secret meeting of the Scandinavian Defense Ministers on October 15, the Committee's secret directives were decided upon.[110] Long before that, however, on June 3, 1948, Truman was advised by Secretary of State Marshall to let the Swedes know that the Americans were disturbed by Sweden's inability to see that what was at stake in the Cold War was more than the simple Great Power struggle between the United States and the Soviet Union—that it was in fact a "'question of survival of nations which believe in freedom and democratic processes.'"[111] Moreover, Matthews was one of the State Department's European experts who had been instrumental in trying "to shift the discussion to a catalog of Russian unilateral actions in Germany, when Clay and Murphy demanded that pressure be brought upon France [...]."[112]

The American resistance sprung partly out of a firm conviction that Sweden was a natural member of the Atlantic community, but mostly from the perceived risk of Sweden, via a neutral Scandinavian pact, dragging Norway and Denmark away from the West. However, the Chief of NOE Benjamin Hulley acknowledged that the Swedish obligations in such a Scandinavian arrangement "appeared to be a long step away from past Swedish policies."[113] More importantly, although all the three Scandinavian countries were of strategic value to the United States, Norway and Denmark were more important for the formation of NATO than was Sweden because of their position on the North Atlantic, and because of their possession of Spitzbergen (Norwegian) and Greenland (Danish). Washington had, in accordance with its Great Power status, a desire to establish a network of bases all around the globe, from where it could conduct military operations. In this strategy Greenland was

109 Ibid., p. 3.

110 *Had There Been A War...*, pp. 61–62.

111 Lundestad, G., *America, Scandinavia, and...*, pp. 102, 200–201, 224, quote on p. 208.; Cole, P. M., *Neutralité du jour:...*, p. 200.

112 Gimbel, J., *The Origins of the Marshall Plan*, p. 129. Robert D. Murphy was the United States Political Advisor in Berlin.

113 Lundestad, G., *America, Scandinavia, and...*, pp. 242–243, quote on p. 243.

seen as falling within the American sphere of influence created by the Monroe Doctrine. Norway and Denmark were part of the so-called stepping stone countries, i.e. they were considered indispensable for the successful defense of Europe, and without them the membership of the United States in a European alliance was unthinkable. Hence, the American insistence on enlarging the number of Brussels Treaty nations.[114]

A looser arrangement could thus be accepted for Sweden than for Norway and Denmark, writes Lundestad.[115] The Americans were far from the only critics of the SDU proposal; an equally harsh criticism came from Swedish diplomats and military officials.[116] But the message relayed by Swedish officials to the American counterparts was not a coherent one. While the Foreign Minister publicly came out *against* alignment with the West, Defense Minister Allan Vougt assured Ambassador Matthews in Stockholm that "if the time were ever to come when Sweden 'must definitely choose sides' there was no doubt that Sweden would join the West."[117] It was clear that the Swedish government was hedging its bets, trying to convince the Americans that the official rhetoric did not in fact imply what it seemed to imply. At this point, however, the Americans seem to have been unable to grasp the subtleties of Swedish diplomacy. "Sweden's policy", writes Cole, "did not make sense to U.S. officials [...]."[118]

At about the same time as Undén hesitantly proposed a SDU to Norway and Denmark, Defense Minister Vougt ordered the Chief of the Swedish Defense Staff General Swedlund to embark on a far-reaching military cooperation scheme with Norway. This almost immediately, i.e. in June 1948, became known to the Americans, and it is from American sources that we know that Swedlund had been ordered:

> [...] to cooperate to the fullest practicable extent with the Norwegian Command. And on the Norwegian side, General BERG of the Norwegian Defense Staff was given similar orders by Defense Minister HAUGE.[119]

114 Kaplan, L. S., *The Long Entanglement:...*, pp. 15, 33, 224. See also: Lundestad, G., *America, Scandinavia, and...*, pp. 63–76, 249, 340–341, 345–346.

115 Lundestad, G., *America, Scandinavia, and...*, p. 249.

116 For this, see: Molin, K., *Omstridd neutralitet...*, pp. 66–80.

117 Cole, P. M., *Neutralité du jour:...*, p. 148.

118 Ibid., p. 184.

119 Memorandum on Swedish-Norwegian Joint Military Planning, dated June 17, 1948, p. 1, in: RG 59; BNA, 1941–1953; SF, 1941–1953; South Africa, Union of, Policy – Sweden, External Political Affairs: U.S.; Box No. 18; (Folder: External Political Affairs: Foreign Relations – U.S.); NARA II, College Park, MD. To this

This was part of the gestating plans for a Scandinavian Pact, and Sweden planned to support Norway with major fighting units in the Tröndelag area.[120] The Americans and the British were kept well informed about the whole thing by the Swedish Military Attaché in Oslo, who had "no objection to being quoted as source since he wanted both AMERICA and ENGLAND to know what was going on."[121] That summer, Swedlund, Berg, and some Norwegian Staff officers went on an extensive reconnaissance trip around Norway including visits to Stavanger, Bergen, Trondheim, Narvik, and the Tröndelag area. After this trip it was resolved that two Staff studies should be conducted. The subject of the first study was road and rail communications within Norway and between the two countries; the second study was to do with the signal communications in Norway, with special reference to international links within the system.[122]

One of the issues to be dealt with in the first study was to make "recommendations as to the most suitable roads to be used and developed in the event of Swedish Forces operating on Norwegian soil." The reason for conducting the studies was that the Swedish government lacked information about these specific issues regarding Norwegian defense. Once this information had been obtained "the Swedes would be in a position to make strategical and tactical appreciations on which Norwegian/Scandinavian defense plans could be based."

The studies were expected to take three months to finish, but neither the Swedes nor the Norwegians thought that there was any point in approaching the Swedish government until after the elections in September. The studies should however be ready:

> [...] so that they may be included in any Swedish/Norwegian strategic and tactical defense plans for presentation to the new Swedish Government, who, it is hoped by the Swedish Defense Forces and the Norwegian Government, will be more willing to talk on defense matters than is the present Swedish Government.[123]

---

date there has been no evidence in either Swedish or Norwegian archives of such cooperation ever being prepared for, much less exercised.

120 Ibid., p. 1. For information regarding Swedish-Norwegian military cooperation, see: Petersson, M., *"Brödrafolkens väl"*..., pp. 233–290. Petersson, however, did not include any information about the plans discussed above in his dissertation, probably because they predate his investigation which starts in 1949.

121 Attachment to Memorandum on Swedish-Norwegian Joint Military Planning, dated June 17, 1948, p. 1, in: RG 59; BNA, 1941–1953; SF, 1941–1953; South Africa, Union of, Policy – Sweden, External Political Affairs: U.S.; Box No. 18; (Folder: External Political Affairs: Foreign Relations – U.S.); NARA II, College Park, MD.

122 Ibid., pp. 1–2.

123 Ibid., p. 2. It is not clear whether the Swedish military expected that the Social Democrats would lose power in the election, or whether it was thought that the

According to the Swedish Military Attaché von Horn there would be cooperation between Norwegian and Swedish forces in the northernmost part of the two countries. The Swedish elite force Fjälljägarna in Kiruna, would at that time be ordered up to the junction of the borders between Norway, Sweden, and Finland where they would:

> [...] cooperate with General Dahl's forces in the Lyngefjord area and the training of both will be directed in coordination.[124]

The author of the memorandum, probably the American Military Attaché in Oslo, then ended the document with a reflection of his own mind:

> It does give definite promise of effective Scandinavian military cooperation, and so far as I have been able to learn the Swedish Military Command is definitely not wedded to the strict neutrality program presently advocated by the UNDEN faction. Rather they believe that after this falls' election the Swedish Government may give considerable ground.[125]

It is not known if the studies mentioned above were ever finished, or if the coordinated training of ground forces ever took place. But the episode reveals the intentions of the Swedish military establishment in 1948. It has to be assumed that at least parts of the Swedish government were informed about the project considering the instructions given to Swedlund to cooperate to the fullest practicable extent with the Norwegians. It seems clear that the studies were to form a foundation for further discussion on the SDU between the two governments in the fall.

Initially, the American State Department was split over the SDU issue, with a majority opposing the formation of a neutral Scandinavian Bloc, and a minority that saw certain merits in it, although they were not necessarily enthusiastic about it. The majority position was personified by Under Secretary of State Robert Lovett, the Chief of the Office of

---

government would be more willing to discuss these matters after the elections anyway. Intuitively, it would make more sense to have this expectation if it was believed that the right would be in power after September, but on the other hand it was the Social Democratic government that had told Swedlund to cooperate to the fullest extent with the Norwegians. Perhaps the military reasoned that the left government needed to be injected with the confidence that an election victory would give, in order for them to dare to go to the end of the line.

124 Ibid., p. 3.

125 Ibid., p. 3.

European Affairs (henceforth: EUR) John Hickerson, the Chief of NOE Hulley, and Ambassador Matthews. The minority view was argued most strongly by George Kennan of the PPS, with Secretary of State Marshall vacillating between the two positions. There was also a more moderate group who would be satisfied with only Norway and Denmark in NATO.[126]

In the middle of this, on July 15, an American C-47D carrying American Air Attaché Colonel Donald L. Hardy, Captain G. S. Tham, Chief of Swedish Military Intelligence, and Captain R. C. Håkanson of the Swedish Defense Staff, departed from Bromma Airport in Stockholm and, via Frankfurt, landed on Erding Air Force Base near Munich, Germany. The reason for the trip was for Håkanson and Tham to attend a conference at the USAFE headquarters.[127] The Swedes spent a week in Munich and were flown back again on July 21.[128] Tham and Håkansson did not know that General Curtis Le May would be in Munich, and apparently without preparations told him that "they wished to start discussions with the General 'on matters relating to mutual defense in the event of hostilities.'" Matthews was surprised at this approach, and stated that:

> It is hard to believe that if the approach is as serious as indicated, they would have made it without the approval of [Supreme Commander] General Jung and possibly the Defense Council [Matthews was perhaps referring to the Swedish Defense Staff], on which the Swedish Foreign Office is represented. If they have in fact made the approach with that approval, it is extremely interesting as an indication of how far the Swedes have moved in their thinking [...].[129]

126 Gabriel, Jürg Martin, *The American Conception of Neutrality after 1941* (Hampshire: The MacMillan Press Ltd., 1988) pp. 94–105; for the views of the majority, see also: Lundestad, G., *America, Scandinavia, and...*, pp. 238–249. Matthews was an ardent cold warrior who had been Director of EUR between 1945 and 1947, before he was assigned to Stockholm in July, 1947. From the Embassy he directed a State Department sanctioned war against Sweden's neutrality policy. It made little difference that this policy had a broad public and parliamentary support in the country that hosted him; he would openly tell Swedish officials that the Swedes "had to take their stand now" and choose which side they wanted to join in the coming war (Lundestad, G., *America, Scandinavia, and...*, p. 201).

127 "Air Operations Report, 1 July–30 July 1948 for aircraft C-47D, 44-77204" from U.S. Air Attaché in Stockholm to Director of Intelligence Headquarters, USAF, Washington D.C., p. 2, in: RG 341; DCSO; DI; DDCD; COD; AAB; AOR, 1948–51; Singapore to Union of South Africa; Box No. 156; NARA II, College Park, MD.

128 Ibid., p. 3.

129 Letter from Matthews to Llewellyn E. Thompson, Jr., dated August 5, 1948, p.

In this, Matthews expressed the same view as NOE had done in a letter to the Deputy Director of EUR Llewellyn E. Thompson, only a few days after the Swedes had returned from Munich. The unknown author wrote that it was unlikely that this approach was unknown to the Swedish Foreign Ministry. "[T]he possibility exists", said the letter, "that Sweden is preparing a step towards the West through Scandinavian defense cooperation." This possibility ought to be fully exploited, according to the author. The recommendation from the NOE was thus:

> I suggest therefore, that the invitation be accepted and discreetly utilized as an opportunity to propagandize the Swedes and to listen to their proposals, while otherwise maintaining a non-committal attitude.[130]

Thompson, after reading these recommendations, subsequently wrote a letter to Matthews in Stockholm, where he informed him that military elements in the Swedish Defense Staff wanted to invite General Le May to a hunting trip in Sweden, at which time they wished to conduct informal deliberations on defense matters. Thompson and Bohlen at EUR had a somewhat different take on the deal than the people in the NOE, feeling that while "this approach should not be rebuffed", they "did not think it should be accepted at least at this time."[131] The risk was, declared Thompson, that the Swedish military once they had received this encouragement, would lose their interest in pressing the Swedish government to change its position. The EUR was anxious to have the Ambassador's view before giving a definite answer to the USAF.[132] In his reply to Thompson Matthews agreed with this, stating that

> It would be interesting to find out what the Swedes have in mind and whether they are really serious. Once that is done, however, I believe that they should be told in the most friendly fashion that we are sorry but we cannot engage in such conversations with neutrals. On the assumption that the approach is serious and therefore that it was made with the knowledge of at least some higher Swedish

1, in: RG 84; SWSE; TSGR; 1943–1949; Box No. 1; (Folder: 1948); NARA II, College Park, MD.

130 Letter to Llewellyn E. Thompson from unknown author, dated July 23, 1948, p. 1, in: RG 59; BNA, 1941–1953; SF, 1941–1953; South Africa, Union of, Policy – Sweden, External Political Affairs: U.S.; Box No. 18; (Folder: External Political Affairs: Foreign Relations – US); NARA II, College Park, MD.

131 Letter from Llewellyn E. Thompson, Jr. to Matthews, dated July 24, 1948, p. 1, in: RG 84; SWSE; TSGR; 1943–1949; Box No. 1; (Folder: Ambassador Matthews – Cumming Sweden Vs. NATO); NARA II, College Park, MD.

132 Ibid., pp. 1–2.

authorities, the word would undoubtedly be reported back and would undoubtedly stimulate Sweden's evolution toward the West [...].[133]

The hunting trip theme seems to have been a popular cover for conducting discussions on matters pertaining to common military planning, for some months later an identical invitation was offered to the Director of Intelligence at the SAC, a Colonel Bently. The purpose would have been to give Bently an opportunity to talk things over with his Swedish counterpart.[134] Most likely, this invitation was also turned down by the Americans.

Until there was further and more tangible proof of a Swedish westward plunge, the Swedes would be left to simmer. The Americans feared that if they opened up discussions with Sweden at this point, the Swedes would stay up in their neutral ivory tower and never come down. This was not the kind of consent to U.S. hegemony that the American officials were working so hard to get—indeed, such behavior did not deserve to be labeled consent at all. There were thus different opinions within the State Department concerning this issue. The result seems to have been a sort of compromise. The Americans found the Swedish feeler extremely interesting, but the invitation to a hunting trip was not acted upon. Instead other ways were chosen to sound the Swedes out. The State Department policy became one of inquiring curiosity, while keeping the Swedes at arm's length.

On September 22, 1948 the two First Secretaries of the Swedish Embassy Per Lind and Sverker Åström had lunch with Charles E. Rogers of the NOE. Åström told Rogers that there had recently been a notable shift away from isolated neutrality in Sweden. He continued with the remarkable statement that he did not preclude the possibility of Sweden joining a Western defense pact if such a thing became a reality, even going so far as saying that it might be desirable for Sweden to partake in the establishment of such an organization. The Scandinavian defense talks were a stepping-stone in that direction, said Åström. A rather surprised Rogers commented that this was "the first time Aström [Åström] has indicated a chink in his normally scrupulous defense of Swedish neutrality." During this critical time, intimated Åström, when the shift was taking place, it was extremely important for the United States not to exert too overt pressure upon Sweden.[135] Åström

133 Letter from Matthews to Llewellyn E. Thompson, Jr., dated August 5, 1948, pp. 1 and 3, in: RG 84; SWSE; TSGR; 1943–1949; Box No. 1; (Folder: 1948); NARA II, College Park, MD..

134 Weinberger, H., "The Neutrality Flagpole:..." in M. Thad Allen and G. Hecht (eds), *Technologies of Power:...*, p. 317.

135 Memorandum of Conversation between Rogers, Lind and Åström, dated

assured Rogers that he was not referring to anything that had happened, although the American believed "he had in mind the withholding of arms licenses."[136] In an earlier conversation with Rogers, Åström had apparently abandoned all arguments for neutrality except one, asking rhetorically "'What else can we do when you only have two divisions ready for combat?'" This attitude, remarked Rogers, reflected rather less complacency than had usually been shown by the Swedes.[137] The pressure applied by the United States on the Scandinavians was mounting, and Hickerson had told the Norwegians in late August that unless Norway joined her neighbors and other states in a collective defense arrangement, her request for arms deliveries from mid-April could not be favorably viewed by the U.S. government. If Norway were to join in the common cause, however, Hickerson said, the Norwegian request would be reconsidered.[138]

While the statement that Sweden would be prepared to join a Western defense pact was clearly exaggerated, and must have been recognized as such by the American, it is equally clear that this effort at calming American fears was an effort to downplay the reverberations of the SDU, an effort officially sanctioned by the Swedish government. And while the plea to the Americans to refrain from 'too overt pressure' might have been received with some sympathy even in Washington, it failed to change U.S. pressure policy in any fundamental way. But this was, perhaps, never Åström's intention.

The Americans felt that in order to permanently sway the Swedish government to give up its policy of subjective neutrality, they had to coordinate policies with their closest trans-Atlantic ally, Britain. Britain had much closer relations with Sweden, at this time, than did the U.S., and it has been stated that the American pressure politics against Sweden were jinxed to some extent by lack of coordination with the British, who had no similar apprehensions about supplying the Swedes

---

September 22, 1948, pp. 1–2, in: RG 84; SLE; CONF 1948; 020-844; Box No. 6; (Folder: 800); NARA II, College Park, MD.

136 Ibid., p. 2.

137 Memorandum of Conversation between Rogers and Åström, dated March 31, 1948, p. 2, in: RG 84; SLE; CONF 1948; 020-844; Box No. 6; (Folder: 800); NARA II, College Park, MD.

138 "Summary of oral statement, made August 7, 1948, by the Department of State, Director John D. Hickerson, Office of European Affairs, to the Chargé d'Affaires ad interim of Norway, at the delivery of the United States Government's answer to the Norwegian Government's Aide Memoire of April 16, 1948, concerning certain military supply requirements for Norwegian defense.", dated August 27, 1948, pp. 1–2, in: HP 24 VI/U.S.A; "Tillstånd till in- och utförsel av vapen, ammunition och krigsmateriel"; 1946, okt.– aug. 1950; Folder 20; Box No. 1353; År: 1946–1952; RIA, Stockholm.

with military materiel.[139] This lack of coordination was an example of British accommodating opposition to American hegemony. Juhana Aunesluoma has written that it was the British need "to integrate both American and Swedish resources to stabilize the north" that lay behind their meddling with the Swedes. In absence of preponderant military and economic strength, the British fell back on their "long tradition of diplomacy and political involvement", in order to secure the sought after collaboration.[140]

At a meeting in late September 1948 the JCS, at General Vandenberg's suggestion, had decided to approach the British Chiefs of Staff and transmit their objection to the sale of Vampire airframes to Sweden.[141] Lovett spoke to the British the same day, and Vandenberg told the JCS that the British Royal Air Force (RAF) had come out against the sale, but that the RAF's hands were tied because a decision on the matter was up to the Ministry of Supply. The RAF had advised that the objection would carry much more weight if it came directly from the JCS to the COS.[142]

General Gruenther, the Deputy Chief of Staff for Plans, was charged with approaching the representatives of the COS in Washington. The JCS spelled out what further pressures should be brought to bear on the Swedes, and made clear that they were greatly concerned over the sale of such aircraft to the RSAF. The British response was to state their belief in Sweden's ability to defend itself, and that the orders were accepted "in the interest of the British aircraft industry." They added, however, that all contracts with foreign powers contained an escape clause and that if any "country having higher priority than Sweden [...] needed aircraft the escape clause could be invoked".[143]

Hickerson and Thompson were the main actors at the State Department, and they had approached the USAF on two occasions,

139 Lundestad, G., *America, Scandinavia, and...*, p. 231. For Britain's policy towards Sweden in this period, see: Aunesluoma, J., *Britain, Sweden and the Cold War....*

140 Aunesluoma, J., *Britain, Sweden and the Cold War...*, p. 155.

141 Memorandum for the Record "Prospective Sale of Vampire Aircraft to Sweden", dated September 22, 1948, in: RG 218; CDF 1948–50; 381 (10-2-46) B.P. Pt. 2 to 381 (8-14-47) Sec. 3; Box No. 184; (Folder: Scandinavia – General); NARA II, College Park, MD. It is worth pointing out that although the Vampire affair has been dealt with by other authors (cf. Aunesluoma, J., *Britain, Sweden and the Cold War...*, pp. 9, 67, 89, 98–100, 105.), no one has made use of the sources that I base my account on. My contribution is thus the American view of the matter.

142 Ibid.

143 Memorandum for General Gruenther "Sale of British De Havilland Vampire Jet-Propelled Fighters to Sweden, dated September 23, 1948, pp. 1–2, in: RG 218; CDF 1948–50; 381 (10-2-46) B.P. Pt. 2 to 381 (8-14-47) Sec. 3; Box No. 184; (Folder: Scandinavia – General); NARA II, College Park, MD.

September 21 and 22, stressing the urgency of the issue.[144] Foggy Bottom's attitude was that the British sale hindered the American policy of forcing Sweden to change its current position and they wished to see a coordination of policies vis-à-vis Sweden between them.[145] A memorandum stated that the reason for the American sense of urgency regarding this matter was that:

> Sweden's strategic importance, from the point of view of both geographic location and industrial potential, and its position of leadership among the Scandinavian states make it important that increased efforts be made to induce Sweden to alter its neutrality policy [...].[146]

Thus, Sweden's strategic importance was once more emphasized by the United States. The American Embassy in London had discussed the issue with the Northern Department of the Foreign Office,[147] and the State Department had asked General Vandenberg "to bring what pressure he could on the military level to bear against the proposed sale [...]."[148] The USAF was apparently worried that the sale would not strengthen Sweden's defenses enough to be justifiable, stating that these aircraft risked falling in the hands of an enemy of the United States.

> The U.S. Department of State is conducting negotiations looking toward firm orientation of Sweden toward Western Union [...] The Department of State feels that so long as Sweden can obtain military supplies [...] she will maintain her position of neutrality.[149]

144 Ibid., p. 1.

145 Aide Memoir, dated September 24, 1948, p. 1, in: RG 59; BNA, 1941–1953; SF, 1941–1953; South Africa, Union of, Policy – Sweden, External Political Affairs: U.S.; Box No. 18; (Folder: External Foreign Affairs: Foreign Relations – UK); NARA II, College Park, MD.

146 Ibid., p. 2.

147 Memorandum for General Gruenther "Sale of British De Havilland Vampire Jet-Propelled Fighters to Sweden, dated September 23, 1948, p. 2, in: RG 218; CDF 1948–50; 381 (10-2-46) B.P. Pt. 2 to 381 (8-14-47) Sec. 3; Box No. 184; (Folder: Scandinavia – General); NARA II, College Park, MD.

148 Memorandum for the Record "Prospective Sale of Vampire Aircraft to Sweden", dated September 22, 1948, in: RG 218; CDF 1948–50; 381 (10-2-46) B.P. Pt. 2 to 381 (8-14-47) Sec. 3; Box No. 184; (Folder: Scandinavia – General); NARA II, College Park, MD.

149 Memorandum for General Gruenther "Sale of British De Havilland Vampire Jet-Propelled Fighters to Sweden, dated September 23, 1948, p. 2, in: RG 218; CDF 1948–50; 381 (10-2-46) B.P. Pt. 2 to 381 (8-14-47) Sec. 3; Box No. 184; (Folder: Scandinavia – General); NARA II, College Park, MD.

American frustration with the Swedes, and with the British, becomes utterly manifest above. The U.S. policy of treating Swedish requests after those of the Brussels Treaty nations, as outlined in NSC 28/1, "cannot be fully effective" it was said in a document, "unless it is supported by Great Britain".[150] In view of this the United States was "seriously disturbed" by British policy, and claimed that the Vampire affair in effect nullified the U.S government's effort to influence Swedish policy.[151] It was hoped that the British government would give these arguments urgent consideration, and that it would be possible to hold up execution of the sale of jet planes to Sweden pending further consultation between Britain and the United States.[152]

Hickerson and Lovett acted again, this time with JCS backing, and made clear to the British Ambassador Hoyer-Millar on September 24 just how perturbed the United States was.[153] The official reply came in from the British Embassy in Washington a few days later, and it was not what the Americans had hoped for. The British claimed that their industry could not be kept alive merely on the orders from the Brussels Treaty nations and other allies alone. The airframes would furthermore not interfere with deliveries to the allies since it was engines that were the limiting factor. [154]

The Embassy explained that "the British Government cannot afford to forego export in the present state of their economy unless it is really essential to do so." According to the British government the present policy was yielding good results, and it "does not therefore consider that they should alter their present policy […]." It remained their hope, the Embassy said, that the United States would accept this view.[155] This attitude irritated the living daylights out of Ambassador Matthews, who despised the British search for quick profits, stating "'why the hell should we spend money to rearm Brussels Pact countries when they are selling jets to neutrals'".

150 Aide Memoir, dated September 24, 1948, p. 4, in: RG 59; BNA, 1941–1953; SF, 1941–1953; South Africa, Union of, Policy – Sweden, External Political Affairs: U.S.; Box No. 18; (Folder: External Foreign Affairs: Foreign Relations – UK); NARA II, College Park, MD.

151 Ibid., p. 5.

152 Ibid., pp. 5–6.

153 Memorandum of Conversation between Hickerson, Lovett, and Millar, dated September 24, 1948, in: RG 84; SWSE; TSGR; 1943–1949; Box No. 1; (Folder: External Foreign Affairs: Foreign Relations – UK); NARA II, College Park, MD.

154 "Supply to Sweden of Vampire Aircraft", dated September 30, 1948, p. 1, in: RG 84; SWSE; TSGR; 1943–1949; Box No. 1; (Folder: External Foreign Affairs: Foreign Relations – UK); NARA II, College Park, MD.

155 Ibid., p. 2.

According to him, the British oozed ignorance of Swedish psychology, and the short-sighted interest in the survival of their domestic aircraft industry was jeopardizing any chance of getting Sweden over to the western camp.[156] The British, who since June had been at pains to follow Bevin's instructions to try and persuade the United States not to put so much pressure on the Swedes, were very distressed by Matthews' tirades. Hankey at the Foreign Office instructed Harold Farquhar, at the British Embassy in Stockholm, to please make Matthews stop making so much trouble.[157] The Americans in Washington were not happy about this intransigent attitude either, and took their time replying to it. But they were not as perturbed as Matthews in Stockholm. Hickerson stated that while the United States:

> [...] have not accepted their position, but it seems useless to pursue the matter further now. The exchange of views had some value in letting the British know how we felt.[158]

This paragraph seems to imply that the United States policymakers did not take the issue all that seriously after all, except as a matter of principle in dealings with the British. This is surely an important point, because what was going on in this interchange was an American struggle to gain British consent to U.S. hegemony. The British, on the other hand, expressed their dissent by accommodating opposition.

However, while the British opposed U.S. pressure to withhold the airframes, it seems as if they, in the end, partly succumbed to the U.S. demands to coordinate policies, and delayed a number of deliveries to and contracts with the Swedes. This becomes obvious from a conversation that the American Air Attaché in Stockholm Colonel Hardy had with Colonel Gustaf Adolf Westring, the Chief of Staff of the RSAF, in late September 1948. According to Westring, the British were dragging their feet in several matters, e.g. refusing to fulfill aircraft engine contracts, even though the engines had been manufactured and only had to be shipped. In addition, said the Swede, the British were procrastinating in negotiations concerning airborne radar to the point where Westring was beginning to think that they had no intention of concluding the deal.[159] And even though there were no signs of a change in the British line with

156 Aunesluoma, J., *Britain, Sweden and the Cold War...*, p. 66.

157 Ibid., pp. 66–67.

158 Letter from Hickerson to Matthews, dated December 14, 1948, in: RG 84; SWSE; TSGR; 1943–1949; Box No. 1; (Folder: External Foreign Affairs: Foreign Relations – UK); NARA II, College Park, MD.

159 Message from the U.S. Air Attaché in Stockholm, dated September 30, 1948, p. 1–2, in: RG 84; SWSE; TSGR; 1943–1949; Box No. 1; (Folder: External Foreign Affairs: Foreign Relations – UK); NARA II, College Park, MD.

regard to the airframes,[160] Hardy summed it up and stated that all of this was "evidence that United Kingdom is to some degree implementing our policy [...]."[161]

The Swedes knew all about the American tactics because they had asked the British whether they had been pressured to deny the airframes to Sweden by the United States, and had been told that the Americans had vociferously voiced their discontent about the Vampire sale. While the British had not fully complied this time, there was no telling how things would develop in the future.[162] Juhana Aunesluoma gives an excellent example of the lingering British non-compliance in this field in his account of how the Swedes, in November 1949, wanted to buy an additional 60 de Havilland Vampire airframes from the British. In London, the government quickly found a loophole enabling them to go through with the sale despite U.S. opposition. Empty airframes could not be said to violate Atlantic Pact requirements, mused the British, and Aunesluoma writes that a "consensus was rapidly reached in London that 'there is no need to tell the Americans about the sale of Vampire airframes to Sweden'." Writes a perceptive Aunesluoma:

> This highlighted the strong feeling held in Britain, that ultimately only they themselves should be in charge of arms exports policy, if it did not explicitly violate agreements with the US.[163]

This is a classic, not to say paradigmatic, illustration of accommodating opposition. The British, like a dour kid brother who sees his chance to have it his way when big brother is not looking, snuck the deal under the nose of the Americans. It is only all the more interesting that this struggle involved a former hegemonic nation and its successor state. This was, moreover, one of the few issues that the Americans and the British did not see eye to eye on. Consensus had been reached between them on other, and perhaps more important, issues. British support for the idea that a truly non-aligned Sweden would not get automatic assistance from the Western powers, for example, was one of these.

The strategy of denying arms to Sweden also became known to the Swedish government through Boheman's reporting from Washington, who pointed out Matthews and his cohorts as being behind this policy.[164] The United States had gained partial, but not full, cooperation from the British, and the issue of British aircraft sales to Sweden would, as we will

160 Ibid., p. 2.

161 Ibid., p. 1.

162 Silva, C., *Keep Them Strong, Keep Them Friendly:...*, p. 251.

163 Aunesluoma, J., *Britain, Sweden and the Cold War...*, p. 98.

164 Silva, C., *Keep Them Strong, Keep Them Friendly:...*, p. 251.

see below, surface again later in the year. While this was surely annoying to Washington, nothing would, or perhaps could, be done about the situation at the moment. In the meantime, the Americans had other fish to fry.

The reason why the United States persevered in its pressure on Sweden was that such pressure was part and parcel of the new policy towards Sweden from the fall of 1948. This policy, expressed in NSC 28/1, stated that the United States should try to induce Sweden to give up the neutrality position and to align itself more clearly with the West.[165] The United States wanted more tangible proof of Sweden's orientation. The NSC paper further stated that the Scandinavian countries were of strategic importance to the United States, and listed the fact that these countries were situated along the shortest flight path to target areas in the Soviet Union.[166] Matthews, at the Embassy in Stockholm, reacted with some apprehension to this, and urged the Director of EUR Hickerson not to tell the Swedes that the United States considered Sweden to be of strategic value to them, because then, foresaw Matthews, it would be quite impossible to influence the Swedes away from their position of subjective neutrality. Matthews insisted:

> [...] it is vitally important, if we are to accomplish our objective of influencing Sweden to abandon its "subjective neutrality" [...], that the Swedes do not repeat not become aware that the United States Government considers Sweden "strategically important to the United States" [...]. If the Swedes do discover these facts, you can kiss goodbye to any chance of getting Sweden to line up with the West.[167]

What Matthews expressed in his letter to Hickerson was nothing less than the essence of hegemony-building. The United States was to influence a basically friendly nation to take a clearer stand in the Cold War, and it was deemed central to this effort that the Swedes be made aware of this wish, in a not too overt sense, so that they themselves would want to effect the change of attitude. Matthews was managing what he saw as the best strategy to manufacture Swedish consent to American hegemony. The Ambassador stated further that the Swedes had to be made to realize that their post-conflict adherence to the West

165 *Had There Been A War...*, p. 101. Mind that it did *not* say that the ultimate goal was Swedish open membership in a Western defense alliance.

166 Ibid., p. 100.

167 Letter from Matthews to Hickerson, dated September 28, 1948, p. 1, in: RG 84; SWSE; TSGR; 1943–1949; Box No. 1; (Folder: Ambassador Matthews – Cumming Sweden Vs. NATO); NARA II, College Park, MD.

would bring the supplies that they would not get otherwise. In order to take "the Westward plunge", wrote Matthews,

> *They must be made to wonder whether our strategic plans include some form of defense of Sweden or whether they stop at the Norwegian frontier.* The recent approaches to General Le May and to Colonel Hardy are signs that the military here [in Sweden] at least would like very much to know the answer.[168] [Italics added.]

What is revealed in the italicized sentence above? Is it perhaps evidence that the American war planning did *not* in fact stop at the Norwegian border, but that it actually did treat the Scandinavian Peninsula as a strategic whole, and thus included Sweden? There seems to be little reason for Matthews to formulate it this way otherwise. Indeed, if the U.S. military plans did not cover Sweden, it would have been much more to the point to stress to Hickerson the importance of relating the true state of things to the Swedes. No, the Ambassador seems to have been more at pains to hide the true state of things from the Swedes—i.e. that Sweden was considered strategically important and that American war plans covered the neutral as well.

Boheman left for Washington a week later, as the new Swedish Ambassador, and Matthews urged Hickerson to arrange for someone "at the top level, preferably [Acting Secretary of State] Bob Lovett" to implement the recommendations of NSC 28/1.[169] Hickerson did not waste any time beating around the bush. On October 8, 1948, he subsequently wrote a letter to Lovett in which he forwarded Matthews' proposal that the new Swedish Ambassador ought immediately to be talked to bluntly.[170] In the letter, Hickerson provided further evidence for my conclusion above that Sweden was included in U.S. strategic plans at the time. Referring to Matthew's letter, Hickerson wrote that the United States should not:

> [...] allow Sweden to believe it is strategically so important to us, that we will come to its defense regardless of its neutrality policy, or that our own security interests are tied up with Swedish freedom from Soviet domination. *He reasons that the Swedes are not certain of our attitude on these points, and that if we let them know of our*

168 Ibid., p. 2.

169 Ibid., p. 3.

170 Letter from Hickerson to Lovett, dated October 8, 1948, p. 2, in: RG 84; SWSE; TSGR; 1943–1949; Box No. 1; (Folder: Ambassador Matthews – Cumming Sweden Vs. NATO); NARA II, College Park, MD.

interest, they will bank on it and continue to cling to neutrality.[171] [Italics added.]

Lovett never got a chance to act upon Matthews' instructions for he, for some reason, never met with Boheman. Instead, Hickerson himself did. When he and the Chief of NOE Hulley met with Boheman on October 12, they made abundantly clear to the Swede the American mind-set regarding neutrality. Hickerson let the official Swedish representative know that the United States was "worried, distressed and shocked by the adherence of Sweden to the policy of neutrality, a word which was really offensive to us."[172] Boheman interpreted Hickerson's attitude as aggressive and later he remarked that Hickerson, who was one of the few within the State Department who uttered such ideologically tainted critique, was not a very constructive interlocutor.[173]

This scolding, however, proved too much for Secretary of State Marshall, who, once he got wind of the conversation, rebuked Hickerson in a telegram from Paris, an unusual thing on his part. The Secretary's concern was that such overly apparent pressure tactics might do more harm than good at the moment.[174] Matthews, on the other hand was amused by the whole incident, and felt that Hickerson had handled Boheman "just right."[175]

Boheman, who seemed less shocked than did Marshall, expressed his understanding for the American stance, and explained to them that neutrality was a bad choice of words and that he had no illusions that Sweden could stay out of a future war in Europe. Boheman also stated that if the ongoing Scandinavian military talks "should have any tangible result, that could only mean that Sweden had in fact abandoned its neutrality."[176] Hickerson acknowledged Marshall's worries, and

171 Ibid., p. 2.

172 "Memorandum of Conversation between Boheman, Hickerson, and Hulley", dated October 12, 1948, p. 1, in: RG 84; SWSE; TSGR; 1943–1949; Box No. 1; (Folder: 1948); NARA II, College Park, MD. Also quoted in: Lundestad, G., *America, Scandinavia, and...*, p. 264.

173 Stridsman, J., "Erik Boheman..." in K. Zetterberg (ed.), ...*Hot and Cold...*, p. 115.

174 Lundestad, G., *America, Scandinavia, and...*, p. 264. Lundestad makes a big deal out the fact that Marshall rebuked Hickerson, but it is worth noting that he explicitly referred to the present situation in his critique. Marshall was not really concerned about pressure politics in general, but only that it might backfire in this case and at that instant. He also made the distinction between harm and good, implying that he endorsed such pressure tactics as long as they generated good dividends.

175 Letter to Hickerson from Matthews, dated November 15, 1948, p. 2, in: RG 84; SWSE; TSGR; 1943–1949; Box No. 1; (Folder: 1948); NARA II, College Park, MD.

176 "Memorandum of Conversation between Boheman, Hickerson, and Hulley",

stated that such conversations should be put on hold for the time being. But he also pointed out that Boheman seemed to have agreed with everything he had said, and that the conversation had been held in a cordial, informal, and friendly atmosphere. Hickerson repeated Matthews' recommendation that the heat should be kept up under the Swedes.[177] The reference to keeping up the heat shows just how seriously (i.e. not at all) Hickerson and the other officials really took Marshall's objections.[178]

Despite the U.S. government's negative attitude towards Sweden's security policy, and its SDU initiative, contacts were slowly improving on the ground with the American forces in Europe. On October 20 the same C-47 that had been used in July to fly the Swedes to Munich and back was employed by the Stockholm Embassy to transport Colonel W. C. Bentley and Captain Hutchins of USAFE to Stockholm for a conference with the Swedish Defense Staff. The Americans stayed for three days and returned home to Wiesbaden on October 23.[179] The content of these discussions is not known, but the topic might have been informal defense discussions or Swedish-American intelligence cooperation.[180]

However, such signs of improving relations between Sweden and the U.S. in Europe did not stop Lovett and Hulley from turning the heat up under Boheman in Washington, in accordance with Matthews' instructions, when they held another meeting with the Swedish Ambassador on October 26, 1948. Boheman now repeated what he had said to Hickerson some days before, but qualified his statement, adding that the term neutrality was "perhaps something of a blind which did not correctly designate the Swedish attitude."[181] The Americans were

---

dated October 12, 1948, p. 2, in: RG 84; SWSE; TSGR; 1943–1949; Box No. 1; (Folder: 1948); NARA II, College Park, MD.

177 Letter from Hickerson to General Marshall S. 'Pat' Carter (forwarded to Matthews), dated October 23, 1948, pp. 1–2, in: RG 84; SWSE; TSGR; 1943–1949; Box No. 1; (Folder: 1948); NARA II, College Park, MD.

178 For this point, see also: Aunesluoma, J., *Britain, Sweden and the Cold War...*, p. 67.

179 "Air Operations Report, 1 October–30 October 1948 for aircraft C-47D, 44-77204" from U.S. Air Attaché in Stockholm to Director of Intelligence Headquarters, USAF, Washington D.C., pp. 1–2, in: RG 341; DCSO; DI; DDCD; COD; AAB; AOR, 1948–51; Singapore to Union of South Africa; Box No. 156; NARA II, College Park, MD.

180 I thank Robert Dalsjö for pointing the latter subject out to me.

181 "Memorandum of Conversation between Boheman, Lovett, and Hulley, dated October 26, 1948, p. 2, in: RG 59; BNA, 1941–1953; SF, 1941–1953; Country (Sweden, External Political Affairs: U.S.S.R.-Sweden, Wood Pulp); (Folder: Military and Naval Affairs: Western Union); Box No. 19; NARA II, College Park, MD. See also: Molin, K., *Omstridd neutralitet...*, p. 69.

interested in what Boheman had to say, but wanted clarification as to the meaning of what he was telling them. A little over a week later Hickerson himself, together with Hulley, again saw Boheman, and this time the Swedish Ambassador went even further in his effort to elucidate the Swedish position. Boheman, now at least for the third time, assured him that neutrality was not a good word to describe the Swedish position since "all defense planning and action had only one possible purpose—defense against Russia. No thought was ever entertained that there was any need to defend against the Western countries. Clearly," he said, "'neutrality' was not a word to describe this attitude."[182]

Karl Molin has suggested that Boheman was acting on his own when he described Swedish neutrality in the way he did, writing that when Boheman came to Washington he "sought to portray the defense union rather as being the beginning of a cooperation with the Western Powers than as the enlargement of neutrality which it was to the government."[183] However, I suggest that there is enough evidence to assume that Boheman acted with (at least) the tacit consent of the Swedish government. Furthermore, it does seem as if this way of portraying Sweden's foreign policy had Undén's expressed agreement. Upon being informed about Boheman's way of presenting Swedish policy in Washington, Undén stated that, for him, it was "self evident that the argumentation for Sweden's standpoint ought to be adjusted to the political and psychological environment in the country where the Diplomat was working."[184] It is thus clear, concludes Jacob Stridsman in his essay on Erik Boheman, that the Foreign Minister gave his Ambassador considerable leeway in his dealings with the Americans, leaving Boheman to decide more or less for himself how it was most suitable to describe Sweden's foreign policy.[185] The Embassy in Stockholm was not impressed by Boheman's endeavors

182 "Memorandum of Conversation between Boheman, Hulley, and Hickerson", dated November 5, 1948, p. 2, in: RG 59; BNA, 1941–1953; SF, 1941–1953; South Africa, Union of, Policy – Sweden, External Political Affairs: U.S.; Box No. 18; (Folder: External Political Affairs: Foreign Relations –U.S.); NARA II, College Park, MD.

183 Molin, K., *Omstridd neutralitet…*, p. 68. "sökte han framställa försvarsförbundet snarare som början till samarbete med västmakterna än som den utvidgade neutralitet som den var för regeringen."

184 Stridsman, J., "Erik Boheman…" in K. Zetterberg (ed.), *…Hot and Cold…*, p. 127. Although this was in the spring of 1949 there is really no reason to think that Undén would have reacted differently had he known about Boheman's statements to the Americans in 1948. Indeed, Undén probably was aware of it and tacitly accepted it.

185 Ibid., p. 128. In correspondence with Boheman, Undén also acknowledged that Sweden would probably rethink its policy of neutrality should the Soviet Union occupy Finland, and he did not argue against Boheman's statement that Sweden would be drawn into a war if the Scandinavian neighbors were attacked (ibid. p. 127).

to make his government's policies more palatable to the Americans, and urged Hickerson not to "let him get away with this rather obvious bit of propaganda".[186]

Perhaps the Ambassador's approach worked, because there can be little doubt that the conversations Hickerson had had with Boheman contributed to him taking a fairly positive attitude towards the sugestion of Sir Robin Hankey's, of the Northern Department of the British Foreign Office, of a special arrangement for the Scandinavian countries. The idea was to solve the problem of the defense of Scandinavia by establishing two interlocking treaties, where Norway and Denmark would be firmly attached to an Atlantic Pact and Sweden tied in through a separate Scandinavian arrangement. This was the so-called 'Hankey Plan', which in turn was based on a proposal in May 1948 by the British Permanent Under Secretary Sir Orme Sargent.[187] A crucial part of this scheme was that Sweden should be obliged to assist Norway and Denmark if they were attacked, and vice versa, and this latter scenario "'would involve the other members of the N. Atlantic Security Pact [...].'"[188] In Hankey's, and the Foreign Office's, view this was an ingenious way of tying Sweden to the western camp. Aunesluoma quotes Hankey as saying:

> By making Sweden feel that they were backing on to something fairly solid, viz, Norway and Denmark with the full support of the members of the N. Atlantic Security Pact, it would give the Swedes confidence, while they would only be expected to intervene if their own chance of national safety or survival were most gravely compromised by the threat of Soviet encirclement.[189]

The Hankey Plan was presented to the Americans on October 12, 1948. It was dutifully studied on a working level within the State Department, and the reply came a month later in a letter from Hickerson addressed to the British Ambassador to Washington Hoyer Millar. Hickerson explained that the State Department thought it to be a good arrangement to tie a Scandinavian defense union to the NAT even if

186 Letter to Hickerson from Matthews, dated November 24, 1948, p. 1, in: RG 84; SWSE; TSGR; 1943–1949; Box No. 1; (Folder: 1948); NARA II, College Park, MD.

187 For the Hankey Plan, see: Aunesluoma, J., *Britain, Sweden and the Cold War...*, pp. 69–73. Our sources are only partly overlapping. For Hickerson's views we make use of the same document, i.e. a letter from Hickerson to British Ambassador Hoyer Millar on November 15. Aunesluoma has used the copy from the British Public Record Office (PRO), while I have used the copy from NARA.

188 Ibid., p. 70.

189 Ibid., p. 71.

Norway and Denmark became members of the said organization. Wrote Hickerson:

> We agree with much of the substance of the paper and particularly with the view that every effort should be made to induce Norway and Denmark to become full parties to a North Atlantic Security Treaty. If these two countries do become parties, the problem of Sweden might well be solved if that country were prepared to enter into a satisfactory Scandinavian Defense Treaty. [...] Provided that Norway and Denmark were parties to an Atlantic Treaty, it would be desirable to have the three Scandinavian countries parties to a separate but substantially similar Scandinavian Defense Treaty.[190]

However, the United States had some problem with the British suggestion that an attack on Sweden should automatically include a response from all members of an Atlantic Pact. According to Hickerson, the view on the American side was that this would:

> [...] appear to come too close to a one-way guarantee of Sweden by the parties to the North Atlantic Treaty [...]. [...] If the Norwegians and Danes heard of it they might well also hope to obtain a similar one-way guarantee.[191]

This risk of the other Scandinavian countries becoming jealous of Sweden was thus the real reason for opposing the British plan, and there is a good chance that the Americans would have accepted the arrangement if such a guarantee would not have spelled problems with Norway and Denmark. At this delicate time caution was necessary in order to cement the Atlantic Alliance. Instead, proposed Hickerson, "[a]n armed attack upon Sweden, assuming it were not a party, would merely require consultation."[192] Hickerson then presented an idea of what the articles in a treaty of interlocking pacts (a Scandinavian and a Western) would look like:

> Provision for individual and collective effort to strengthen the capacity of the parties to resist aggression (Article 4).
>
> Provision for establishment of agencies for implementation of the Treaty (Article 12).

190 Letter from Hickerson to Hoyer Millar, dated November 15, 1948, p. 1, in: RG 59; ReEUR; SFREDA, 1948–1954; Lot File 55D258; Box No. 2 of 4; (Folder: Scandinavia - General); NARA II, College Park, MD.

191 Ibid., p. 1.

192 Ibid., p. 1.

> Delineation of the area within which an armed attack would bring into effect the mutual assistance provisions (Article 7).
>
> Provision that an attack on any party within the delineated area is regarded as an attack on each and that accordingly each is obligated to assist in meeting the attack (Article 5).
>
> A provision for consultation in the event of an armed attack outside the area or any other event indirectly strengthening the parties (Article 10).[193]

In order to illustrate how the American version of the arrangement would work in practice, Hickerson spelled out how the situation would be under the proposed Treaty:

> An attack on Belgium would require full assistance from Norway and Denmark but not from Sweden. An attack on Norway or Denmark would require assistance from all parties to the Atlantic Treaty and from Sweden. An attack on Sweden would require assistance from Norway and Denmark but not from the other parties to the Atlantic Treaty unless, as a result of consultation, they chose to give it. No one-way guarantee would be given.[194]

The difference in the last sentence, compared to the British version, was obvious. Hickerson wanted to know what the people in the Foreign Office thought of his proposition, and he said that if they could come to terms they should sound out the Norwegians and the Danes fairly soon. Regarding Sweden, however, he urged caution; "[w]e will want to think very carefully about what we say to the Swedes and I don't think anything should be said to them yet."[195] Aunesluoma writes that Hankey commented that if Sweden only got consultations from the arrangement then there was no hope of getting her to take any "'risks at all on our behalf.'" Nevertheless, the British Foreign Office came to accept that the one-way guarantee had to be dropped, since the proposal had to be acceptable to the Norwegians if anything was to come out of it.[196]

In the meantime, the U.S. unwillingness to deal with the Scandinavian countries on equal terms was being felt in Stockholm. Matthews reported a conversation he had had with Supreme Commander General

193 Ibid., p. 2.

194 Ibid., p. 2.

195 Ibid., p. 3.

196 Aunesluoma, J., *Britain, Sweden and the Cold War...*, p. 72.

Jung on November 15 to Washington. The Ambassador stated that Jung had expressed considerable "uneasiness" over the U.S. talks with the Norwegians and Danes, "and by implication our failure to have similar talks with the Swedes."[197] This was certainly good news for the Americans, busy as they were with trying to stitch together a workable scheme for Scandinavia. While Hickerson discussed how to best make Sweden do its part for the defense of Scandinavia, after repeatedly having queried Boheman regarding the meaning of Swedish neutrality, a representative of the Swedish Defense Staff was perhaps offering some answers to Hickerson's questions to the Americans on the European mainland. American Air Attaché Hardy accompanied Håkanson of the Swedish Defense Staff to Wiesbaden, on November 15, for yet another conference, this time with USAFE and EUCOM. This time Håkanson stayed for eight days, returning to Stockholm on November 22.[198] Once again, the agenda of the meeting is not known. The contacts between the Swedish military and the Americans in Wiesbaden were apparently handled by Officer Stone Christopher of the USAF. He was stationed in Germany between 1948 and 1952, and arranged for Swedish officers to travel discreetly to Wiesbaden for meetings with their American counterparts. He later moved on to Oslo, from where he helped USAF generals, dressed as civilians, to get across the Swedish-Norwegian border for conferences with the Swedes.[199]

Despite the tacit collaboration between the Swedes and the USAFE, the U.S. government announced its attitude towards the SDU to the Scandinavian countries on November 26, 1948. Ambassador Matthews informed the Secretary General of the Swedish Foreign Ministry Johan Beck-Friis that a neutral SDU could not count on any American support, nor could it count on deliveries of military supplies until the needs of the Atlantic Pact members had been taken care of. The Ambassador was careful to point out that the United States wished "to avoid any appearance of pressure" and added, disingenuously, that the American

197 Letter to Hickerson from Matthews, dated November 15, 1948, p. 1, in: RG 84; SWSE; TSGR; 1943–1949; Box No. 1; (Folder: 1948); NARA II, College Park, MD.

198 "Air Operations Report, 1 November–30 November 1948 for aircraft C-47D, 44-77204" from U.S. Air Attaché in Stockholm to Director of Intelligence Headquarters, USAF, Washington D.C., p. 1, in: RG 341; DCSO; DI; DDCD; COD; AAB; Aircraft Operations Reports [henceforth: AOR], 1948–51; Singapore to Union of South Africa; Box No. 156; NARA II, College Park, MD.

199 Thorsell, Staffan, *Sverige i Vita Huset* (Stockholm: Bonnier Fakta, 2003), pp. 57–58. Christopher was a good friend of both Marcus Wallenberg and Stig Wennerström, who became the Swedish Air Attaché in Washington in the spring of 1952. Wallenberg was, according to Christopher, very influential and had a hand in almost everything that went on. The Director of SEB, Karl Arvid Norlin, had been sent to Wiesbaden in the late 1940s, because Wallenberg felt the Swedish government was not moving fast enough (ibid., p. 58–59).

government was "really not much interested in Swedish defense policy." The Ambassador also told Beck-Friis that similar messages would be given to the Norwegian and the Danes.[200]

Östen Undén met with his Scandinavian counterparts later that day to discuss the American message. The Norwegian Foreign Minister Halvard Lange read the statement that had been read to him earlier in the day. Undén stated that he thought it dissatisfying that the United States should dictate Scandinavian foreign policy by stipulating preconditions for arms deliveries, and suggested that if this was to be the Americans' attitude then the Scandinavians should tell them that they would look to the East for armaments. This suggests that Undén thought that the American refusal had serious implications for Sweden. The Swedish Foreign Minister's proposal, however, was too much for Lange, who said that such actions would be totally unacceptable to the Norwegian public. Regarding the issue of connections to the West, Undén warned against a situation where the United States decided over Scandinavian territory, and where the USSR would feel that the easternmost boundary of the U.S. was at the Norwegian-Russian border. To this Lange snapped that the Norwegian public would be relieved if the Americans considered their border to end at the said point. The difference between the Norwegian and the Swedish position could perhaps not have become more clearly expressed, and Undén stressed that there was no need for Norway and Denmark to rush a decision on whether to join the Atlantic Pact.[201]

The withholding of this vital technology tipped the scales, for once it became clear that no American military aid would be forthcoming, Norway and Denmark agreed to join the NAT.[202] This hardline bargaining was aimed toward Sweden, and not towards Norway or Denmark, and the United States actively tried to get Britain to assent to its embargo on military sales to the Swedes,[203] although, as we have seen, with a less unambiguous result.

200 FRUS: *Foreign Relations of the United States*; 1948; Vol. III, *Western Europe* (Washington, D.C.: United States Government Printing Office, 1974), pp. 295–296. See also: Telegram from Beck-Friis to the Embassies in Washington, London, Paris, Oslo, and Copenhagen, dated November 26, 1948, pp. 1–2, in: HP 24 VI/U.S.A; "Tillstånd till in- och utförsel av vapen, ammunition och krigsmateriel"; 1946, okt.– aug. 1950; Folder 20; Box No. 1353; År: 1946–1952; RIA, Stockholm.

201 Untitled memorandum written by Undén, dated November 26, 1948, pp. 1–3, in: HP 24 VI/U.S.A; "Tillstånd till in- och utförsel av vapen, ammunition och krigsmateriel"; 1946, okt.– aug. 1950; Folder 20; Box No. 1353; År: 1946–1952; RIA, Stockholm.

202 Kaplan, L. S., *The Long Entanglement:...*, p. 32. Regarding the American refusal to provide arms, see also: Silva, C., *Keep Them Strong, Keep Them Friendly:...*, p. 255.

203 Lundestad, G., *America, Scandinavia, and...*, p. 346.

Perhaps by happenstance, Hugh Cumming arranged a dinner in his home on the very same day, i.e. November 26. Attending the dinner were a number of Swedish officials, among them Thord Bonde of the Swedish Defense Staff. It is not known whether Cumming's guests knew about Matthews' message, but it can be assumed that they did. A conversation between Cumming, Bonde, and Folke Wennerberg (former Counselor of the Swedish Legation in Washington) reveals the Swedish sense of its plight. Wrote Cumming:

> Mr. Wennerberg remarked that as part of the process of strengthening the Swedish defense system to ensure Swedish neutrality Sweden should build a large number of airfields which could be utilized by military aircraft in the event of a threatened attack on Sweden. I asked, "Utilized by whom?"
> Mr. Wennerberg said, "By American planes, of course."[204]

Considering that the memorandum was written three days after the event, one should probably not make too much of the fact that Wennerberg seemed to imply that in order to protect the nation's neutral position, Sweden should construct airfields which the Americans could make use of *before* an actual attack had occurred. Perhaps Cumming's memory was playing a trick on him? Be that as it may, Cumming, who was just as devoted a 'hardliner' as Matthews, was not impressed by this expression of goodwill, and quipped that he thought:

> [...] it more likely that Sweden might wake up one day to discover Mr. Wennerberg's airfields occupied by Soviet aircraft.[205]

Bonde sadly agreed with Cumming and the two split off from the rest and continued the discussion between themselves. Bonde said that he understood that the United States was disappointed in Sweden, but that the Americans should not be overly pessimistic since the Swedes were "moving in the general direction he understood we desired." What Bonde then said is worth quoting in full:

> He said that [...] assuming that we had a direct strategic interest in Norway which we did not have in Sweden (he said this looking at me in a questioning way) and assuming further that the Scandinavian Peninsula was a strategic entity he thought that the ideal solution would be for us to encourage the formation of a Scandinavian

204 "Conversation with Thord Bonde, of the Swedish Defense Staff.", dated November 29, 1948, p. 1, in: RG 84; SLE; CONF 1948; 020-844; Box No. 6 (Folder: 800); NARA II, College Park, MD.

205 Ibid., p. 1.

defense union and to give such a union a unilateral and secret guarantee of support in the event of attack or threatened attack by Russia. Without waiting for my comment, he immediately went on to say that he understood that there might be some difficulties in the way of such a unilateral guarantee but he felt that the same purpose could be accomplished if we would, first, encourage or at least not discourage the formation of a defense union, secondly, make such provision as we could for the supply of military equipment to all three Scandinavian countries and, thirdly, embark on very confidential staff talks looking to the correlation of Swedish and American war plans.[206]

To this more than open-hearted plea for American recognition of the Swedish situation Cumming replied that:

> Sweden's continued adherence to neutrality lessened our [the U.S.] interest in Sweden's defense plans busy as we were in discussions of the North Atlantic Pact, etc.[207]

The message was engraved upon Bonde by giving him "the usual line" that those "who were not prepared publicly to associate themselves" with the West would be put at the "tailend" of applicants for American military equipment. Cumming continued his rebuttal:

> [...] as a former second lieutenant it seemed to me that secret staff conversations with a country which openly declared that it was prepared only to defend itself and not to go to the aid of other countries if attacked would be fruitless from our standpoint.[208]

There can be no doubt as to the effectiveness of Cumming's approach. According to his notes, Bonde took all of this in a very good way, but:

> [...] remarked a little sadly that if what I had said was true and he had no doubt that it was that a unilateral guarantee of a Scandinavian union could not be forthcoming and that we could not secretly or otherwise help Sweden with arms and staff conversations, "Then my country is indeed in a difficult position."[209]

206 Ibid., p. 2.
207 Ibid., p. 3.
208 Ibid., p. 3.
209 Ibid., p. 4.

The irony in all of this, of which Cumming was probably not unaware, was of course that Bonde's suggestion was almost identical with certain aspects of the Hankey Plan. It is hard to believe that this was simply yet another coincidence. Bonde had perhaps been informed of Hankey's proposal by the British, and might have tested the grounds with the American. Cumming feigned innocence to further the higher political objective, i.e. American hegemony in Scandinavia as a whole. As we saw above, Hickerson was a bit more candid about the true reasons why Hankey's suggestion was not considered feasible by the United States than was Cumming.

A few days later, on November 30, Boheman met Hickerson at a dinner at which time he inquired what the reason for the American declaration was. The American replied that the statement had been made in response to Swedish requests regarding this matter to Ambassador Matthews in Stockholm. Boheman then asked why, if this was the case, the same statement had been made to the Norwegians and the Danes. Hickerson had then, disingenuously, replied that they had not. Instead, he said, the American Embassies in Oslo and Copenhagen had been notified about the message delivered to the Swedish government by Matthews. The Norwegian and Danish governments, said Hickerson, had known about the American position for a long time.[210] We know from other sources, as we have seen above, that this was untrue. But why would Hickerson tell such a flagrant lie, when he surely must have understood that the Swedes knew that he was being dishonest? There is no way of knowing the answer to this question. Perhaps he was just trying to avoid opening up a can of worms at a dinner party.

On December 3, Beck-Friis inquired of Matthews what exactly the American policy meant for Sweden. The American Ambassador reported to Washington, with some satisfaction, that the Swede had "evinced considerable anxiety over the breadth and meaning of the word 'aid' from which Sweden would be disqualified until the requirements of members of the North Atlantic Pact were met." Apparently, Beck-Friis tried to reason with Matthews stating that he could not find any specific statement in the Vandenberg resolution that said that a neutral country or group of countries would be deprived of American aid, but the U.S. Ambassador quickly retorted that whether this was explicitly stated in this resolution or not, "it had now been given him as the official interpretation of the American Government."[211] It is worth noticing

210 Telegram from Boheman to the Foreign Ministry, dated December 1, 1948, in: HP 24 VI/U.S.A; "Tillstånd till in- och utförsel av vapen, ammunition och krigsmateriel"; 1946, okt.– aug. 1950; Folder 20; Box No. 1353; År: 1946–1952; RiA, Stockholm.

211 FRUS: *Foreign Relations of the United States*; 1948; Vol. III, *Western Europe*, pp. 299–300.

that Matthews did *not* say that Sweden had to join the Atlantic Pact, only that it had to meet the requirements put on the members in the alliance.

In his report to Washington, Matthews made clear that the American policy of denying Sweden arms was yielding the desired effects. Wrote Matthews:

> Foregoing illustrates the growing anxiety of Swedish Government and military circles over the whole question of the Atlantic Pact and its effect in isolating Sweden. I had Department's views [...] conveyed orally to General Jung, Chief of Swedish Defense Forces, and General Nordenskiold, Chief of Swedish Air Force.[212]

Erlander felt that this message from the American government was "a new [and] perhaps serious element in the war of nerves against our neutrality." According to the Swedish Prime Minister, it was Matthews' biased and uninformed, indeed "foolish", reporting from Stockholm that was the cause of this attitude. The government of the United States, wrote Erlander in his diary, did hence did not know the true feelings and thoughts of the Swedish government on this question.[213] Erlander was perhaps being unfair, because Matthews simply restated to the Swedes the policy entailed in NSC 28/1.[214] This war of nerves was the U.S. government's way of telling the Swedes that they needed to toe the line, and consent to American hegemony. A neutral Scandinavian union went directly against the American plans for the postwar European order, and was thus not acceptable to them.

There is more evidence that the Swedes considered the American attitude to be a serious threat against the formation of the SDU. Boheman met with Hickerson again on December 7 at which time he read out a four-page official response from the Swedish government to him. The reply was just as much a plea for understanding of the Swedish and Scandinavian situation as it was a forceful argument and a protest. It began by stating diplomatically that the Swedish government was happy to hear that the United States did not wish to affect Sweden's foreign policy, and said that Sweden would defend herself without hesitation and with all her power if she was attacked. At the same time the point was made that the Swedish government considered the SDU to be in complete harmony with the U.N. Charter, and that American discrimination against Sweden's efforts to arm and defend

212 Ibid., p. 300.

213 *Tage Erlanders dagböcker, 1945–1949*, p. 276. "[...] ett nytt måhända allvarligt element i nervkriget mot vår neutralitet." "[...] dåraktigt [...]."

214 FRUS: *Foreign Relations of the United States*; 1948; Vol. III, *Western Europe*, p. 233.

herself would lead the government to believe that Sweden was being put in a less favorable position for political reasons. Hickerson replied that the American statement was not intended to contain any new information, and should thus not to have come as a complete surprise to Sweden. The intention was only to inform Sweden of the content of the Vandenberg resolution. Boheman then stated that the view in Stockholm was probably that Matthews' message was intended to blow the SDU out of the water, and argued that the Americans instead ought to encourage this cooperation. The American, however, was unmoved by Boheman's colorful speech.[215] The U.S. refusal to supply weapons to the SDU was hence seen as a threat to the entire Scandinavian endeavor.

Earlier research has suggested that despite the important differences between Norway and Sweden, the Norwegians accepted the Swedish position at the time. Magnus Petersson has written that the Norwegian government "[...] seems to have had an understanding for the Swedish policy."[216] Without questioning the accuracy of Petersson's conclusion over a longer period of time, at the time the Norwegians voiced some strong opposition to the Swedish line in conversations with the Americans, and in fact urged them to put additional pressure on Sweden.

The Norwegians, who wished to have Sweden with them in a defense pact backed by the United States, were understandably frustrated by the Swedish attitude. So frustrated were they in fact, that Sivert A. Nielsen, of the Norwegian Embassy in Washington, called on Rogers at NOE on December 8, 1948, to discuss the matter of the intransigent Swedes. Nielsen said that it was becoming increasingly evident that Sweden was not on the same wavelength as Norway and the rest of the Western European countries, and it "appeared that Sweden was as unwilling as ever to compromise its policy of neutrality." The Norwegian specifically referred to the SDU negotiations where Sweden, he said, "behaved in a very non-committal fashion." According to Nielsen, the Swedish Embassy people in Washington were now convinced that their relations with the United States had taken a turn for the better, and that the Americans understood the grounds for the Swedish policy and recognized Sweden's "special position."

The Norwegians were not pleased with this complacent attitude, and wanted the United States to correct it. According to Rogers:

215 Telegram from Boheman to the Foreign Ministry, dated December 7, 1948, pp. 1–2; transcript of the Swedish note attached to Letter from Boheman to Åström, dated April 11, 1949, pp. 1–4, in: HP 24 VI/U.S.A; "Tillstånd till in- och utförsel av vapen, ammunition och krigsmateriel"; 1946, okt.– aug. 1950; Folder 20; Box No. 1353; År: 1946–1952; RIA, Stockholm. "att torpedera"

216 Petersson, M., "*Brödrafolkens väl*"..., p. 99. "[...] tycks ha haft förståelse för den svenska politiken."

> I gathered that this kind of talk irritated the Norwegians in Washington and Nielsen suggested that *it might be well to correct any erroneous impressions the Swedes might have.* I replied that the Swedish Embassy people might not be fully up to date on all developments since I was quite sure that the Swedish Government in Stockholm could be under no uncertainty as to our position.[217] [Italics added.]

Thus it is clear that even Sweden's Scandinavian brothers wished to see increased pressure put on them to force a friendlier posture towards the West. The issue was of some importance to the Norwegians, since they were painfully aware of Sweden's importance to their national defense even if Norway became a member of an Atlantic Pact.

The next day, on December 9, 1948, the Americans and the British met to discuss the issue of interlocking pacts once again. Hoyer-Millar related to the Americans the Foreign Office's new position, i.e. that it had become increasingly aware that there was little likelihood of Sweden abandoning its neutrality in the near future.[218] The conference concluded that the British and U.S. governments had no objection to a Scandinavian Pact if the parties were free to enter other regional defense associations, and that they should forget Sweden in the formation of the Atlantic Pact. The two countries now agreed that:

> 2. We cannot offer any unilateral guarantee to any country.
> 3. We should not urge Norway and Denmark to accept a pact less advantageous to their national interest than Sweden is offered.
> 4. There is little present hope of bringing Swedes in by interlocking pacts.[219]

This decision was made even easier after the Norwegians had firmly rejected the idea on the grounds that the political price of granting Sweden such a unique position, while Norway still had to accept reciprocal responsibilities within the Atlantic Treaty, was too high. It would, writes Aunesluoma, "have meant that an obstinate policy of

217 Memorandum of Conversation between Rogers and Nielsen, dated December 8, 1948, in: RG 59; Records of the Bureau of European Affairs [henceforth: REEUR]; Subject Files Relating to European Defense Agreements [henceforth: SFREDA], 1948–1954; Lot File 55D258; Box No. 2 of 4; (Folder: Sweden); NARA II, College Park, MD.

218 Memorandum of Conversation, "British and US Views in Relation to Scandinavian Countries and North Atlantic Pact", dated December 9, 1948, p. 1, in: RG 59; REEUR; SFREDA, 1948–1954; Lot File 55D258; Box No. 2 of 4; (Folder: Scandinavia - General); NARA II, College Park, MD.

219 Ibid., p. 2.

neutrality had paid bigger dividends than opting for a closer association with the Western cause."[220]

It apparently also caused some change in Hankey's attitude toward Sweden, at least temporarily. In a conversation with Gerald Kieth, of the American Embassy in London, the head of the Foreign Office's Northern Department had said that with regard to Sweden, he felt that the British and American position was essentially the same, i.e.:

> [...] we should not give Sweden defence equipment until the Swedes have clearly stated where they stand as to their cooperation with countries in the Atlantic Pact. [...] Hankey then continued by saying that he would tell me very confidentially that only recently the Swedes had requested an opportunity to train some of their airmen in night flying with the British Forces. Hankey [...] inquired how they [the airmen] would be used if, for example, a British military plane were compelled to fly over Sweden some time. Hagglof [Hägglöf, the Swedish Ambassador to London] responded by remarking that a British plane would not be flying over Sweden. Hankey then proceeded to produce a map of the Polar Regions and the Northern Part of Europe to explain to him how Sweden might lie in the course of such necessary flights. Hankey reports that he told Hagglof in no indecisive manner that it was absurd to make such a request when Sweden would not take its stand which would make evident how they might use those fliers whom the British had helped train. He said the Swedish request was most definitely refused.[221]

This illustrates the contradictory policies followed by the British at this time. Firstly, it seems to contradict Aunesluoma's conclusion that the British had decided that they would work to relieve the pressure the Americans were putting on Sweden. Secondly, it does not really fit well with British actions regarding the de Havilland Vampire airframes when they clearly disobeyed U.S. wishes. Thirdly, Hankey's point regarding the absurdity of the Swedish request can be expanded onto an area that has bearing on this thesis, namely why would the West supply guided missiles to Sweden if there was even the slightest chance of them being used against Allied aircraft flying over Sweden? It is inconceivable that the Americans would have supplied Sweden with guided missiles without having asked the same question. The British position was qualified somewhat about a year later at a meeting between the Chief of

220 Aunesluoma, J., *Britain, Sweden and the…*, pp. 72–73.

221 Memorandum of Conversation between Kieth and Hankey, dated January 5, 1949, p. 2, in: RG 319 AIDF; From ID 523271 To ID 523460; Box No. 3392; (ID: 523455; NARA II, College Park, MD.

the Air Staff Westring and his British counterpart Air Marshal Sanders. The Air Marshal had explained to Westring that the refusal was made wholly on the part of the Foreign Office, and did not represent British military sentiments.[222]

Matthews was, despite the intransigence of the British, very pleased with the response that the American policy had generated in Sweden. At the end of December 1948, with Swedish participation in the Marshall Plan secured, and with Sweden having indicated that it might be willing to abide by the trade embargo within the near future, he reported to Foggy Bottom that the policy of withholding arms from Sweden had indeed been a great success. Matthews expressed his hope that this policy would stimulate increasing concern in Swedish government circles.[223] He also took the opportunity to caution Hickerson that Boheman, while "intelligent and very pleasant to deal with", was "a very smooth operator."[224] It seems as if Matthews thought that you could never really trust the Swedes.

The SDU Committee presented its findings on January 14, 1949, which included such matters as the preferred, and required, assistance from the Western Powers, general prerequisites for Scandinavian military cooperation, and "to clarify in what areas such cooperation could be carried out and what forms it should take."[225] On the same day the Americans restated, this time publicly, that the members of the SDU, if not associated with the United States, would not be eligible to receive military supplies from them. Similar statements were made again a few days later through the American Ambassador in the respective countries.[226] The message was being hammered in, and the adverse

222 Memorandum, dated October 19, 1949, pp. 2–3, in: HP 24 VI/U.S.A; "Tillstånd till in- och utförsel av vapen, ammunition och krigsmateriel"; 1946, okt.–aug. 1950; Folder 20; Box No. 1353; År: 1946–1952; RiA, Stockholm.

223 Letter from Matthews to Hickerson, dated December 29, 1948, p. 1, in: RG 84; SWSE; TSGR; 1943–1949; Box No. 1; (Folder: Ambassador Matthews – Cumming Sweden Vs. NATO); NARA II, College Park, MD.

224 Ibid., p. 2.

225 *Had There Been A War…*, pp. 61–69, quote on p. 69. Cole cites a book written by Norwegian Foreign Minister Halvard Lange in which Lange states that Undén also insisted upon the investigation being open-ended, and that the study "would not exclude association in one form or another to the broader western security cooperation […]." However, Lange argued that the Swedish interpretation of this was very different from the Norwegian, and effectively precluded association with any body of states except the United Nations (Cole, P. M., *Neutralité du jour:…*, pp. 187–188).

226 FRUS: *Foreign Relations of the United States*; 1949; Vol. IV; *Western Europe* (Washington, D.C.: United States Government Printing Office, 1975), p. 774. See also: Skottsberg Åhman, Brita, "Scandinavian Foreign Policy, Past and Present" in Henning Friis (ed.), *Scandinavia: Between East and West* (Westport, Connecticut: Greenwood Press, Publishers, 1979 (originally published by

American attitude towards the Scandinavian union, and implicitly against Sweden's security policy, could not escape the Swedish government. This was, at least, the fourth time that the Scandinavian countries were told that the SDU would not get access to American weapons.

Important differences remained between the Scandinavian countries, and the positions did not change, during the meeting of the Prime, Foreign, and Defense Ministers in the Swedish city of Karlstad on January 5–6, the submitting of the Committee's report on January 14, and the conference in Copenhagen on January 22, 1949.[227] What had changed, though, was the international situation. For on Christmas Eve 1948 the Norwegian and Danish Ambassadors in Washington received information that their countries would very soon be offered membership in NAT.[228] This obviously put the SDU in a completely new light for Norway and Denmark. The Swedish government felt this too, of course, and decided to act swiftly. Therefore, at the meeting in Karlstad the Swedish government offered the Norwegians and the Danes a defense union that was to be concluded even before their defenses had reached parity with the Swedish (something which up till then had been an important precondition for Sweden).[229]

Given the situation in which the Swedish government now found itself, the Swedish Prime Minister Erlander thought that there was no alternative but to make this offer.[230] It was by now evident that there would not be a formal arrangement between the Scandinavian countries, and the following gathering in Oslo on January 29–30 was merely a funeral wake for the SDU which had, once again, been declared dead.[231] The Swedish refusal to contemplate even the most informal link to the Western Powers has been cited as the main reason for the failure of the SDU.[232] That is a strange, if not erroneous, conclusion considering what has been said above. Indeed, Sweden stated that such an informal link was an absolute prerequisite for Swedish participation. Instead, it must be concluded that the main reason for the failure was the U.S. refusal to provide support for a neutral Scandinavian defense bloc.

Cornell University Press, 1950)), p. 301; Ekman, J. G., *Förhandlingarna om Skandinaviskt…*, pp. 36–37.

227 *Had There Been A War…*, p. 75.

228 Ekecrantz, S., *Hemlig utrikespolitik…*, p. 86.

229 Ibid., pp. 86–87; Skodvin, M., *Nordic or North Alliance?…*, p. 70.

230 Ekecrantz, S., *Hemlig utrikespolitik…*, p. 88.

231 *Had There Been A War…*, p. 75.

232 Riste, Olav, "'Janus Septentrionalis'? The Two Faces of Nordic Non-Alignment" in J. Nevakivi (ed.), *Neutrality in History*, p. 321; Logue, J., "Sweden" in S. V. Papacosma and M. R. Rubin (eds.), *Europe's Neutral and…*, pp. 87–88.

The American refusal to supply a neutral SDU with weapons has been known for a long time,[233] but I argue that prior research has not given this factor the weight that it deserves.[234] By withholding arms deliveries the Americans ensured that Norway and Denmark could not possibly choose the Scandinavian option, and this factor was just as important as the Swedish insistence on a neutral union, for the failure of the SDU—they were two sides to the same coin.

Even though the idea of a formal Scandinavian defense pact was no longer viable, there was a second issue that the SDU Committee had been ordered to investigate, and that was an option involving partial secret defense cooperation. One important conclusion in the report had been that "it has to be strongly emphasized that a defence union, cooperation otherwise prepared in advance between the three Scandinavian states, does not eliminate the necessity of external assistance."[235] While even Undén probably realized this all along, the Swedish government's official position did not allow for such connections to be made in the open. In late January 1949 the Swedish Supreme Commander Helge Jung provided the Swedish government with a paper in which he emphasized the importance of rapid military build-up of Norway and Denmark, insisting also that if a defense union should prove impossible, peacetime cooperation between the Scandinavian countries should still be carried out. The Supreme Commander argued most forcefully that Western assistance was still necessary, and that "such assistance could be made more effective and employed more rapidly, had mutual peacetime preparations been made."[236]

The same theme was reiterated by the Chief of the Defense Staff Nils Swedlund in a memorandum to Prime Minister Erlander in February 1949. Swedlund suggested that a series of measures should be taken by the three countries, including the establishment of wire and radio communications, the development of plans to coordinate air surveillance, air defense command and control, air force search and rescue, and military weather service. He also recommended that naval cooperation should be provided for, as well as establishing joint

233 Cf. Silva, C., *Keep Them Strong, Keep Them Friendly:...*, pp. 254–255; Molin, K., *Omstridd neutralitet...*, p. 66; Wahlbäck, K., "Norden och blockuppdelningen 1949–49" in *Internationella Studier, 1973:B*, pp. 50, 53–58; Ekman, J. G., *Förhandlingarna om Skandinaviskt...*, pp. 34–37; Dalsjö, R., *Life-Line Lost:...*, p. 49; Skodvin, M., *Nordic or North Alliance?...*, p. 73.

234 The only possible exception is Ekman, who has pointed out that Norway and Denmark's dependence upon weapons deliveries from the U.S. was used to pressure them into joining an Atlantic Alliance of some form (Ekman, J. G., *Förhandlingarna om Skandinaviskt...*, p. 34), but does not make this an important factor in his concluding remarks.

235 *Had There Been A War...*, p. 72.

236 Ibid., p. 74.

terminology, reporting and identification procedures in general. The government gave a green light to most of these efforts, and they were overseen by Swedlund and the Defense Staff.[237] Swedlund informed the Chief of the Norwegian Defense Staff General Ole Berg about the Swedish government's green light in November 1949, and Berg gave his o.k. to the plan. Both sides were to provide contact men, and the planning was conducted by small groups of Norwegian and Swedish staff officers under the supervision of the respective Defense Staff. This cooperation, commonly called DNS,[238] was based on personal contacts and as little as possible (even though it did leave traces in the archives) was put down on paper; instead the result was materialized in practice.[239]

The Swedish-Norwegian defense cooperation fitted perfectly with the American war plan Dropshot, which had been presented to the JCS on January 31, 1949. Both Norway and Sweden were seen as strategically important from offensive and defensive standpoints, since these countries could provide air bases 800–1100 kilometers (500–700 miles) closer to the Soviet Union than those in the British Isles. It was also deemed important to deny such bases to the Russians, in order to make enemy operations in, and over, the North Sea more difficult. Sweden's considerable industrial capacity and natural resources (e.g. iron ore and uranium) were also part of the reason for why the Americans wished to deny this area to the USSR. Dropshot argued forcefully for a concerted defense of Scandinavia, and, as Hans Weinberger has commented, if the ideas contained in the plan were communicated to Sweden or Norway it must have provided a strong incentive for closer military cooperation between the two neighbors.[240] Even though Dropshot was never formally approved by the JCS it remained a summary of U.S. strategy until 1951.[241]

On February 7, 1949, Boheman talked to the Chief of the NOE Hulley and Robert McClintock, from United Nations Affairs, and discussed matters relating to the imminent creation of an Atlantic Pact. The Swedish Ambassador was very outspoken and said that "Sweden could not be neutral in any future war", and agreed that although they were not to formally enter the Atlantic Pact "[Sweden] could, through secret staff conversations, be placed in a position to act in its own and our [the U.S.] behalf."

237 Ibid., pp. 126–128, 137–147.

238 Denmark, Norway, Sweden. For a good summary of these efforts, see: Dalsjö, R., *Life-Line Lost:...*, pp. 148–156.

239 Petersson, M., *"Brödrafolkens väl"...*, p. 241; Agrell, W., *Den stora lögnen...*, pp. 82–83.

240 Weinberger, H., "The Neutrality Flagpole:..." in M. Thad Allen and G. Hecht (eds), *Technologies of Power:...*, pp. 312–113.

241 Ross, S. T., *American War Plans 1945–1950*, pp. 131–132.

Reflecting upon this, Hulley thought that there might "be considerable profit to be derived from a secret diplomatic understanding between the Swedish and American Governments." Although he acknowledged that "Sweden would not qualify to receive overt military assistance [...]", he reasoned that if the United States "could assist the Swedes financially and in an economic way and at the same time have staff talks leading to standardization of weapons and basic war planning we [the U.S.] might get many of the benefits now contemplated by the more formal Atlantic Pact."[242]

On February 9, 1949, Boheman informed Hulley and Secretary of State Acheson that the Swedish government had reached a decision "to enter a defense alliance with Norway and Denmark". This decision had been reached "with the consent of all important parties", and it represented "a radical change from previous Swedish policy", said Boheman.[243] The Ambassador was referring to the offer made to the Danes and Norwegians in Karlstad in early January, although it seems strange that he would present the offer in this way *after* the breakdown of the SDU negotiations, when this option had already been rejected by Norway and Denmark. A plausible explanation would be that the Ambassador was being less than candid with Hulley and Acheson, trying to convince the Americans that Sweden had been prepared to shoulder some responsibility in Scandinavia. By this show of good will on the part of Sweden, Boheman might have hoped to receive some sympathy from the United States.

Perhaps surprised at this, and curious as to the effect this would have on the current talks on an Atlantic Pact, Acheson inquired if this precluded military staff talks between Sweden and the Western Powers. To this direct query Boheman now answered that such talks "could not be formally arranged but, of course, there was such a thing as suggestions offered by our Service attaches in the northern capitals."[244]

242 "Conversation with Swedish Ambassador on the Atlantic Pact" dated February 8, 1949, p. 1, in: RG 59; BNA, 1941–1953; SF, 1941–1953; South Africa, Union of, Policy – Sweden, External Political Affairs: U.S.; Box No. 18; (Folder: External Political Affairs: U.S.); NARA II, College Park, MD. See also: *Had There Been A War...*, p. 148.

243 Memorandum of Conversation between Boheman, Hulley, and Acheson, dated February 9, 1949, p. 1, in: RG 59; REEUR; SFREDA, 1948–1954; Lot File 55D258; Box No. 2 of 4; (Folder: Sweden); NARA II, College Park, MD. Boheman and Acheson were personal friends. Boheman had met Acheson, then a young lawyer, in Stockholm in 1928 (Thorsell, S., *Sverige i Vita Huset* (Stockholm: Bonnier Fakta, 2003), p. 38).

244 Ibid., p. 2. A similar conversation between Boheman and Acheson is referred to by the Neutrality Policy Commission. The Secretary asked if a Scandinavian Pact would "'preclude potential secret staff cooperation", to which Boheman responded that this was not possible since it could hardly be kept secret, but that it should not be too difficult for the Americans to "get a general idea of

The phrase 'our Service attachés' referred to the secret collaboration schemes offered Sweden by the American representatives in Norway, Sweden, and Denmark. Boheman's statement can reasonably be assumed to have been the expressed hope of the Swedish government that some sort of secret arrangement could be made on the side with the United States.

Acheson's response was that this was hardly a strong enough connection, and he remarked that in his opinion this Scandinavian grouping would lack sufficient strength to defend itself. The Secretary then asked:

> [...] whether the alliance would be based on neutrality in the sense that none of its members could have any military connection with other countries. He [Boheman] confirmed that this was the case.[245]

Boheman had bluffed, and to the Ambassador's chagrin Acheson had called him. Presented with a direct question it was impossible for Boheman not to tell the truth. Acheson and Hulley probably felt that their fears had been confirmed by Boheman's answer; a neutral Scandinavian Pact with Norway and Denmark outside a broader Atlantic security arrangement was not palatable to the United States. Acheson then provided Boheman with a chance to save face by wondering what would happen if the Soviet reaction to the SDU "took a much more vigorous line" than was expected by the Swedes. Boheman then stated that "in that case the Scandinavian countries would have to apply for admission to the Atlantic Pact."[246] This answer at least gave him the opportunity to show that Sweden would not succumb to threats from the USSR.

But even if Acheson was hesitant he nevertheless felt that the matter should be properly investigated, and sent a letter to Secretary of Defense James Forrestal to which he attached several questions which he felt should be submitted to the JCS, to be addressed by them as soon as possible. In discussions with the Ambassadors of the Brussels Treaty countries it had already been agreed that they would put the same questions urgently before the military committee in London; the Canadians would do the same to their military. The attachment deserves a full quotation:

> What strategic role is anticipated for Scandinavia in the event of war? Would our strategic objectives be accomplished by Scandinavian

---

where and how assistance could and should be implemented." (*Had There Been A War...*, p. 148).

245 Ibid., p. 2.

246 Ibid., p. 2.

non-belligerence or neutrality? In the light of these factors what are the relative advantages to the North Atlantic Powers of:

a) a Norwegian and Danish membership in the North Atlantic Pact with a neutral Sweden unwilling to contribute in any way to the defense of Norway or Denmark unless it is itself attacked.
b) A Scandinavian defense pact committing Norway, Sweden, and Denmark to go to war in the event of an armed attack on the metropolitan territory of any of them, but precluding any association, either by treaty or through military conversations, with the parties to the North Atlantic Pact?

Would it be in the interest of the parties of the North Atlantic Pact to furnish arms for Scandinavia under (b)?[247]

Forrestal replied the same day that he had asked the JCS to consider these questions "as a matter of urgency", and that he would advise Acheson as soon as he got the views from the military Chiefs.[248] The JCS had in fact already been given a report on this matter by the Joint Strategic Survey Committee (JSSC) on December 15, 1948. This was in response to another request from Forrestal in late November, and the report restated the importance of the base areas in Norway, and on Spitzbergen and Greenland. The Spitzbergen Archipelago was not of so much strategic importance to the United States as it was to the Soviet Union; therefore it was in the interest of the United States to deny the Soviets bases there. It was deemed to be in the security interest of the U.S. that Sweden, Norway, and Denmark remain free of Soviet domination, and it was concluded that:

> A purely Scandinavian defense pact might […] deny base rights and communication facilities to the United States; […] Unless the Scandinavian countries participate in a defense alliance as a part of a North Atlantic security system, it is to the advantage of the U.S. military position that they refrain from undertaking a purely Scandinavian pact of either neutrality or defense […].[249]

247 Letter from Acheson to Forrestal, dated February 9, 1949, p. 2, in: RG 59; ReEUR; SFREDA, 1948–1954; Lot File 55D258; Box No. 2 of 4; (Folder: Scandinavia – General); NARA II, College Park, MD.

248 Letter from Forrestal to Acheson, dated February 9, 1949, in: RG 59; ReEUR; SFREDA, 1948–1954; Lot File 55D258; Box No. 2 of 4; (Folder: Scandinavia – General); NARA II, College Park, MD.

249 J.C.S. 1929/4, "The Military Implications to the United States of a Scandinavian Pact", dated December 15, 1948, pp. 2–3, in: RG 218 Records of the U.S. Joint Chiefs of Staff [henceforth: RG 218]; Central Decimal File [henceforth: CDF] 1948–50; 381 (10-2-46) B.P. Pt. 2 to 381 (8-14-47) Sec. 3; Box No. 184; (Folder:

The JSSC thus effectively put an end to any chance of American military support for the SDU. The Chiefs had the interests of the United States before their eyes when they offered these views, but at they same time they seem to have equated the interest of the United States with that of Scandinavia as a whole. This attitude corresponds well with a strategy of hegemony and the kind of foreign policy that the Americans were undertaking at the time.

That the Americans were getting able help in their efforts to influence the Swedish government, and to convince it to assume a more amicable attitude vis-à-vis the United States, has been known for a long time. On its side, the United States had a sizable portion of the Swedish diplomatic and military communities.[250] Supreme Commander Helge Jung was responsible, according to Molin, for laying down clear guidelines as to what the military's attitude should be.[251] Jung had confirmed this in a conversation with Ambassador Matthews in November 1948. He also stated that a change in the Swedish attitude was on its way, and asked if American weapon deliveries would be forthcoming if the Swedish policy developed in the right direction.[252]

The Norwegian Foreign Minister Halvard Lange went to the United States in early February 1949, in order to "seek information about the North Atlantic Pact and to discuss questions of Scandinavian defense."[253] Once there, the Norwegians corroborated Boheman's statement from two days earlier. Norwegian Under Secretary of Defense Dag Bryn told the Americans that all three Scandinavian countries were in agreement that any kind of Scandinavian defense pact was predicated on armaments support from the West. "The Swedish Government [was] in full agreement with this concept", stated Bryn, "and in fact did not desire an isolated Scandinavian military pact unless Norway and Denmark were assured of U.S. assistance."[254] Bryn and Lange's effort was due to a directive from the Norwegian government, which stated that they should try to create an understanding on the American side

Folder: Defense of Scandinavia (Norway, Denmark and Sweden); NARA II, College Park, MD.

250 Molin, K., *Omstridd neutralitet…*, pp. 37–45.

251 Ibid., p. 39.

252 Ibid., p. 41.

253 Letter from Acheson to Forrestal, dated February 9, 1949, p. 1, in: RG 59; ReEUR; SFREDA, 1948–1954; Lot File 55D258; Box No. 2 of 4; (Folder: Scandinavia – General); NARA II, College Park, MD. See also: Wahlbäck, K., *Norden och blockuppdelningen, 1948–49* in *Internationella Studier, 1973:B*, p. 52.

254 "Memorandum of Conversation", dated February 11, 1949, p. 1, in: RG 59; ReEUR; SFREDA, 1948–1954; Lot File 55D258; Box No. 2 of 4; (Folder: Norway); NARA II, College Park, MD. See also: Wahlbäck, K., *Norden och blockuppdelningen, 1948–49* in *Internationella Studier, 1973:B*, pp. 54–55; Ekman, J. G., *Förhandlingarna om Skandinaviskt…*, p. 36.

for the Scandinavian situation, so that new possibilities opened up for a regional arrangement without overt connections to an Atlantic Pact.[255]

In fact, it had been known to the Americans at least since September 1948 that the Swedes "hoped for unilateral military guarantee from US which if forthcoming would insure adherence three northern countries to western bloc."[256] It is of course hard to imagine that Erlander or Undén would ever have said that such a guarantee would ensure that the Scandinavian countries would support the West, but it is perhaps not so hard to think that, for example, Boheman could have gone so far as to promise this to the Americans. Anyway, Ambassador Matthews commented that this was most certainly not to be taken seriously, saying that if such a guarantee were given that would ensure that Sweden would not adhere to the Western Bloc at all; it would, according to him, be contrary to everything that the Embassy had been working for.[257]

The State Department's reaction to the Swedish position, as related to them by Bryn and Lange, was negative, and displayed the apprehensions about neutrality that were so prevalent within the new administration. Charles E. Bohlen and the Director of EUR Hickerson wrote a letter for Secretary of State Acheson after the conversation with Bryn. In the letter they advised Acheson to tell the Norwegians that "a Scandinavian pact of the only type Sweden would be willing to conclude would [not] have the blessing of this country".[258] Hickerson (as did Hulley and, we may suspect, Bohlen) belonged to the majority position within the State Department which did not look benignly upon a neutral Scandinavian bloc. There is certainly also a chance that they did not approve of the apparent 'free riding' entailed in the Swedish attitude, even if the Swedes had now decided to opt for secret cooperation with the West. Why should the United States commit to a nation that was unwilling to align itself more clearly with the West?

But this negative reaction was not shared by everyone. At least there seemed to be an interest on the American side in inquiring where this might lead. Rogers at NOE brought the issue of Sweden's decision to enter into an alliance up with Arne Gunneng, of the Norwegian Foreign Office, on February 14. Rogers, who was a bit hesitant as to whether the information convened a few days earlier was really correct, asked Gunneng if this really was the Swedish position, which the Norwegian

255 Wahlbäck, K., *Norden och blockuppdelningen, 1948–49* in *Internationella Studier, 1973:B*, p. 52.

256 FRUS: *Foreign Relations of the United States*; 1948; Vol. III, *Western Europe*, p. 252.

257 Ibid., p. 252.

258 "Letter from Bohlen and Hickerson to Acheson", dated February 11, 1949, p. 1, in: RG 59; ReEUR; SFREDA, 1948–1954; Lot File 55D258; Box No. 2 of 4; (Folder: Norway); NARA II, College Park, MD.

official could only confirm.[259] Having been reassured, Rogers then stated that this seemed rather untenable since the Swedes did not permit staff talks "or other coordinating measures" to take place. Gunneng agreed to this, adding "but they are definitely drifting".[260] "It seems to me", wrote Rogers, "that this last point may be important."[261] This judgment was no doubt colored by the fact that Rogers was generally more positive in his attitude toward Sweden's claims of readiness to defend itself than were other State Department officials.[262] He then offered his view as to the current, and future, position of Sweden:

> The Swedes have cast themselves loose from the traditional position of pure neutral isolation and will find it difficult to determine upon a logical stopping point from here on.[263]

Rogers was clearly acknowledging what was going on, and had indeed summed it up very neatly. For his prognosis for Sweden turned out to be accurate. Once Sweden had embarked on the westward journey, and especially before an acceptable way of masking the discrepancy between the official and unofficial policies had been found, the government always had to worry that they might travel too far, hence the stop and go motion in the late 1940s.

Boheman saw Hulley again (this time accompanied by Hickerson) on February 25, 1949, and presented the Americans with an offer that they could not refuse. Boheman said that Swedish Military Attaché Colonel Gunnar Möller had returned from Sweden primed with full knowledge regarding Sweden's defense plans and that he would like to set them before the Americans, without any demands of reciprocity.[264] In fact, as Robert Dalsjö has pointed out, this was part of a lobbying campaign on the part of Swedish diplomats and military officers to convince the Americans of Sweden's strategic importance and of its

259 Memorandum of Conversation between Rogers and Gunneng, dated February 14, 1949, p. 1, in: RG 59; ReEUR; SFREDA, 1948–1954; Lot File 55D258; Box No. 2 of 4; (Folder: Norway); NARA II, College Park, MD.

260 Ibid., p. 1–2.

261 Ibid., p. 2.

262 Silva, C., *Keep Them Strong, Keep Them Friendly:...*, p. 272.

263 Memorandum of Conversation between Rogers and Gunneng, dated February 14, 1949, p. 2, in: RG 59; ReEUR; SFREDA, 1948–1954; Lot File 55D258; Box No. 2 of 4; (Folder: Norway); NARA II, College Park, MD.

264 "Memorandum of Conversation between Hickerson, Hulley, and Boheman", dated February 25, 1949, p. 1, in: RG 84 Foreign Service Posts of the Department of State [henceforth: RG 84]; Sweden; Stockholm Embassy [henceforth: SWSE]; Top Secret General Records [henceforth: TSGR]; 1943–1949; Box No. 1; (Folder: 1949); NARA II, College Park, MD.

will and ability to defend itself.[265] The former objective was certainly an unnecessary undertaking, since the Americans already considered Sweden to be strategically important; and if they had not, nothing that Möller could have said would have changed U.S. policy in the area. What the presentation could have been hoped to achieve, however, was to convince the United States of the latter.

In early March a conference was held in the Pentagon, and Military Attaché Möller and the Senior First Secretary of the Swedish Embassy Olof Rydbeck descended upon the five-pointed edifice ready to reveal Sweden's order of battle. The meeting was presided over by General Courtlandt van R. Schuyler (who became Chief-of-Staff at NATO's Supreme Headquarters Allied Powers Europe (SHAPE) in 1953),[266] and the talk appeared both sincere and legitimate to the American participants—25 officers from the Plans and Operations staff representing the Air Force, Army, Navy, and State Department.[267] One of the attendants remarked that the presentation was very similar to one recently given to the American Service Attachés in Stockholm, and said that the briefing "was frank and covered the whole picture [...]." He continued:

> The defense plans described reflected joint Scandinavian planning, the importance of U.S. military assistance to Norway and Denmark under the Atlantic Pact being stressed as the crucial factor.[268]

265 Dalsjö, R., *Life-Line Lost:...*, p. 156. See also: Silva, C., *Keep Them Strong, Keep Them Friendly:...*, pp. 275–280.

266 Schuyler was handpicked by Supreme Allied Commander Europe (SACEUR) General Alfred M. Gruenther (Pelletier, George Eugene, "Ridgeway: Trying to Make Good on the Promises", in Robert S. Jordan, *Generals in International Politics: NATO's Supreme Allied Commander, Europe* (Lexington: The University Press of Kentucky, 1987), p. 39).

267 "Swedish Defense Plans" dated March 11, 1949, in: RG 59; BNA, 1941–1953; SF, 1941–1953; Country (Sweden, External Political Affairs: U.S.S.R..-Sweden, Wood Pulp); Box No. 19; (File: Military Affairs: Defense); NARA II, College Park, MD. Letter regarding Colonel Möller's presentation, dated April 12, 1949, in: RG 59; BNA, 1941–1953; SF, 1941–1953; South Africa, Union of, Policy – Sweden, External Political Affairs: U.S.; Box No. 18; (Folder: External Political Affairs: U.S.); NARA II, College Park, MD. Incoming message from the American Military Attaché in Stockholm, dated April 21, 1949, in: RG 59; BNA, 1941–1953; SF, 1941–1953; South Africa, Union of, Policy – Sweden, External Political Affairs: U.S.; Box No. 18; (Folder: External Political Affairs: U.S.); NARA II, College Park, MD. See also: Cole, P. M., *Neutralité du jour:...*, pp. 301–302; Dalsjö, R., *Life-Line Lost:...*, p. 156; *Had There Been A War...*, pp. 149, 253.

268 "Swedish Defense Plans" dated March 11, 1949, in: RG 59; BNA, 1941–1953; SF, 1941–1953; Country (Sweden, External Political Affairs: U.S.S.R..-Sweden, Wood Pulp); Box No. 19; (File: Military Affairs: Defense); NARA II, College Park, MD.

Apparently, General Schuyler was "'extremely impressed' by what he had heard, and learned a great deal, and thought Moller's [Möller's] exposition had been excellent." The General

> [...] was impressed by the political remarks and implications in Moller's [Möller's] presentation and felt they reflected right thinking. He was also impressed by the initiative taken by the Swedes and considered that our attitude has been too negative towards them in the past.

Finally, the conclusion was drawn that these appraisals differed considerably from that of the Service Attachés in Stockholm. They had viewed the parallel presentation as sheer propaganda and had "condemned in very strong terms what they regarded as the smug ignorance of the Swedish military."[269] This difference in appreciation represented an increasingly large rift between the judgments of the Pentagon officers and the Service Attachés at the Embassy in Stockholm. The Embassy had problems recognizing that the Swedes were moving towards a tacit consent to American hegemony.

The generosity expressed by Möller's exposé of Sweden's defense plans, writes Cole, was part of a Swedish governmental effort to repair the damage done by Vougt's statement in early February, at which time he had said that Sweden would fire upon Western aircraft flying over Swedish territory.[270] It is doubtful, however, whether this interpretation does Möller's presentation justice. Perhaps it was more an expression of the Swedish need for assistance, and a way to neutralize Matthews' critique of Swedish policy.[271] Möller told his audience that preparations were being made for a strategic plan to aid Denmark with the RSAF, and for cooperation between Swedish and Danish navies. The Swedish Military Attaché said that:

> [i]f the Scandinavian Pact had become a reality, Sweden would have had to help her neighbors immediately. Without that alliance, we are not forced to do so, but our opinion is that under such circumstances [a Russian attack on Norway and Denmark leaving Sweden out of the war] it will be impossible for us not to be mixed up in the war sooner or later. At the latest we will be forced into a war when the Russians demand of us permission to transport supplies through

269 Ibid. See also: Agrell, W., *Den stora lögnen...*, p. 100.

270 Cole, P. M., *Neutralité du jour:...*, p. 301.

271 *Had There Been A War...*, p. 149. I thank Robert Dalsjö for pointing this out to me.

Swedish territory to Russian forces. No Swedish Government will ever agree to such a demand.[272]

Möller was cautiously, but unmistakably, pointing out to the Americans the importance of Sweden to the defense of Norway and Denmark, and that by having worked against a neutral SDU the United States had missed out on an automatic Swedish response to a Soviet attack against them. While this kind of hint was surely a gamble, Möller was probably not unaware of the importance that the British attached to Sweden in defending Norway. The British COS held the view that it would be undesirable to have Norway and Denmark in NATO without defense arrangements with Sweden, and had repeatedly intimated to the Swedes that there was a possibility for a middle way whereby Sweden could enter into such arrangements with the Atlantic Pact countries.[273] Möller had also been careful enough to add that Swedish assistance would probably be forthcoming anyway. But in the end what was insinuated was that by assisting Sweden, the United States would also be assisting the other Scandinavian countries.

Möller was probably more uncertain about the American attitude, which, as he no doubt realized, was the one that really mattered. He would probably have been calmer had he known about the new NSC policy paper on Scandinavia NSC 28/1, which had been approved by President Truman on September 4, 1948, which stated that Sweden was after all considered to be of strategic importance to the United States. But neither Möller, Rydbeck, nor the Swedish government were privy to this information in March 1949, so the Military Attaché was sent to try to convince the United States' policymakers that Sweden was a nation worth supporting. According to the Americans, Möller had claimed that Sweden wanted to build up its defenses and act as a bastion for NATO.[274]

Möller could not have taken an initiative to present the Swedish defense secrets to foreign powers on his own without at least informing the Defense Minister (and probably also the Foreign Minister). To this, we can add the fact that the Swedish Military Attaché in London provided

272 "Transcript of Möller's presentation" dated March 24, 1949, in: RG 59; BNA, 1941–1953; SF, 1941–1953; South Africa, Union of, Policy – Sweden, External Political Affairs: U.S.; Box No. 18; (Folder: External Political Affairs: U.S.); NARA II, College Park, MD.

273 Aunesluoma, J., *Britain, Sweden and the Cold War…*, cf. pp. 78–80.

274 Incoming message from the American Military Attaché in Stockholm, dated April 21, 1949, in: RG 59; BNA, 1941–1953; SF, 1941–1953; South Africa, Union of, Policy – Sweden, External Political Affairs: U.S.; Box No. 18; (Folder: External Political Affairs: U.S.); NARA II, College Park, MD.

similar information to the British in May 1949.[275] Möller's presentation perhaps set something of a precedent for the future, and it is hard not to draw a parallel to Vougt's memorandum to the British almost two years later. Basically, there are too many examples of diplomats and military representatives that expressed the same view, for it to be a probable interpretation that these were all rogue elements. Indeed, with this view, the only actors that did *not* portray Swedish policy as being essentially a Potëmkin neutrality, were the Prime and Foreign Minister, who for obvious reasons could never be expected to voice any other opinion. True, Erlander could not afford to do away with all the dissenters, but it seems as if he reasoned that these people could be of good use as long as they kept themselves within certain boundaries.

According to the memo quoted above, Sweden's plans were essentially based on the precondition offered to Norway and Denmark by Erlander for Swedish participation in the SDU—i.e. that joint Scandinavian defense presupposed U.S. backing of Norway and Denmark. It seems as if Möller's talk convinced some of the American officials that it was possible to establish some kind of cooperation with the Swedes. A memorandum written some days after the Pentagon conference spoke of the potential establishment of a "silent partnership" between Sweden and the United States.[276] The fact that an official from the Embassy, i.e. Rydbeck, had joined Möller at the Pentagon was seen by the Americans as a confirmation that the talk was "somewhat more than 'purely technical'."[277] Indeed, the Swede had intimated at the Pentagon that Sweden wanted to build itself up as a bastion for NATO in the north.[278]

A few weeks later Rydbeck and Per Lind, both First Secretaries of the Swedish Embassy in Washington, offered further evidence of the change in the Swedish attitude in a conversation with Rogers. On a direct question regarding the issue of overflights of Sweden by Western aircraft in wartime, both answered that allied planes would be waved through and that they were positive the government had discussed this

275 *Had There Been A War...*, p. 149.

276 "U.S. Policy Toward Sweden", dated March 30, 1949, in: RG 59; BNA, 1941–1953; SF, 1941–1953; South Africa, Union of, Policy – Sweden, External Political Affairs: U.S.; Box No. 18; (Folder: External Political Affairs: U.S.); NARA II, College Park, MD.

277 Letter regarding Colonel Möller's presentation, dated April 12, 1949, in: RG 59; BNA, 1941–1953; SF, 1941–1953; South Africa, Union of, Policy – Sweden, External Political Affairs: U.S.; Box No. 18; (Folder: External Political Affairs: U.S.); NARA II, College Park, MD.

278 Message from U.S. Military Attaché in Stockholm, dated April 21, 1949 in: RG 59; BNA, 1941–1953; SF, 1941–1953; South Africa, Union of, Policy – Sweden, External Political Affairs: U.S.; Box No. 18; (Folder: External Political Affairs: U.S.); NARA II, College Park, MD.

and reached such a decision.[279] The fact that they were positive that this had been done was hardly reassuring to Rogers. The Americans could not be expected to base the planning of their air war strategy on the assumptions of two Swedish diplomats no matter how 'positive' they might have been that the government had come to this conclusion.

The tacit contacts between the U.S. military in Europe and their Swedish counterparts were slowly, but definitely, progressing during the spring of 1949. In April the RSA sent a group of three officers to act as observers at a EUCOM, the unified military command for the U.S. military in Europe, maneuver on the European continent. The request had been put before EUCOM unofficially by the RSA, and no objections were raised by the Americans. A formal invitation would of course not be issued; only an answer that their presence was welcome would be presented to Sweden. This raised some problems with regard to how to handle the issue with respect to the United States' NATO allies Norway and Denmark. It was apparently decided that the Norwegians and the Danes were also to be notified and allowed to send observers, so as to counter any impression that the United States put Sweden, a non-signatory to the Atlantic Pact, before them.[280] This did not mean that the Americans granted every Swedish request to observe such exercises. For example, the Swedes were not allowed to send representatives to the larger exercises which were planned to take place in the fall. The refusal was later stated as being on the grounds that since new American materiel would be displayed, certain people in Washington felt that only NATO countries should be permitted to send observers.[281]

279 Memorandum of Conversation between Rogers, Rydbeck, and Lind, in: RG 59; BNA, 1941–1953; SF, 1941–1953; South Africa, Union of, Policy – Sweden, External Political Affairs: U.S.; Box No. 18; (Folder: External Political Affairs: U.S.); NARA II, College Park, MD. See also: *Had There Been A War...*, pp. 148f. Although the Commission made use of this document, it offered no comment or conclusion based on what the First Secretaries had said. No analysis regarding why they said what they did, or how they could possibly have gotten the (presumed erroneous) impression that the government had arrived at this decision. This period is also foregone completely in the Commission's general conclusions regarding Sweden's contacts with the Western Powers (*Had There Been A War...*, pp. 170–172, 261–268).

280 Message regarding Swedish observers to EUCOM maneuver, dated April 9, 1949, in: RG 59; BNA, 1941–1953; SF, 1941–1953; South Africa, Union of, Policy – Sweden, External Political Affairs: U.S.; Box No. 18; (Folder: External Political Affairs: U.S.); NARA II, College Park, MD; Message regarding Swedish observers to EUCOM maneuver, dated April 19, 1949, in: RG 59; BNA, 1941–1953; SF, 1941–1953; South Africa, Union of, Policy – Sweden, External Political Affairs: U.S.; Box No. 18; (Folder: East-West Trade); NARA II, College Park, MD.

281 Memorandum, dated October 19, 1949, p. 1, in: HP 24 VI/U.S.A; "Tillstånd till in- och utförsel av vapen, ammunition och krigsmateriel"; 1946, okt.–aug. 1950; Folder 20; Box No. 1353; År: 1946–1952; RiA, Stockholm.

On November 22, 1949, Dag Hammarskjöld and Boheman met with George W. Perkins, Assistant Secretary for European Affairs and Hulley of NOE. After discussing matters related to CEEC, Germany's place in the future Europe, and the plans for a Scandinavian customs union, Boheman opened up a whole other field and stated that he assumed that the NAT's Northern European Regional Planning Group was formulating defense plans, "and he wondered if the question of relations with Sweden had come up."[282] Perkins answered that he had not seen anything related to the progress of these discussions. Hammarskjöld then interjected that there would necessarily exist a limited technical cooperation with Norway and Denmark, even though these countries were now members of the North Atlantic Pact. This obviously seemed to contradict what Swedish leaders had said concerning this issue, and Perkins wished to have matters clarified regarding what the Swedish position actually was:

> Queried as to the meaning of recent public statements by the Swedish Prime and Foreign Ministers to the effect that Sweden would be unable to cooperate militarily with its two neighbors, Mr. Hammarskjold said that limited technical cooperation was of course going on but that for political reasons public announcements would necessarily take the line which the Swedish Ministers had followed in their speeches [Erlander and Undén in the Riksdag on October 26 and 27]. Such cooperation was secret and there would be no announcement that it is going on. The Ambassador said that the difficult problem for Sweden as a country outside the group would be to keep abreast of technical developments. He hoped that when the Swedes tried to buy military supplies here [in the U.S.] it would be possible for them to do so.[283]

Boheman and Hammarskjöld were thus doing the best they could to convince the United States that Sweden wanted to contribute to the defense of Scandinavia. Once again, the Swedish Ambassador, in the hope that the Americans would come to assume an amiable attitude to Swedish requests, addressed the question of Swedish access to American military technology. Perkins, however, could not offer any consolidating promises.[284]

282 Memorandum of Conversation between Hulley, Perkins, Hammarskjöld, and Boheman, dated November 22, 1949, pp. 2–3, in: RG 59; BNA, 1941–1953; SF, 1941–1953; South Africa, Union of, Policy – Sweden, External Political Affairs: U.S.; Box No. 18; (Folder: External Political Affairs: Miscellaneous); NARA II, College Park, MD.

283 Ibid., p. 3.

284 Ibid., p. 3.

What the Americans wanted was not a Sweden that was a formal member of NATO; what was important was that Sweden should give its consent to American hegemony. Once this overarching goal had been reached the United States was willing to sell military material to Sweden. This was made evident in a memorandum from Assistant Secretary of State James Webb to James Lay, Jr., the Executive Secretary of the NSC, in November 1950. The memorandum discussed the policy aims of NSC 28/1 and stated that the objectives with regard to Norway and Denmark had been reached; both countries had aligned themselves with the Western Powers and economic aid and military equipment were being furnished to them. Sweden, however, was another story. It was pointed out that the American position had been made clear to the Swedes, and applications for licenses to export military supplies were considered in the light of the policy spelled out in this document. It was acknowledged that both Norway and Denmark wished to see a militarily strong Sweden, and it was concluded that:

> [...] it is in the overall interest of the United States in the area that the views of these countries be given due weight. We propose no modification of the present policy of having Sweden to go to the end of the line for arms, *but after this condition is met there should be sufficient leeway to permit commercial exports of military equipment.*[285]

The use of the phrase "sufficient leeway" clearly shows that what U.S. policymakers had in mind was not Swedish membership of the Atlantic Alliance. What was sought was a firm commitment by the Swedes to adhere to the West, and to make sure that American military materiel was not re-exported eastward. If the Americans had seen Swedish participation in the alliance as 'the end of the line', then there would have been no point in speaking about 'leeway'. Sweden would then have had the same rights as any other member of the alliance.

In the end, the second burial of the SDU in January 1949 marked the birth of a highly secret cooperation scheme between Sweden, Norway, Denmark, and the United States and Britain along the lines that had been part of the instructions to the SDU Committee. The Swedish military establishment had perhaps assisted the government in reaching this conclusion. If nothing else, the military confirmed to the Government that the line chosen was the right one from a strategic point of view.[286] Karl Molin has noted that this was a clever solution to the quandary in

285 Memorandum for James S. Lay, Jr., from James Webb, dated November 14, 1950, pp. 1–2, in: RG 218; CDF 1948–50; 381 (10-2-46) B.P. Pt. 2 to 381 (8-14-47) Sec. 3; Box No. 184; (Folder: Defense of Scandinavia (Norway, Denmark and Sweden)); NARA II, College Park, MD.

286 Agrell, W., *Den stora lögnen...*, p. 67.

which the government had found itself in late 1948, namely: once the SDU proposal had received a positive response at home, thereby denying the opposition the possibility of hi-jacking this issue in order to win votes in the upcoming elections (in September), the government had to take responsibility for its realization. The tricky part was that success could easily turn into a failure if Norway and Denmark flatly rejected the Swedish neutrality provision. On the other hand, the government could not afford to tie Sweden openly to the West by agreeing to a SDU that had to be associated with the Western Powers in order to survive[287]—it seemed like a catch 22 situation. The obvious solution, once it became apparent that Norway and Denmark (not to mention the Americans) would never accept a non-aligned neutral SDU, was to opt for secret military cooperation with Norway, Denmark, and the United States on the lines of the partial defense cooperation envisioned by the Committee's report in January 1949. This indeed provided a convenient solution to the predicament in which the government found itself all through the rest of 1948.

The aerial part of this collaboration was formalized in 1960 and designated SVENORDA[288], a system that included a joint communications network that could be connected to NATO's European network. Moreover, communication in the SVENORDA network was conducted in English, something that inadvertently "gave RSAF personnel training in English terminology."[289] Perhaps unsurprisingly, the unofficial side of the Swedish-American security cooperation is not always easy to equate with the knowledge that Prime Minister Erlander claims to have had about the extent of the cooperation with the Scandinavian neighbor states after the breakdown of the SDU negotiations. The most intimate source we have regarding Erlander's knowledge about these affairs is his diaries, published by his son Sven Erlander. On March 28, 1950, Jung and Swedlund briefed Erlander, Vougt, and Undén in Erlander's chambers regarding the cooperation that had been undertaken so far with the Danes and the Norwegians. Erlander wrote in his diary:

> Can one trust Jung and Swedlund? I hope so. In that case the statement they gave yesterday to Undén and me regarding Scandinavian cooperation was fully satisfying. Nothing other than such things, that have to be viewed as perfectly natural for us has been done according to the Generals. No staff negotiations, no joint planning. […] The contacts that have been made have apparently strictly toed the line; not to undertake anything that can be used to compromise our foreign

287 Molin, K., *Omstridd neutralitet…*, p. 63.

288 This was an acronym for: SVErige, NORge, DAnmark.

289 *Had There Been A War…*, pp. 126–128.

> policy. Let's hope no subordinates have gone further! At least, we have now done what we can in order to make clear to the military that we are serious about our policy of neutrality.[290]

Before drawing any conclusions based on this statement, it must be remembered that these diaries cannot be taken at face value; there are plenty of reason to remain cautious, and even skeptical, regarding Erlander's honesty and how much he really confided to his diaries. It is not know whether this was Erlander's true thoughts on the matter.[291]

Moreover, if Erlander did not trust his Supreme Commander to provide him with an accurate picture of the ongoing defense collaboration with Norway and Denmark, then he should have replaced him with someone who would. It is true, however, that although Swedlund had been preferred as Supreme Commander by the Social Democrats because it was expected that he would be less politicizing than his competitor for the post, Chief of the RSA C. A. Ehrensvärd, this turned out to be a false hope, for Swedlund was to be everything but easy to handle for the government.[292] Even so, Swedlund regularly briefed the Prime, Foreign, and Defense Ministers about the Scandinavian cooperation on at least a half a dozen occasions between 1951 and 1961.[293] It is thus probably safe to assume that Erlander knew more about the actions undertaken than he pretended to do.

290 *Tage Erlanders dagböcker, 1950–1951*, Published by Sven Erlander (ed.) (Hedemora: Gidlunds Förlag, 2001), pp. 71–72. "Kan man lita på Jung och Swedlund? Jag hoppas det. I så fall var den redogörelse som de lämnade igår inför Undén och mig om det skandinaviska samarbetet fullt tillfredsställande. Inget annat än sådana saker, som måste betraktas som helt naturliga för oss har enligt generalerna utförts. Inga stabsförhandlingar, inga gemensamma planer. [...] De kontakter som tagits har tydligen strikt hållit sig på linjen; icke företa sig någonting som kan användas för att kompromettera vår utrikespolitiska linje. Bara nu inte några underhuggare har gett sig längre ut! Nu har vi i alla fall gjort vad vi kunnat för att hos militärerna inskärpa att vi menar allvar med vår neutralitetspolitik."

291 Robert Dalsjö has, in a conversation with the author, remarked that Erlander was often very anxious to convince himself that he had made the right decision. The diary was a safety valve, according to Dalsjö, with which Erlander could relieve himself of much of the pressure placed upon him.

292 Zetterberg, K., "Nils Swedlund" in Gunnar Åselius (ed.), *Svenska officersprofiler under 1900-talet* (Stockholm: Militärhögskolan & Axel Abrahamsons Förlag, 1996), pp. 499–500. Zetterberg writes that Swedlund's insistence that Sweden should acquire tactical nuclear weapons drove a wedge between him and the government, and contributed to the splitting of the Social Democratic Party over the issue. The Swedish defense went through a growth in power that was unprecedented in the 20th century, either before or since, not least from a qualitative point of view compared to the surrounding countries (ibid., p. 500).

293 Agrell, W., *Den stora lögnen...*, p. 89.

However, there certainly were differences in views between the political and military leaderships at this time, but prior research suggests that this difference was perhaps more a matter of degree than one of substance. There are, however, plenty of examples that suggest that it was not uncommon for Swedish foreign policy to be formulated, and initiatives undertaken, by factions within the Foreign Ministry or the military.[294] The government consisted of different factions that pulled in opposite directions. Erlander, who was not undivided himself, was probably honest when he worried about something being done that could be used against the government both domestically (by the opposition) and abroad (by the Soviet Union). But it is plausible to assume that what Erlander feared the most was not cooperative measures in themselves, but measures that he and the government did not know about. Such undertakings, if they became known, could not be readily defended by the Prime Minister as being consonant with the policy of neutrality.

The line between what Erlander wished to know about, and what he did not wish to know about, due to plausible deniability, was a fine line indeed. There also seems to be little doubt that Erlander wished to keep some distance between the military establishment and the politicians. For example, he made derogatory remarks about the Defense Minister, stating that Vougt was all too fond of hanging out with the Generals and the Colonels.[295] He even went so far as to say, when discussing

294 Ibid., pp. 77–80. Agrell writes that one of the reasons why the resumption of the neutrality line after 1949 went so smoothly was the fact that the parties within the Foreign Ministry and the military that had been pushing a dissenting agenda stopped doing so (ibid., p. 80). Presumably, the Swedish decision to stay out of NATO when it was formally established in the spring of 1949 confirmed to possible dissenters that this was no longer a viable choice, but that it was better to work under the guise of neutrality and to make as much as possible of the situation.

295 Ibid., p. 118. Vougt was also averse to Torsten Nothin's chairmanship in the Defense Committee of 1949, indeed to the existence of the Committee as a whole, and was pretty open about it. This was not appreciated by Erlander, who in the fall of 1950 wondered whether he would not have to "give some serious thought to what kind of Defense Minister I have" (ibid., p. 121; "allvarligt fundera på vad det är för slags försvarsminister jag har"; for more on Erlander's criticism of Vougt, see: pp. 141, 155, 165, 181, 254–255, 262, 272). Apparently, Vougt did not carry much clout with the Supreme Commander Helge Jung (who was about to resign, and be replaced by Swedlund), who, according to Nothin, wished to see Vougt leave his post. Jung was willing to work with Erlander, Nothin had said, but not with Vougt. Such a move, said Nothin, would very much increase the military's trust in Erlander, to which Erlander quipped that he had almost too much trust in these circles but that, unfortunately, Nothin was right about what he had said (ibid., p. 131; for more on the military's lack of confidence and respect for Vougt, and the conflict between them, see: pp. 141, 147, 181–183). In the end the conflict between foremost Jung and Vougt got so bad that Erlander concluded, in December 1950, that both of them "had to disappear." (ibid., p. 184; "måste försvinna.")

the composition of the government after the upcoming elections in the fall of 1951, that "Vougt is a mistake, that never ought to have been included [...]."[296] Torsten Nilsson subsequently replaced Vougt after the September elections.[297]

How far the collaboration with Norway and Denmark actually went is still not entirely clear, but there is ample evidence that it was very profound in scope. In October 1953 the Swedish Prime, Defense and Interior Ministers were informed by General Swedlund that:

> 'the Atlantic Pact commanders' [...] had agreed that all contacts with Sweden should be made via Norway and that the Norwegians had been ordered to provide the Swedes with all information without any restrictions and without demands for reciprocity.[298]

Conclusion

In 1946 the American government withheld the delivery of P-51s to Sweden. This was the earliest example of the U.S. using technology as a tool to achieve a political objective against Sweden in the postwar period. The purpose of the action was to make the Swedish government cancel, or at least withhold, deliveries of armaments to Argentina, a country which had been blacklisted by the Americans for trading with the enemy during World War II. However, the decision to withhold the aircraft, which had been made by Acheson and the State Department, conflicted with the objectives of the War Department. The latter wished to gain Swedish agreement to the aerial photography project *Casey Jones*, and it was thought that the State Department's pressure had caused the Swedish government to turn the War Department's request down. Acheson and the State Department eventually decided to release the P-51s, but this did not change the Swedish position on the *Casey Jones* project. It is currently unknown whether Casey Jones was ever carried out according to plan in Sweden, but the government did delay Bofors' arms deliveries to Argentina.

Even before the articulation of American hegemony, and before the Cold War had gained momentum, there was limited military cooperation between Sweden and the United States. However, what little that there was soon ebbed away and was replaced by a harsh American attitude towards Swedish requests for export licenses for armaments. The Swedish policy of neutrality demanded, for reasons of credibility, that Swedish defenses should be strong enough to repel an attack or a violation of its neutral territory in wartime. To do this, Sweden opted

296 Ibid., p. 254.

297 Ibid., pp. 346, 350.

298 Ibid., p. 146. See also: *Tage Erlanders dagböcker, 1953* Published by Sven Erlander (ed.) (Hedemora: Gidlunds Förlag, 2001), p. 179.

for supplying its armed forces to a great extent from domestic sources. Moreover, the lack of foreign currency and foreign technical assistance in the guided missile area made it seem only natural to the Swedes that guided missile development should also be undertaken domestically; everyone agreed that missiles were the way of the future.

The lack of access to classified data pertaining to foreign developments made necessary a concentration of scarce resources on a few, very important, projects. But even though this was realized from very early on, and explicitly stated by people within the Swedish community responsible for developing guided missiles, concentrated development was nonetheless hard to achieve. The latter half of the 1940s and the early 1950s were characterized by a disparate development effort where valuable resources were scattered out on many different projects. In fact, it was not until about the mid-1950s that the development was narrowed down to only three projects—RB 304, RB 321, and RB 322. There was thus a connection between the many guided missile projects initiated and the small results, in terms of operational guided missiles, they yielded.

The personnel and financial problems experienced by the GMB became a factor from the very beginning. Only under circumstances where the Swedes did not have access to American and British know-how, and where the GMB was heavily dependent upon the success of the domestic development program, could this become such a disturbance. The inability to purchase guided missiles and know-how from the United States necessitated a diverse and intensive domestic research effort that became a *cul-de-sac*. There were some tiny exchanges with the surrounding world during this period, among them Britain and Norway, with some kind of (more or less) regular barter trade program being established between Norway and Sweden in early 1947. This did not do much to help Sweden's development projects, however.

Initially, the Swedish government had great difficulty in making the Americans understand that they had nothing to fear from a neutral Sweden, or even from a neutral Scandinavia. The Swedish policymakers had trouble in understanding that it would take real tangible steps to acquire the trust of their U.S. counterparts. The SDU proposal was seen as a direct challenge to U.S. security interests in Scandinavia; indeed it threatened the success of American hegemony in the region. Scandinavia as a whole, but especially Norway and Denmark (due to their geostrategic location as well as their island possessions Svalbard and Greenland), was too strategically important for the Americans to sit idly by and watch as these countries were sucked into a neutral Scandinavia. The SDU challenged the United States policy objective of securing Norwegian and Danish participation in NATO, and hence the effectiveness of the Atlantic Alliance as a whole. The response by the U.S.

was to use military technology as a lever to achieve its policy objectives in the region; a neutral SDU would not be considered for arms deliveries from the United States. Norway and Denmark thus abandoned the SDU and signed up for membership of the Atlantic Alliance instead.

While this fact has been mentioned by every study of the SDU negotiations, no one has so far viewed it as instrumental for the failure of the Scandinavian option. Instead, prior research has given precedence to the Swedish government's unwillingness to let go of the neutrality provision. However, technology was an important tool for the U.S. when establishing its hegemony in Western Europe. It was an important lever because a lot of governments depended upon it. Western Europe was not expected to last long without the support of the United States to build up its defenses. It therefore deserves to be pointed out that the American refusal to provide arms effectively doomed the alliance. If the Americans had accepted the SDU provisions, it is hard to imagine that Norway and Denmark would have sought membership of the NAT. Rather, they would have chosen an informal tie to the West. But a NATO without Denmark and Norway was deemed undesireable by the American government so the pressure was kept up.

The aftermath proved to be not so that bad for Sweden after all, and a secret collaboration was initiated with the Western Powers, with Norway acting as proxy for NATO and the U.S. But even so, all through the period dealt with in this chapter, the Americans continued to press the Swedes to take a firm stand in the East-West conflict. The assurances of Boheman and other diplomats were not enough for the Americans; they wanted more tangible evidence of Sweden's position. The period dealt with in this chapter, i.e. 1945-1948/49, was a time when American hegemony had not yet been fully acknowledged by the Swedish government. By 1947, and the declaration of the Truman Doctrine, the hegemonic project had been articulated by the U.S. government, which was prepared to take up the reins after the British. The Swedish government, however, had yet to offer its unambiguous consent to American hegemony. The Swedes certainly had an understanding of the importance of American assistance for the defense of Scandinavia, and saw U.S. involvement in Scandinavian security affairs as desirable, as long as this could be accomplished on terms acceptable to the Social Democratic government. Chapter 4 below will show how the government's consent was finally offered.

Chapter 4

The Consent to U.S. Hegemony: Sweden and the ERP, COCOM, and the MDAA Agreement, 1948–1952

It was as if a gigantic magnet had somehow come into existence, compelling most states, often even movements and individuals within states, to align themselves along fields of force thrown out from either Washington or Moscow. Remaining uncommitted, in a postwar international system that seemed so compulsively to require commitment, would be no easy matter.

(John Lewis Gaddis)[1]

[T]hey [the Swedes] can still kid themselves that ERP is only economic and not political or military.

(Matthews to Hickerson, March 16, 1948)[2]

Introduction

Below, I intend to provide a conclusive analysis of the three steps that were absolutely essential to Sweden getting access to military (and to a certain extent also civilian) technology on basically the same terms as the NATO countries. Just like the following one, this chapter is guided by the first set of questions formulated in Chapter 1. The chapter describes the material *forms* in which the Swedish government expressed its consent to American hegemony, as well as the *methods* and *tactics* employed by the United States to obtain this consent.

Sweden and the Marshall Plan

In September 1946, a trade agreement, including a five-year SEK 1 billion credit had been concluded between Sweden and the Soviet Union. The Americans reacted strongly and feared that the agreement would take up such a large part of Sweden's production resources that it would tie the Swedish economy to the Soviet Union for many years to come, and thereby neglecting its perceived obligations in rebuilding war-torn

1 Gaddis, J. L., *We Now Know:...*, p. 26.

2 Letter from Matthews to Hickerson, dated March 16, 1948, p. 1, in: RG 84; SWSE; TSGR; 1943–1949; Box No. 1; (Folder: 1948); NARA II, College Park, MD.

Western Europe. The Swedish government was presented, in the midst of negotiating the agreement, with a note from the U.S. government in which the Americans expressed concerns that Swedish trade would become tied to the Soviet Union. In the note the Americans emphasized that the agreement would not be in accordance with American free trade policy.[3]

The U.S. determination to establish an international system of free trade was matched only by an equal determination to deny the Soviet Bloc access to that very system. The free trade dogma could only be stretched so far. Hence the Americans had no problem erecting trade embargo walls against the Soviet Union, while at the same time urging their Western allies to open up their markets. But the Truman administration was honestly concerned about the proliferations of bilateral trade agreements among both friends and foes.[4] Adding to the American difficulties was the fact that there was as yet no hegemonic apparatus tying Sweden to the American version of free trade.

It is not unreasonable to assume that John Foster Dulles might have had this agreement (and others) in mind when, at the foreign ministers conference in Moscow in April 1947, he warned Secretary of State Marshall that the economy of Central Europe, *including Scandinavia,* was being integrated into and drained into that of the Soviet Union. The same argument had been used to justify the denial of reparation payments from the western zones in Germany to the Russians.[5] The fear of Russian penetration of Western European economies surfaced repeatedly through the years, as, for instance, in 1952 regarding Finland's growing dependence on imports from the USSR.[6] From the American

3 Karlsson, B., *Handelspolitik eller politisk handling…*, pp. 44–48. Only half the credit was actually used (or SEK 505 million). The reasons for this were legion, but it should be mentioned that the Swedish government only promised to approve export licenses to a value of SEK 1 billion, without promising to guarantee deliveries to that amount. This was left to the individual companies to negotiate amongst themselves. This gave the Swedish companies the upper hand – there was only to be as much trade as they wished to engage in. The credit was never very popular in Swedish business circles anyhow, and efforts were continually made to avoid deliveries. This was done by alleging that the production resources were already tied up for production of other products, and by charging too much for products that could be delivered (pp. 122–123). The Soviet Union complained that the United States was behind this policy, and stated that pro-American Swedish capitalists, such as the Wallenbergs, had been used to this end (Sevón, C., *Visionen om Europa…*, p. 202). See also: Appelqvist, Örjan, *Bruten Brygga. Gunnar Myrdal och Sveriges ekonomiska efterkrigspolitik 1943–1947* (Stockholm: Santérus Förlag, 2000), pp. 376–385, for the domestic debate concerning the credit agreement, see: ibid., pp. 367–376.

4 Leffler, M. P., *A Preponderance of Power*, p. 107.

5 Ibid., pp. 152–153.

6 Hanhimäki, J. M., *Scandinavia and the United States:…*, p. 51.

point of view, moreover, the Swedish-Soviet trade agreement could not have come at a worse time. As seen in Chapter 2, the British and the Americans had already opted for a partition of Germany, and thereby Europe as a whole, into two rival blocs. And though originally a British initiative, the Bizone was just as much an American creation when it formally became operational on January 1, 1947.[7]

The emerging Cold War frontlines in general, and the American note in particular thus turned the trade agreement into high politics, and the Soviet negotiators were suddenly very eager to conclude the agreement. The Swedish negotiators were then more or less forced to conclude the agreement in order to demonstrate that Sweden's non-alignment policy was not subject to American pressure.[8] Paradoxically then, the American note contributed to the finalizing of the deal, hence producing exactly the opposite result to what was intended. This was an example of the unpredictability of international relations, and goes to show that not even an immensely preponderant nation can get its way all the time. However, time and, perhaps more importantly, money, were now working in favor of the Americans.

Like all European nations at this time,[9] Sweden had quite a substantial balance of payments problem after the war due to a dollar deficit related to the breakdown of convertibility between the American dollar and the British pound sterling. The Swedish problem was structural, built into the pattern of trade. Swedish exports were primarily directed towards the European market (i.e. most of the income from this trade was not in the form of dollar), and at the same time an increasing share of Swedish imports (almost 35 per cent in 1947) now came from the United States.[10] This meant that the discrepancy was slowly but steadily draining the Swedish dollar reserve.

During the summer of 1947 Sweden, represented by State Secretary Dag Hammarskjöld, had renegotiated its 1935 trade agreement with the United States. The new agreement stipulated that the volume of Swedish imports from the United States would be dependent on the Swedish financial situation. These *de facto* import regulations would gradually

7 Deighton, A., *The Impossible Peace:...*, pp. 105–115 and Eisenberg, C., *Drawing the Line: ...*, pp. 233–248.

8 Karlsson, B., *Handelspolitik eller politisk handling...*, p. 48.

9 For this point, cf.: Deighton, A., *The Impossible Peace:...*, p. 174. The dollar gap amounted to $ 8 billion in 1946, and rose to $ 12 billion in 1947 (see: McCormick, T., J., *America's Half-Century...*, p. 74; see also: Hogan, M. J., *The Marshall Plan:...*, pp. 238–292).

10 Silva, C., *Keep Them Strong, Keep Them Friendly:...*, p. 211. A detailed coverage of Swedish-American diplomacy in connection to the Marshall Plan can be found in: ibid., pp. 81–240. For a detailed study of Sweden's economic woes at that time, see: Appelqvist, Ö., *Bruten Brygga...*, pp. 219–446.

be lifted as the Swedish payment balance improved.[11] The problem was that the financial situation did not improve at all, it actually got worse, and the European market, moreover, did not manage to recuperate on its own.

During the course of November 1947 Hammarskjöld repeatedly asked the American embassy in Stockholm for a dollar loan. According to the Chief of the Swedish Central Bank (Riksbanken), Ivar Rooth, Sweden would need a $100 million loan.[12] But the Swedes had requested financial assistance from the Americans at precisely the wrong moment. At this time, the Americans were preparing to launch the Marshall Plan, and did not intend to give Sweden any financial assistance outside this hegemonic apparatus. According to Geir Lundestad, "there could be little doubt that dissatisfaction with Sweden's policy of neutrality represented [a] factor behind Washington's negative attitude."[13]

The larger issue was of course that Sweden could not be granted assistance outside the hegemonic structure that constituted the Marshall Plan. Certainly, even a bilateral agreement could have been made to play the same role, but such an arrangement would have given Sweden a negotiating space that the American wish to attain hegemony could not tolerate. It was important to gain the northern neutral's consent to U.S. hegemony, and to align Sweden with the rest of the Western European nations.[14]

Thomas Jonter touched upon this issue when he argued that the United States wanted to force the Swedes into cooperating economically

11 Appelqvist, Ö., *Bruten Brygga...*, pp. 384–395; Jonter, T., *Socialiseringen som kom av sig...* pp. 145–146.

12 Jonter, T., *Socialiseringen som kom av sig...* pp. 149, 151.

13 Lundestad, G., *America, Scandinavia, and...*, p. 103.

14 Lundestad's point that Washington was prepared to let the Swedes restrict their imports from the dollar area, and even to postpone payments on current dollar transactions (Lundestad, G., *America, Scandinavia, and...*, p. 103), does not contradict these conclusions. Sweden's role in the reinvigoration of Europe involved *intra*-European trade, and not their trade with the United States. Lundestad also argues that there were strong limits to the extent of possible American pressure in the economic area. "Washington made no serious attempt to interfere with normal trade with Sweden", he states, and continues by saying that with "a possible exception for the amounts extended in the second quarter of 1948, the European Recovery Program was generally administered on the basis of economic criteria." (ibid., p. 231). He sums up by stating that the lack of pressure could be explained by the fact that there was little or no point in waging full scale economic warfare against a nation that could not be described as a real threat to U.S. interests (ibid., p. 232). However, while there is certainly merit to his conclusions, Lundestad once again misses the overall aim of Sweden's participation in the Marshall Plan. There would have been little sense in disrupting Swedish trade, since it was a U.S. goal to *increase* this trade in order to reconstitute the Western European economy. A weakened Sweden might turn toward the Eastern Bloc in search for markets.

with the West, and by doing so, securing a unified Western front relative to the Soviet Union.[15] Melvyn P. Leffler writes that American officials were convinced that in order to align Western Europe, West Germany and Japan with the United States "they had to narrow the dollar gap and help their industrial allies sell their goods, earn dollars, and purchase foodstuffs and raw materials in the underdeveloped periphery."[16]

In fact, the prosperity of the Marshall Plan, and with it the American design for a multilateral world economy based on nondiscrimination in trade and full convertibility of currencies, depended on the Americans successfully closing the dollar gap.[17] So for the Americans, then, it was a question not only of getting something out of Europe but also of giving something back. Material benefits were given to the European allies in order to elicit their consent to U.S. hegemony.

Consequently, on January 22, 1948, the American State Department informed its Embassy in Stockholm that Sweden would not be granted the coveted dollar loan outside the realm of the ERP.[18] The reasons for the negative American stance were stated bluntly to the Swedish Embassy in Washington; Sweden could not be given priority outside the Marshall Plan context, and the credit to the Soviet Union had played a role as well.[19] Sweden was clearly receiving not very well concealed pointers as to where her foreign policy allegiances should be if she wanted to be a qualified member of the Western European states.

Nils Andrén has written that Sweden declared "in its own economic interest, that it perceived it [the ERP] as a reconstruction plan, not as a formation of a bloc directed against outside powers."[20] As was made obvious in Chapter 2, this was hardly in full conformity with the political intentions behind the Plan.[21] Ambassador Matthews had been working towards getting Sweden to include itself in the Marshall Plan machinery. In March 1948 he wrote to Hickerson telling him that he thought he had obtained fuller government support for the Marshall

15 Jonter, T., *Socialiseringen som kom av sig…*, p. 149; see also: Silva, C., *Keep Them Strong, Keep Them Friendly:…*, p. 58. Silva states that the Americans drove a campaign to force Sweden to choose sides in the Cold War.

16 Leffler, M. P., *A Preponderance of Power:…*, p. 17; Kolko, G. and Kolko, J., *The Limits of Power:…*, p. 360.

17 Hogan, M. J., *The Marshall Plan:…*, p. 238.

18 Jonter, T., *Socialiseringen som kom av sig…*, p. 155.

19 Sevón, C., *Visionen om Europa…*, p. 225.

20 Andrén, N., "Sweden: Neutrality, Defense and Disarmament", in Hanspeter Neuhold and Hans Thalberg (eds.), *The European Neutrals in International Affairs* (Wien: Wilhelm Braumüller, Universitäts-Verlagsbuchhandlung, 1984), p. 42.

21 For this, see: Andrén, N., "Sweden: Neutrality, Defense and Disarmament", p. 42.

Plan; but only, he added, "because they can still kid themselves that ERP is only economic and not political or military."[22] There can thus be no doubt that the Marshall Plan was used by the Americans as a stepping stone to gain Swedish consent to U.S. hegemony in Europe. In the same letter Matthews said that even if Sweden managed to remain neutral in a coming war, it could only be done by supplying Russia with industrial goods. In a belligerent tone, the Ambassador wrote that in such a case:

> I say, it is inconceivable that the United States with its duty of self-preservation and its presumed obligations to its Western allies would not, albeit with reluctance, be compelled to destroy Swedish industry from the air.[23]

Matthews even urged Hickerson to let the Swedes know about this inherent risk because, according to him, the only way that the Swedish government or the Swedish people could be swayed to abandon the neutrality path would be if it was made clear to them that the "consequences in material harm" would be severe.[24] Matthews continued:

> I want some help at your end. [...] [P]erhaps around some quiet lunch table you could quietly and with proper embellishment drive home the fact that [...] whether she sticks to neutrality or not, Sweden will be either occupied or plastered from the air or both if war comes [...].[25]

Although Foreign Minister Östen Undén was on the whole positive to the American proposal, he nonetheless wished Sweden's acceptance to be on the condition that the assistance ought to be administered through the ECE, where the Swedish economist Gunnar Myrdal had been the new Chief since April. Undén took this line because his trusted friend and ally Rolf Sohlman (Ambassador to Moscow) thought that an unreserved acceptance might increase the Soviets' inclination to include Sweden in the hostile camp, but also because Myrdal himself saw such an approach as necessary if the Russians were to participate.[26] But as we saw in Chapter 2, Russian participation was exactly what the Americans

22 Letter from Matthews to Hickerson, dated March 16, 1948, p. 1, in: RG 84; SWSE; TSGR; 1943–1949; Box No. 1; (Folder: 1948); NARA II, College Park, MD.

23 Ibid., p. 2.

24 Ibid., pp. 1–2.

25 Ibid., pp. 2–3. See also: Dalsjö, R., *Life-Line Lost:...*, p. 52.

26 Kronvall, O., *Östen Undéns sovjetsyn*, pp. 27–28; Silva, C., *Keep Them Strong, Keep Them Friendly:...*, p. 63; Molin K., *Omstridd neutralitet*, p. 40; Kronvall, O., "Rolf Sohlman" in G. Artéus and L. Leifland, *Svenska diplomatprofiler...*, p. 294; Sevón, C., *Visionen om Europa...*, pp. 140–141.

wanted to avoid, even though it had at first been suggested that the erp should be handled through the ECE. One can imagine that Washington must have found this ironic reference to the PPS' secret, and since discarded, recommendation less than amusing.

Undén thus wanted to placate the Soviet government by suggesting that Swedish participation should be conditioned upon ece handling the Marshall Plan, but found himself outmaneuvered by Ambassador Erik Boheman, who was irritated by most aspects of Swedish foreign policy at the time, and who had argued for a non-conditional acceptance at meeting of the Advisory Council for Foreign Affairs on July 7–8, 1947. Boheman had spoken to Bevin in person, something that no doubt lent great weight to his arguments. The only option left for Undén was damage control. In a highly publicized speech in Strömstad he tried to mitigate the consequences of Sweden's participation for the credibility of the neutrality policy by stating that the Swedish decision, as well as the Marshall Plan itself, was not political, i.e. not directed against the Soviet Union.[27] Undén would not go as far, however, as to recommend that Sweden refrain from participating if the Soviets backed out. He and the Norwegian Foreign Minister Halvard Lange had already jointly agreed on this over dinner at the Norwegian Embassy in Stockholm already on June 27 (the day of the Paris meeting).[28]

Whatever possible illusions regarding the American motive behind the Marshall Plan may still have existed within government circles in Sweden in July, 1947, they were soon to be brutally shattered. In January, 1948, Dag Hammarskjöld was summoned to the American State Department where he was told that the American government deemed Sweden's "'whole-hearted cooperation'" in the Marshall Plan as being of the utmost importance. The Plan was not only to promote economic cooperation, he was told, it was also intended to increase political and ideological consolidation in the Western Bloc. Hammarskjöld was told that the United States thought that the Swedish government seemed hesitant about such collaboration.[29]

27 Kronvall, O., *Östen Undéns sovjetsyn*, pp. 27–28; Silva, C., *Keep Them Strong, Keep Them Friendly:...*, pp. 62–63; Molin K., *Omstridd neutralitet*, p. 39–40; for Boheman's arguments, see also: Stig Ekman (ed.), *Sven Grafström. Anteckningar 1945–1954* (Stockholm: Kungl. Samfundet för utgivande av handskrifter rörande Skandinaviens historia, 1989), p. 815; Sevón, C., *Visionen om Europa...*, p. 187. Boheman had also, according to Grafström said that the ECE was not a possible forum, not only because of the Russian presence, but also because of Myrdal who was a suspicious persona in the eyes of both London and Paris (Stig Ekman (ed.), *Sven Grafström. Anteckningar...*, p. 815). Örjan Appelqvist has shown that Mydal was, due to his economic policies, a controversial person also within the Swedish government (see: Appelqvist, Ö., *Bruten Brygga...*).

28 Sevón, C., *Visionen om Europa...*, p. 166.

29 Karlsson, B., *Handelspolitik eller politisk handling...*, p. 156.

This was connected to the negotiations on a new bilateral trade agreement between Sweden and the United States. Sweden's trade balance difficulties demanded a fast remedy, and only two options seemed possible—restricting import from the dollar area or a dollar loan. After protracted negotiations the Americans now demanded non-conditional Swedish support for the Marshall Plan policies, and a moral-political break with the Soviet Union in the foreign policy area. Hammarskjöld understood the message as meaning that what was necessary was that the government assured the Americans of the importance that European reconstruction had to Sweden. He also drew the conclusion that a firm Swedish commitment was a *de facto* precondition for all economic cooperation with the United States and Western Europe in the future. Undén subsequently telegrammed Hammarskjöld, who in his turn forwarded the same message to Foggy Bottom, and permitted him to confirm the Swedish government's unwavering support for the program intended to rebuild the European continent.[30]

The Swedish decision to join the program had profound implications for Sweden. The Russian Foreign Ministry intensely criticized the decision, and stressed how the United States had feverishly searched for ways to include Sweden in the anti-Soviet military-political bloc that was being created in Western Europe. The Soviets referred to American pressure upon Sweden in connection with the credit agreement in 1946, and said, correctly, that the problems in the Swedish trade balance had been used to influence Sweden's policies.[31]

Swedish Foreign Minister Östen Undén was the first Scandinavian official to make a public statement regarding Bevin's Western Union speech (see p. ? above).[32] Undén reacted unfavorably to Bevin's address, and stated that the Swedish policy of neutrality would remain unchanged. Sweden would refuse to openly take sides in the way that Bevin wanted. The Danish response had also been negative; the problem, as far as Sweden was concerned, was the Norwegians. Norway was positive to both Bevin's suggestion and the Marshall Plan, and did not rule out membership of a pact. The Norwegian Foreign Minister Halvard Lange intimated to the Swedish Ambassador in Oslo, Johan Beck-Friis, that

30 Sevón, C., *Visionen om Europa...*, pp. 215–224.

31 For the Russian reaction, see: Ibid., p. 202.

32 Cole, P. M., *Neutralité du jour:...*, p. 137. Cole expresses his indignation over the Swedish wish to form a neutral Scandinavian Bloc by quoting Machiavelli's *The Prince*: "It is always the case that the one who is not your friend will request your neutrality [...]." (ibid., p. 136). This quote was taken out of context; Machiavelli was speaking about a situation where the one requesting your neutrality had himself *started* the war.

his country soon "had to make a choice.",[33] making it painfully clear to Undén where Norway was heading.

Sweden signed up for participation in the ERP on July 3, 1948, almost exactly three months after congressional approval of the Plan. Sweden received a total amount of $107.5 million between 1948 (in November) and 1951 out of which $87.1 million came as *conditional aid*, which meant that Sweden would get paid in dollars when exporting to other participating countries. All of this was handled through the European Payments Union (EPU).[34] This conditionality "gave the US direct leverage over Sweden, despite the fact that the latter was not a member of the Western Alliance [...]."[35]

Charles Silva has suggested that the Swedish participation in the Marshall Plan was the turning point in the generally negative American attitude towards Sweden and the Swedish neutrality policy which now became a more positive one.[36] I argue that this is only partly true, since the American judgment of Sweden was intrinsically tied to the embargo issue. Silva also states, correctly I think, that "the technique of the informal unilateral declaration" pioneered with the Swedish participation in the Marshall Plan, provided Sweden and the United States with a precedent on which future cooperation could be based. This arrangement, he says, "proved particularly useful in security policy arrangements."[37] This seem plausible, but from this it is also clear that the alleged change in the American attitude towards Sweden came *post facto*, after a prior *de facto* change in Sweden's attitude. The United States never wavered on the subject of getting Sweden's consent to American hegemony. From this point of view it was Sweden's change of course when navigating through the precarious Cold War waters and finally arriving in the Western port that effected this perceived change in American attitude.

The Marshall Plan was a hegemonic apparatus and with it came more than just money, i.e. American power and influence. Among these things was to be found section 117 (D) of the Foreign Assistance Act of 1948. With this paragraph the United States was able, to some extent, to

33 Lundestad, G., *America, Scandinavia, and...*, p. 200; see also: Aalders, G., *Swedish Neutrality and the Cold War, 1945–1949*, pp. 62, 64–65, quote on p. 65.

34 Jonter, T., *Socialiseringen som kom av sig...*, pp. 160–161; Silva, C., *Keep Them Strong, Keep Them Friendly:...*; for the functions of the EPU, see: Hogan, M. J., *The Marshall Plan:...*, pp. 320–328.

35 Dohlman, E., *National Welfare and Economic Interdependence:...*, p. 71.

36 Silva, C., *Keep Them Strong, Keep Them Friendly:...*, pp. 110–112. Silva sets the total amount at $118.4 million, of which $20.4 million came as a non-conditional loan.

37 Silva, C., *Keep Them Strong, Keep Them Friendly:...*, p. 345.

control the foreign trade of the participating countries.[38] Sweden was no exception in this case. This was only the first step in the cumulative process of establishing the hegemonic cooperation that was to make Sweden a trusted, secure nation, and make it eligible to receive American technology—including guided missile technology.

The amount of $107.5 million that Sweden got through the Marshall Plan was, essentially, the very same amount that Sweden had tried to obtain from the United States all along, but with the crucial difference that Sweden was no longer the sole master of its foreign trade policy. Granting the loan outside the Marshall Plan might have carried with it unpleasant consequences for the Americans, such as, perhaps, the Swedes 'wasting' their dollars on imports from the United States. Letting Sweden buy a lot of goods in the U.S. might have made sense in the short term, from a myopic and self-interested point of view. However, it would have been bad as a long-term hegemonic strategy, because it would not have contributed to the larger U.S. policy objectives in Europe. This would, in the long run, also have been bad for the United States. By granting the loan within the Marshall Plan framework the U.S. government could make sure that Sweden would participate in the reconstruction of the European economy, and, as a consequence, a multilateral global trade order.

Be that as it may, the Americans ensured that this no longer was a viable option for Sweden, or any other Western European nation for that matter. In order to correct the payment balance problem Sweden was forced to engage in commercial activities with Western Europe, and at the same time contribute to the rebuilding of its war-torn neighbors. A sober assessment of Sweden's decision to partake in the ERP was offered by the ECA Special Mission to Sweden in February 1951:

> Thus it is that Sweden has become increasingly devoted to the objectives of ERP, and to its instruments for its accomplishment such as trade liberalization, increasing economic integration of Western Europe, etc. Circumstances have so operated as to intensify Sweden's interest in the European organizational institution vested with the responsibility for ERP – the OEEC. It is the Mission's view that Swedish support for ERP objectives and for OEEC also serve the objectives of the West. ERP and OEEC have accomplished a degree

38 Adler-Karlsson, G., *Western Economic Warfare…*, p. 23. Section 117 (D) instructs the Marshall Aid Administrator to refuse aid to any nation that receives American goods "'which go into the production of any commodity for delivery to any non-participating European country which commodity would be refused export licences to those countries by the United States in the interest of national security'." (Ibid., p. 23.)

> of economic cooperation between Sweden and the West that could not be achieved by any other device. The Mission believes that not only is this cooperation important in the economic field, but that it is also important in the political field. Economic cooperation today is, in a large sense, political cooperation; the manifold activities of OEEC permit the Swedes to have a peripheral political relationship with Western Europe and particularly the United States which could otherwise not be realized because of Sweden's adherence to neutrality. [...] The Mission will continue to urge Swedish support of ERP objectives, of the OEEC, and considers itself to be an instrument for influence in Swedish policy on matters economic in origin but which may have political connotations and which can be most appropriately handled through the instrumentality of the Mission, in close collaboration with the Embassy.[39]

It was furthermore stated that since the Marshall Plan was considered to be a very non-controversial subject in Sweden "and cannot be construed to infringe upon her neutrality, it offers a unique method of gaining support for general U.S. objectives."[40] There was an apparent awareness in the Mission that it, and the Embassy, were instruments for the establishment, and strengthening, of American hegemony in Sweden. The subtlety of the Marshall Plan's political effects was seen as the perfect way in which to encourage increasing Swedish adherence to the West. The main reason for the usefulness of the Marshall Plan was that it did not seem to infringe upon Swedish neutrality.

Sweden, COCOM, and the American Trade Embargo

Trade, or rather the restriction of trade, quickly became a means for the American administration to conduct the Cold War,[41] and it was deemed to be in the United States' interest for Sweden to remain inside the Western hemisphere when the trade embargo wall towards the East was erected. This trade embargo, and Sweden's adaptation to it, was to be of the utmost importance for Sweden's ability to obtain American high technology (something which was also decisive in the survival of the Swedish guided missile program).

39 Mission Action Program for the ECA Special Mission to Sweden, dated February 2, 1951, p. 6, in: RG 469 Records of the U.S. Foreign Assistance Agencies [henceforth: RG 469], 1948–61; Special Representative in Europe [henceforth: SRE]; Defense Production Division; SF, 1950–52; Programs: Economic, UK-Military, UK; Box No. 19: (Folder: Sweden – Military); NARA II, College Park, MD.

40 Ibid., p. 6.

41 For a short and excellent introduction to this subject, see: McGlade, Jacqueline, "COCOM and the Containment of Western Trade and Relations" in J. Eloranta and J. Ojala (eds.), *East-West Trade and the Cold War*, pp. 47–61.

The first American step towards the trade embargo was taken on December 17, 1947, when the NSC decided that all exports to Europe should be subjected to licensing. The decision was made with the explicit purpose of cutting off the flow of strategic commodities to the eastern bloc. Only ten days later, on December 27, the British concluded a trade and financing agreement with the Soviet Union. The American reaction to this challenge to their hegemonic aspirations was swift and decisive. On December 31 the U.S. Department of Commerce issued a new export licensing regulation calling for improved destination controls of goods, which were already under licensing, and on January 15, 1948, a second licensing regulation was issued. The latter required *all* commercial shipments to Europe to have individual licenses after March 1, 1948.[42]

American officials deeply resented, and feared, bilateral dealings with the Soviet Union which clashed with U.S. objectives for Europe and the non-Communist world.[43] The neutral countries could seriously have impaired the American embargo policy, had they stuck to their neutral stance. The fact is that without them the embargo would essentially have been a true paper tiger. Even though the neutrals did cooperate with the United States, some of the most influential authors on the subject have held that the embargo policy was in essence ineffective, and was indeed never intended to be effective other than in rallying the Western front against its main adversary: the Soviet Union.[44] If this is true, this actually supports the theory of hegemony since it was a strategy intended to elicit the consent of the European nations more easily.

The neutral countries had already acted as transit countries for embargoed goods during already during World War I, and more so during World War II, and the United States had suffered first hand the consequences of the German embargo policy. Swedish companies, controlled by the Wallenberg family, contributed a great deal to this by acting as fronts for German companies in the United States.[45] The Americans had no intention of allowing Sweden to become a great big hole in their net surrounding Eastern Europe, and were hence quick to apply pressure on Sweden to join in the embargo. In October 1948

42 Adler-Karlsson, G., *Western Economic Warfare...*, pp. 22–23; Dobson, Alan P., *US Economic Statecraft for Survival 1933–1991: Of sanctions, embargoes and economic warfare.* (London: Routledge, 2002), p. 89.

43 Leffler, Melvyn P., *A Preponderance of Power:...*, pp. 162–163.

44 Adler-Karlsson, G., *Western Economic Warfare...*, passim; Dobson, Alan P., *US Economic Statecraft:...*, passim.

45 Olsson, U., *At the center of development...*, pp. 208–210; Olsson, U., *Furthering a fortune...*, pp. 227–253 and p. 258. See also: Aalders, G. and Wiebes, C., *The Art of Cloaking Ownership:...*

negotiations were initiated, and the Swedes were presented with lists of goods that were to be covered by the American trade blockade.[46]

According to Matthews, it was the Swedes that had asked for the list,[47] but if this is correct then it is probable that it was due to American pressure on them to comply with the strategic embargo and in this way align themselves with the other Western European countries. Matthews remarked that the level of Swedish compliance was still indeterminate. The amount of success varied between the European nations, said Matthews, and was dependent upon e.g. "our [U.S.] leverage, namely the degree countries [were] dependent on ECA aid [and] economic importance of trade particularly imports from east [...]."

Matthews wrote that the fact that Sweden had only received a small amount of Marshall Plan funds, and only in the form of loans, "warrant belief we shall get less of our objectives here [...]." The Swedish government had, however, "expressed informally willingness to conduct joint technical examination to appreciate impact on two way trade with east." Matthews showed unusual restraint and understanding for the Swedish position, and noted that the issue was one of great importance to the Swedes especially with regard to Poland which, according to Matthews, supplied "eighty percent of Sweden's steam coal."[48]

Matthews was speaking in plain hegemonic terms when he noted that the extent of U.S. leverage would determine the outcome, i.e. how much of their political objectives could be realized. While the result was still hard to forecast in the Swedish case, the government had now been properly served a notice that acquiescence to this U.S. hegemonic scheme was expected in Washington. American leverage might have been less vis-à-vis Sweden than compared to other countries in Europe that were more dependent upon the Marshall Plan funding and less reliant on imports from the east, but Sweden nonetheless showed that they were willing to consent to U.S. hegemony.

In the middle of January 1949, Swedish Minister Without Portfolio Dag Hammarskjöld[49] took an active part in the discussions at the

46 Memorandum entitled "US Pressure on Sweden to Stop Trade with the East", dated October 13, 1948, in: RG 59; BNA, 1941–1953; SF, 1941–1953; South Africa, Union of, Policy – Sweden, External Political Affairs: U.S.; Box No. 18; NARA II, College Park, MD.

47 Telegram from Matthews to Harriman, dated November 23, 1948, p. 1, in: RG 469, 1948–1961; SRE (1948–53); Office of the Director of Administration Communications and Records Unit [henceforth: DACRU]; Country Files [henceforth: COUNF], 1948–51; Sweden P-T; Box No. 55; (Folder: Swed: Trade East-West); NARA II, College Park, MD.

48 Ibid., p. 1.

49 Hammarskjöld was responsible for East-West trade-related issues in the contacts between the Swedish and the American governments. He was elected to take over from Norwegian Trygve Lie as Secretary General to the United

French Foreign Ministry Quai d'Orsai in Paris.[50] The purpose was to coordinate the trade embargo policy against the Eastern Bloc.[51] The French chairman emphasized the importance of the issue of the definition of contraband and the transit trade.[52] This question was to haunt the Swedes over the next year and a half, because the Americans were determined to sway the Swedish government into giving their consent. Indeed, the United States had at their disposal the leverage they needed to accomplish this task—a technological monopoly.

John Haskell, the ECA Mission Chief in Sweden, immediately started to work on the difficult Swedes, recommending that since the "Swedes have not yet signified their intention to cooperate in restriction of exports to east [...] question advisability of offering to them items already denied to East since we have no assurance relative to reexport."[53] The United States' special representative to the ECA in Europe Averell Harriman agreed that the Swedes should be pressed for more complete information regarding the controls, and specifically recommended that "licenses for three multispindle Gridley screw machines be denied if Swedes refuse to cooperate."[54] This was an explicit demonstration to the fact that the United States used technology as a tool in their efforts to get Sweden's consent to American hegemony. These tactics did not of course go by unnoticed, but the Americans feigned innocence and issued official denials. In mid-March 1949 Rogers at NOE sent a memorandum to Acheson advising the Secretary of State on how to respond to such charges. Rogers wrote:

> You may be asked for comment on reports that the United States is "pressing" Sweden to limit trade with the East. You may wish to

Nations in April 1953, despite some initial resistance from the Soviet Union. If the Soviets knew about the role that Hammarskjöld played in Sweden's subordination to American hegemony, it is easy to imagine why they might have felt uncomfortable about him becoming General Secretary.

50 Adler-Karlsson, G., *Western Economic Warfare...*, p. 51; an article in English summarizing the prior research regarding Sweden and the trade embargo (of which most is in Swedish) can be found in: Nilsson, M. and Stenlås, Niklas, "Cold War Neutrality and Technological Dependence: Sweden's Military Technology and the East-West Trade" in J. Eloranta and J. Ojala (eds.), *East-West Trade and the Cold War*, pp. 133–151.

51 Adler-Karlsson, G., *Western Economic Warfare...*, p. 6.

52 Ibid., p. 51.

53 Telegram from Haskell to Harriman, dated January 21, 1949, in: RG 469, 1948–1961; SRE (1948–53); DACRU; COUNF, 1948–51; Sweden P-T; Box No. 55; (Folder: Swed: Trade East-West); NARA II, College Park, MD.

54 Telegram from Harriman to Secretary of State, dated January 21, 1949, in: RG 469, 1948–1961; SRE (1948–53); DACRU; COUNF, 1948–51; Sweden P-T; Box No. 55; (Folder: Swed: Trade East-West IA-IB Lists); NARA II, College Park, MD.

say that the United States has not in fact brought pressure to bear on Sweden.[55]

The message also contained a brief background on the matter for Acheson, but said nothing of the hints at the pressures revealed in the telegrams quoted above.[56] It is hard to say if Acheson was in the know or not, but the Secretary of State probably knew as much as he needed to know in order to perform his duties; plausible deniability by the heads of government was most certainly sought after and Rogers' memo provided for just that.

On February 28, 1949 a new Export Control Act was enacted in the United States, and on August 12, the same year, the Department of Commerce issued a 'master list' of all goods requiring an export license. The Consultative Group and the Coordinating Committee (henceforth: COCOM) were created on November 22, 1949, and COCOM formally started operating on January 1, 1950.[57] According to Gunnar Adler-Karlsson, COCOM:

> [...] had no foundation in any formally binding international treaty, but has been considered as a 'gentlemen's agreement' only. In spite of this it has in reality functioned as an international treaty with binding effects on the participating nations. [...] The United States concluded bilateral agreements with more than 50 nations in the West, concerning restrictions on trade with the communist nations.[58]

At the following meeting in October 1950, Hammarskjöld reported that Sweden could not formally take part in such actions, since that would compromise the policy of non-alliance.[59] The reason was that, officially, it would not be possible for Sweden to discriminate against Eastern trade the way a strategic embargo required. Informally, however, this could be achieved by extending the definition of contraband. Directly after the first meeting Hammarskjöld suggested to Cabinet Minister Johan Beck-Friis that Sweden should make a commitment to keep strategic goods on the domestic scene, and pointed out that a broadening of the term 'contraband' ought to be considered. All items that were regarded as strate-

55 Memorandum from Rogers to Secretary of State, dated March 17, 1949, in: RG 59; BNA, 1941–1953; SF, 1941–1953; South Africa, Union of, Policy – Sweden, External Political Affairs: U.S.; Box No. 18; (Folder: East-West Trade); NARA II, College Park, MD.

56 Ibid.

57 Adler-Karlsson, G., *Western Economic Warfare...*, pp. 6, 25, 51; Dobson, A. P., *US Economic Statecraft:...*, pp. 96–99.

58 Adler-Karlsson, G., *Western Economic Warfare...*, p. 6.

59 Karlsson, B., *Handelspolitik eller politisk handling...*, pp. 161–162.

gic and not viable for re-export to the East were labeled contraband and it was thus automatically forbidden to export them under Swedish law. The Americans were initially very pleased with this arrangement, which they regarded as highly efficient,[60] but the Swedes were soon to discover that the U.S. would not remain pleased for very long.

As in the case of the participation in the Marshall Plan, the Swedish government had officially interpreted its tacit participation in COCOM as "a purely economic matter that could be reconciled with a policy of neutrality."[61] But Hammarskjöld's actions show that this was nothing but plain lies—the Swedish government was painfully aware of the harm this cooperation could do to foreign confidence in Swedish neutrality and non-alignment. There would have been no reason to keep it secret otherwise. Strange as it may seem, the international trading order that was established by the Americans in the immediate postwar period did not make any exception for the neutrals. This was due to the fact that neutrality was seen as being purely a political matter, "which had few, if any, economic implications [...]."[62] This assumption, writes Ebba Dohlman, meant that Sweden was able to cooperate in the effort to depoliticize trade in the immediate postwar period.[63] But the American initiative to restrict trade with the Eastern Bloc made it blatantly obvious that this assumption was nothing but an illusion.

However, the initial Swedish decision to follow COCOM's lists was really just a unilateral promise, and no documents were signed tying Sweden to the trade embargo. This arrangement soon proved, after the outbreak of the Korean War, to be insufficient. Now the American regulations got stricter still, and during the fall of 1950 Swedish industry was really starting to suffer the consequences of their government's refusal to abide less ambiguously by the embargo rules. In April, 1950, a number of American export licenses for Swedish companies were stalled and withheld. The Korean conflict was a point after which the United States reinforced its effort to get the Western European allies in line. This sharpening of American policy was directed particularly against the European neutrals.[64]

An American Policy Change in 1950?

Charles Silva has argued that there was a change in U.S. policy in early 1950, stating that "[i]n February 1950, the US simply abandoned its

60 Ibid., pp. 171–173.

61 Af Malmborg, M., *Neutrality and State Building...*, p. 152.

62 Dohlman, E., *National Welfare and Economic Interdependence:...*, p. 41.

63 Ibid., p. 59.

64 Karlsson, B., *Handelspolitik eller politisk handling...*, pp. 211–212; Silva, C., *Keep Them Strong, Keep Them Friendly:...*, p. 306.

objective of including Sweden in the Western alliance."[65] The reason for this, said Silva, has not been adequately explored by scholars, who have not devoted enough attention to the issue.[66] Simon Moores also adopted this view, writing that this shift led the U.S. to "renounce its commitment to the idea of Sweden joining NATO […]."[67] The conclusion that the United States abandoned its objective of getting Sweden to join NATO is puzzling, all the more so since one may question whether there ever really was a U.S. policy objective of tying Sweden formally to the West. In NSC 28/1 from September 1948, the first official policy document dealing with Sweden, Swedish membership of the Western alliance was not stated as an aim of American policy. It said simply that the objective of the United States was to:

> […] influence Sweden to abandon this attitude of subjective neutrality and *look toward eventual alignment* with other Western Powers *in such form as may be found collectively acceptable* […]. [T]he United States should explore with the members of Western Union the problem of increasing the security of Norway, Denmark, and, *if appropriate*, Sweden, through their integration into *or association in some form* with the Brussels Treaty system or otherwise.[68] [Italics added.]

As we saw in Chapter 2, Sweden (and Italy) had been sorted into the third-tier category during the discussions at the Atlantic Pact exploratory talks in Washington in early July 1948, with a recommendation that she could be *associated* with the Atlantic alliance "under such special arrangements as might be necessary owing to their geographical position or to their international obligations […]."[69] The paragraph in NSC 28/1 was certainly a little ambiguous, but what is clear is that nowhere is it said that it was the objective of U.S. policy to have Sweden formally join a Western alliance. It is also clear that with regard to Sweden, not much was hoped for in September 1948. It was not even certain that it would be appropriate to discuss Swedish association with the Brussels Treaty nations, probably because the Americans realized that getting these countries to accept responsibility for a non-member would be a delicate matter indeed. What was sought was to get Sweden to lean towards the West more clearly, and the phrase 'or otherwise' implied that the

65 Silva, C., *Keep Them Strong, Keep Them Friendly:…*, p. 280.

66 Ibid., pp. 280–281. Silva specifically mentions Geir Lundestad (ibid., p. 280).

67 Moores, S., *'Neutral on Our Side':…*, p. 31.

68 FRUS: *Foreign Relations of the United States*; 1948; Vol. III, *Western Europe*, p. 233.

69 Lundestad, G., *America, Scandinavia, and…*, p. 274–276, quote on p. 276. See also above: p. 54.

United States kept other options open, with the possible *exception* of an outright membership of the alliance. The eventual alignment they were after was to be in a form that was collectively acceptable, hinting that they were well aware of the popular attitudes in Sweden, and of the minute chance of the Swedes actually joining formally.

Granted that this should be able to pass for an expressed American wish to include Sweden in a Western alliance, one would expect the next policy document to be substantially different on this point. That, however, is not the case. The spirit of NSC 121 which superseded NSC 28/1 in January 1952 is basically the same, stating that although:

> primarily because of the advantage to the organization of Scandinavian defense, it would be to our interest to have Sweden in NATO, we must for the predictable future accept as a political fact Sweden's policy of avoiding great power military alliances.[70]

It is very hard to conclude that these formulations represent the volte-face, or revolutionary reversal, of the policy contained in the preceding document that these scholars contend. If one so desires one can find much more obvious references in NSC 121 to the desirability of eventual Swedish adherence to a Western alliance. For example, in the first sentence saying that it would be in the American interest to have Sweden in NATO. This interest was never so unambiguously stated in NSC 28/1. That is quite the opposite of what Moores and Silva argue. The formulation in this policy document was no doubt influenced by the signs of increased cooperation that Sweden had already shown, such as the participation in the Marshall Plan in the summer of 1948, an event whose importance even Silva himself acknowledges.[71]

It might well be that Acheson and certain other members of the State Department had all along had as their policy objective to include Sweden in NATO; however, Acheson could not dictate the terms to the other departments and walk all over them in the NSC when it came to formulating overall U.S. policy. But if one looks at the documents, it is hard to find support for the view that even the State Department was seeking to include Sweden formally in the Western alliance.

But perhaps it is unfair to demand that NSC 28/1 should express any wish for Sweden to become a member in the Western alliance? At the time that this document was prepared the United States was itself unwilling to associate itself more firmly with the Brussels Treaty nations. It wanted the European association to broaden its membership to include all the

70 FRUS: *Foreign Relations of the United States*; 1952–1954; Vol. VI, *Western Europe and Canada* (Washington, D.C.: United States Government Printing Office, 1986), p. 1759.

71 Silva, C., *Keep Them Strong, Keep Them Friendly:...*, pp. 110–112.

Atlantic nations. Was there thus any change in American attitude once the North Atlantic Treaty was being contemplated? To try and answer this we can look at the discussions regarding the role of the Scandinavian countries in the NAT, and what kind of association was sought for them. There is an obvious difficulty in assessing what the Americans were actually striving for, because the word 'association' has two meanings: on the one hand it can designate a connection or relationship, and on the other it can refer to a union or alliance. The same is true of the term 'alignment'.

The content of NSC 28/1 was repeated in an aide-mémoire attached to a letter from Hickerson to Lovett on September 21, 1948, i.e. just over two weeks after the document was approved by President Truman. Here it was said that the United States government would "endeavor to influence Sweden toward eventual alignment with other Western Powers in such a way as may be found collectively acceptable [...]."[72] In other words this was just a repetition of what was said in the policy document. The goal to achieve the "integration into or association in some form", of the Scandinavian countries, "with a North Atlantic security system or the Brussels Treaty system" was also reiterated. Military supplies would first accrue to those nation which "have given indication of intention to cooperate with the United States or Brussels Treaty signatories in security arrangements."[73] I argue that while the Americans opted for 'integration into' with respect to Norway and Denmark, they chose 'association in some form' with respect to Sweden.

It is true that Hugh S. Cumming, Jr. from the American Embassy in Sweden used the word associate interchangeably when referring to all Scandinavian countries in his statements made before the PPS (see Chapter 3, p. 122), but that was in May 1948—i.e. before NSC 28/1 was approved.[74] Moreover, Cumming's focus was more on the apparent inability of Sweden to make a moral distinction between the United States and the Soviet Union.[75] It would seem that there is little evidence of the United States seeking to do anything else than to influence Sweden to associate herself, in the looser sense of the word, more clearly with the West.

72 Aide-Memoire attached to letter from Hickerson to Lovett, dated September 21, 1948, p. 3, in: RG 59; BNA, 1941–1953; SF, 1941–1953; South Africa, Union of, Policy – Sweden, External Political Affairs: U.S.; Box No. 18; (Folder: External Political Affairs: Foreign Relations – U.K.); NARA II, College Park, MD.

73 Ibid., p. 4.

74 "Summary of Remarks by Hugh S. Cumming, Jr. before the Policy Planning Staff of the State Department", dated May 4, 1948, p. 3, in: RG 84; SLE; CONF 1948; 020-844; Box No. 6; (Folder: 800); NARA II, College Park, MD.

75 Ibid., p. 3.

Then what about Ambassador Matthews? Surely one can find plenty of documentary evidence to corroborate the view that the United States was seeking Swedish membership in a Western alliance prior to February 1950? Well, in early 1948 he urged Hickerson to make the Swedes understand that they had to join with the West immediately, no doubt referring to membership of some form of alliance.[76] But again, this is *prior* to NSC 28/1. In light of this it is interesting to view Matthews' comments on that very document which he provided in a letter to Hickerson on September 28, 1948. Matthews did not object to the stated policy, but said that he was "on the whole very pleased with the document."[77] He did, however, state that "I might have made some of the language a little firmer",[78] but apart from that, Matthews seems to have been perfectly satisfied with the new official policy of the United States with regard to Sweden. He even ended his letter (see Chapter 3, p. 137) with a recommendation that someone at the State Department should duly proceed to "implement" the policy entailed in NSC 28/1 when the new Ambassador Boheman came to Washington.[79]

Further support for my argument can be found in the discussion referred to above pertaining to the Hankey Plan and the possibility of some kind of interlocking pact arrangement with the Scandinavian countries. In these discussions it was agreed that the United States and Britain should forget Sweden in the formation of the North Atlantic Pact.[80] It is equally clear that only Norway and Denmark would be actively encouraged to seek pact membership.[81] It is hard to find a clearer indication of American policy at the time immediately after the adoption of NSC 28/1.

Paul Cole has also observed this fact, saying that by January 1949 it was "confirmed that Norway and Denmark were more important than Sweden in U.S. planning for NATO."[82] As also noted by Cole, Sweden

76 Cf. Letter from Matthews to Hickerson, dated March 16, 1948, pp. 1–3, in: RG 84; SWSE; TSGR; 1943–1949; Box No. 1; (Folder: 1948); NARA II, College Park, MD.

77 Letter from Matthews to Hickerson, dated September 28, 1948, p. 1, in: RG 84; SWSE; TSGR; 1943–1949; Box No. 1; (Folder: Ambassador Matthews – Cumming Sweden Vs. NATO); NARA II, College Park, MD.

78 Ibid., p. 1.

79 Ibid., p. 3.

80 Memorandum of Conversation, "British and US Views in Relation to Scandinavian Countries and North Atlantic Pact", dated December 9, 1948, p. 1, in: RG 59; REEUR; SFREDA, 1948–1954; Lot File 55D258; Box No. 2 of 4; (Folder: Scandinavia – General); NARA II, College Park, MD.

81 Letter from Hickerson to Hoyer Millar, dated November 15, 1948, p. 1, in: RG 59; REEUR; SFREDA, 1948–1954; Lot File 55D258; Box No. 2 of 4; (Folder: Scandinavia – General); NARA II, College Park, MD.

82 Cole, P. M., *Neutralité du jour:...*, p. 289.

was subsequently not sounded out regarding membership of NATO in early 1949.[83] If it was U.S. policy at this time to get Sweden to join NATO, then surely such an overture would have been made. Certainly, the hope was still entertained that Sweden would eventually decide to apply for membership,[84] but as we saw above the same thing was expressed much later, in NSC 121 in 1952. But a hope is not the same as state policy. If one would follow such logic, it would still have been a policy goal in 1952, which is something that nobody has ever argued.

My analysis is also confirmed by Acheson's letter to Secretary of Defense James Forrestal on February 9, 1949, where he spelled out the alternatives pertaining to the role of the Scandinavian countries in the event of war to be evaluated by the JCS (see chapter 3). Neither of the two scenarios outlined by Acheson included Swedish membership of the North Atlantic Pact;[85] a clear indication that this was not the objective of U.S. policy at that time. A State Department Policy Statement on Sweden from August 15, 1949, stated unambiguously that "[...] while we recognize the importance of Sweden for our own security and that of our allies, it is against our policy to exert pressure on Sweden to join the North Atlantic Pact."[86] All of this clearly shows that if there was a reversal of American policy with regard to the desirability of Swedish membership in NATO it happened much earlier than February 1950. Moreover, it implies that if it ever was U.S. policy to try and get Sweden into NATO, then it was a very brief endeavor since NATO had only been around since April 1949.

I would instead suggest another explanation of why the American attitude toward Sweden improved in early 1950. To understand this we have to turn our attention to Sweden's progressive consent-giving to American hegemonic policy in Western Europe; most specifically concerning East-West trade. The fact is that the perceived change in general American policy towards Sweden at this time was paralleled by encouraging developments with regard to the trade embargo issue. In December 1949, the Swedish government had issued new regulation pertaining to export of military items. The government had at the same time broadened the definition of such items, something which, in addition to earlier undertakings, according to Birgit Karlsson gave Sweden the "means to formulate an export control policy that for the

83 Ibid., p. 291.

84 Ibid., p. 294.

85 Letter from Acheson to Forrestal, dated February 9, 1949, p. 2, in: RG 59; ReEUR; SFREDA, 1948–1954; Lot File 55D258; Box No. 2 of 4; (Folder: Scandinavia – General); NARA II, College Park, MD.

86 FRUS: *Foreign Relations of the United States*; 1949; Vol. IV; *Western Europe*, p. 775.

moment satisfied the Western Powers",[87] and Hammarskjöld briefed representatives from the State Department and the ECA on the new Swedish regulations.[88]

Jussi M. Hanhimäki has observed that Truman's declaration in late February 1952 that Sweden was eligible for military assistance on a reimbursable basis was connected to the Swedish decision to join the strategic embargo. According to Hanhimäki, the Swedish agreement to participate in export controls was based on a tacit *quid pro quo*; in return the United States made military assistance available to Sweden.[89] Cole, who is Hanhimäki's source for this statement, quotes a document saying that:

> After intensive negotiations by the Embassy, the Swedish Government agreed to adopt east-west trade policy equivalent to that of NATO countries. [...] As a result, U.S. EXCON [Export Control] policy applied to Sweden has been placed on the same basis as for NATO countries [...]. In principle Sweden also regarded by U.S. as eligible for priority assistance on defense materials *within certain defined limits*.[90] [Italics added.]

The italicized phrase was of importance for the Swedish guided missile program, because such material was *not* included in the package. However, as I will show later, this provision was not something applying only to the Swedes and thus had nothing to do with the official policy of neutrality. I argue that getting Sweden to join a Western alliance stopped being a policy objective for the Americans no later than August 1948, as is made evident in NSC 28/1. It was thus not so much the American policy that changed in February 1950, and which then made possible extended collaboration on a whole range of issues, but Swedish unilateral measures that convinced the U.S. that Sweden had consented to American hegemony. The trade embargo was a crucial part of the American hegemonic project, and increasing abidance by this hegemonic apparatus resulted in a more relaxed relationship between the United States and Sweden. The reason that scholars have not devoted much

87 Karlsson, B., *Handelspolitik eller politisk handling...*, p. 172. "medel att utforma en exportkontroll-politik, som för tillfället tillfredsställde västmakterna."

88 Ibid., pp. 172–173.

89 Hanhimäki, J. M., *Scandinavia and the United States:...*, p. 48.

90 Cole, P. M., *Neutralité du jour:...*, p. 352. Cole's source is a telegram from the Embassy in Stockholm to the State Department dated December 28, 1951; this led Hanhimäki to conclude that it was not until then that Sweden actually agreed to participate in the trade embargo (Hanhimäki, J. M., *Scandinavia and the United States:...*, p. 48). This is incorrect. The telegram most likely refers to the agreement signed little more than six months earlier.

attention to this issue is thus perhaps that there was no reason to do so—there was in fact no change in U.S. policy. It is true that there was a clear change in the U.S. *attitude* towards Sweden, but that is another matter completely since this was only a consequence of the change in Sweden's policy.

Silva has specifically referred to Perkins as being a part of a group of new officials within the State Department that took over the handling of issues regarding Sweden. This group had a more positive experience of Sweden, writes Silva, and Hickerson and Thompson were phased out. This was supposedly part of a large reorganization of the State Department and as a consequence decisions relating to Sweden were removed from EUR and instead handled by the new Office of British, Commonwealth and Northern European Affairs (henceforth: BNA). Perkins had also been the one to suggest a review of NSC 28/1 in February 1950.[91]

However, it is doubtful that what Perkins had in mind at this point was a reevaluation of U.S. policy vis-à-vis Sweden. This becomes apparent from Perkins' own wording, quoted by Silva, when recommending a review of the document. Perkins wrote: "'We doubt that any amendment to the basic policy set forth in NSC 28/1 is necessary [...].'"[92] True, a new policy document was eventually produced, but not until 1952 when NSC 121 was approved by the President. Moreover, the reorganization within the State Department referred to by Silva had begun more than a year earlier, in January 1949.[93] This too makes one doubt the correctness of regarding this particular event as the decisive moment.

Thus it seems untenable to regard Perkins' call for a review in order to achieve an agreed interpretation of the existing policy in February 1950 as the point when the State Department turned around and started to view Sweden more favorably. It is more likely that it was the compounding effect of Swedish actions (i.e. a change in Sweden's behavior) that produced the perceived turnabout on the part of the United States. As we shall see later, it is only if one connects the issue of military deliveries to Swedish overall consent to American hegemony, symbolized by formal abidance by the COCOM trade embargo, that one can understand the alleged 'change' in U.S. attitude toward Sweden.

The reason why Moores and Silva must insist on February 1950 as a line of demarcation in U.S.-Swedish relations is that without the hegemony perspective they fail to see the connection between Swedish tacit adherence to the trade embargo and the improving relationship between the two governments. Moreover, their insistence upon a change

91 Silva, C., *Keep Them Strong, Keep Them Friendly:...*, p. 281.

92 Ibid., p. 281.

93 Ibid., p. 281.

in American policy objectives follows automatically from the fact that Sweden did not join NATO. Thus, if Sweden did not change its policy, then the United States must have, since relations took a turn for the better. Once hegemonic theory is applied as a lens through which this relationship can be interpreted, it becomes evident that from the fall of 1948 there were no major changes made in American policy objectives with regard to Sweden. The direct connection between the improved U.S. attitude toward Sweden, and Sweden's adherence to the trade embargo, was alluded to by Perkins in April 1952, when he wrote:

> [...] the cooperation thus far obtained from the Swedish Government in the matter of East-West trade. Mr. Butterworth cautions, however, that this should be done in such a way as to encourage continued cooperation which, after all, is in Sweden's best interest and, incidentally, has resulted in a more favorable attitude on our part in the matter of export licenses and allocations to Sweden.[94]

The relationship between the Marshall Plan, the MDAP, and the COCOM trade embargo was illustrated in a memorandum on June 19, 1950. This document also plainly states that the Americans were conscious of how American hegemony should be implemented in order for the policy not to be opposed by the United States' allies. The memorandum, written by Marcy M. Dupre, Jr., for Henry J. Tasca, Averell Harriman's Special Assistant for Staff Planning at the Office of the Special Representative to the ECA in Europe, was intended "as a possible basis for discussions on this subject which may arise from time to time, especially during [Tasca's] current mission to Europe." It deserves an extensive quotation:

> In the interest of world peace, as well as its own security and that of the non-Soviet community, the United States has embarked upon two comprehensive and extensive programs of assistance to the countries of Western Europe. These are the European Recovery Program and, under the aegis of the North Atlantic Treaty, the Mutual Defense Assistance Program (MDAP). As an additional security measure, the United States commenced some two years ago to severely restrict the flow of strategic commodities to the Soviet Bloc and since that time has been endeavoring to persuade the countries participating in the Marshall Plan to apply similar parallel controls. *These efforts*

94 758.13/4-1052, Letter from Perkins to Acheson, dated April 10, 1952, pp. 1–2, in: RG 59; Decimal File [henceforth: DF] 1950–54; from 758.13/1-852 to 758.5MSP/12-1554; Box No. 3773; NARA II, College Park, MD. For the same, see: 758.13/4-152, Letter from Butterworth to Matthews, dated April 1, 1952, p. 1, in: RG 59; DF 1950–54; from 758.13/1-852 to 758.5MSP/12-1554; Box No. 3773; NARA II, College Park, MD.

> *have been partially successful.* [...] *Experience in the implementation of these three programs has revealed clearly the strong impact which each of these activities has on the other, and the consequent necessity for close integration and coordination of effort with respect to the whole.* Unquestionably this necessity will deepen as time goes on and the activation of the whole becomes a reality under what might be termed the total diplomacy concept. [...] Ambassador V. Averell Harriman's new appointment involves assisting the President in the coordination of all these various elements. [...] It seems probable that a great contribution in this respect could be made through General [Omar] Bradley's Standing Group in Washington, *especially in scanning and screening carefully the entire program of Military Assistance* [...] *with respect to the East-West trade problem and the European Recovery Program in addition to its relationship to their main function—the strategic implications.*[95] [Italics added.]

After this very frank summary of the connection between the Marshall Plan, the MDAP, and the East-West export controls, and the importance of a close integration of these programs in the future, Dupre proceeded to issue some recommendations for American policy toward the Western European countries:

> It appears also essential that the United States exert a more forceful and dynamic leadership in the planning area than is now being done. This is not a criticism of the personalities involved for they are loyally conforming to our national policy of *avoiding the appearance* of dominating the European countries. Our policy of *avoiding the appearance* of dominating the European countries appears wise. Therefore, the question is posed as to whether we can step up the tempo of planning by exerting a more vigorous leadership in the field and still *avoid the appearance* of domination.[96] [Italics added.]

It is perhaps hard to find any document that more clearly expresses the awareness amongst the contemporary American actors of the hegemonic character of the United States' relationship with the Western European countries. Hegemonic leadership is indeed all about dominating while at the same time avoiding the appearance of dominating. That is how consent is enticed from other nations—if these countries felt that they

95 Memorandum from Dupre to Tasca, dated June 19, 1950, pp. 1–3, in: RG 469, 1948–1961; SRE (1948–53); Office of the Special Assistant for Staff Planning; Subject Files of Henry J. Tosca [sic: Tasca, the name is misspelled on the box], Special Assistant, 1949–52; Box No. 1; NARA II, College Park, MD.

96 Ibid., p. 3.

were being dominated by an outside force then their consent would not be forthcoming.

Dupre thus duly recognized that it was the policy of the United States to tacitly dominate the European countries; but at the same time the American official also saw a need for a more forceful leadership in order to more fully realize the strategic goal of the three inter-related programs. There were a great deal of resistance to the American export control regulations among the European countries, especially among those who were dependent on imports from the Eastern Bloc and stood to lose financially from the embargo. Dupre apparently felt that this resistance was threatening to overturn the overall aim of American hegemony, perhaps even in the end threatening American hegemony as such.

It is worth pointing out that the memorandum was written before the outbreak of hostilities in Korea, and that Dupre seemed anxious to make sure that these considerations would figure in the minds of Tasca (and Harriman) as they planned and executed ECA policy in Europe. This points to the fact that there was a feeling brewing amongst the American officers in Europe that the United States was perhaps losing a part of its political momentum. From this it is also evident that Sweden's decision to partake in the strategic embargo, and to implement the same policy as the COCOM countries, was something that was of great importance to the United States at this point. It contributed significantly to the ability of the United States to fulfill the strategic implications of official U.S. policy.

It is now even easier to understand how this decision could be a catalyst for granting Sweden access to military supplies on a reimbursable basis. The three American assistance programs converged at this crucial juncture, the leverage that the United States had on Sweden through the Marshall Plan was limited by the fact that Sweden received only a small amount of loans (and no grant aid). This pushed to the fore the export control regulations since with these the Americans controlled Sweden's access to military technology, as well as dual-use technology. Swedish participation in the MDAA was unthinkable without prior reassurance that these items would not end up in the hands of the enemies of the United States.

But at the same time, as we know from hegemonic theory, the United States needed its allies and had to make sure that material rewards would be available once consent to American leadership had been given. Hence, even though the MDAP was really a program designed to rearm the NATO countries, Sweden was made eligible to procure defense material on the same basis as the NATO allies within the framework of that program.

This process was in itself a fascinating display of the flexibility of American hegemony, something that will be more thoroughly probed later in this thesis.

The Final Consent to U.S. Hegemony

However, by 1951 it had become clear to U.S. officials that Sweden's improvised way of complying with the embargo was utterly impractical. The American Ambassador to Stockholm Walton Butterworth contacted Dag Hammarskjöld early that year, and informed him that the United States was no longer satisfied with the Swedish case-by-case insurance—the Americans now wanted a general guarantee that strategic commodities would not be re-exported to the Eastern Bloc. Hammarskjöld could not agree to formal multilateral discussions, but he was willing to discuss on a bilateral basis. Sweden did indeed enact regulations concerning their foreign trade policy's most sensitive areas. Out of these regulations it was possible to infer that most of the commodities that the Americans had been apprehensive about were no longer to be exported to the Eastern Bloc.[97]

On June 15, 1951, Hammarskjöld and Butterworth signed the so-called Stockholm Agreement [Stockholmsöverenskommelsen], under which Sweden would, if possible, avoid exporting listed commodities to the East. The American embassy would be kept posted about Swedish undertakings on the subject. Birgit Karlsson has argued that this indicates that the United States got Swedish cooperation, if not to the letter, then in the spirit of the American wishes. For the United States *the content* was what mattered while to the Swedes the important thing was *the packaging* of the content.[98] The United States concluded an identical

97 Karlsson, B., *Handelspolitik eller politisk handling…*, pp. 176–177.

98 Ibid., pp. 178–179. For extensive information about the agreement: Ahlström, Christer, *Stockholmsöverenskommelsen rörande den svenska öst-väst-handeln – ett folkrättsligt bindande avtal?* [The Stockholm Agreement Concerning the Swedish East-West Trade] (Stockholm: Utrikespolitiska institutet, 1996). The deal made with the Americans regarding COCOM's export control was similar to an agreement made in the fall of 1945. Gunnar Skogmar has shown how, in the summer of that year, the Americans had concluded that Sweden's uranium assets were among the third or fourth largest in the world, and the only one of any significance in Western Europe. On top of that, the Swedish ones were the only of its kind not under the control of the United States or Britain. Negotations were initiated with the Swedish government to convince it to let the Americans and the British control the Swedish uranium. In the end, Sweden promised not to export uranium, and to inform the U.S. and British governments of any requests by foreign powers for uranium-bearing materials in Sweden. As in the COCOM-case, the Swedish government had rejected American demands for more formal arrangements, e.g. that Sweden could not grant the United States and Britain the right to veto Swedish export of uranium or guarantee them the right to purchase the uranium if the Swedes were ever to decide to export it. However, by 1948 the issue had lost much of its significance, and did not to an

agreement with Switzerland, the so-called Hotz-Linder agreement, the same month. The Swiss government had been 'persuaded' to conclude the agreement subsequent to the U.S. blocking 40 import demands from Swiss companies.[99]

American legislation was strengthened in a series of steps during 1951, the last of which was the Battle Act. The official name of the Battle Act was The Mutual Defense Assistance Control Act of 1951, and Congress approved it on October 26, 1951. The Battle Act was in turn tied to the Mutual Defense Assistance Act of 1949, and was the most important legislative document connecting the American foreign aid program with specific demands upon the aid-receiving nations to conform to American regulations in their external commerce with Communist nations.[100] It stipulated that the aid to any nation that exported strategic commodities included in the COCOM NO. 1 list[101] to the Eastern Bloc be cut off immediately.[102] The United States could do this only because of its hegemonic position, since this domestic legislation would have been useless if the Western European nations had not been so dependent on the American economy. But which were the implications of the Battle Act for Sweden? Well, not much at the time it was passed, because Sweden was not yet eligible for military assistance in the eyes of the Americans, although this was about to change as a corollary to the signing of the Stockholm Agreement.

Swedish Guided Missile Development in the Wake of COCOM and Korea

The outbreak of the Korean War in June 1950 had consequences for the Swedish guided missile development effort, because of the tightened American foreign trade regulations that followed it.[103] The pressure

appreciable degree affect the U.S.-Swedish relationship (Skogmar, Gunnar, *De nya malmfälten. Det svenska uranet och inledningen till efterkrigstidens neutralitetspolitik* (Stockholm: Forskningsprogrammet *Sverige under det kalla kriget Arbetsrapport nr 3*, 1997)).

99 Ammann, K., "Swiss Trade with…" in J. Eloranta and J. Ojala (eds.), *East-West Trade and the Cold War*, p. 126.

100 Adler-Karlsson, G., *Western Economic Warfare…*, p. 28; see also: Kaplan, Lawrence S., *A Community of Interests: NATO and the Military Assistance Program, 1948–1951* (1980), p. 161. The Battle Act got its name from Congressman Laurie C. Battle from Alabama who chaired the House Committee on Foreign Affairs which presented the bill to Congress.

101 There were three COCOM-lists: list NO. 1 dealt with totally embargoed goods; list NO. 2 handled quantitatively embargoed goods and list NO. 3 contained goods to be kept under observation only.

102 Karlsson, B., *Handelspolitik eller politisk handling…*, p. 167.

103 "NR 8/1950" ,"Protokoll från FRR sammanträde den 15. december 1950.", in KFF, FRR, Serie: A I, "Protokoll från sammanträden 1948–1957", År: 1948–1957,

applied by the United States on Western Europe to adopt trade restrictions against the Communist nations "multiplied in strength" after the outbreak of the war,[104] when the United States moved to consolidate its hegemonic status versus its allies.

As I showed in Chapter 3,[105] the Americans kept a close eye on guided missile developments in Sweden, and an evaluation of the American Air Attaché Colonel Donald L. Hardy's reporting to Washington from the fall of 1949 concluded that the intelligence "[r]eports concerning the RSAF have in general been excellent to include those covering the subject of guided missiles."[106] The official government attitude toward Hardy was, according to him, "unusually cordial", and he made the observation that "[n]o unusual courtesies were received during the current month from the members of any government other than the Swedish."[107]

In September Colonel Hardy and a representative group from the U.S. House of Representatives, the Interstate and Foreign Commerce Committee, were authorized by the Swedish government to visit aircraft and guided missile manufacturer SAAB in Linköping. The company offered the group demonstrations of the plant's activities, assisted by SAAB's "top managerial and operational personnel". What really impressed Hardy, however, was that the Swedish government granted the group the unprecedented permission to fly to Linköping and land at the factory airfield. "This was an unusual courtesy," wrote Hardy, since to his knowledge, no other landings by foreign military aircraft had been permitted at this field during his over two year long stay in Sweden.[108] It was obvious that the Swedes wanted to keep up a good relationship

---

VOL. 1, KrA, p. 1. For information about Sweden prior to the Korean War, see: Mörth, Ulrika, *Sverige och Koreakriget: ett förkrigsskede* (Stockholm: Försvarsdepartementet, 1988); For domestic debate on the issue, see: Holmström, Barry, *Koreakriget i svensk* (Stockholm: Rabén & Sjögren, 1972); Björklund, F., *Samförstånd under oenighet…*; Bjereld, Ulf and Demker, Marie, *Utrikespolitiken som slagfält. De svenska partierna och utrikesfrågorna* (Stockholm: Nerenius & Santérus, 1995); for a summary of the domestic debate, see: Ottoson, Sten, *Den (o)moraliska neutraliteten…*, pp. 155–161; Sweden's official position on foreign policy during the Korean War is dealt with in: Lödén, Hans, *"För säkerhets skull"…*, pp. 78–82.

104 Adler-Karlsson, G., *Western Economic Warfare…*, p. 26.

105 See above, pp. 113–115.

106 Letter to Hardy, dated August 2, 1949, p. 3, in: RG 341; DCSO; DI; DDCD; COD; AAB; StF, 1948–49; Spain to Sweden; Box No. 67; (Folder: Sweden-Miscellaneous); NARA II, College Park, MD.

107 Letter from Hardy to Director of Intelligence, USAF, dated October 31, 1949, p. 2, in: RG 341; DCSO; DI; DDCD; COD; AAB; StF, 1948–49; Spain to Sweden; Box No. 67; (Folder: Sweden-Miscellaneous); NARA II, College Park, MD.

108 Letter from Hardy to Director of Intelligence, USAF, dated September 30, 1949, in: RG 341; DCSO; DI; DDCD; COD; AAB; StF, 1948–49; Spain to Sweden; Box No. 67; (Folder: Sweden-Miscellaneous); NARA II, College Park, MD.

with the USAF. While there was certainly more than one reason for that, the possibility of getting information for the guided missile program probably figured among them.

The Swedish openness toward Hardy ricocheted back at him at one occasion, however. In the summer of 1949 Hardy had a very interesting and revealing conversation with General Nordenskiöld, then the Chief of the Swedish Air Staff. Nordenskiöld told Hardy that he had spoken to Defense Minister Vougt and Foreign Minister Undén about the necessity of allowing American overflights in time of war, and that he had gotten their express agreement to such overflights. In fact, not only had Undén and Vougt said that such violations of neutrality would be allowed but, according to Nordenskiöld, they had said that Sweden would encourage such flights. Hardy reported:

> He [Nordenskiöld] hastened to assure that Swedish interference with such flights was utterly inconceivable. Public statements by Defense Minister Vough [Vougt] which might be interpreted to the contrary were merely political eyewash. He Nordenskiold, had shown to Vought the absolute necessity for such overflights by our aircraft and Vought had agreed that Sweden would not only permit but encourage such flights. Nordenskiold had further within the past month discussed the same problem with the Swedish Foreign Minister Unden [Undén], who had likewise agreed to Swedish noninterference with American overflight. Nordenskiold had then urged Unden to so inform our Secretary of State or other high State Department representative on an informal basis that such was the case. Unden had replied that much as he would like to, he could not bring himself to utter any statement incompatible with his official stand on neutrality.[109]

This is of course extraordinary information, if it was true. Nordenskiöld is known to have been very friendly to the Western Powers, and it is no secret that he did not approve of the Swedish government's policies at the time. The question is how far the General would go in trying to convince the Americans that Sweden was safe, and that supplying equipment to the RSAF would not constitute a threat to U.S. interests. In other words, how much was he prepared to stretch the truth in order to secure the survival of the RSAF in wartime? There is no way to provide a satisfactory answer to this question.

109 Memorandum of Conversation between Hardy and Nordenskiöld, dated July 8, 1948, p. 1, in: RG 59; BNA, 1941–1953; SF, 1941–1953; South Africa, Union of, Policy – Sweden, External Political Affairs: U.S.; Box No. 18; NARA II, College Park, MD. For the same, see also: Weinberger, H., "The Neutrality Flagpole:..." in M. Thad Allen and G. Hecht (eds), *Technologies of Power:...*, pp. 329–330.

Hardy also noted that:

> [t]he Swedish military hierarchy sense growing indifference to Sweden's welfare on the part of the U.S. This is desirable; otherwise the Swedes will convince themselves an attempt to impress the other Scandinavian powers that adherence to the Atlantic Pact is an unnecessary assumption of obligations.[110]

The feeling purveyed in this statement is indeed not that the Swedes were entirely wrong in assuming that Sweden's welfare was important to the U.S., but only that awareness of this fact would lessen the leverage held by the Americans in their interaction with the Swedes. But even though Nordenskiöld had been very reassuring, he cautioned against drawing too large conclusions from Nordenskiöld's statements, but he does not seem to have been sceptical toward them. Hardy reminded the people in Washington that:

> [d]espite Nordenskiolds' [sic] assurance, we must not forget that in the event of American overflight of Sweden, it will be the Swedish in power at that time that will decide what action to take. [...] The Swedish military may be expected to exert themselves, particularly through military channels and through the good offices of their Scandinavian neighbors, to press the issue that from a military standpoint, Sweden is a logical, desirable, and necessary partner in combined but unofficial planning to counteract the moves of the USSR.[111]

Hardy's conversation with Nordenskiöld created a considerable stir on the American side. General Norstad reacted strongly to this, as he saw it, unilateral initiative on the part of Colonel Hardy, and he thought any such discussion by any of the U.S. attachés with any official of any foreign country ill advised.[112] It was apparently felt that in the conversation regarding possible Swedish reactions to overflights by the USAF:

> [the] cable indicated that our Air Attache exerted a degree of pressure

110 Memorandum of Conversation between Hardy and Nordenskiöld, dated July 8, 1948, p. 1, in: RG 59; BNA, 1941–1953; SF, 1941–1953; South Africa, Union of, Policy – Sweden, External Political Affairs: U.S.; Box No. 18; NARA II, College Park, MD.

111 Ibid., p. 2.

112 Memorandum for Major General Charles Cabell, dated July 23, 1949, in: RG 341; DCSO; DI; DDCD; COD; AAB; StF, 1948–49; Spain to Sweden; Box No. 67; (Folder: Sweden-Miscellaneous); NARA II, College Park, MD.

> to induce what he thought would be favorable Swedish action, and that he may have initiated that discussions.[113]

The latter point was what upset Norstad the most; no instructions had been issued to Hardy to initiate any conversations on this very sensitive subject. Cables were subsequently sent on August 2 to the Air Attachés in Oslo, Copenhagen, Stockholm, and to the Supreme Commander of the USAF in Europe (CINCUSAFE), instructing them not to engage in discussions on this subject with foreign nationals.[114]

Hardy responded to these worried voices of opposition on August 25, and said that the whole matter was due to a misunderstanding of his first report on the subject. He wrote:

> I regret that the inadequate wording of the opening statement of my message [...] has caused concern. On re-reading it, I can understand how the implication that I had initiated a discussion of this nature could come about. However, in the preparation of the message, I took it thoroughly for granted that such action on my part would not be construed.[115]

The American Air Attaché continued by saying that the subject of overflights had come up repeatedly in his conversations with Nordenskiöld. The Swede had raised this issue in order to seek Hardy's reaction, and Nordenskiöld was apparently concerned about the American's unwillingness to engage in a discussion regarding this matter. According to Hardy, it was Nordenskiöld's "desire that the United States Air Force would precisely know his own position in the matter, and the effort he had made to pry the Swedish Foreign Office away from its firm neutrality stand in at least some commitment as to a possible future action."

Hardy then pointed to an important fact regarding Sweden's possible actions in he event of war, and said that "Swedish reaction to actual overflight would be a decision for the Swedish government in power at the time, *no matter what assurances we were given now*."[116] [Italics added.] This point has bearing upon this thesis in so far as it further

113 Memorandum for the Record, dated August 11, 1949, in: RG 341; DCSO; DI; DDCD; COD; AAB; StF, 1948–49; Spain to Sweden; Box No. 67; (Folder: Sweden-Miscellaneous); NARA II, College Park, MD.

114 Ibid.

115 Letter from Hardy to Director of Intelligence, USAF, dated August 25, 1949, p. 1, in: RG 341; DCSO; DI; DDCD; COD; AAB; StF, 1948–49; Spain to Sweden; Box No. 67; (Folder: Sweden-Miscellaneous); NARA II, College Park, MD.

116 Ibid., p. 1. Hardy regarded the Swedish hope of receiving American arms without a commitment to the West as being nothing but "wishful thinking".

underlines that this issue must reasonably have been satisfactorily resolved by the end of the 1950s, when Sweden was allowed to purchase guided missiles from the United States.

At USAF Headquarters in Washington they were gratified to hear that Hardy had not initiated the discussion, and advised him rather cryptically:

> We are confident that you will handle this subject diplomatically and wish to reemphasize the fact that when Swedish officials mention this subject it is quite proper for you to give immediate verbal reply along lines that would contribute to USAF interests, but that you must be very careful to avoid any implications that we wish to press any particular Swedish action.[117]

Dovetailing this controversy and the question regarding American overflights was an intelligence report sent to Washington in November 1949. The report was a questionnaire on the present and future status of radar in the Swedish air defense system. It had been collected by the Air Technical Liaison Officer Lieutenant Colonel Florian A. Holm of the American Embassy, and had been drawn up for the Americans by 'the Reporting Officer' to the Operations Staff of the RSAF in October. The answers had been supplied by members of the Operations Staff of the RSAF.[118] It is uncertain who the Reporting Officer was, but it reasonable to believe that he was an American national tied to the U.S. Embassy in Stockholm.

The questions ranged from what type of sonic and visual warning system was in operation, what type of communications were used, location of early warning radar and type of equipment, what kind of airborne intercept radar the Swedes had, what type of Identification System (IFF) were used, to questions concerning maintenance, training and availability of spare parts.[119] The RSAF had been very candid in their answers, letting the Americans know that the Swedish early warning system covered the coastline and consisted of a chain of British ER-IIIB stations, similar to the American AMES-6. The Swedes were planning on purchasing further stations from Britain and the United States, e.g. a modified version of the American AN/TPS-1A, as well as AMES-13 and

117 Letter from Department of the Air Force HQ in Washington to Hardy, undated, in: RG 341; DCSO; DI; DDCD; COD; AAB; StF, 1948–49; Spain to Sweden; Box No. 67; (Folder: Sweden-Miscellaneous); NARA II, College Park, MD.

118 "Report on Present and Future Status of Swedish Radar in Air Defense Systems", dated November 2, 1949, p. 1, in: RG 341; DCSO; DI; DDCD; COD; AAB; StF, 1948–49; Spain to Sweden; Box No. 67; (Folder: Sweden-Miscellaneous); NARA II, College Park, MD.

119 Ibid., pp. 2–4.

AMES-14 from the Bendix Radio Corporation. The Americans also wanted to know where the plotting and filter rooms were located, and if there were any associated airfields or anti-aircraft defense, and how the control of air defense and warning organizations was organized.[120]

To this the answer from the Swedes was that "[o]ur systems are exactly the same as the British system", and stated that particular answers to these inquiries would be offered at a later date.[121] With regard to bottlenecks in the system the Swedes had listed four major areas where help was needed to improve the organization. The defense efforts were hampered by "severe lack of competent instructors"; adequate training aids were needed "in order to relieve the very urgent need for better trained operators and maintenance personnel; it was further stated that the Swedish night fighting officers (both staff officers and pilots) needed to go to Britain for training because "they need complete education in proper night fighting tactics."[122]

As we saw in Chapter 3,[123] Hankey had explained to Gunnar Hägglöf, the Swedish Ambassador in London, that such education was quite impossible as long as the British were unsure whether or not the pilots would be used against British aircraft overflying Sweden. Perhaps the RSAF was hoping for American support in this issue to influence the British to allow this training to take place. The number one bottleneck factor that the Swedes had listed was not a personnel issue, however, but one of technology. According to the report, it was of imperative importance:

> [t]o realize actual deliveries of important radar on order from abroad and especially those units ordered from the U.S., as our whole system centers around the TPS-1A deliveries as soon as possible.[124]

This alluded to the problems the Swedes had had with deliveries of radar sets ordered from the Bendix Radio Corporation, i.e. the so-called Bendix stations, which I will deal with below. The Swedes felt that the American questions were of a general character, and not necessarily connected to an American decision regarding the Bendix stations. But there was a twist to this inquiry. The American Military Attaché in Stockholm had flown down to Wiesbaden in June, at which time

120 Ibid., pp. 2–3.

121 Ibid., p. 3.

122 Ibid., p. 4.

123 For this, see p. 152.

124 "Report on Present and Future Status of Swedish Radar in Air Defense Systems", dated November 2, 1949, p. 4, in: RG 341; DCSO; DI; DDCD; COD; AAB; StF, 1948–49; Spain to Sweden; Box No. 67; (Folder: Sweden-Miscellaneous); NARA II, College Park, MD.

representatives of the USAF told him that because of the possibility that the RSAF would be used against American bombers overflying Sweden in wartime, the USAF did not see how it could be to their advantage to grant export licenses for American radars and thereby increase the effectiveness of the RSAF. This said, however, it was also stated that "the positioning of the Swedish radar stations could possibly have a positive effect on the export licensing matter."[125]

This seemed to imply that if the Swedish air defense was organized in a way which benefited the Americans, e.g. with the radar coverage aimed towards the Baltic, the USAF might revoke its refusal. The fact that U.S. aircraft would fly over Swedish territory in wartime was thus well known to the Swedish government for it was recounted in a Foreign Ministry transcript of a memorandum originating with the Swedish military. Colonel Holm had, however, assured Westring that the questionnaire that he had given the Swedes had not been prepared by the USAF Headquarters in Washington but by the Air Materiel Command, Wright Field, who had put the same questions to a great number of countries at the same time.[126] Lagerfelt at the Foreign Ministry chose to believe Holm's version, and commented on the American interest in Swedish radar stations that it did not have "the serious implications that could have been expected, but should rather be regarded as being part of a routine gathering of information."[127]

The Korean Divide

The American attitude towards Sweden and Swedish companies also got much harsher following the outbreak of the Korean War. The Americans felt that the Swedish government had not yet completely complied with the COCOM trade regulations[128]—i.e. the Swedes had not given their unambiguous consent to American hegemony. Captain Oxenstierna argued that it was vital, for Swedish guided missile development, that

125 "P.M. angående amerikansk inställning i vissa militära frågor berörande Sverige", dated October 19, 1949, p. 3, in: HP 24 VI/U.S.A; "Tillstånd till in- och utförsel av vapen, ammunition och krigsmateriel"; 1946, okt.– aug. 1950; Folder 20; Box No. 1353; År: 1946–1952; RIA, Stockholm. "grupperingen av de svenska radarstationerna kunde påverka lösningen av exportlicensfrågan i förmånlig riktning."

126 "PM nr 2 angående Bendix radar m.m.", dated October 1, 1949, in: HP 24 VI/U.S.A; "Tillstånd till in- och utförsel av vapen, ammunition och krigsmateriel"; 1946, okt.– aug. 1950; Folder 20; Box No. 1353; År: 1946–1952; RIA, Stockholm.

127 Letter from Lagerfelt to Hägglöf in Washington, dated October 21, 1949, p. 1, in: HP 24 VI/U.S.A; "Tillstånd till in- och utförsel av vapen, ammunition och krigsmateriel"; 1946, okt.– aug. 1950; Folder 20; Box No. 1353; År: 1946–1952; RIA, Stockholm. "icke haft den allvarliga betydelse, som man kunnat frukta, utan snarast voro att betrakta som del av ett rutinmässigt uppgiftsinhämtande."

128 Karlsson, B., *Handelspolitik eller politisk handling…*, p. 183.

emergency purchases [beredskapsköp] of certain components be made, in order to ensure that the ongoing guided missile tests could be continued; Sweden's ability to produce operational guided missiles was in the balance.[129]

The Americans, however, had no intention of granting Sweden access to classified material, especially information pertaining to guided missiles. In March 1950 a group of three Swedish scientists planned a two-month trip to the U.S. to visit numerous military establishments. The group was interested in matters relating to chemical and biological warfare, radiation protection measures, and guided missiles. It was agreed on the American side that the 'usual' formula should be applied, i.e. "admit them to material no more highly classified than restricted."[130]

In the absence of governmental consent to American hegemony it was obviously seen as vital to get as much as possible from the Americans before the curtain closed. The situation demanded an intensified research and testing activity, and use of the opportunities that the war presented of following the foreign development of counter-measures against new weapons.[131] In April 1950 the American Air Attaché submitted an intelligence report containing all available information regarding the Swedish guided missile development program to Washington. The information had been related to the Americans by personal contacts (it is not clear if they were Swedes or Americans) of some of the officers of the Office of the Air Attaché at the American Embassy in Stockholm.[132]

The report supplied as detailed information as possible concerning air-to-surface missiles RB 301, RB 302, RB 303; surface-to-surface missiles RB 310, RB 311, RB 312, RB 315; air-to-air missile RB 321; and ground-to-air missile RB 322. It was stated that the emphasis was presently being applied to the surface-to-surface missiles but that this might change in the future.[133] Captain Oxenstierna had 74 people from all the services under his direction, the report said, and it continued to list all the testing, manufacture and development facilities that Sweden had

129 "NR 8/1950" ,"Protokoll från FRR sammanträde den 15. december 1950.", in KFF, FRR, Serie: A I, "Protokoll från sammanträden 1948–1957", År: 1948–1957, VOL. 1, KrA, p. 1.

130 Memorandum from Caspar D. Green, dated March 8, 1950, in: RG 59; BNA, 1941–1953; SF, 1941–1953; South Africa, Union of, Policy – Sweden, External Political Affairs: U.S.; Box No. 18; (Folder: Administrative: Visitors); NARA II, College Park, MD.

131 "NR 2/1951" ,"Protokoll från FRR sammanträde den 6. februari 1951.", in KFF, FRR, Serie: A I, "Protokoll från sammanträden 1948–1957", År: 1948–1957, VOL. 1, KrA, p. 1.

132 "ID 662497" from the Air Attaché in Stockholm to Washington, dated April 6, 1950, pp. 1–2, in: RG 319 AIDF; From ID 662461 To ID 662710; Box No. 4268; NARA II, College Park, MD.

133 Ibid., p. 2.

at its disposal.[134] In its entirety, the document contained 15 pages, but the reporting USAF Officers Air Technical Liaison Officer Lieutenant Colonel Florian A. Holm and Assistant Air Attaché Major H. B. Walsh offered no analysis of their own regarding the Swedish development.

It seems as if the United States was engaged in an effort to acquire more technical intelligence on foreign guided missile development at this time. The American Committee on Guided Missiles stated in December 1950 that it "in planning its guided missile program is being handicapped by insufficient technical intelligence information." This was especially true with regard to the Soviet Union.[135]

The Swedish development efforts were eventually concentrated on three projects—namely air-to-surface guided missile RB 304, the ship-based anti-ship guided missile RB 315, and the air-to-air guided missile RB 321. The RB 321 guided missile, intended for the RSAF, had the highest priority of them all.[136] Despite the omnipresent troubles haunting Swedish guided missile development the Swedes, at least until the mid-1950s, still expected it to render operational guided missiles, although purchasing from abroad was considered as well. The Swedes were very eager to obtain technical information on guided missiles from the United States, and tried to induce the Americans to give out some information in exchange for Swedish material. In the summer of 1951 the Director of the Research Department of the Swedish Armed Forces Colonel Schmidt approached the Americans and said that he could offer them all patent descriptions and test results involving a newly developed recoilless grenade rifle. In return, the Swedes wanted "such information as the United States government is willing to release concerning rocket launchers and guided missiles, particularly in the field of anti-aircraft."[137]

In further conversations with U.S. Army Attaché in Stockholm Lieutenant Colonel Roy W. Cole, Jr., Schmidt almost begged the Americans to give Sweden "at least a method of approach to the problem

134 Ibid., pp. 2–3.

135 "GM 29/4, Technical Intelligence regarding Foreign Guided Missiles Development.", dated December 20, 1950, in: RG 156; EOHB; AGMP, 1940–62; GM 29/Foreign Intelligence through GM 78 Consolidation Navy AAM Pro; Box No. 12; NARA II, College Park, MD.

136 "NR 7/1951", Protokoll från FRR sammanträde den 5 november 1951.", in KFF, FRR, Serie: A I, "Protokoll från sammanträden 1948–1957", År: 1948–1957, VOL. 1, KrA, pp. 1f.

137 Letter from Army Attaché in Stockholm to the Assistant Chief of Staff, G-2, U.S. Army, dated July 26, 1951, p. 1, in: RG 319 Army-Intelligence Project Decimal File [henceforth: AIPDF] 1951–1952; From 335.11 Sweden to 676.3 Sweden, Horby, Box No. 179; (Folder: 472.-474.6 Sweden); NARA II, College Park, MD.

of designing and producing a satisfactory anti-aircraft guided missile."[138] Cole thought that the Swedish grenade rifle warranted the United States sending a qualified specialist to Sweden to test-fire the weapon. He also recommended that technical data of low classification regarding field artillery rocket launchers should be provided to Schmidt, and perhaps because he took pity on the Swedish Colonel, suggested that:

> [c]onsideration be given to releasing information which will permit Swedish guided missile anti-aircraft research to crystallize and facilitate the development phase. Should this last recommendation be considered favorably, it is believed that best channel to initiate same will be through Captain Bo Westin, who will shortly arrive in Washington as Assistant Army Attaché. Captain Westin is currently Vice-Chief of the Swedish guided missile program and is highly qualified.[139]

Although the Office of the Chief of Ordnance agreed to the suggestion that the grenade rifle should be investigated, it could not concur with the barter trade recommended:

> The United States is interested as a possible customer for the weapon; however, the exchange of rocket and/or guided missile information should not be a part of any agreement when this weapon is being discussed.[140]

Thus disappeared any hope that the Swedes might have had of using the new grenade rifle as a pawn in their bargaining with the United States. It is clear that Schmidt divulged important information to Cole and the Americans when being so frank about the Swedish development problems. The strategy of denying Sweden access to such pertinent information as to let them get their guided missile program, literally, off the ground was part of the American hegemony-building tactic of using advanced technology as a tool of persuasion. However, the guided missile technology was only indirectly part of the effort to sway the Swedes to consent to American hegemony, because there was a general

138 Letter from Army Attaché in Stockholm to the Assistant Chief of Staff, G-2, U.S. Army, dated August 31, 1951, p. 1, in: RG 319 AIPDF 1951–1952; From 335.11 Sweden to 676.3 Sweden, Horby, Box No. 179; (Folder: 472.-474.6 Sweden); NARA II, College Park, MD.

139 Ibid., p. 2.

140 Letter from the Assistant Chief of Staff, G-2, U.S. Army to Army Attaché in Stockholm, dated December 21, 1951, p. 1, in: RG 319 AIPDF 1951–1952; From 335.11 Sweden to 676.3 Sweden, Horby, Box No. 179; (Folder: 472.-474.6 Sweden); NARA II, College Park, MD.

restriction in force with regard to this technology that denied all the allies this important piece of air defense equipment. The United States was as yet not ready to let the guard down; not until the end of the 1950s were the Western European countries offered this technology.

A report sent to Colonel Torsten Schmidt of the Swedish Defense Staff in 1952 dealt with how Sweden ought to handle the United States in matters concerning guided missile procurement. The report reflected the limited knowledge that the Swedes had of guided missile development in the United States at the time. It considered Sweden to be on an essentially equal stance with the United States on guided missile research and development (at a time when the Americans were conducting test flights with Sidewinder missile[141] although it stressed the necessity of providing the Americans with such information so as to make them interested in a mutual exchange of information on the subject. Echoing an awareness of the lack of necessary collaboration in this field, the report stated that the Americans would probably not be willing to release any such information at all to Sweden at this time and therefore Sweden ought to expose, and lay bare entirely, the domestic situation relating to guided missile development. This information should, at least initially, be provided without reciprocity.[142]

> However, the information mentioned above constitutes a valuable starting point for ensuing discussions. The person who is entrusted to give out this information would hereby create confidence not only for ensuing discussions but above all for his person. During the conversation or those discussions where this information is provided, an enquiry must be made concerning the USA's interest in contributing to Swedish guided missile development so as to pave the way for continuing discussions on the matter. [...] The ensuing discussion can hence ensue from more equal conditions between Sweden and the USA and one can therefore demand that the provided technical information is equaled by the USA – a barter should be able to be established. [143]

141 Westrum, R., *Sidewinder...*, pp. 95–97.

142 "RH A7:7", in KFF, Försvarets robotvapenbyrå [henceforth: FRB], Serie: F III A, "Inkomna och avgångna skrivelser ordnade efter klassifikationssystem", A 01-M 212, År: 1952, VOL. 8, KrA.

143 Ibid. "Ovannämnda upplysningar utgöra emellertid ett värdefullt utgångsmaterial för följande diskussioner. Den person, som bemyndigas utlämna sagda upplysningar bör härigenom skapa förtroende icke blott för fortsatta diskussioner utan framför allt för sin person. Vid det samtal eller de diskussioner där dessa upplysningar utlämnas måste USA [sic] intresse för att bidraga till svensk robotutveckling sonderas så att förutsättningar skapas för vidare diskussioner i ärendet. [...] Den fortsatta diskussionen kan alltså utgå från en förutsättning av större jämspelthet mellan Sverige och USA och man kan därvid påfordra att

This approach had several priors, e.g. the talk held by Military Attaché Möller at the Pentagon in March 1949 (see Chapter 3, p. 162–166). It should be kept in mind that this was just after the exchange of diplomatic notes that confirmed Sweden's participation in the MDAA (see below, p. 231), and it must have been considered valuable to make use of this new possibility for exchange with the Americans as soon as possible. A prerequisite for this was presumed to be that Sweden had something to offer in return. Sweden's participation in the MDAA provided a forum for realizing this strategy. It has been observed that the agreement did not lead to an immediate extensive import of military equipment from the United States, and that the Americans did not provide Sweden with classified technical information to an appreciable degree.[144]

The advice contained in the report was heeded in conjunction with Admiral Felix Johnson's three-day visit to Stockholm between December 9–11, 1952. Johnson, who was formerly the Chief of the U.S. Naval Intelligence, met with Supreme Commander Swedlund, Chief of the Swedish Defense Staff Major General Richard Åkerman, and Chief of the Army Colonel Bonde, and the purpose of his visit was "to establish vital contacts with the Swedish military and return to Washington with important information on the Swedish defence posture."[145] Swedlund, Åkerman, and Bonde took the opportunity to inquire whether it would be possible for Swedish researchers to meet with their American counterparts and discuss pertinent questions, "especially the question of developing guided missiles."[146] In accordance with the report, it was suggested that the Swedes take the first step towards such collaboration by divulging non-sensitive information relating to their own development. Moores states:

> Washington was to be given a clear picture of the kinds of guided missiles and the types of devices and motors currently under development *so that it could see for itself that it would benefit from entering into such discussions.* [The hope was also expressed that] Sweden would learn that it might have overcome certain technical obstacles better than Washington had done.[147] [Italics added.]

It was obviously felt that it would be possible to establish collaboration on the basis of the new hegemonic apparatus, but these

lämnade tekniska upplsysningar [sic] från svensk sida motsvaras av liknande från USA – en byteshandel bör kunna etableras."

144 *Had There Been A War...*, p. 123.

145 Moores, S., *'Neutral on Our Side':...*, pp. 124–125.

146 Ibid., p. 125.

147 Ibid., p. 125.

notions were premature, however reasonable they might have seemed at the time. But as the last quote shows, there was no lack of confidence within the Swedish military establishment, and there was a clear sense of pride over the domestic guided missile development.

The content of the report was communicated in a memorandum to the Foreign Ministry on December 12, the day after Johnson had left Stockholm. Nothing was said of the fact that the military had already begun to implement the strategy outlined in the memorandum.[148] The Defense Staff seems to have been trying to keep the government informed, and to get its approval, while at the same time keeping it partly in the dark.

The Bendix Impasse

In late November 1949 Hammarskjöld and Boheman discussed the best time for the Swedes to present the Americans with an application for export licenses regarding twenty radar sets purchased from the Bendix Radio Corporation during his visit to Washington. The order was worth about $1 million and the radar stations were being produced by Bendix. The first station had been completed, and was being tested, and the remaining nineteen sets were expected to be ready for delivery in April 1950.[149] Little did the Swedes know that the Bendix radar affair would be a long haul indeed.

The issue of the twenty Bendix radar sets came to the fore in early 1950, and is interwoven with the alleged policy change dealt with earlier in this chapter. Simon Moores has written that the change in U.S. policy vis-à-vis Sweden was emphasized by this affair, and that it "highlighted the gaping chasm between the State Department's views, on the one hand, and those of Matthews, on the other regarding Sweden's future treatment."[150] Sweden had been denied an export license for the ordered sets and was desperately trying to get them released. The episode revealed some intriguing aspects of the American policy to get Swedish consent to U.S. hegemony.

148 "PM angående önskade upplysningar från USA inom robotvapenområdet", dated December 12, 1952, pp. 1–2, in: HP 24 VI/U.S.A; "Tillstånd till in- och utförsel av vapen, ammunition och krigsmateriel"; 1950 sept.–1952; Folder 21; Box No. 1353; År: 1946–1952. See also: Moores, S., *'Neutral on Our Side':...*, p. 125. Moores had no knowledge about the report upon which this memorandum was based.

149 Letter from Boheman to Lagerfelt, dated December 1, 1949, pp. 1–2, in: HP 24 VI/U.S.A; "Tillstånd till in- och utförsel av vapen, ammunition och krigsmateriel"; 1946, okt.– aug. 1950; Folder 20; Box No. 1353; År: 1946–1952; RIA, Stockholm.

150 Moores, S., *'Neutral on Our Side':...*, pp. 35–36. Moores bases this estimate on Silva (Silva, C., *Keep Them Strong, Keep Them Friendly:...*, pp. 281–285).

On January 9, 1950, Bendix submitted a munitions export application for consignment to the RSAFB, who in turn let Ambassador Boheman present this application to the U.S. government on January 12.[151] Echoing the USAF's statement about the possible importance of the positioning of the radar sets, Boheman emphasized that these radars would be placed along the Swedish east coast, and that they were of great importance for Sweden's defense.[152] According to Moores, Matthews now wanted clarification regarding the official U.S. policy toward Sweden, stating that if it was still the desire of the United States' government that Sweden ultimately join the Atlantic Alliance, then the licenses should be denied. If, on the other hand, it was now felt that Swedish neutrality was beneficial to the United States then they should approve the application. The exchange originated in an insecurity pertaining to the interpretation of NSC 28/1 and its application to Sweden, and has been interpreted by Charles Silva as the Secretary and the Ambassador going "head to head" over the issue.[153]

The 'gaping chasm' between the State Department and Matthews became, according to Moores, evident in Acheson's reply, where he said that Foggy Bottom believed that withholding the radars would have a negative impact on the defense of Norway and Denmark. While it was still the desire that Sweden relinquish its neutrality, the inclusion of Sweden in NATO was no longer a policy objective, and the State Department was therefore prepared to agree to Sweden's request. Even so, concedes Moores, the Department remained wary about this stance.[154] As we remember, this policy had been laid down in a State Department Policy Statement in mid-August 1949.[155] Apparently, Matthews had not read it.

However, for Ambassador Matthews, the Bendix issue provided a long sought after opportunity to squeeze consent out of the Swedes, and his message to Acheson contained much more information than revealed by Moores. Matthews specifically cited the Bendix question as especially pertinent to sort out, because this case was, he said:

151 758.56 1/8-2450, Bonbright to Matthews, p. 3, in: RG 59; DF 1950–54; from 758.52/6-2551 to 758.713/10-3052; Box No. 3774; NARA II, College Park, MD.

152 Letter from Boheman to Hammarskjöld, dated January 12, 1950, p. 1, in: HP 24 VI/U.S.A; "Tillstånd till in- och utförsel av vapen, ammunition och krigsmateriel"; 1946, okt.– aug. 1950; Folder 20; Box No. 1353; År: 1946–1952; RIA, Stockholm.

153 Silva, C., *Keep Them Strong, Keep Them Friendly:...*, pp. 281–282, quote on p. 282.

154 Moores, S., *'Neutral on Our Side':...*, p. 36. The same interpretation has been made by economic historian Birgit Karlsson, see: Karlsson, B., *Handelspolitik eller politisk handling...*, p. 183.

155 FRUS: *Foreign Relations of the United States*; 1949; Vol. IV; *Western Europe*, p. 775.

> [...] regarded in Swedish military and political circles as the vital test case of Sweden's ability to purchase in US much needed electronic and similar modern equipment not available here and will be followed by other applications before long. This radar case therefore assumes in Swedish circles an importance far greater than relatively small dollar value involved: it is the test of how Sweden's position is regarded by the Atlantic Pact powers in general and the US in particular.[156]

The Ambassador reminded Acheson that even if the radar would help fill a gap in Sweden's defense, and provide early warning to the Norwegians and Danes "[t]he radar could, of course, be used against us in the case of overflights of American planes if the Swedish Government should wish to hinder such overflights to protect Swedish neutrality."[157] Matthews then came to his main point. There was one other factor that the United States' government ought to consider when deciding upon this matter, he said; a factor which:

> [...] might modify my recommendation that the license be granted if we want a neutral Sweden *and that concerns our control objectives on east-west trade*. As Department is well aware Swedish position on such controls while possibly showing some slight improvement is far from satisfactory and our leverage to obtain better results seems very small. With the Department soon assuming responsibility *in the field of east-west trade controls I should point out that the withholding of the radar license, pending some further assurances from Sweden that it does not intend to obstruct west military objectives by strengthening the USSR military potential, might be more effective than any other step.*[158] [Italics added.]

Matthews thus connected the issue of the release of the Bendix radar stations with a Swedish promise to abide by the trade embargo. He thus saw withholding this vital technology as the crucial lever that could sway the Swedes to align unambiguously with the West. Matthews was here, albeit unbeknownst to himself, expressing the essence of hegemonic thinking.

Acheson and the State Department were on the other hand, as stated by Moores, worried about Norway and Denmark. The State Department, said Acheson, was more inclined "to weigh heavily fact Norway desires radar net in Sweden regardless of accuracy of Nor

156 FRUS; *Foreign Relations of the United States*, 1950, Volume III, *Western Europe* (Washington, D.C.: US Government Printing Office, 1977), p. 15.

157 Ibid., p. 16.

158 Ibid., p. 17.

estimate of its importance."[159] This was connected to concerns over difficulties to fulfill U.S. undertakings to arm its Norwegian NATO ally. As far as the State Department knew, releasing the radars in question would not entail release of classified material, and the members of the Atlantic Pact did not want it either. "If so," concluded Acheson, "withholding license wld [would] clearly be act of pressure with adverse consequences in Nor and Den and possible exploitation by Soviet propaganda."

This was the reason why the State Department was in favor of approving the application. It was not intended as a reward for good Swedish behavior, but as a reassurance toward Norway (regardless of if it was effective or not). The underlying rationale seems to have been that if Sweden wished to pay for radar sets that would give the Norwegians a sense of security, then let them do it. The question of defending Norway had become a concern once the discussion began on the implementation of the NAT. The arms shipments that did reach Norway were too limited and too protracted to satisfy the Norwegians, writes Silva. This contributed to the increasing appreciation of Sweden's strategic value as a buffer in the defense of Norway.[160] Acheson was also clearly worried about the effects overt pressure upon Sweden might have. This worry was caused by the effect of an article in the *Washington Post* written by the American journalist Marquis Childs (which had also been carried by the Swedish media in late 1949). Childs had argued that the American Embassy in Stockholm was going too far in its campaign against Sweden, and Matthews had considered this coverage highly damaging to the Embassy.[161]

Acheson also remarked that the issue had not been discussed with the DOD, nor fully cleared within the State Department, and the Secretary intended to give Matthews an opportunity to comment upon the issue.[162] It is clear from this that what has been designated a turn in the State Department's policy had not even been cleared within its own ranks. It is evident that the proposed action, i.e. to approve Sweden's request, hardly constituted a definite decision on Acheson's part, and much less an overall change of policy towards Sweden. Further testimony to this effect is provided by the fact that no export license was approved at this time.

In his reply, Matthews agreed that this would indeed constitute an act of pressure, but considered it to be worth the risk, since the effect of approving release could be to demonstrate to the Norwegians and

159 FRUS; *Foreign Relations of the United States,* 1950, Volume III, *Western Europe,* p. 20.

160 Silva, C., *Keep Them Strong, Keep Them Friendly:...*, p. 283.

161 Ibid., p. 284.

162 FRUS; *Foreign Relations of the United States,* 1950, Volume III, *Western Europe,* p. 20.

the Danes that benefits accrue to non-members as well and try to lure them away from NATO. Sweden, noted Matthews "would leave no stone unturned" if it thought that there was even the slightest possibility of enticing these countries away from the West. Matthews then concluded his reply by reiterating what he considered to be the most important point, namely:

> I assume from this that larger considerations involved in radar license outweigh the leverage we might obtain in securing greater Swedish cooperation in controlling I-A and I-B list shipments to the East. *If such control is important I must repeat, however, we are relinquishing one of the few real cards we hold.*[163] [Italics added.]

Matthews thus took the opportunity offered to him by his boss to make sure his recommendations were heard. International diplomacy is not a simple game of push and shove, and even though the United States was very powerful and had much influence in world affairs it could not subject potential allies to too obvious pressure. That lack of influence can be coupled with great power is one of the great paradoxes of international relations. Consequently, Matthews regarded the radar licenses as one of the few real cards that the United States government held in their relation with Sweden; here was a real chance to influence Swedish policy, and the Ambassador recommended that since this card had been dealt, it should also be played.

The whole episode seems strange in retrospect. Why would Acheson go head to head with Matthews over this issue? Was there even a conflict between them? Matthews did not have any formal leverage on the matter, and Acheson did not need the Ambassador's approval in order to make a decision regarding the State Department's policy regarding the Bendix radars. There seems to be little foundation for the claim that this exchange of opinions highlighted the change in U.S. official policy towards Sweden. Furthermore, when Undén and Boheman spoke to Acheson in October, the Secretary did not let on that he felt embarrassed by the American withholding of the radar sets. He simply said that he knew about the issue, but he could not tell the Swedes whether he was in favor of a release or not.[164]

The matter was subsequently sent over for evaluation to the DOD. In late March 1950 Major General J. H. Burns of the U.S. Army submitted a report to Deputy Secretary of Defense Dean Rusk in which the view of the JCS on the refusal to release the radar sets was spelled out. The

163 Ibid., p. 24.

164 Letter from Boheman to Erlander, dated October 3, 1950, pp. 1–2, in: HP 24 VI/U.S.A; "Tillstånd till in- och utförsel av vapen, ammunition och krigsmateriel"; 1950 sept.–1952; Folder 21; Box No. 1353; År: 1946–1952.

JCS' recommendations would not have been happy reading for any proponent of the Swedish cause in Washington, or elsewhere. The report stated:

> From a military point of view, they [JCS] recommend that export licenses to Sweden covering the following items be denied:
>
> a. Items of substantial military significance whether munitions or munitions-producing which could be used against us in the event Sweden were subjected to direct or indirect domination by the Soviets.
> b. Items which could jeopardize or delay the strengthening of the North Atlantic Treaty nations or of the United States itself.
> c. End items of a technological nature which, if made available to enemy technicians, would enhance the enemy's military knowledge or techniques.
>
> Manufacturing licenses or production facilities for the types of items described in a and c above.[165]

There was thus little chance for Boheman to affect the present state of things when he met with Hulley and George W. Perkins, Assistant Secretary for European Affairs, in mid-April. But Boheman did his best, stating that Swedish acquisition of these radar stations would strengthen the defense of West, and this would be accomplished at the expense of Sweden rather than the NATO countries. The Ambassador also took the opportunity to assure the Americans that with regard to American overflights in peacetime, of which there had been some lately (this was of course ironic considering the Swedish lack of air defense radars), the United States did not have to worry a great deal about them. The Swedish government, said Boheman, understood that such things were bound to happen.[166] This rather desperate statement serves to underline the importance the Swedish government attached to the radar issue at this time.

It does indeed seem as if Matthews' recommendations in the telegram to Acheson were heeded by at least certain parts of the State Department, for the connection between Sweden's policy and the

165 From Burns to Rusk, dated March 25, 1950, in: RG 330 Records of the Office of the Assistant Secretary of Defense [henceforth: RG 330], International Security Affairs [henceforth: ISA], Office of Military Assistance [henceforth: OMA]; Project Decimal File [henceforth: PDF] Apr 1949–May 1953; 333 Spain to 091.3-523 Thailand (File: 091-523 Sweden 1950+51); Box No. 77; NARA II, College Park, MD. See also: Moores, S., *'Neutral on Our Side':...*, p. 37.

166 Memorandum of Conversation between Hulley, Perkins, and Boheman, dated April 18, 1950, in: RG 59; BNA, 1941–1953; SF, 1941–1953; South Africa, Union of, Policy – Sweden, External Political Affairs: U.S.; Box No. 18; (Folder: External Political Affairs: U.S.); NARA II, College Park, MD.

American reluctance to export radar equipment was explicitly stated to Boheman by Llewellyn Thompson in April. The United States was not at all happy with the Swedish political attitude, said the State Department official.[167] The matter now seems to have remained dormant for some time and it was not until August, after the outbreak of hostilities in the Far East, that it again left any documentary trace in the archives. Now the Truman administration informed Stockholm that it would withhold the license because the equipment was needed by the Department of the Air Force.[168]

The crux of the matter was that the radar sets had already been paid for by the Swedes, i.e. they had paid Bendix Radio Corporation, so the Americans now had to buy them back from the Swedes.[169] There was thus a very strange situation were the U.S. government wished to deny Sweden access to radar sets which the Swedes had already legally purchased. The only way the United States could possibly do this was by withholding the export licenses. The Americans now tried to come up with a more viable reason for not approving the Swedish request; what they found was that a basis for this action could be created by referring to a need to include the sets in the MDAP.

The strained effort to justify the American action is manifest in a letter from the Chief of the Munitions Department at Foggy Bottom John C. Elliot to Colonel C. K. Moffatt, Acting Chief of the Economic Division at the Office of International Programs. Elliot stated that the equipment:

> […] even if modified, is not desired for inclusion in the present procurement plans for the Department of the Air Force. However, the total number available of the subject equipment can be utilized by the Department of the Air Force for use in the Mutual Defense Assistance Program […].[170]

Thus the Department of the Air Force would contact the Swedish Ambassador "to attempt to purchase this equipment from that country […]."[171] The phrasing does not convey any particular urgency, and suggests that Elliot was indeed making an effort in his search for a way

167 Cole, P. M., *Neutralité du jour:…*, pp. 338–339.

168 Moores, S., *'Neutral on Our Side':…*, p. 37.

169 758.56 1/8-2450, Elliot to Moffatt, dated August 14, 1950, in: RG 59; DF 1950–54; from 758.52/6-2551 to 758.713/10-3052; Box No. 3774; NARA II, College Park, MD; Cole, P. M., *Neutralité du jour:…*, p. 339.

170 758.56 1/8-2450, Elliot to Moffatt, dated August 14, 1950, in: RG 59; DF 1950–54; from 758.52/6-2551 to 758.713/10-3052; Box No. 3774; NARA II, College Park, MD.

171 Ibid.

of justifying the American decision. This becomes even clearer in a letter from Bonbright at EUR to Matthews in late August. Bonbright said that he would let the Swedish Chargé d'Affaires ad interim Hubert de Besche, at the Swedish Embassy, know that it 'had been found' that the sets were needed elsewhere.[172]

Furthermore, Bonbright suggested, he would tell the Swede:

> [...] that in view of present international situation, including its obvious effect upon the supply situation, no assurances can be given that a license would be issued for replacements at a later date, as suggested by Ambassador Boheman. I would add that we know the Swedish Government, in this connection, will appreciate our obligation towards the countries which have entered into defense agreements with us.[173]

The last sentence underlined that the whole episode was intended to be a lesson for Sweden; Bonbright would in a diplomatic way tell de Besche that the Americans knew that this caused a lot of anxiety in Sweden, and the Swedes would sit and take it because they had no choice. Here, once again, the tactic of using technology as a hegemonic tool becomes glaringly evident. According to Paul Cole, Bonbright told de Besche this "in order to send Sweden a clear message that countries associated with the Mutual Defense Assistance Program would receive priority for U.S. military material."[174]

Upon hearing this decision the governments of Sweden, Norway, and Denmark stressed to the Americans that because of this the RSAF now found itself at a major disadvantage because it had no air defense radar, and that a militarily strong Sweden was something that all governments wished for. Undén actually expressed his understanding for the American action in a meeting with Hayden Raynor, the Chief of BNA, and Acheson in October. The Swedish Foreign Minister did not share the view that the RSAF necessarily was at a disadvantage because of this; of course Sweden, as a non-aligned nation, would be far down on the list of recipients of American military technology.[175] The American intransigence continued all through 1950, until the export license was

172 758.56 1/8-2450, Bonbright to Matthews, in: RG 59; DF 1950–54; from 758.52/6-2551 to 758.713/10-3052; Box No. 3774; NARA II, College Park, MD.

173 Ibid.

174 Cole, P. M., *Neutralité du jour:...*, p. 339. Cole includes a comment that borders on the comic, saying that "This policy was not intended to create undue pressure on Sweden." (ibid., p. 339.) This only makes sense if Cole thought that the pressure was not *undue* because Sweden deserved what they got.

175 Moores, S., *'Neutral on Our Side':...*, pp. 37–38.

finally granted over half a year later in May 1951.[176] The reason for this was, I argue, the Swedish promise to abide by the trade embargo regulations (I will get back to this below).

The disingenuousness of the American claim that the equipment was needed by the NATO countries for use in the MDAP was clear from the outset. Acheson even referred to this fact in his telegram to Matthews in February 1950, stating that "[o]ur understanding is that equipment covered by present application involves no problem of release of classified material, and so far as we are aware is not desired by NAT members."[177] Matthews confirmed that this was the case.[178] This was the very reason that Acheson thought that withholding the license would constitute overt pressure on Sweden that would have adverse consequences for Norway and Denmark, and Matthews concurred with this view as well, but he thought that granting the license would have even worse consequences.[179] In fact, the Swedes knew that this was not entirely true. When he handed over the export license application to Mr. Labouisse at the State Department on January 12, Boheman had himself stressed the point that the radar sets were technically different from American sets because they had been made especially for the RSAF. This, said Boheman, probably meant that the radar stations could not be used by the Americans or the NATO powers.[180]

Vougt tried to lobby the Americans to treat Swedish requests for defense materiel favorably. In September 1950 he met in Oslo with Generals Scott, new Chief of Office of Military Assistance (OMA) in Washington, and Biddle, Chief of the Military Assistance Program Advisory Group (MAPAG) in London, as well as with the Chief of the Military Assistance Advisory Group (MAAG) and the Embassy representative in Oslo. Vougt indicated to the U.S. representatives that Sweden might renew the attempt to purchase equipment from the United States, and referred especially to radar. In an effort to reassure the Americans he stated that "in case of war Sweden will of course

176 Ibid., p. 42; Karlsson, B., *Handelspolitik eller politisk handling…*, p. 146.

177 FRUS; *Foreign Relations of the United States*, 1950, Volume III, *Western Europe*, p. 20.

178 FRUS; *Foreign Relations of the United States*, 1950, Volume III, *Western Europe*, p. 24.

179 For Acheson's view, see: FRUS; *Foreign Relations of the United States*, 1950, Volume III, *Western Europe*, p. 20; for Matthews' view, see: ibid., p. 24.

180 Letter from Boheman to Hammarskjöld, dated January 12, 1950, p. 1, in: HP 24 VI/U.S.A; "Tillstånd till in- och utförsel av vapen, ammunition och krigsmateriel"; 1946, okt.– aug. 1950; Folder 20; Box No. 1353; År: 1946–1952; RIA, Stockholm.

attempt remain neutral, but if such not possible, 'there is no doubt where it will stand'."[181]

Despite this, the issue lingered on within the inter-Departmental bureaucracy in Washington, and the DOD was searching for a legitimate reason not to grant the export license. During late 1950 and early 1951 it was again stated that the USAF had no need for the radar sets either in MDAP or in the domestic organization. A Department memorandum from February 1950 stated that a check performed on the technical specifications of the sets ordered by Sweden revealed that they were "substantially different" from the sets which had been decided upon "as standard for MDAP and the NATO countries", and thus confirmed what Boheman had said to Labouisse a month earlier. The parts were not interchangeable, and the DOD consequently "doubted whether it would be wise to have 20 non-standard sets included in the MDAP." With this option officially dead, two other equally improbable options were thrown out there. It was instead said that there was an 'urgent' need for them in Formosa, or perhaps the Navy "might have some use for them."[182]

On November 25–26, 1950 the Swedish newspaper *Sydsvenska Dagbladet Snällposten* carried some articles which did not make it any easier for the Swedish government to get the Bendix sets released. The newspaper wrote that a U.S. radar set had almost been installed in a Polish ship being renovated at the Kockums ship wharf in Malmö, but had been stopped in the last minute due to American intervention. According to one of the articles, an American source within the U.S. Embassy thought that the effort was an unbelievably brazen maneuver. Hammarskjöld had spoken to Butterworth about the incident a few days later, on November 28, at which time the Ambassador had said that he did not know about the affair, and did not believe that the paper had been in touch with anyone inside the Embassy. If that had been the case, said Butterworth, he would have known about it. Hammarskjöld assured him that the Swedish government would look in to the matter, but he could confidently say that Swedish authorities had not been involved since granting such reexport went against the government's policies. According to Hammarskjöld, Butterworth thanked him for this, and it did not seem as if the Ambassador thought the issue to be especially important.[183]

181 758.5MAP/9-2650, Telegram from U.S. Embassy in Oslo to State Department, in: RG 59; DF 1950–54; from 758.13/1-852 to 758.5MSP/12-1554; Box No. 3773; NARA II, College Park, MD.

182 758.56 1/2-851, Memorandum for the Files, in: RG 59; DF 1950–54; from 758.52/6-2551 to 758.713/10-3052; Box No. 3774; NARA II, College Park, MD.

183 Memorandum by Hammarskjöld, dated November 28, 1950, pp. 1–3; in: HP 24

However, it turned out that the Ambassador had been wrong. A Mr. Christensen called the Foreign Ministry later the same day and said that he was the source that *Snällposten*'s reporter had spoken to. Apparently, Butterworth had been very unhappy with Christensen for not having informed him about this, and he now wished to set the record straight. During the conversation the American was told by the Foreign Ministry that it had been SJ, the Swedish state owned railway company, that had wished for a new navigation radar to be installed in the Polish ferry. The ferry, called Kopernik, had been built in 1903 and was transporting railway cars between Sweden and Poland. Kockums had, through an SJ engineer, contacted Swedish radar suppliers and had found a company prepared to supply a radar. The company had had no idea that the radar would be used on a Polish ship, and had, as soon as this fact had become clear to it, immediately refused the request. The Foreign Ministry told Christensen that there had been no application for export of the radar set, and if such had been forthcoming it would naturally have been denied. The same had been stated in a press release on November 28. The American seemd pleased with this response, and said that it appeared as if *Snällposten* had made a mountain out of a mole hill.[184] The Swedes thought that the episode illustrated how well the export control regulations worked in Sweden.[185] Quite possibly, that the Americans were of another opinion.

Adding to the trouble with the Swedish sets was the fact that the Swedes had paid substantially more for them, due to extensive engineering work and the small amount of sets ordered. Because of this it was uncertain whether the DOD could "appropriately pay the same price [...] since engineering costs per unit on its other procurements will be appreciably less." The option that seemed open if the sets were to be continually withheld from Sweden seemed to be to requisition them. However, this had not been considered, and it was not known just how much authority there was for attempting it.[186] It was due to this fact that the Americans had tried to get the Swedes to give the radar sets

VI/U.S.A; "Tillstånd till in- och utförsel av vapen, ammunition och krigsmateriel"; 1950 sept.–1952; Folder 21; Box No. 1353; År: 1946–1952.

184 Letter from Hammarskjöld to Bagge in Washington, dated December 1, 1950, pp. 1–2; "P.M. angående samtal med Mr. Christensen", dated November 29, 1950, pp. 1–2; "Pressmeddelande den 28 november 1950", dated November 28, 1950, in: HP 24 VI/U.S.A; "Tillstånd till in- och utförsel av vapen, ammunition och krigsmateriel"; 1950 sept.–1952; Folder 21; Box No. 1353; År: 1946–1952.

185 "P.M. ang. samtal med amerikanske handelsrådet, Mr. Tuthill", dated November 30, 1950, p. 2, in: HP 24 VI/U.S.A; "Tillstånd till in- och utförsel av vapen, ammunition och krigsmateriel"; 1950 sept.–1952; Folder 21; Box No. 1353; År: 1946–1952.

186 758.56 1/2-851, Memorandum for the Files, in: RG 59; DF 1950–54; from 758.52/6-2551 to 758.713/10-3052; Box No. 3774; NARA II, College Park, MD.

up volontarily by selling them. When Boheman had spoken to George Perkins on December 1, 1950, Boheman had put some pressure on him by stating that, he could not avoid drawing the conclusion that the American refusal had been a political decision, Perkins was manifestly disturbed, but confirmed that the decision had been made on political grounds, saying that the U.S. government did not want to grant export licenses to a nation that had not declared itself. This made the Swedish Ambassador utter that a free country apparently had to be a part of an alliance in order for it not to be hindered in its efforts to build up its defense. Boheman consequently told the American official that because of the American attitude, there was no telling if the Swedish government would assent to the U.S. demand that the Swedes should sell the materiel back to Bendix. Perkins had then ardently stressed that the Swedes should sell, and that it be a serious mistake not to do so. In his report back to the Foreign Ministry, Boheman stated that the Embassy's ability to affect the matter further had now been drained, and said that it was perhaps best to sell the stations back to the Americans.[187]

The case with the Polish ferry and the American radar came back to haunt the Swedish government, however, despite Christensen's statement that the matter had been blown out of proportion. On December 7 Boheman had met with Assistant Secretary of Defense Lovett, at which time the American had said that the Bendix-issue had almost been resolved, but the latest incidents had made it impossible for the U.S. government to grant the licenses. Lovett deplored the situation, but told the Swedes not to despair and that a solution might soon be found to the problem.[188]

So how do we explain the protracted conflict over these unclassified radar sets that were not needed by the Americans? It seems that we cannot get a satisfactory answer from earlier histories. As I showed above, prior research has contended that there was a change in the American attitude (or at least the State Department attitude) toward Sweden in the spring of 1950.[189] While it is true that the belligerent Matthews was replaced by the more Sweden-friendly Butterworth, I will argue that there are a number of factors that warrant a slightly nuanced interpretation.

187 Telegram from Boheman to the Foreign Ministry, dated December 1, 1950, pp. 1–2; Memorandum by Lagerfelt, dated December 2, 1950, pp. 1–2, in: HP 24 VI/U.S.A; "Tillstånd till in- och utförsel av vapen, ammunition och krigsmateriel"; 1950 sept.–1952; Folder 21; Box No. 1353; År: 1946–1952.

188 Memorandum, dated December 7, 1950, in: HP 24 VI/U.S.A; "Tillstånd till in- och utförsel av vapen, ammunition och krigsmateriel"; 1950 sept.–1952; Folder 21; Box No. 1353; År: 1946–1952.

189 Moores, S., *'Neutral on Our Side':...*, pp. 31, 35; Silva, C., *Keep Them Strong, Keep Them Friendly:...*, pp. 281–285; Karlsson, B., *Handelspolitik eller politisk handling...*, pp. 146–147. For a discussion, see above, pp. 191–202.

Firstly, there is the fact that the radar sets were still withheld for over a year following this momentous change in attitude. There was still a huge aversion to giving Sweden access to advanced military (as well as civilian) technology both within the State Department, and most definitely within the DOD. With such adamant opposition in Washington it must be considered justified to question the contention that the date for the 'turnabout' in U.S. policy was February 1950.[190] Indeed, Moores himself cites a State Department document drafted in March 1950 in which Sweden's "neutrality policy was referred to as 'dangerous and impractical'."[191] This glaring inconsistency in an otherwise diligent study demands an explanation that neither Moores, nor anyone else, has been able to provide. Why this continuing American suspicion? In a similarly offhand manner, Birgit Karlsson simply acknowledged that "[t]he radar licenses were not approved until May 1951."[192] This, too, without any effort to provide an explanation as to why the granting process took so long.

Secondly, we have seen that the radar impasse paralleled the negotiations between the Swedish and American governments concerning Swedish adherence, and abidance by, the COCOM trade embargo. The outbreak of hostilities in Korea in the summer of 1950 only reinforced American anxieties that Sweden would act as a transit country for technology exports to the Eastern Bloc. Oral commitments were no longer considered good enough, and in early 1951 the United States renewed its campaign for a more unambiguous Swedish stance regarding East-West trade. The prolonged negotiations that ensued ended with the signing of the Stockholm Agreement in June 1951.

I consequently argue that only if we take the issue regarding East-West trade into consideration are we able to account for these expressions of doubt and fear. The report from the JCS to Assistant Secretary of Defense Dean Rusk in March 1950 (see above, p. 220–221) explicitly stated that advanced military technology should not be released to Sweden due to the risk of it ending up in the hands of the Russians. The risk involved in possible transit trade was expressed in the recommendation that the United States should refrain from exporting to Sweden "items of

190 Moores, S., *'Neutral on Our Side':...*, p. 31. Moores even speaks about "a de facto volte-face in US thinking" as well as a "U-turn" in American policy (ibid., p. 31.).

191 Ibid., p. 36. Moores does not make any effort to explain how this statement, which totally contradicts his thesis concerning a volte-face in U.S. thinking, was produced.

192 Karlsson, B., *Handelspolitik eller politisk handling...*, p. 146. "Radarlicenserna beviljades inte förrän i maj 1951."

a technological nature, which if made available to enemy technicians, would enhance the enemy's military knowledge or techniques."[193]

So does this alone provide enough evidence to conclude that these issues were connected? No, more tangible evidence would be required for such a statement to be made without uncertainty. Moores explain the eventual granting of export licenses by once more referring to the Secretary of State. According to Moores, Acheson explained to Secretary of Defense Marshall that the Pentagon's rigid policy was causing embarrassment to the administration, and states that General Dwight D. Eisenhower, at that time Supreme Allied Commander for Allied Forces in Europe (henceforth: SACEUR) intervened and urged Pentagon officials to review their policy. This resulted in the State Department informing Boheman in mid-May that "owing to a relaxation in the supply situation, the radar licenses were to be finally granted."[194]

While these events certainly took place, Moores is still unable to explain why Acheson chose to urge Marshall at this particular time. Acheson gave his views to Marshall on April 20, 1951.[195] Had something happened with regard to the East-West trade issue that could have warranted action on Acheson's part? Well, during the spring of 1951 Sweden had shown tangible proof of its will to adhere to the embargo policies, and by early April Sweden had finally given the United States general guarantees against re-export of many of the items that the Americans had demanded. The Swedes had also supplied lists of Sweden's total exports to the Eastern Bloc, assuring the Americans that, ball-bearing exports aside, only about one percent of this trade involved embargoed items covered by the COCOM lists.[196] Sweden had basically complied with U.S. demands; that was why the American refusal was now becoming an embarrassment to the U.S. government.

Further evidence to this effect is given by the fact that on the same day that Boheman was informed about the granting of the licenses, the Embassy in Stockholm telegrammed to Washington recommending Acheson to explicitly link the radar issue to the paralleling negotiations regarding East-West trade between the American Ambassador and Dag Hammarskjöld in Stockholm.[197] Armed with this advice the Director of BNA Hayden Raynor issued instructions to George Perkins at EUR to prepare the latter for his meeting with Boheman later that day. Raynor referred to a letter from Secretary of Defense Marshall on May 11 where

193 From Burns to Rusk, dated March 25, 1950, in: RG 330; ISA; OMA; PDF Apr 1949–May 1953; 333 Spain to 091.3-523 Thailand (File: 091-523 Sweden 1950+51); Box No. 77; NARA II, College Park, MD.

194 Moores, S., *'Neutral on Our Side':...*, p. 42.

195 Ibid., p. 42. see document source in note 27.

196 Karlsson, B., *Handelspolitik eller politisk handling...*, pp. 176–177.

197 Ibid., pp. 146–147.

he had stated that the DOD was now prepared to withdraw its objections to the issuing of export license for the 20 radar sets. The instructions given by the Director of BNA to Perkins are illuminating regarding the true reasons for not granting Sweden a license and deserve to be quoted in full:

> Although the Swedish Ambassador has on several occasions expressed to various officers of the Department [of State] his personal belief that our refusal to issue the export license has been based on "political" grounds, we have maintained the position that our action was dictated by our need for the MDAP. It is, therefore, suggested that in explaining to the Swedish Ambassador our decision to grant the license you should state that our action is the result of an improvement in the requirements and supply situation which could not be foreseen at the time of our earlier discussions with him.

With regard to the Stockholm Embassy's recommendation to tie the issues of East-West trade and the granting of the license together, Raynor advised against it:

> It is believed that in view of our previous position outlined above, any direct connection of the two subjects would be unwise. However, should the occasion arise in your discussion with Mr. Boheman, it might be well to mention as a separate subject the growing congressional pressure for limitation of exports to the Soviet Bloc and our interest in continuing and increasing Swedish cooperation.[198]

Perkins followed Raynor's advice and made sure to mention the importance of the embargo question to the United States.[199] It is clear from Raynor's letter that the earlier references to the use for the sets within MDAP were only a pretext to lend an air of legitimacy to the

198 Raynor to Perkins, dated May 14, 1951, in: RG 59; Miscellaneous Office Files of the Assistant Secretaries of State for European Affairs [henceforth: MOFASSEA], 1943–1957; Box No. 27; (Folder: Sweden 1948–1955); NARA II, College Park, MD.

199 Memorandum of Conversation between Perkins and Boheman, dated May 14, 1951, in: RG 59; MOFASSEA, 1943–1957; Box No. 23; (Folder: Memorandums of Conversations – George W. Perkins 1951); NARA II, College Park, MD. This reference was apparently made with Acheson's blessing (Karlsson, B., *Handelspolitik eller politisk handling…*, p. 147). Interestingly, Boheman did not mention this in his report to Hammarskjöld, see: Letter from Boheman to Hammarskjöld, dated May 18, 1951, pp. 1–2, in: HP 24 VI/U.S.A; "Tillstånd till in- och utförsel av vapen, ammunition och krigsmateriel"; 1950 sept.–1952; Folder 21; Box No. 1353; År: 1946–1952.

withholding of the export license. If they were not, then there would have been no reason for him to urge Perkins to refer to this when explaining to Boheman of how the same radar sets could now suddenly be available again. Indeed, if the Swedish Ambassador had been totally off target when 'personally' referring to political grounds for the refusal, then he would have said so explicitly. The fact that he did not has to be taken as a confirmation that politics was indeed the real reason. Likewise, it was due to the earlier explanations offered by the Americans that Perkins was advised not to make any overt connection of the two issues. The implications of Perkins' 'separate' reference to the embargo matter could hardly have escaped Boheman, and might well have contributed to the conclusion of the negotiations in Stockholm, and the signing of the Stockholm Agreement, only a month later.

The MDAA Agreement—U.S. Confirmation of Sweden's Consent

On February 28, 1952, President Truman declared that Sweden, under the provisions of the Mutual Defense Assistance Act of 1949 (henceforth: MDAA), was eligible to purchase military equipment from the United States. Through an exchange of notes between the two governments in the summer of 1952 Sweden became a participant in the MDAA.[200] The negotiations prior to the conclusion of the agreement were a display of the plasticity of U.S. policy, and of what could be achieved with respect to Sweden's official policy of neutrality once the American government had received satisfying confirmation of Swedish consent to its hegemony.

On the American side, the decision to make Sweden eligible for military assistance on a reimbursable basis had effectively been made already on January 8 as a concomitant of the new official policy paper on Scandinavia and Finland, NSC 121, which Truman signed just over a week later on January 17. Sweden was now, finally, a step away from becoming an entrusted nation with the right to buy American military technology. But as such, Sweden was required to abide by American legislation, i.e. the prescriptions of the Battle Act. One has to understand that these separate legislations were necessary instruments to keep the Western countries from exporting high-technology commodities to the East, since COCOM, being a hegemonic apparatus, was only an organizing committee and could not take legal action against dissenters—this would have to be done by the hegemonic state itself. However, before Sweden could benefit from the American confirmation of the Swedish consent

200 *Had There Been A War...*, pp. 122–123.; Hanhimäki, J. M., *Scandinavia and the United States:...*, p. 48. For the notes, see: *Sveriges överenskommelser med främmande makter* (henceforth: SÖ) 1954:74.

to its hegemony, Sweden had to qualify for arms shipments by agreeing to the provisions of the MDAA.[201]

Since late 1951 it had been known to the Americans that the Swedish Prime Minister Tage Erlander would come to the U.S. in April 1952, at the invitation of the Swedish immigrant city of Rockford, Illinois, for the celebration of its centennial. Butterworth had recommended in December that the State Department should instruct him to call on Erlander and express on the behalf of the U.S. government the hope that the Prime Minister would find time to visit Washington as well. Butterworth described Erlander as being a "very open-minded and able man", but that he knew too little of the world outside the Swedish borders and would thus "be enormously impressed and permanently affected by a view of Amer[ican] scene." It was clear that the Ambassador saw a chance to further the relations between the two countries by having Erlander visit the American capital. Wrote Butterworth:

> Since there is every Prospect Erlander will be Prime Minister of Swed for many a year I know of no comparable action that our govt cld take which wld confer upon it corresponding benefit in its relations with Sweden.[202]

As it turned out, the Swedes beat the Americans to it. While the State Department was considering the issue, Ambassador Boheman approached Perkins at EUR, and informed him that the Prime Minister was himself planning on visiting Washington and wished to have an opportunity to see Acheson and Truman. Perkins subsequently recommended that Butterworth's request should be granted, and he agreed with the Ambassador that such a visit "would be beneficial to our interests". However, he did not think that it was necessary to issue a formal invitation to Washington since Erlander himself had stated that the reason for his trip was to visit Rockford, and Perkins was also "doubtful whether Mr. Erlander himself would wish to make his visit here an official one." The White House quickly approved Erlander's wish and Acheson cabled Butterworth that he should inform the Prime Minister that the President "will be pleased if during his visit to US he can arrange schedule to be in WASH for informal conversations with US officials during week APR 13." It was expected that the President would give a lunch for Erlander, and host a conference in the Cabinet.[203]

201 Silva, C., *Keep Them Strong, Keep Them Friendly:...*, p. 337; *Had There Been A War...*, pp. 122–123.

202 758.13/1-852, telegram from Butterworth to Secretary of State, dated January 8, 1952, in: RG 59; DF 1950–54; from 758.13/1-852 to 758.5MSP/12-1554; Box No. 3773; NARA II, College Park, MD.

203 "will be pleased...": 758.13/1-852, telegram from Acheson to Butterworth, dated

While this matter brewed, the issue of Sweden's eligibility to receive reimbursable military aid came up. The Swedes did not know about Truman's late-February declaration until March, when the Department of the Army sent a letter to notify the Swedish Military Attaché Möller. Möller was puzzled by what the letter seemed to imply, and had Ragnvald Bagge, Minister-Counselor at the Swedish Embassy, query the Americans as to its meaning and implications. In the written response it was said that the necessary documents formalizing the authorization for Sweden were about to be signed, and that negotiations to obtain Swedish assurances were in progress. Upon realizing that the letter was referring to the President's decision of February 28, the Officer in Charge at NOE Andy Ronhovde cabled Butterworth instructing him to let the Swedes know that their status had officially changed. A few days later the statements in the letter had been confirmed to the Swedes by a Colonel Burgess at the Pentagon. Ronhovde explained that:

> [d]epending upon the sort of assurances given [...] Sweden would be authorized to purchase military equipment in certain security classification categories.

In connection with this, Bagge asked if the Embassy could be provided with a copy of a model agreement required to obtain the aid, whereupon the Americans promised to send them a copy of the draft exchange of notes that had already been dispatched to Stockholm for the discussions between the Americans and the Swedish government. Ronhovde then stated that the agreement could be kept secret and did not have to be made public immediately; however, in Bagge's view, it might be better to publish it directly to avoid that the matter being dragged out in the press, although this was of course a decision the Swedish government would have to make.[204] The Americans actually

---

January 23, 1952, in: RG 59; DF 1950–54; from 758.13/1-852 to 758.5MSP/12-1554; Box No. 3773; NARA II, College Park, MD. "would be beneficial...", "doubtful whether...": 758.13/1-2352, telegram from Perkins to Secretary of State, dated January 11, 1952, in: RG 59; DF 1950–54; from 758.13/1-852 to 758.5MSP/12-1554; Box No. 3773; NARA II, College Park, MD. See also: 758.13/1-1752, Letter from Perkins to Secretary of State, dated January 17, 1952, pp. 1–2, in: RG 59; DF 1950–54; from 758.13/1-852 to 758.5MSP/12-1554; Box No. 3773; NARA II, College Park, MD. See: also: 758.13/1-2852, Note regarding Erlander's visit, dated January 24, 1952, and Memorandum from Perkins to Secretary of State, dated January 28, 1952, in: RG 59; DF 1950–54; from 758.13/1-852 to 758.5MSP/12-1554; Box No. 3773; NARA II, College Park, MD.

204 "[d]epending upon...": 758.5/3-1452, Memorandum of Conversation between Ronhovde, Hilliker, and Bagge, dated March 18, 1952, pp. 1–2, quote on page 2, in: RG 59; DF 1950–54; from 758.13/1-852 to 758.5MSP/12-1554; Box No. 3773; NARA II, College Park, MD. See also: 758.5/3-1452, Memorandum of Conversation between Ronhovde and Bagge, dated January 14, 1952, in: RG 59;

delivered on their promise, and nothing was mentioned about the agreement in American legislation for 1952, even though the Swedes had several transfers pending.[205]

Bagge thus thought it better to make the agreement official at once, because it would avoid it being widely discussed in the press. This points to one of the paradoxes of democracy: while much, but far from all, information is available to anyone who takes time to search for it, little is actually known to the general public. In order to keep it out of the newspapers, it should be made public by the government. The rationale seemed to be that if the existence of an agreement was not hidden, no one would pry into the issue to know more about it. In this episode it was also revealed that the United States would demand further assurances from Sweden before an agreement could be signed; the nature of which I will return to below. Boheman telegrammed to the Foreign Ministry that it now seemed as if the Americans would let Sweden purchase military materiel under the provisions of the MDAA, without giving up either neutrality or non-alignment.

The draft of the note to the Swedes was included in a letter to Butterworth on March 21, in which he was instructed to inform the Swedish government of the American decision, which he also did. The Ambassador was to explain to the Swedish government that certain formal assurances, entailed in the enclosed draft, would have to be forthcoming from them prior to the conclusion of any contracts. The assurances would take the form of an exchange of notes, which would constitute an agreement. The draft was a standard form and had been used in making similar agreements with other states under the provisions of the MDAA, and Butterworth was authorized to conclude the agreement should the Swedish government prove willing to accept the draft. However, if they wished to make changes to the wording of the draft, then the matter had to be referred back to Washington for a decision to be made there. He would also instruct the Swedish government that in case of a Swedish rejection, any questions pertaining to export licenses, allocations or priorities, individual cases would continue to be considered in the light of the existing supply situation.[206] The last point was probably thought

---

DF 1950–54; from 758.13/1-852 to 758.5MSP/12-1554; Box No. 3773; NARA II, College Park, MD.

205 Silva, C., *Keep Them Strong, Keep Them Friendly:...*, p. 338.

206 758.5-MSP/3-2152, Letter from Bonbright to Butterworth, dated March 21, 1952, pp. 1–2, in: RG 59; DF 1950–54; from 758.13/1-852 to 758.5MSP/12-1554; Box No. 3773; NARA II, College Park, MD. Telegram from Boheman to the Foreign Ministry, dated March 17, 1952, in: HP 24 VI/U.S.A; "Tillstånd till in- och utförsel av vapen, ammunition och krigsmateriel"; 1950 sept.–1952; Folder 21; Box No. 1353; År: 1946–1952. See also: *Tage Erlanders dagböcker, 1952* [The Diaries of Tage Erlander, 1952], published by Sven Erlander (Hedemora: Gidluns Förlag, 2002), p. 57.

to act as an incentive for the Swedes to sign, considering the licensing difficulties that were still fresh in the mind of Swedish government officials.

Butterworth's announcement caused Prime Minister Erlander some concern because of the connection that could be made between this decision and his planned trip to the United States. Erlander does not say much about this in his diary, but wrote, without being more specific, that the motives behind the American decision were discussed with Hammarskjöld. The last entry for this day reads: "Now we're obviously to be [unreadable] into the trap."[207] Whatever this meant, it seems as if Erlander was not uncritically pleased about the timing of the American display of amicability. He clearly sensed that something would have to be given up in exchange, and that it was a matter of a *quid pro quo* and not a unilateral favor on the part of the United States.

Already at this point, however, the American willingness to compromise became visible; the State Department told Butterworth that:

> [...] should the Swedish Government reject the idea of an agreement in the form of an exchange of notes, the Department is prepared to consider the alternative of a unilateral declaration of assurance embodied in a single note from the Swedish Government. A draft of such a note [...] is enclosed.[208]

Even so, it was still considered preferable for the deal to be sealed with an exchange of notes, and Butterworth was ordered not to tell the Swedes about the back-up option before he had received an answer from the Swedes regarding the exchange.[209] The American proposal was discussed by the Swedish government on April 1, and parallels were drawn with the Swiss acceptance of the same agreement. Hammarskjöld had been informed by Max Petitpierre, the Swiss Foreign Minister during the OEEC meeting in Paris of what Switzerland had achieved. Thus, once again the Swiss experience was to be essentially the same as the Swedish one. The Americans seemed to treat the two neutrals in the same way. It appeared, said Petitpierre, that the two neutrals were put in a category of their own, and that there should not be any risk of them being confused with allied or quasi-allied states. The main point that Hammarskjöld took with him seems to have been that the Swiss

207 *Tage Erlanders dagböcker, 1952.*, p. 57. "Nu ska vi tydligen [oläsligt] in i fällan."

208 758.5-MSP/3-2152, Letter from Bonbright to Butterworth, dated March 21, 1952, p. 2, in: RG 59; DF 1950–54; from 758.13/1-852 to 758.5MSP/12-1554; Box No. 3773; NARA II, College Park, MD.

209 Ibid., p. 2.

had taken out all references to American legislation from the text, thus making it more palatable to the Swiss parliament. Undén thought that even if this was done by the Swedes as well, the agreement still contained parts that were not unobjectionable to him. The matter ought not to be forced, said the Foreign Minister, and apparently Erlander agreed on this. The best thing would be an even more "watered down" agreement or, better yet, no agreement at all but simply oral commitments put in writing.[210]

That the matter was handled on a bilateral basis was not strange; the idea behind MDAA was that bilateral arrangements were the only way to secure reciprocal assistance from participating nations.[211] The two drafts were basically the same; they both featured six articles which were formulated in more or less the same way. In the first article, Sweden would agree to use the assistance furnished in a way that corresponded to the framework of the Charter of the United Nations; in the second, Sweden promised not to undertake acts of aggression but to use the material solely to maintain internal security; in the third, Sweden swore not to relinquish the material to a third party without the consent of the United States; in the fourth, Sweden guaranteed to safeguard any article, service, or information acquired through the MDAA; in the fifth, Sweden acknowledged the American right to terminate any transaction prior to delivery; and in the sixth and final article, Sweden accepted the terms and conditions of payment.[212]

Since the two drafts were so similar it is hard to see why the Swedish government would have a problem with the exchange of notes, but not with the unilateral assurance. There was, however, one crucial difference in the formulation of the first article. In the exchange draft (and only in this draft), the second half of the first article read:

210 *Östen Undén. Anteckningar 1918–1952* [Östen Undén: Notes, 1918–1952], Published by Karl Molin (Stockholm: Kungl. Samfundet för utgivande av handskrifter rörande Skandinaviens historia, 2002), p. 399. "urvattnat [...]." See also: Silva, C., *Keep Them Strong, Keep Them Friendly:...*, p. 338. For Hammarskjöld's conversation with Petitpierre, see: Memorandum by Hammarskjöld, dated April 1, 1952, pp. 1–2, in: HP 24 VI/U.S.A; "Tillstånd till in- och utförsel av vapen, ammunition och krigsmateriel"; 1950 sept.–1952; Folder 21; Box No. 1353; År: 1946–1952. For more regarding the similar treatment of Sweden and Switzerland, see: Letter from Boheman to Undén, dated March 18, 1952, pp. 3–4, in: HP 24 VI/U.S.A; "Tillstånd till in- och utförsel av vapen, ammunition och krigsmateriel"; 1950 sept.–1952; Folder 21; Box No. 1353; År: 1946–1952.

211 Kaplan, L. S., *A Community of Interests:...*, p. 38.

212 See both drafts in: 758.5-MSP/3-2152, Letter from Bonbright to Butterworth, dated March 21, 1952, p. 2, in: RG 59; DF 1950–54; from 758.13/1-852 to 758.5MSP/12-1554; Box No. 3773; NARA II, College Park, MD.

> The Government of _______ further agrees to furnish equipment and materials, services, or other assistance, consistent with the Charter of the United Nations, *to the United States or to and among other nations eligible for assistance under the Mutual Defense Assistance Act* to further the policies and purposes of this Act, as set forth above, and as may be mutually agreed hereafter.[213] [Italics added.]

It is perhaps not very hard to imagine that the Swedish government, whose official foreign policy was guided by the policy of neutrality and non-alignment, could find this paragraph slightly problematic. The section was a pledge to provide assistance to these nations in times of peace *and* war, and it is difficult to understand how the State Department could even begin to think that the Swedish government would perchance acquiesce to the suggested agreement.

But it is instructive that the U.S. government wished Butterworth not to tell the Swedish negotiators that the United States could consider another way of granting the Swedes the same rights, only without the concomitant obligations spelled out in the second half of Article 1. Clearly, what the Americans hoped for was that the Swedish government would agree to the exchange; perhaps goaded by the reference to the elusive 'supply situation'. The Americans were prepared to take their chances with the proposed agreement, all the while making sure to have another alternative ready for presentation. In the interest of hegemonic policy the United States did not want to push a potential tacit ally into a corner where the Swedes would feel forced to back out into a reinforced stance of neutrality. There was also a sentence in Article 2 in both drafts which stated that one of the ways in which the assistance received could be used was to permit it "to participate in the defense of the area of which it is a part", but this was stated in connection with other possible scenarios which did not allude to any regional commitments whatsoever, including U.N. collective security arrangements.[214] The latter sentence had been copied *verbatim* from the amended MDAA, and would in fact become an issue later.[215]

At the same time Erlander was getting ready for his trip to the U.S., and would leave Sweden on April 2. In his diary, the Prime Minister again confessed to be very annoyed by the timing of the American

213 "Draft Exchange of Notes" enclosed in: ibid., p. 2 (page number refers to the Draft).

214 See both drafts in: ibid., p. 2.

215 "Public Law 621—81st Congress, Chapter 491—2D Session, S. 3809, An Act to Amend the Mutual Defense Assistance Act of 1949", p. 4, in: HP 24 VI/U.S.A; "Tillstånd till in- och utförsel av vapen, ammunition och krigsmateriel"; 1950 sept.–1952; Folder 21; Box No. 1353; År: 1946–1952.

decision to make Sweden eligible for procuring military equipment on a reimbursable basis from the United States. Erlander wrote:

> Damned Americans, who had to bring up the question regarding our possibility of procuring weapons right now. We have no special matters at the moment. And we definitely have no interest in tying ourselves to America more firmly than necessary *at the moment*. Not much is necessary. That is why we could have done without the latest declaration of affection.[216] [Italics added.]

The whole matter was clearly seen as a nuisance; but it is pertinent to note the italicized phrase; a closer tie between Sweden and the United States was thus seen as desirable, just not at that precise moment. Erlander was not averse towards establishing closer ties eventually, when attention was aimed in another direction.

The President was scheduled to have lunch with the Swedish Prime Minister on April 15, and Butterworth wrote to Acheson that he had resisted the temptation to recommend that he himself should be in the U.S. to greet Erlander, since this "would create the unwarranted and embarrassing implication that his visit was official and designed to achieve a concrete political result."[217] Butterworth emphasized that Erlander was a "totally non-doctrinaire Social Democrat", and added that:

> [a]lthough I have no reason to believe that Erlander is not in accord with Sweden's so-called alliance-free policy, he is much more sympathetic and understanding of our aims and purposes than is Undén and I am sure that he is willing to go quite far to cooperate with us, particularly if it can be done without publicity, for he does not wish to fly in the face of Swedish public opinion which, it must be admitted, supports current Swedish foreign policy.[218]

Butterworth also said that Erlander had told him over dinner that although he did not wish to bring up anything particular, he was:

216 *Tage Erlanders dagböcker, 1952.*, p. 61. "Djäkla amerikaner, som skulle aktualisera frågan om våra möjligheter till vapenköp just nu. Vi har inga särskilda ting aktuella nu. Och vi har absolut intet intresse att *just nu* binda oss till Amerika hårdare än vad som är nödvändigt. Nödvändigt är inte mycket. Därför hade vi kunnat få vara ifred för den senaste kärleksförklaringen." [Italics added.]

217 758.13/4-152, Letter from Butterworth to Acheson, dated April 1, 1952, p. 1, in: RG 59; DF 1950–54; from 758.13/1-852 to 758.5MSP/12-1554; Box No. 3773; NARA II, College Park, MD.

218 Ibid., p. 2.

> [...] looking forward to a frank discussion of the general world situation with particular reference, of course, to Europe. I underline the word, "frank", because he did.[219]

Butterworth also said he had discussed "one or two matters" which Acheson might wish to have in mind when he and Matthews spoke to the Swedish Prime Minister.[220] Butterworth had in fact written Matthews a letter as well, in which he briefed the former Ambassador to Sweden on a conversation that Marshall Green, First Secretary of the Embassy in Stockholm, had had with Sverker Åström some weeks earlier. During the exchange Åström had said:

> '[...] As to covert strategic relationships with the West, certainly there are some and they can be developed to a certain extent. But the extent has its limits. For *at some point the covert becomes too generally known or at least threatens to become known*. Short of this point covert cooperation must cease.'[221][Italics added.]

The italicized phrase obviously made a great impression on Butterworth, and every other American officer who heard of it, for it was perpetuated several times as it traveled up the hierarchies in Washington, even all the way up to the White House. Åström was interpreted as having said that the sentiments in Sweden had moved considerably westward, and this seemed to bode well for the future with regard to U.S.-Swedish security collaboration.[222] These words were echoed in a memorandum for President Truman signed by Acheson and intended to prepare the President for the meeting with Erlander a few days later. The memorandum stated the United States' relations with Sweden were:

> [...] strongly influenced under present circumstances by that country's determination to avoid involvement in any military bloc or alliance. Nevertheless, there is an official Swedish desire, even

219 Ibid., p. 2.

220 Ibid., p. 2.

221 758.13/4-152, Letter from Butterworth to Matthews, dated April 1, 1952, p. 2, in: RG 59; DF 1950–54; from 758.13/1-852 to 758.5MSP/12-1554; Box No. 3773; NARA II, College Park, MD. See also: Memorandum of Green's conversation with Åström, dated March 20, 1952, pp. 1–2, in: RG 84; SWSE; Classified General Records [henceforth: CGR] 1953–1963; 321.3 German Assets 1954 370; Box No. 2; (Folder: 350 Sweden); NARA II, College Park, MD.

222 758.13/4-152, Letter from Butterworth to Matthews, dated April 1, 1952, p. 2, in: RG 59; DF 1950–54; from 758.13/1-852 to 758.5MSP/12-1554; Box No. 3773; NARA II, College Park, MD.

in matters involving East-West tension, for inconspicuous [deleted word] cooperation with Western countries, *up to the point where there is danger of such cooperation becoming generally known.* [...] For these reasons, the Swedish Government understandably would not wish to have the Prime Minister's visit interpreted publicly as a step toward closer association with the United States [...].[223][Italics added.]

In the attached Briefing Paper there was included a reference to the ongoing deliberations regarding Swedish assurances prior to the signing of the MDAA agreement. At the same time it was stressed that the Swedish government had so far "shown no inclination to enter into military talks that would endanger its public position of adhering to the 'alliance-free' policy."[224] It should be acknowledged that this statement was not considered to contradict what had been said above regarding Sweden's will to cooperate up to a point where the cooperation became generally known. With regard to East-West trade the Briefing Paper said:

Because of its unwillingness to align itself with any great power "bloc" Sweden has chosen to follow what it calls an "autonomous" foreign trade policy. The controls by which its policies are implemented are well enforced and at least as effective as those of other Western European countries.[225]

The matters for discussion mentioned by Butterworth to Acheson were included in a letter to Acheson from Perkins at EUR ten days later, in the middle of Erlander's visit, and four days before he was scheduled to meet Truman and Acheson. Perkins concurred with the recommendations offered by Butterworth, and recommended that:

[...] when Mr. Erlander calls you might express appreciation of the cooperation thus far obtained from the Swedish Government in the matter of East-West trade. Mr. Butterworth cautions, however, that this should be done in such a way as to encourage continued

223 758.13/4-1152, Memorandum for the President, dated April 11, 1952, pp. 1–2, in: RG 59; DF 1950–54; from 758.13/1-852 to 758.5MSP/12-1554; Box No. 3773; NARA II, College Park, MD.

224 Briefing paper "Sweden: Policy Regarding Collective Security", p. 1, attached to: 758.13/4-1152, Memorandum for the President, dated April 11, 1952, in: RG 59; DF 1950–54; from 758.13/1-852 to 758.5MSP/12-1554; Box No. 3773; NARA II, College Park, MD.

225 Briefing paper "Sweden: East-West Trade", attached to: 758.13/4-1152, Memorandum for the President, dated April 11, 1952, in: RG 59; DF 1950–54; from 758.13/1-852 to 758.5MSP/12-1554; Box No. 3773; NARA II, College Park, MD.

> cooperation which, after all, is in Sweden's best interest and, incidentally, has resulted in a more favorable attitude on our part in the matter of export licenses and allocations to Sweden. I should add that it would be particularly unfortunate at this time to permit the Swedish Government to assume a complacent attitude, since further discussions of export control policies are planned with Sweden for the near future.[226]

I ask the reader to note with extra care the reference to the importance of Swedish cooperation on the export control issue for the more lax attitude towards Sweden on the part of the U.S. government. The upcoming negotiations alluded to by Perkins were probably the question of SKF's bearing exports to Czechoslovakia, Hungary, and Poland. This goes to show that total Swedish acquiescence was not deemed necessary in order for U.S. policymakers to consider Swedish consent given. As long as a basic adherence was achieved, the Americans were prepared to grant Sweden its long sought material rewards—in the form of the MDAA agreement. Perkins, however, advised the Secretary that he should not raise the question of Swedish neutrality policy "by direct reference." If Erlander was to introduce the subject, it was recommended that Acheson explain that while it was not the policy preferred by the United States, the American government did respect it.[227]

There seems to be little reason to doubt that no issues relating to Swedish-American military and security relations were dealt with during Erlander's visit, at least not including Erlander himself.[228] However, Erlander's notes in his diary point toward a conclusion that the visit was not purely a matter of courtesy on his part. For example, when on January 26 Butterworth came to see Erlander and brought word from Acheson that he was welcome to see Truman in early April he remarked "What have I gotten myself into?", at the same time noting that Butterworth seemed delighted.[229] This was before Erlander knew

226 758.13/4-1052, Letter from Perkins to Acheson, dated April 10, 1952, pp. 1–2, in: RG 59; DF 1950–54; from 758.13/1-852 to 758.5MSP/12-1554; Box No. 3773; NARA II, College Park, MD. For the same, see: 758.13/4-152, Letter from Butterworth to Matthews, dated April 1, 1952, p. 1, in: RG 59; DF 1950–54; from 758.13/1-852 to 758.5MSP/12-1554; Box No. 3773; NARA II, College Park, MD.

227 758.13/4-1052, Letter from Perkins to Acheson, dated April 10, 1952, p. 2, in: RG 59; DF 1950–54; from 758.13/1-852 to 758.5MSP/12-1554; Box No. 3773; NARA II, College Park, MD.

228 Moores, S., *'Neutral on Our Side':...*, pp. 44–45, note 31; Silva, C., *Keep Them Strong, Keep Them Friendly:...*, pp. 338–339. Whether something was discussed between e.g. Boheman and the Americans in connection with the visit is not known, but there seem to be no sources in existence to indicate that this was the case.

229 *Tage Erlanders dagböcker, 1952.*, p. 21. "Vad har jag egentligen gett mig in på?"

that an American offer to Sweden to join the MDAA was forthcoming, and that this decision would collide with his trip to the United States. There could hardly have been much reason for the Prime Minister to express such worries over a simple trip to see some relatives in the United States, but on the other hand this is not enough to assume that something big was going to be negotiated during the brief sojourn on the other side of the Atlantic.

But once the merging of these two events was clear to him, it seems as if Erlander was feeling a bit uneasy about going to Washington to see the President. This becomes clear from an entry on April 14, the day Erlander saw Truman and Acheson, when he wrote:

> Now to Acheson and Truman today. Useful to whom? It will probably be just a courteous bow without that significant a meaning.[230]

The note was clearly made before he met with the President and his Secretary of State, and a worry is abundantly evident in these sentences. Erlander seems to have looked upon the meeting as being of no real use to either side since nothing important was to be discussed, probably foreseeing trouble on the horizon when thinking about the MDAA agreement and possible reactions to the timing of the visit. Perhaps he would much rather have abstained from the meeting altogether. Such action would not have been very diplomatic, however, and could very well have caused harm to the relations between the two countries. There was also the question of Swedish domestic politics where the coalition with the Agrarian Party, with which the Social Democrats shared a whole-hearted support for the neutrality policy, was perhaps yet another reason for why Erlander felt a bit awkward.[231] Erlander thought that Acheson, whom he thought had been very interesting in the morning, must have been greatly disappointed with him because the

230 Ibid., p. 67. "Nu till Acheson och Truman idag. Nyttigt för vem? Det blir väl bara en hövlig bugning utan så allvarlig mening." (I have benefited from Simon Moores' translation of the last sentence, which he quotes in a footnote, see: Moores, S., *'Neutral on Our Side':...*, p. 45, note 31.)

231 Regarding the Agrarians' support for the policy of neutrality, see: Silva, C., *Keep Them Strong, Keep Them Friendly:...*, p. 336. The coalition government had come about because the Social Democrats had run up against increasing difficulties in winning support for the government's economic policies in the Riksdag during the winter and spring of 1951. Despite a common view on domestic policy, the Social Democrats and the Agrarians were both firm supporters of the official policy of neutrality. In return for their support for Erlander's economic and social policies, the Agrarians obtained benefits in the field of agrarian policy (Ruin, Olof, *I välfärdsstatens tjänst. Tage Erlander 1946–1969.* (Stockholm: Tidens Förlag, 1986), pp. 89–90.

Secretary seemed very "[...] cold and disinterested" later in the day.[232] Perhaps Acheson had wished to get some kind of concrete assurance from the Swedish Prime Minister regarding the tie-in of Sweden into the Western defense system.

Hammarskjöld had received the draft proposal for an exchange of notes between the two governments from Butterworth on April 1, and although he had the impression that there was no immediate Swedish need for equipment he would consult with the Swedish military authorities and get back to the Ambassador. Butterworth thought it best to leave the matter with the Swedes and let Hammarskjöld bring the matter up for discussion.[233] This might, considering the content of the proposal, have added to Erlander's frustration, and led him to curse the American timing, as quoted above. Whether the Prime Minister saw the drafts on April 1 has not been established, but if he did, they could certainly have caused him some concern. By the end of April the Swedish government had given its preliminary consideration to the American proposals, and Hammarskjöld told Butterworth that his government understood full well that the legal requirements of the MDAA had to be fulfilled, but at the same time he expressed his hope that Sweden's situation would be taken into account. The Swedish government had a host of objections to the texts as they were, and wished to have certain references, which according to the Swedes did not apply to them, omitted.

The Swedes wanted to remove all references to the MDAA as such and to Section 408 (e) of the Act, since according to them, Sweden was not involved in the mutual defense arrangements that were the main purpose of the Act.[234] The reason was that if the assurances offered by the Swedish government included references to American legislation,

232 *Tage Erlanders dagböcker, 1952.*, p. 68. "[...] kylig och ointresserad."

233 758.5/4-1652, telegram from Butterworth to Acheson, dated April 16, 1952, in: RG 59; DF 1950–54; from 758.13/1-852 to 758.5MSP/12-1554; Box No. 3773; NARA II, College Park, MD.

234 758.5/4-2652, telegram from Butterworth to Acheson, dated April 26, 1952, in: RG 59; DF 1950–54; from 758.13/1-852 to 758.5MSP/12-1554; Box No. 3773; NARA II, College Park, MD. See also: Silva, C., *Keep Them Strong, Keep Them Friendly:...*, p. 338. The amended version of Section 408 (e) stipulated that the President could transfer military materiel to "(A) nations eligible for assistance under title I, II, or III of this Act, (B) a nation which has joined with the United States in a collective defense and regional arrangement, or (C) any other nation not eligible to join a collective defense and regional arrangement [...], but whose ability to defend itself or to participate in the defense of the area of which it is a part, is important to the security of the United States [...]." (See: "Public Law 621—81st Congress, Chapter 491—2D Session, S. 3809, An Act to Amend the Mutual Defense Assistance Act of 1949", p. 4, in: HP 24 VI/U.S.A; "Tillstånd till in- och utförsel av vapen, ammunition och krigsmateriel"; 1950 sept.–1952; Folder 21; Box No. 1353; År: 1946–1952.)

“they might compromise Sweden’s policy of non-alignment.”[235] The Swedish objection proves that it was the draft of an exchange of notes that the Swedish government was commenting on (since the other draft text did not include any reference to the MDAA). This was quite natural, since this was the only draft that the Swedes had any knowledge of at this point. In light of that, the Swedish objections, as we shall see, were quite remarkable. The government also stated that in order for it to sign the note, it would require omission of the phrase in Paragraph 2 (quoted above) referring to the defense of the area of which the signatory was a part. This phrase was, it was said, obviously inapplicable to Sweden’s situation. Furthermore, the Swedish government did not wish to undertake a bilateral exchange of notes with a nation which required it to agree not to take aggressive action against another state.[236]

This might seem strange at first, but it was explained by the fact that Sweden did not wish to set a precedent to be used by the Soviet Union in a possible future suggestion by them of some form of non-aggression treaty. Hammarskjöld assured Butterworth that Sweden certainly did not have any aggressive intentions, and that if U.S. law required that such a promise be given then the Swedish government was prepared to issue a unilateral declaration as it had in the case of East-West trade. Butterworth now requested further instructions from Washington on where to go from here.[237] From Butterworth’s cable it is not possible to discern whether the Swedes had any other objections, but it is most likely they did not. If they had had any further protestations, he would surely have included them in the cable as well. Nothing was said about the paragraph in Article 1, which said that a receiving nation *had* to provide assistance to the United States and other MDAA countries to further the policies and purposes of the MDAA.

While this was going on, Ambassador Butterworth sent a telegram to Acheson on April 30, 1952, in which he elaborated on how the Swedish-American defense cooperation ought to be arranged. Butterworth cited Åström’s words concerning how far the Swedes were willing to take the collaboration scheme. Wrote Butterworth:

> […] it would be highly desirable that Sweden coordinate defense plans with West as basis coop in event of war and that such coop shld be carried out with secrecy and at least tacit consent of Swed Govt. […]

235 *Had There Been A War…*, p. 123.

236 758.5/4-2652, telegram from Butterworth to Acheson, dated April 26, 1952, in: RG 59; DF 1950–54; from 758.13/1-852 to 758.5MSP/12-1554; Box No. 3773; NARA II, College Park, MD. See also: *Östen Undén. Anteckningar 1918–1952*, p. 399.

237 758.5/4-2652, telegram from Butterworth to Acheson, dated April 26, 1952, in: RG 59; DF 1950–54; from 758.13/1-852 to 758.5MSP/12-1554; Box No. 3773; NARA II, College Park, MD.

> There is a barrier in developing even covert strategic relationships between Swed and West beyond which Swedes will not (rpt not) go *lest the covert becomes known or threatens to become known.* This barrier cannot (rpt not) be considered as static: It moves forward and backward in accordance with Swed estimates of urgent need for defense collaboration, possibility of secrecy being maintained and secondary factors. [...] As to promote this relationship, the final recommendation in my desp[atch] 1227 of June 15, 1951 stated that: "The US (and SHAPE) shld consider encouraging the Nor and Dan Govts to cooperate, or even invite, Swed overtures re joint planning and strategy. The Swed mil[itary] fully understands the need for such coop; [...]. Swed is likely to prefer that all such planning be arranged on a clandestine basis and wld be more likely to respond favorably to the arrangements handled directly between Scandinavian mil auths [authorities] [...]."[238] [Italics added.]

Butterworth reasoned that any collaboration between the Scandinavian countries would be much easier for the Swedish public to swallow if the preparations should become known, and that since the Soviets thought that much more was going on that there actually was (although he confessed to be a bit uncertain as to the extent of the cooperation himself), this left some leeway for the Swedes before the Soviets could object. He was sure that the Swedes must appreciate this point. It was logical, said the Ambassador, that joint planning efforts between Sweden and the West should be undertaken through the instrumentality of the Norwegians.[239] He continued by stating:

> Swedes will, of course suspect that any covert defense coordination with Nors will be reported to NATO. At present Swedes cannot (rpt not) be expected, at best, to move more than cautiously forward in extending area of collaboration, but to do even this wld require

238 758.5/4-3052, Telegram from Butterworth to Acheson, dated April 30, 1952, p. 1, in: RG 59; DF 1950–54; from 758.13/1-852 to 758.5MSP/12-1554; Box No. 3773; NARA II, College Park, MD. The whole document has been cited *in extenso* by Paul Cole (Cole, P. M., *Neutralié du jour:...*, pp. 358–362), and has been referenced by the Neutrality Policy Commission (*Had There Been A War...*, pp. 145f), as well as by Wilhelm Agrell (Agrell, W., *Den stora lögnen...*, pp. 113–114), and Mats Berdal (Berdal, M., R., *The United States, Norway and...*, pp. 137, 140). Berdal stated that Butterworth sent the telegram in April 1953, which is incorrect; this error sprung out of the fact that he based his narrative on Paul Cole, whose footnotes are very confusing in this case (see: Berdal, M. R., *The United States, Norway and...*, p. 137; and: Cole, P. M., *Neutralié du jour:...*, pp. 359–362).

239 758.5/4-3052, Telegram from Butterworth to Acheson, dated April 30, 1952, p. 2, in: RG 59; DF 1950–54; from 758.13/1-852 to 758.5MSP/12-1554; Box No. 3773; NARA II, College Park, MD.

> making it clear to them that they wld not (rpt not) be tied to NATO in anyway [sic]. The details of Swed-Nor defense coordination must. Therefore, remain a matter to be worked out largely between those two countries. This may be frustrating at times to US and UK mil planners but I think it offers the best initial solution for getting maximum defense collaboration from the Swedes in their present frame of mind. [...] Then as Swedes find that they can extend the degree of covert coop with Nors without leaks or public repercussions they will cease to be fearful of dealing with Nor acting, as she gradually will be, as the agent of NATO.[240]

Intentionally or not, the cooperation between Sweden, Norway and NATO seems to have been handled in great concordance with the strategy laid out by Butterworth in this telegram.[241] The intimate cooperation between Sweden and Norway on issues related to NATO, took on almost ridiculous proportions in some instances. For example, a Swedish officer, Sven Hasselrot, stated to the Neutrality Policy Commission that he at one point in the 1950s [unclear exactly when] partook in a NATO conference on radar issues *disguised* as a Norwegian.[242]

Butterworth had furthermore stressed the imperative issue of standardizing Sweden's military equipment so that eventual collaboration in wartime could be facilitated. He said that the Swedes were aware of this need, and that the Chief of Weapons of the RSA as well as the Artillery Director of Bofors had approached the Embassy independently of each other, wishing to discuss this issue.[243] This fit well with what we know regarding American objectives in Scandinavia. As Mats Berdal has shown, most evidence suggest that "it was official US policy to encourage the highest degree of interdependence in terms of equipment, force posture and planning between Norway and Sweden" at this time.[244]

In the meantime, the negotioations regarding the MDAA agreement continued unabated. The State Department had no problem accepting the first two Swedish objections to the American draft, but it did state that the last one would not be so simply met, since renunciation of aggression

240 Ibid., pp. 2–3. See also: Cole, P. M., *Neutralié du jour:...*, p. 360.

241 For a good overview of the defense cooperation with Norway, see: Petersson, M., *"Brödrafolkens väl"...*, pp. 223–290. Petersson refers to Butterworths telegram, but only through secondary sources (see, ibid., p. 244).

242 "Neutralitetspolitikkommissionen – samtal med Sven Hasselrot", 1993-04-02, in: Box No. Regeringsbeslut 1995-06-29; NPCA; Riksdagens arkiv- och dokumentcenter.

243 758.5/4-3052, Telegram from Butterworth to Acheson, dated April 30, 1952, p. 3, in: RG 59; DF 1950–54; from 758.13/1-852 to 758.5MSP/12-1554; Box No. 3773; NARA II, College Park, MD.

244 Berdal, M. R., *The United States, Norway and...*, p. 137.

was statutory, i.e. required by U.S. law, and something along those lines thus had to be included in the exchange of notes. The Department said, however, that any reference to this phrase would of course be omitted when the exchange of notes was published and registered with the United Nations. On the American side they preferred an exchange of notes on the matter, since there was no guarantee that even a unilateral statement by the Swedish government would not at some point become known (even a unilateral statement might have to be registered with the U.N.). Acheson included the proposed changes to the agreement in a message to Butterworth on May 6.[245]

So what had happened to the phrase in Article 1, which implied that Sweden had to furnish assistance to the other MDAA signatories including the United States? Well, in the proposal, any explicit reference to the MDAA had been omitted, but Article 1 was otherwise unrevised: logically, since the Swedes had not said anything about that clause. In Article 2 the part which spoke of 'the defense of the area of which it is a part' had been taken out, in accordance with the wishes of the Swedish government. All references to the MDAA had been replaced by the phrase 'this agreement'; thus the spirit of the agreement was the same because 'this agreement', as far as the Americans were concerned referred to the MDAA.[246] Hammarskjöld regarded the most important victory after this first round to be the fact that the Swedes had been able to excise all references to U.S. law. He believed, however, that the new American proposal could be watered down even more. The non-aggression issue could perhaps be solved by referring to the U.N., and making a pledge not to attack another nation in connection to that.[247]

The Swedes turned a new proposal over to Butterworth in late May. The controversial phrase contained in Article 1 in the U.S. draft had now been removed. The Ambassador promised to forward the draft to Washington, and he said that he thought that the Swedish text ought to be acceptable to the U.S. government. After further negotiations it was revealed that the Swedes wished to exclude the statement contained in the original draft that the exchange of notes would constitute an

245 758.5-MSP/5-652, telegram from Acheson to Butterworth, dated May 6, 1952, p. 1, in: RG 59; DF 1950–54; from 758.13/1-852 to 758.5MSP/12-1554; Box No. 3773; NARA II, College Park, MD. Memorandum by Hammarskjöld, dated May 19, 1952, p. 1, in: HP 24 VI/U.S.A; "Tillstånd till in- och utförsel av vapen, ammunition och krigsmateriel"; 1950 sept.–1952; Folder 21; Box No. 1353; År: 1946–1952.

246 Ibid., pp. 1–2. That the United States regarded the exchange of notes as basically the same as a formal signing of agreements, has been shown by Silva (Silva, C., *Keep Them Strong, Keep Them Friendly:…*, p. 338).

247 Memorandum by Hammarskjöld, dated May 19, 1952, p. 2–3, in: HP 24 VI/U.S.A; "Tillstånd till in- och utförsel av vapen, ammunition och krigsmateriel"; 1950 sept.–1952; Folder 21; Box No. 1353; År: 1946–1952.

agreement. The State Department thought that this was doable, but remarked that it could not guarantee that it would not be referred to as such at a later time, e.g. possibly in connection with the registration of the exchange with the U.N. The Swedish government also suggested that the exchange be altered so that the American reply to the Swedish note would only include a simple acknowledgment.[248] On June 13 Butterworth told Hammarskjöld that the Swedish counter-proposal had been accepted by Washington.[249]

The exchange of notes took place on June 30, with a Swedish note signed by Foreign Minister Östen Undén, and on July 1, with a short American acknowledgement signed by Butterworth.[250] The Swedish note did not refer to the Articles in the draft proposal, but contained assurances in the spirit of the draft. For example, the Swedish government stated that the materials and services acquired would only be used to maintain Swedish internal security, and to permit it to partake in U.N. collective security measures, and also that Sweden would not relinquish its possession of these materials without the explicit consent of the U.S. government. It also said that Sweden's foreign policy would continue to be guided by its obligations as a member of the U.N., and Sweden would thus not undertake any act of aggression against another state.[251]

Secretary of State Marshall had perceived the military assistance program to be the most efficacious method of building up the defense of the European landmass. The alternative would have been an increasing U.S. military presence in Europe, something which American budgetary restraints, coupled with other global military undertakings, did not permit. In 1948, President Truman had perpetuated the gap between

248 758.5-MSP/6-952, telegram from Acheson to Butterworth, dated June 9, 1952, pp. 1–2, in: RG 59; DF 1950–54; from 758.13/1-852 to 758.5MSP/12-1554; Box No. 3773; NARA II, College Park, MD. Letter from Hammarskjöld to Boheman (and attached draft note), dated May 30, 1952, pp. 1–3, in: HP 24 VI/U.S.A; "Tillstånd till in- och utförsel av vapen, ammunition och krigsmateriel"; 1950 sept.–1952; Folder 21; Box No. 1353; År: 1946–1952.

249 Telegram from the Foreign Ministry to the Swedish Embassy in Washington, dated June 13, 1952, in: HP 24 VI/U.S.A; "Tillstånd till in- och utförsel av vapen, ammunition och krigsmateriel"; 1950 sept.–1952; Folder 21; Box No. 1353; År: 1946–1952.

250 *Had There Been A War…*, p. 123; SÖ 1954:74. See also: Notes by Undén and Butterworth, dated June 30 and July 1 respectively, in: RG 330; ISA, OMA; PDF Apr 1949–May 1953; 333 Spain to 091.3-523 Thailand (File: 091-523 Sweden 1950+51); Box no. 77; (Folder: 113 Reimbursements and transfers Sweden 1952); NARA II, College Park, MD. Letter from Undén to Butterworth, dated June 30, 1952, pp. 1–2, in: HP 24 VI/U.S.A; "Tillstånd till in- och utförsel av vapen, ammunition och krigsmateriel"; 1950 sept.–1952; Folder 21; Box No. 1353; År: 1946–1952.

251 SÖ 1954:74. See also: Jonter, T., "Det amerikanska spåret…", p. 145.

means and ends when he endorsed NSC 20/4, a policy document that outlined the central strategy and tenets of U.S. national security policy.[252] This document was tied to, and acted as political guidance for, the American war plan Offtackle, and established U.S. political objectives toward the USSR. It envisioned the political breakup of the Soviet state, and a postwar order where Russian power and influence had been reduced to a level which did not threaten the peace and stability of the world. Offtackle demonstrated how difficult it was to attempt to respond to such ambitious policy objectives with inadequate forces. Despite complaints from the JCS the limited budget continued to haunt the Pentagon until the outbreak of the Korean War, and the concommitant approval of NSC 68 in September 1950. Military spending soared from $14.258 billion in FY 1950 to $53.208 billion in FY 1951 and to $65.992 billion in FY 1952.[253]

The American Minster to Iceland Edward B. Lawson also made a small contribution to the effort to bring about closer ties between Sweden and NATO in early September 1952. The Foreign Ministers of Norway, Denmark, Sweden, and Iceland had their annual meeting in Reykjavik, and Lawson took the opportunity to propagate for the American cause with Swedish Foreign Minister Undén. According to Lawson, who had met Undén once before, this time around the Swedish Foreign Minister was "much more affable and showed a much better disposition to discuss quite freely subjects which I raised."[254] With the double purpose of affording normal courtesies and to stimulate Undén's interest in NATO and the advantages of collective defense, the American Minister arranged a visit to the Keflavik Air Field through the Commander of the Icelandic Defense Force General Brownfield. If one is to believe Lawson, Undén "expressed [...] very great appreciation."[255]

It is worth noticing that NATO planning for a general war in Europe at this time, just like the American Dropshot plan (see Chapter 3, p. 156), stressed the importance of defending Scandinavia as a whole entity. On December 9, 1952, the North Atlantic Military Committee (henceforth: NAMC) approved the NATO Standing Group's strategy guidance document MC 14/1, which contained a detailed study of the problems of the defense of Scandinavia. In this document it was stated that:

> [t]he defense of Norway and Denmark must take account of Sweden,

252 Leffler, M. P., *A Preponderance of Power:...*, pp. 264–265; see also: Ross, S. T., *American War Plans 1945–1950*, pp. 79–80.

253 Ross, S. T., *American War Plans 1945–1950*, pp. 103, 111–112, 139.

254 Memorandum of Conversation between Lawson and Undén, dated September 4, 1952, p. 1, in: RG 59; MOFASSEA, 1943–1957; Box No. 27; (Folder: Sweden 1948–1955); NARA II, College Park, MD.

255 Ibid., p. 2.

which lies in the path of any attack against Scandinavia from the east, particularly as Russia cannot ignore the possibility of Sweden joining the Allies should another Scandinavian country be attacked.[256]

Exactly what was implied by this cryptic formulation is of course hard to say with any certainty. What did 'take into account' actually mean in this context? Did it imply that NATO force requirements in the area did not have to be as substantial as they would have to be if NATO could not count on support by Swedish forces in the defense of Norway and Denmark? What is clear is that NATO did not consider it an impossibility that Sweden might come to the assistance of its neighbors if they were attacked; even if Sweden was *not* attacked simultaneously, *or prior to* this. The Norwegian and Danish representatives to the NAMC must thus have been confident enough to assume that the cooperation that was going on between the three countries suggested that there was a good possibility that Sweden would assist them, in order for them not to voice any serious opposition to the assumptions of MC 14/1.

The document continued to say that because of this:

> [...] it is clear that Scandinavia *must* be defended as a whole [...]. Plans for the defense of Norway and Denmark, *and also of Sweden should this be possible*, must therefore be integrated. These plans should be based on the need for the utmost flexibility, to permit concentration to deal with surprise attacks before they can be reinforced.[257] [Italics added.]

The MC 14/1's view of Scandinavia's role in NATO strategy had its roots in the general reappraisal, and increased significance, of the area in U.S. military planning in 1950. The United States now envisaged using NATO's southern and northern flanks as launching pads for the defence of central Western Europe.[258] The paragraph above contains several intriguing factors, regarding the planning for the defense of the area. Swedish assistance was considered by NATO to be a necessity for the defense of the rest of the Scandinavian Peninsula, as implied by the use of the word 'must'. The uncompromising tone of this sentence was slightly challenged by the more loosely formulated following sentences.

256 MC 14/1 (Final), dated December 9, 1952, "Decision on M.C. 14/1: A Report by the Standing Group on Strategy Guidance", p. 19, accessed at: http://www.nato.int/docu/stratdoc/eng/a521209a.pdf (2007-03-06); see also: Weinberger, H., "A Small/Medium-sized Great Power in the North..." in K. L. Kleve (ed.), *The Cold War, Military Power...*, p. 55; Tamnes, R., *The United States and the...*, pp. 144–145.

257 Ibid., pp. 20–21. See also: Dalsjö, R., *Life Line Lost:...*, p. 60.

258 Tamnes, R., *The United States and the...*, pp. 63–89.

Defense plans were only to integrate Sweden to the extent this was deemed possible. Apparently, it was not a total disaster if defense plans had not been coordinated when Sweden joined up with Norway and Denmark. By the time MC 14/1 was approved, Swedish consent to American hegemony had been secured by Washington, and Swedish assistance in the defense of Scandinavia more or less taken for granted by the Americans. But the document was not unambiguous, because it was also stated that even though Sweden, due to her strategic location and her economic, industrial, and military strength, could play an important part in the defense of Scandinavia she was not a member of NATO and was "likely to remain neutral unless attacked."[259]

The latter was identical to a statement made by Norway's permanent representative to NATO Dag Bryn in Februari 1952, when he had said that Sweden would not *fight* unless attacked.[260] The Norwegian's apprehensions thus seems to have made its way into MC 14/1. Bryn's statement was not original, however. The Americans had made the same assumption in the Dropshot plan in 1949.[261] This clearly contradicted the assumption made in the same document (in fact on the following page) that the Soviets had to count on Sweden joining the Allies even if she was *not* attacked. It does seem, however, as if the former statements laid closer to the true expectations upon Sweden. Raynor at BNA commented, with regard to Sweden remaining neutral unless attacked, that he thought that this underestimated Sweden's possibilities, and that he believed that she would probably fight if one or more of her northern neighbors were attacked.[262]

259 MC 14/1 (Final), dated December 9, 1952, "Decision on M.C. 14/1: A Report by the Standing Group on Strategic Guidance", p. 18, accessed at: http://www.nato.int/docu/stratdoc/eng/a521209a.pdf (2007-03-06).

260 Petersson, M., "'She would not *fight* unless attacked'..." in *Kungl krigsvetenskapsakademiens handlingar och tidskrift*, No. 3, 2002, p. 87. Bryn continued by saying that Sweden would tie down considerable Russian forces anyway, and hence in practice act as a flank guard for Norway and, to a lesser extent, Denmark. One has to assume that what he meant was that a neutral Sweden would *indirectly* bind considerable Russian forces because of the Soviet need to spare divisions in case Sweden decided to fight, even if she was not attacked (ibid., p. 87). Rolf Tamnes has shown that the Norwegians were instrumental in persuading the Alliance of Sweden's ability and will to defend itself (Tamnes, R., *The United States and the...*, p. 83).

261 Ross, S. T., *American War Plans 1945–1950*, pp. 119–120. The same had been assumed with regard to Ireland, Spain, Switzerland, Greece, Turkey, Pakistan, and the Arab League members (ibid., p. 120).

262 "Comments on Draft of Report by the International Planning Team to the Standing Group on Strategic Guidance", Raynor to Wolf, dated October 8, 1952, pp. 1–2, in: RG 59; Miscellaneous Lot Files; Lots 57 D 321, 323, & 324; Office Files of Leonard Unger, 1951–1956; Box No. 4; (Folder: NATO - Problems); NARA II, College Park, MD.

Raynor's comments (as well as the assumptions of MC 14/1) were consistent with the conclusions of the CIA's National Intelligence Estimate 49 (henceforth: NIE 49), presented on March 20, 1952, concerning Sweden's position in the East-West conflict. As Jerker Widén and Simon Moores have shown, the study was a joint product of the intelligence organizations of the State Department, the Army, the Navy, the Air Force, the JCS, and the CIA, and thus represented the view of the U.S. intelligence and defense communities.[263] It was expected that Sweden might join the NATO powers if the USSR attacked Norway, even if Sweden was bypassed by the Russians, if the Swedes thought that large-scale assistance could be received from the West. Furthermore, the NIE 49 stated that "Sweden would certainly fight if attacked.[264] The U.S. intelligence community's confidence in Sweden was thus very high.

The same supposition was made in NATO's "Capabilities Plan of the Allied Command Europe 1957" prepared in late 1953 and early 1954.[265] The Americans had reason to feel optimistic regarding the possibility of getting assistance from the Swedes. The Chief of the RSAF General Nordenskiöld intimated to Colonel Bernt Balchen, Special Assistant on Arctic Affairs to the Secretary of the Air Force, around this time, that in the event of an emergency, seven Swedish airbases could be made available for NATO in Sweden.[266] The Americans knew of course that there were no guarantees to this promise coming true, but it was an important indicator of the Swedish military establishment's sentiments. Balchen had met Nordenskiöld, together with Major General Westring, for the first time in November 1951, at which time Nordenskiöld had expounded on the possibilities of a secret joint defensive war plan between the Swedes and the Norwegians. The Chief of the RSAF had said that such a plan could be integrated with NATO planning, and that Swedish troops should be used on Norwegian soil in the Finnmark to hold the Boden-Finnmark area in the north.[267]

At the time, however, Balchen's optimism with regard to the possibility of realizing these ideas was not shared by either the USAF

263 Moores, S. and Widén, Jerker, "Sverige under det kalla kriget – den amerikanska underrättelsetjänstens perspektiv, mars 1952" in *Kungl Krigsvetenskapsakademiens handlingar och tidskrift*, 2003:5, pp. 78–92.

264 Ibid., pp. 84, 89, quote on p. 84.

265 SGM 600-54, "Capabilities Plan Allied Command Europe 1957", dated September 10, 1954, p. 5, in: North Atlantic Treaty Organization [henceforth: NATO]; Standing Group Records Section [henceforth: STAGRES]; NATO Archives, Brussels. I thank Hans Weinberger for making this document available to me. See also: Weinberger, H., "A Small/Medium-sized Great Power in the North..." in K. L. Kleve (ed.), *The Cold War, Military Power...*, p. 55.

266 Berdal, Mats, R., *The United States, Norway and the Cold War, 1954–60* (Basingstoke: Macmillan Press Ltd., 1997), p. 138.

267 Tamnes, R., *The United States and the...*, pp. 84–85.

or the American Embassy. Nevertheless, Balchen kept coming back to Scandinavia several times in the course of 1952–1954 to confer with the Scandinavians, and to sound out the prospects of secret Scandinavian defense cooperation.[268]

Conclusion

Two processes stand out when we assess how Sweden gave its consent to American hegemony during the late 1940s and early 1950s; these two processes were the participation in the Marshall Plan and the trade embargo of the Eastern Bloc. This chapter has studied the material forms in which the Swedish government expressed its consent to American hegemony, as well as how the U.S. went about securing it. I also looked more closely at the event that was at the same time a confirmation of U.S. acceptance of Swedish policy and a further consent to the United States as the preponderant power in world affairs: the MDAA agreement of June/July 1952.

Through the Marshall Plan Sweden partook in the political, economic, and military formation of Western Europe in 1948–1951, thus showing the American government that Sweden wished to adhere to the Western camp in the future. The Swedish government, however, tried to portray the Marshall Plan in purely economic terms, something which was ridiculed by Ambassador Matthews. According to him, the Marshall Plan had both political and military implications and the Swedes were now part of this parcel. Through her participation in the ERP Sweden consented to a curtailment of its autonomy in her foreign trade, and thereby affirmed, through its objective effects, her close ideological affinity to, and dependence upon, the West. Sweden took upon herself to actively support its Western European neighbors in their efforts to rebuild their war-torn countries, and since the Marshall Plan had been cunningly engineered to exclude the Soviets this also meant that there was to be no equal hospitality granted to the countries of Eastern Europe. Sweden's participation in the ERP was an absolute necessity if the U.S. government was to accede to Sweden buying military materials from the United States.

Even more important, however, was the Swedish government's agreement to adhere to the COCOM trade embargo. This question of hindering re-export of strategic materials to the East was very important to the United States government, and Sweden, with its policy of neutrality, risked becoming a hole in the embargo wall toward the Soviet Union and Eastern Europe. Sweden could not be allowed to procure defense materials until the Americans could be sure that such equipment would not be exported eastward. Sweden (along with Switzerland) turned out to be the most disciplined ally out of all the Western European nations,

268 Ibid., p. 85.

and was very firm in its implementation of the embargo rules. This was done by extending the term 'contraband' to cover goods on the COCOM lists.

Swedish guided missile development was running ever deeper into a *cul-de-sac* by the early 1950s and no help was forthcoming from the Americans. The difficulties encountered by the GMB in its efforts to realize an operational Swedish missile were directly connected to the American technological monopoly. However, Sweden did not gain access to American guided missile know-how even though the Swedish consent to American hegemony was recognized by the U.S. government by 1952. But I argue that without the Swedish consent, Sweden would not have gotten access to guided missiles even once the Americans decided to share this technology with its allies in the late 1950s.

The Americans kept a close eye on the Swedish development efforts through the reporting of the Military Attachés, and the Swedes tried widely different approaches in their efforts to get information from the Americans, reaching from suggestions of bartering one technology for another to more or less desperate calls to expose the state of Swedish R&D to the Americans without any demands for reciprocity in the hope that this would inspire reciprocal measures in the future. None of it worked. However, this was not due to any Swedish idiosyncrasy, such as the policy of neutrality, but was consistent with overall American policy with regard to guided missile technology, as will be shown in the next chapter.

In early 1952 President Truman declared that Sweden was now eligible to procure defense materials in the United States on a reimbursable basis, and this process culminated in the exchange of notes that summer which made Sweden a participant in the MDAA. Here, too, the Americans proved very flexible; indeed, the fact that there was only an exchange of notes and not a formal signing of agreements testified to this. The Swedes were also granted latitude with regard to the express content of the agreement. The Swedes were allowed to remove any explicit reference to the exchange as an agreement, as well as any explicit reference to American legislation.

All this was made with reference to the importance of non-alignment in order to preserve the option of neutrality in wartime, and thus points to the importance of this policy (or more correctly: the political effects a breach of this policy would have for the government) for the construction of the secret collaboration with the United States during this period. The Americans tried to get the Swedish government to sign a note which would have resulted in Sweden becoming an integrated part of the American defense effort around the globe, and they nearly succeeded. Even though the Swedes insisted upon considerable changes to the American draft, a crucial paragraph remained unaltered after the first round of negotiations, namely the one specifying that materiel and information purchased by Sweden with reference to the exchange of notes should be supplied upon request to the United States or any other signatory of the MDAA. The Americans had prepared two drafts, one of which was basically the same as the final version, but the American Ambassador was instructed not to tell the Swedes about this option before they had offered their views on the first draft. Clearly it was hoped that the Swedish government would accept the more stricter version of the agreement. However, when it became evident that the it would not, the U.S. government was willing to compromise, and in the end accepted a note drawn up by the Swedes. For the Americans the end result was essentially the same, although there is a good chance that the Swedes did not see things this way. Since the exchange of notes lacked an explicit reference to the MDAA, the Swedes probably did not consider themselves bound by it.

We have also seen that NATO thought that Scandinavia had to be defended as a whole, and that there was a good chance that Sweden would fight alongside the Allies if one of her neighbors was attacked, even if Sweden had both been attacked prior to this. While MC 14/1 was certainly a bit ambiguous in its statements, Raynor at the State Department considered this a likely possibility as well. Clearly, if the Swedish government was trying to persuade foreign governments that Sweden would remain neutral unless she was attacked it was not doing a very good job. Considerable doubt about the Swedish position seems to have existed at least within the Western camp. Prior research has shown that the USSR was not at all convinced by Sweden's policy either.[269]

269 Molin K., "The Central Issues…", in M. Gehler and R. Steininger (eds.), *The Neutrals and the European Integration*, pp. 271–274.

Chapter 5

The Turning Tide:

How Sweden Gained Access to Guided Missiles, 1953–1958

[i]f we do not now [...]provide our allies with the newer weapons in being in these years, our relative deterrent and fighting capabilities will be diminished to a serious degree. *If we should fail to make this provision, our allies would be keenly aware of this changed situation, and the impact thereof on their political attitudes could be drastic.* Also, the deterrent effect of our defenses on the Soviet forces would be weakened. [Italics added.]

(Memorandum for the Chairman of the JCS, March 12, 1956)[1]

[the] number one priority [of domestic Swedish missile R&D] has been, and still is, to acquire information and experiences [...]. *Hereby we educate ourselves to become competent buyers of fully developed foreign missile systems* [...]. [Emphasis in original.]

(Major Hans Neij, September 28, 1960.)[2]

Introduction

This chapter addresses the second set of questions introduced in Chapter 1. If we want to correctly understand the Swedish difficulties in getting access to American guided missiles during the 1950s it is essential that we study the general U.S. disclosure policy in this field, as well as the developments within NATO during this period. As it turned out, the American decision to grant Sweden access to this technology had very little, if anything, to do with the idiosyncrasies of Swedish foreign and security policy. But first we look briefly at a Swedish effort to contribute very concretely and materially to the U.S. and NATO's defense of Western Europe.

1 Memorandum for the Chairman of the JCS, dated March 12, 1956, p. 6, in: RG 218; Chairman's File [henceforth: CHF]; Admiral Radford [henceforth: AR], 1953–57; 471.6 (1 Aug 56) to 500; Box No. 44; (Folder: 471.6 (1956)); NARA II, College Park, MD.

2 *Vestmanlands Läns Tidning*, "Svensk robotpolitik", September 28, 1960.

Sweden and the Off-Shore Procurement Program

In the spring of 1953 the Americans and the Swedes had found a way for Sweden to contribute to the European defense effort. The United States wished to see Swedish participation in NATO's Off-Shore Procurement Program (henceforth: OSP), whereby production of military material for NATO and U.S. forces in Europe would be placed in Western Europe instead of in the U.S. The Swedes were more than willing to assist, for both economic and political reasons.

By April 1953 the American Embassy was arguing forcefully for placement of prime OSP contracts in Sweden. The Counselor for Economic Affairs DuWayne G. Clarke wrote to Grant Hilliker at BNA, telling the latter how dependable Sweden was as a source of supply:

> Sweden, while not a NATO country, is probably one of the best prepared European countries, prepared in the sense that it has an excellent Airforce, a small but efficient Navy, and a well trained Army. There need be no question about the Swedes being willing and prepared to defend their country against any form of aggression, and [...] invasion of this country would present difficulties far outweighing any advantage to be derived. In other words, as I see it, *as a source of supply*, while the Swedes may not be within the NATO area proper, *they certainly are in a dependable and comparatively secure position.*[3][Italics added.]

According to Butterworth it was quite clear that the Swedes were keenly interested in participating in the OSP, particularly as prime contractors.[4] According to the U.S. Mission to the NATO and European Regional Organizations (henceforth: USRO) in Paris, American policy required there to be a political, economic or military advantage to the United States if prime contracts were to be placed there, and it was believed that problems could arise if the Swedish government was to become too optimistic as to the possibility of receiving such contracts. But Sweden took part in the program in other ways; USRO stated that several "non-MDAP OSP contracts have already been placed there for U.S. forces".[5] Certain prime contracts had also been awarded the

3 758.5/4-2453, Letter from Clarke to Hilliker, dated April 24, 1953, pp. 1–2, quote on p. 2, in: RG 59; DF 1950–54; from 758.13/1-852 to 758.5MSP/12-1554; Box No. 3773; NARA II, College Park, MD.

4 Telegram from USRO to the Secretary of Defense, dated April 27, 1953, p. 1, in: RG 469, 1948–61; USRO; Executive Secretariat [henceforth: EXSEC]; COUNF, 1952–55; Spain-Turkey; Box No. 22; (Folder: Sweden); NARA II, College Park, MD.

5 Ibid., p. 2.

Swedes, who then in turn had subcontracted to certain NATO countries, particularly to Norway for explosives and propellants.[6]

In this way, Sweden "could indirectly contribute to the defense of NATO by producing military end items for NATO forces [...]."[7] However, it was not thought that any major contracts could be placed in Sweden since political and economic factors worked against it. It was believed that the NATO countries, as well as interests in the United States, would not like it very much if funds targeted for production of defense materiel within NATO were siphoned off to a non-NATO producer. There was still a worry, despite the assurances from the Stockholm Embassy, that a neutral nation might not be a reliable source of supply in time of war. Sweden's exposed position, wrote USRO, "raised the strategic question whether in time of war it could effectively supply NATO nations other than Norway from production fostered through OSP."[8]

That Sweden was planning to supply NATO with military hardware, and to a certain extent was already doing this, has previously been unknown to the scholarly community, and raises a host of questions. How could the Swedish government even begin to contemplate such a scheme, when it was otherwise so sensitive to possible negative publicity with regard to its policy of neutrality? What kind of items had Sweden already produced for the American forces? These questions are not likely to be answered in the near future, but they are worth bearing in mind when discussing Sweden's postwar behavior. It was, however, recommended that the Stockholm Embassy should evaluate the advantage to the United States of specific types of OSP production in Sweden, and then advise Washington regarding the result of the evaluation.[9]

The Swedish government delegated to the Swedish Export Association (Sveriges Allmänna Exportförening), a quasi-official organization helping Swedish companies find markets abroad,[10] to act as an intermediary on matters regarding OSP affairs. This was done because, as the U.S. Embassy in Stockholm explained, the organization was headed by Torsten Vinell, a retired Swedish Commercial Attaché, and it was thus "uniquely qualified to advise re [regarding] production

6 Telegram from USRO to U.S. Embassy in Stockholm, dated July 18, 1953, p. 4, quote on p. 2, in: RG 469, 1948–61; USRO; EXSEC; COUNF, 1952–55; Spain-Turkey; Box No. 22; (Folder: Sweden); NARA II, College Park, MD.

7 Telegram from USRO to the Secretary of Defense, dated April 27, 1953, p. 2, in: RG 469, 1948–61; USRO; EXSEC; COUNF, 1952–55; Spain-Turkey; Box No. 22; (Folder: Sweden); NARA II, College Park, MD.

8 Ibid., p. 2.

9 Ibid., p. 3.

10 This organization still exists; now called the Swedish Trade Council (or: Exportrådet).

possibilities in Sweden if it can be informed as to types and kinds of materials on which Sweden might bid." The Export Association had already sent a representative to the Procurement Office in Frankfurt in November 1952, but with a negative result. Wrote Butterworth:

> Association desires identification of organization or organizations in Paris or elsewhere through which it can determine possible off-shore procurement orders and to which it can arrange submission of bids by Swedish industries.[11]

The Swedish government knew full well what the OSP was about. In early March 1953 the Foreign Ministry sent a memorandum containing official U.S. statemens regarding the purpose of the program to the Export Association.[12] There can thus be no doubt that the Swedish government was seeking, in 1952–53, to contribute materially to NATO's defense of Western Europe by means of active participation in the OSP. While this was surely being done with an eye to the potentially very lucrative contracts involved, and argued for in terms purely of the national economy, it would seem to have been a big step away from a credible policy of neutrality, regardless of what the government did or did not intimate in its declarations on foreign policy at the time,[13] considering that the government did not offer the USSR the same courtesy (at least in the eyes of the Russians). As in the case of Sweden's participation in the Marshall Plan, it seems as if this action was rationalized by the assumption that economic choices did not entail political consequences. However, just as when it participated in the ERP, the Social Democratic government was in fact expressing its consent to American hegemony in Western Europe.

The disingenuousness of this reasoning becomes obvious if one imagines what it would have meant for the credibility of Swedish neutrality in the West if Sweden would have participated in a program designed to boost Soviet war potential by producing and selling weapons *only* to them. A reason why it is not usually seen as a problem to export defense-related items is the idea that trade is somehow devoid of political content. Once the tables are turned it becomes apparent that this

11 Telegram from Butterworth to Washington, dated May 4, 1953, p. 1, in: RG 469, 1948–61; USRO; EXSec; COUNF, 1952–55; Spain-Turkey; Box No. 22; (Folder: Sweden); NARA II, College Park, MD.

12 "P.M. ang. Officiella uttalanden om offshoreprogrammet", dated March 10, 1953, in: Sveriges Allmänna Exportförening; Serie: Korrespondens, samlingsserie; Signum: E I a; Box No. 1257; RiA, Arninge.

13 Dalsjö, R., *Life-Line Lost:...*, pp. 70–74. Dalsjö shows that the government's declarations allowed considerably more latitude with regard to military cooperation with foreign powers than has traditionally been thought.

argumentation is just a façade; as long as the trade does not strengthen the enemy it can be viewed as purely economic, but if the opposite is the case then trade becomes very political indeed. In fact, COCOM was essentially a colossal admission that trade had political consequences. The Swedish Foreign Ministry knew about Sweden's participation in the OSP program, for it had on its own initiative approached Butterworth and asked to be kept fully apprized of the matter. The problem now, wrote the Ambassador, was how to arrange for the Export Association to be kept informed regarding opportunities for Sweden to participate in the OSP, and he requested that Washington send some guidance that could be passed along to the Export Association.[14]

Later in the summer of 1953 USRO in Paris stated the problems of placing prime contracts in Sweden in a telegram to the State Department. After conferring with Ambassador Butterworth, USRO stated again that the ongoing evaluation concerning the placement of OSP contracts in Sweden should not be allowed to raise false hopes with the Swedish authorities. Although there were some items which it might be suitable to produce in Sweden, USRO concluded that they were "not (rpt) not optimistic re[garding] placement there of MDAP OSP contracts [...]."[15] The reasons for this were legion and had to do with, for example, the fact that ample capacity was expected to be available within the NATO area during 1954, and the lack of U.S. tax exemption in Sweden.[16] The risk of the NATO countries feeling cheated was, as stated above, another reason for not awarding prime contracts to Sweden. The practice of Sweden's subcontracting to other NATO countries "appears", stated USRO, "as in the past to be Sweden's most practical avenue for participation in the MDAP-OSP program."[17]

On October 27, 1953, Sven Dahlman at the Swedish Embassy in the Hague, informed the Foreign Ministry that the Dutch branch of the American aircraft engine manufacturer Curtiss-Wright wished to contact Swedish companies which could deliver spare parts etc. for the production of jet engines. The request covered everything from nuts and bolts, machine parts, tojet engine accessories. Dahlman wrote that Curtiss-Wright had received large OSP contracts, to a value of $27

14 Telegram from Butterworth to Washington, dated May 4, 1953, p. 1, in: RG 469, 1948–61; USRO; EXSEC; COUNF, 1952–55; Spain-Turkey; Box No. 22; (Folder: Sweden); NARA II, College Park, MD.

15 Telegram from the USRO to U.S. Embassy in Stockholm, dated July 18, 1953, pp. 1–2, quote on p. 2, in: RG 469, 1948–61; USRO; EXSEC; COUNF, 1952–55; Spain-Turkey; Box No. 22; (Folder: Sweden); NARA II, College Park, MD.

16 Ibid., p. 3.

17 Telegram from the USRO to U.S. Embassy in Stockholm, dated July 18, 1953, p. 4, quote on p. 2, in: RG 469, 1948–61; USRO; EXSEC; COUNF, 1952–55; Spain-Turkey; Box No. 22; (Folder: Sweden); NARA II, College Park, MD.

million, at the end of June to be used in the Netherlands and other NATO countries. The company was to produce spare parts for aircraft engines delivered to Europe from the United States.[18] The Swedish companies were thus not intended to provide spare parts to Curtiss-Wright so that it could manufacture engines, the Swedes were being asked to contribute directly by producing such parts for NATO. This intention was abundantly clear to the Foreign Ministry, and there can not have been any doubt as to the purpose of these contracts.

The request was forwarded to SFA, but the company, which had been in touch with Curtiss-Wright before, was unable to assist at this time due to current production for the RSAF.[19] It is not clear if SFA had manufactured parts for Curtiss-Wright, or if the two companies had only corresponded prior to this request. The government had written to the Export Association on November 5 asking it to provide a list of Swedish companies able to satisfy Curtiss-Wright's request. A week later the Export Association replied and included a long list of possible candidates for the different tasks involved. Among the companies listed by the Association were SFA, Bofors, STAL, SAAB, and SKF.[20] Later in November B. A. Odhner at the Foreign Ministry informed Captain af Klint at Bofors regarding American requests for production of ammunition. Bofors let Odhner know that the company was indeed very interested, and Bofors could provide the Americans with NATO-standardized 10.5 centimeter haubits, and 90 millimeter anti-aircraft gun ammunition. The company also specified how many shells per month it could be produce. The Foreign Ministry replied the next day that this information had been forwarded to the Swedish Embassy in Washington, and that Bofors would be notified as soon as a reply had been received from the Americans.[21]

18 Letter from Dahlman to Undén, dated October 27, 1953, pp. 1–3, in: HP 24 VI/U.S.A; "Tillstånd till in- och utförsel av vapen, ammunition och krigsmateriel"; 1953–juni 1955; Folder 22; Box No. 62; År: 1955–1960; RIA; Arninge.

19 Leter from the Director of SFA G. Engellau to the Foreign Ministry, dated November 9, 1953, in: HP 24 VI/U.S.A; "Tillstånd till in- och utförsel av vapen, ammunition och krigsmateriel"; 1953–juni 1955; Folder 22; Box No. 62; År: 1955–1960; RIA; Arninge.

20 Letter from the Export Association to the Foreign Ministry, dated November 12, 1953, pp. 1–2, in: HP 24 VI/U.S.A; "Tillstånd till in- och utförsel av vapen, ammunition och krigsmateriel"; 1953–juni 1955; Folder 22; Box No. 62; År: 1955–1960; RIA; Arninge. For the same, see: Letter from C. F. Gadde at the Export Association to the Foreign Ministry, dated November 26, 1953, in: Sveriges Allmänna Exportförening; Serie: Korrespondens, samlingsserie; Signum: E I a; Box No. 1257; RIA, Arninge.

21 Letter from Arvid Larsson of Bofors to Odhner, dated November 19, 1953; letter from Odhner to Bofors, dated November 21, 1953, in: HP 24 VI/U.S.A; "Tillstånd till in- och utförsel av vapen, ammunition och krigsmateriel"; 1953–juni 1955; Folder 22; Box No. 62; År: 1955–1960; RIA; Arninge.

The production of ammunition and spare parts was not a small or unimportant matter within the OSP. On the contrary, official American statements about the program had made clear that particular emphasis would be given "the production of ammunition and spare parts in Europe", because of the "obvious necessity for close sources of supply of these items." This was well known to the Swedish government because this information was included in the official U.S. statements that the Foreign Ministry had sent to the Export Association on March 10. In fact, the section quoted above had been underlined by the Foreign Ministry.[22] The Export Association had provided the government with a list of companies that could possibly contribute to the NATO arms buildup.[23]

It is not known, at present, how many OSP contracts were awarded Swedish companies during the following years. Already in February 1954 an American source had told Ambassador Boheman that the United States was considering excluding Sweden (as well as Switzerland, Israel and Yugoslavia) from the OSP. Boheman stated that he would bring the matter up with Deputy Secretary of Defense Keys, who was influential in the matter, and argue the Swedish case. He would tell Keys that Sweden wished to retain an effective armaments industry, and also bring up the purely commercial value of these orders to Swedish companies.[24] However, Hägglöf received a negative response from Hughes of the USRO and Tracy Voorhees of the OSP just a couple of days later. Sweden would no longer be eligible for prime contracts within the OSP, said the Americans. Nevertheless, Hägglöf was assured that Swedish businesses could continue participation as subcontractors.[25] The matter surfaced again a year later in May 1955, when a cut in the funding for OSP necessitated a decrease in the number of contractor nations. Secretary of State Dulles wrote to the U.S. Embassy in Stockholm that:

> [...] it was considered necessary generally to limit OSP contracts to NATO countries and certain other countries with which MDAP

22 "P.M. ang. Officiella uttalanden om offshoreprogrammet", dated March 10, 1953, p. 2, in: Sveriges Allmänna Exportförening; Serie: Korrespondens, samlingsserie; Signum: E I a; Box No. 1257; RiA, Arninge.

23 Letter from C. F. Gadde at the Export Association to the Foreign Ministry, dated November 26, 1953, in: Sveriges Allmänna Exportförening; Serie: Korrespondens, samlingsserie; Signum: E I a; Box No. 1257; RiA, Arninge.

24 Leter from Boheman to the Foreign Ministry, dated February 16, 1954, in: HP 24 VI/U.S.A; "Tillstånd till in- och utförsel av vapen, ammunition och krigsmateriel"; 1953–juni 1955; Folder 22; Box No. 62; År: 1955–1960; RiA; Arninge.

25 Letter from Hägglöf to the Embassy in Washington, dated February 18, 1954, in: HP 24 VI/U.S.A; "Tillstånd till in- och utförsel av vapen, ammunition och krigsmateriel"; 1953–juni 1955; Folder 22; Box No. 62; År: 1955–1960; RiA; Arninge.

> agreements were in effect. Under the circumstances it is unlikely that Sweden will receive any contracts except where justified by compelling reasons. It is the Department's understanding that *this policy does not apply to purchases by the United States Armed Forces for their own use or to mdap-osp subcontracts.*[26] [Italics added.]

Thus, the result was the same as before. Business could continue as usual up to this point, i.e. that certain OSP contracts had been given to Sweden and that subcontracting to NATO countries was also going on. It is also clear that even after this point the Swedes were free to contribute to the defense of Western Europe by supplying material to U.S. forces, and to continue the practice of subcontracting. Even though the program was never realized to the extent originally intended, it might well have compromised the credibility of Sweden's neutrality policy if it had become known. It might be argued that the Swedish participation in the OSP was simply a matter of exporting military items to other nations, and could thus not be considered to be problematic from a foreign policy perspective. Sweden was, just as any other nation, certainly allowed to trade in armaments. However, there is a big difference between exporting weapons on a commersial basis to any country willing to pay for them, and systematically excluding some countries for ideological reasons. Within the context of the Cold War the government's decision to let Swedish companies produce military items exclusively for NATO was a political act. It was an affirmation of Sweden's consent to American hegemony, and a concrete manifestation of the Swedish will to strengthen NATO's position on the European continent.

Reaching Westward

At the same time as the OSP negotiations, in April 1953, the Swedish Army Attaché to Washington Colonel Möller wrote a letter to the U.S. Army's Foreign Liaison Officer of the Office of the Assistant Chief of Staff, G-2, to inform him of the Swedish military's intention to relay information about Sweden's guided missile development to American experts. The purpose was to pave the way for further discussions and six months later Möller's assistant, Major Bo Westin, briefed representatives from the U.S. Army on the Swedish development program. As it turned out, almost 85 percent of the area in which the Swedes, through Westin, expressed an interest corresponded to the development performed by the USAF.

26 758.5MSP/5-1755, Message to the U.S. Embassy in Stockholm from Dulles, dated May 17, 1955, in: RG 59; 1955–1959 CDF; From: 758.11/11-1555 To: 758.55361/5-1556; Box No. 3466; (Folder: 758.5/1-755); NARA II, College Park, MD.

It was quickly made evident to Westin, however, that the amount of data or technical assistance that could be made available to the Swedes at this point, due to the current disclosure policy, was of an 'extremely limited' nature. Within G-2, on the other hand, it was apparently felt that exchange of some data, particularly technical and intelligence data, might be beneficial to the United States. Perhaps the Swedes had achieved some significant developments it was said, especially with regard to electronics and guidance systems. Unfortunately, for the Swedes and the U.S. Army, neither the USAF or the U.S. Navy agreed with this conclusion. "[I]t was felt that the US would scarcely benefit from exchanging technical or intelligence information, and so the Swedes were informed that their request had been declined."[27] Apparently, a major reason for the unfavorable attitude was an unwillingness on the American side to discuss anything that had to do with homing devices. The reason for this was largely because of a fear that the wavelengths that these devices operated on might become known to the Soviets, and thus render them ineffective.

After the rebuff in April, 1953, Boheman took over from Möller in trying to persuade the Americans to view Sweden more favorably. He did so by using one of his many highly influential contacts, and friends, within the Eisenhower administration. On January 8, 1954, he met with the newly appointed Director of the CIA (henceforth: DCI) Allen Dulles. Boheman told the DCI that what the Swedes wanted was mainly to arrange for Swedish and American technicians to discuss guided missile development, and to compare the Swedish and American results in the area. The Ambassador stated that this could save the Swedes a lot of money and time. Even American advise regarding developmental paths not worth exploring would be very valuable, said Boheman, and continued by saying that it must certainly be in the interest of the United States for Sweden to be able to defend its territory from intrusions. Dulles agreed, and intimated that it was important to the U.S. that Sweden could serve "common interests" in the North. At the same time the Swedish Naval Attaché Commodore Östberg had brought the subject up with Admiral Espe at the Navy Department. The response was negative, however, and the reason was once again said to have been the highly classified nature of the homing devices. According to the Americans there was no point in conducting talks on the subject of guided missiles if homing devices could not be discussed. However, they had emphasized that this decision was not a political one, and that such information would not be shared with any other country either. Boheman concluded gloomily that if this was true, then all Swedish efforts to get access to such information were hopeless.[28]

27 Moores, S., *'Neutral on Our Side':...*, pp. 125–126; quote on p. 126.

28 Letter from Boheman to Lundberg, dated January 10, 1954, pp. 1–4, quote on

The Ambassador also felt that he was in a very bad position to negotiate with the Americans regarding guided missiles because of his miserably limited technical knowledge, and said that he was prepared to fly home to Sweden if a short course could be prepared for him. Dulles promised, however, after Boheman had made it very clear to him that Sweden was in desperate need of assistance in the guided missile area, to speak to Under Secretary of State General Walter Bedell Smith, who had preceded him as DCI, and give the matter a push in the right direction. The DCI also said that it could be a good idea for Boheman to speak to the former DCI himself. Dulles would, he said, pave the way for Swede by talking to Bedell Smith prior to this.[29] In Stockholm Boheman's message was interpreted with perhaps a lesser sense of despair. The Foreign Ministry instructed the Ambassador that the Swedes were not interested in the construction of homing devices, but in general discussion regarding fruitful and unfruitful guided missile development. In what seems to have been a response to Boheman's complaint about his low technical skills, the Ministry included a list in English of the areas which the Swedes wanted to discuss with their American counterparts. Among these areas were general principles of guidance, engine choices, gyros, autopilots, and tracking devices for missile tests.[30]

Dulles delivered on his promise. When Dulles and Boheman spoke again in early February things were looking a little brighter. The DCI had spoken to Bedell Smith, as well as to some senior members of the USAF, who had not reacted unfavorably to the Swedish inquiries. Boheman himself got to meet with the Under Secretary, who apparently showed great interest in the Swedish cause. So much, in fact, that he was prepared to push the Swedish case with the Chairman of the JCS, U.S. Navy Admiral Arthur W. Radford, to whom Boheman later gave a written version of the Swedish inquiry. Moores writes that in the six weeks that followed Boheman's first meeting with Dulles he met with Bedell Smith on two more occasions, and finally he was given an audience with Donald Quarles, the Assistant Secretary of Defense for Research and

---

p. 2, in: HP 24 VI/U.S.A; "Tillstånd till in- och utförsel av vapen, ammunition och krigsmateriel"; 1953–juni 1955; Folder 22; Box No. 62; År: 1955–1960; RIA; Arninge. "gemensamma intressen." See also: Moores, S., *'Neutral on Our Side':...*, pp. 126–127.

29 Letter from Boheman to Lundberg, dated January 10, 1954, pp. 1–5, quote on p. 2, in: HP 24 VI/U.S.A; "Tillstånd till in- och utförsel av vapen, ammunition och krigsmateriel"; 1953–juni 1955; Folder 22; Box No. 62; År: 1955–1960; RIA; Arninge. "gemensamma intressen." See also: Moores, S., *'Neutral on Our Side':...*, pp. 126–127.

30 Telegram from the Foreign Ministry to Boheman, dated, January 15, 1954, in: HP 24 VI/U.S.A; "Tillstånd till in- och utförsel av vapen, ammunition och krigsmateriel"; 1953–juni 1955; Folder 22; Box No. 62; År: 1955–1960; RIA; Arninge.

Development. Quarles pointed out to Boheman that the discussions would be informal and "that only matters of purely a technical nature would be on the agenda." The American could not guarantee that any data concerning American guided missile development could be related to the Swedes, but he thought that he would be able to advise the Swedes on their program and to help them that way. This, writes Moores, prompted Boheman to advise the people in Stockholm to do as had previously been suggested—to lay all cards on the table for the Americans to see. It was now up to the Swedish experts to get as much out of the discussions as possible.[31]

In the meantime considerable progress had been made with regard to ensuring the survival of the Swedish defense in wartime as outlined in NSC 121. In August 1953 Under Secretary of State Bedell Smith submitted a Progress Report on NSC 121 to the NSC, in which he stated that the problem of reserve stocks of aviation fuel for the RSAF had now been solved. The Departments of State and Defense had decided in January 1953 to permit the build up in Sweden of such reserves to a level of ninety days of estimated combat consumption, in addition to peacetime operating supplies. Wrote Bedell Smith:

> This was an exceptional decision based upon the particular geographic position of Sweden and the relatively small amounts of gasoline required. The possession of adequate reserves of aviation gasoline should be an important factor in encouraging Sweden to use its Air Force to best advantage in the event of hostilities in Northern Europe.[32]

Efforts had also been made to expand port and transportation facilities in Trondheim, Norway, and one or more points in Sweden, in order to ensure the continued supply of necessities to Sweden if the shipping routes in the Skaggerrak and Kattegat were obstructed.[33] The

31 Moores, S., *'Neutral on Our Side':...*, pp. 127–128, quote on p. 128; see also: Letter from Boheman to Lundberg, dated January 29, 1954, p. 1; telegram from Boheman to Lundberg, dated February 6, 1954; Letter from Boheman to Lundberg, dated March 17, 1954, pp. 1–2, in: HP 24 VI/U.S.A; "Tillstånd till in- och utförsel av vapen, ammunition och krigsmateriel"; 1953–juni 1955; Folder 22; Box No. 62; År: 1955–1960; RIA; Arninge.

32 "National Security Council Progress Report by the Under Secretary of State on the implementation of the Position of the United States with Respect to Scandinavia and Finland (NSC 121)", dated August 25, 1953, p. 3, in: Dwight D. Eisenhower; White House Office; National Security Council Staff Papers, 1948–61; Disaster File; Box No. 67; (Folder: Scandinavia 6); Eisenhower Library, Abilene, KS.

33 Ibid. See also: Agrell, W., *Den stora lögnen...*, p. 86.

Swedish Prime Minister expressed surprise regarding this decision, and commented this delivery in his diary on February 12 by saying:

> Strange, but we can hardly do anything else than to say thank you and accept it. Is this the new administration's way of telling us that it intends to continue its Sweden-friendly policies?[34]

It is hard to tell whether Erlander really was so innocent, or if he actually knew, or at least suspected, what the U.S. objectives were. There is certainly a chance that he did know, because the Swedes had been informed of the American decision two days earlier by Marshall Green of the U.S. Embassy. The American official had hinted to Ingemar Hägglöf at the Foreign Ministry that the amount of fuel allocated to Sweden meant that Sweden had received better treatment than other West European countries. This corresponded well, said Green, with the American objective of keeping Sweden militarily strong. He went on to state very frankly that the American decision had been made in the light of the expected difficulty in supplying Sweden with aviation fuel in the early stages of a war, and the U.S. government thus felt that the RSAF should have a decent supply from the outset. This was the meaning of the reference to Sweden's geographical position, wrote Hägglöf.[35]

This was essentially everything short of a an American promise to keep the RSAF flying during wartime. Considering what has been recounted above, the only reason for why Erlander could honestly consider the American decision strange, was if he had not been given all the information possessed by Hägglöf. Perhaps the question if Erlander knew or not was not all that important, as long the Supreme Commander and the Chief of the RSAF got the point. If the RSAF, thanks to an adequate supply of fuel, was able to operate for a prolonged period of time, its activities would serve U.S. and NATO objectives, whether or not Sweden chose to defend only its own territory or to extend this limit to cover its Scandinavian neighbors. The important point was for the Swedish forces to tie down as many Soviet divisions as possible, for as long a period as possible; if the Swedes were able to do that, then the aims of the Western Powers would have been well served. A prior commitment from Sweden was not really necessary to accomplish

34 *Tage Erlanders dagböcker, 1953* Published by Sven Erlander (ed.) (Hedemora: Gidlunds Förlag, 2003), p. 30. "Underligt, men vi kan väl inte göra annat än att tacka och ta emot. Är det den nya administrationens sätt att tala om att den ämnar fortsätta sin Sverige vänliga politik?"

35 Letter from Hägglöf to Boheman, dated February 10, 1953, in: HP 24 VI/U.S.A; "Tillstånd till in- och utförsel av vapen, ammunition och krigsmateriel"; 1953–juni 1955; Folder 22; Box No. 62; År: 1955–1960; RIA; Arninge.

this beneficial result for the Allies. The Americans certainly seemed to reason that their enemy's enemy was their friend.

At a meeting in March 1954 Major Bo Westin of the Defense Staff briefed the GMC about the development of guided missiles in the United States, but it was noted that it was not possible to give a full picture of the American guided missile program because of the rigorous secrecy surrounding it. Westin, who had been Vice-Chief of the GMB before 1951, pointed out that surface-to-air missiles were being produced within all three Services in the United States, an unambiguous proof that the Americans had solved the most pertinent problems with regard to guided missile technology.[36]

In April 1954, Captain Oxenstierna (then still the Chief of the GMB) led a group of Swedish experts on a trip to the United States for a meeting with Quarles, and their American counterparts. The meeting did not really yield as much as was hoped for because the Americans "stayed fully within security restrictions". Besides receiving constructive criticism of the Swedish development efforts, very little else was accomplished. The American Air Attaché in Stockholm USAF Colonel Charles H. MacDonald mused a year later that "the visit could not be considered too profitable from a Swedish viewpoint." Nonetheless, the Americans apparently felt that the Swedish development had a high technological standard, and they also believed that it was viable and the Swedes were urged to continue their domestic development. However, they "were warned that not enough funding was being allocated" to the Swedish program. These words struck a chord with Boheman, who immediately urged the Swedish government to furnish the Swedish guided-missile-developing community with the financial support deemed necessary. If combined with intelligence from abroad, as well as the possibility of procuring missiles in other countries, this could, he felt, produce a guided missile ready for production.[37]

The question of funding was thus seen as extremely pertinent on both sides of the Atlantic. The government's purse was not opening as fast as desired by the GMB, however, and despite the perceived opening in the great American 'guided missile technology wall' there were no such weapons forthcoming to Sweden in the mid-1950s. But this was

36 Annex 1 to "NR 4/1954", Protokoll från FRR sammanträde den 5. mars 1954.", entitled: "Utveckling av robotvapen i Förenta Staterna." in KFF, FRR, Serie: A 1, "Protokoll från sammanträden 1948–1957", År: 1948–1957, VOL. 1, KrA, p. 1.

37 Letter from MacDonald to Abbott, dated April 18, 1955, in: RG 59; 1955–1959 CDF; From: 758.11/11-1555 To: 758.55361/5-1556; Box No. 3466; (Folder: 758.5/1-755); NARA II, College Park, MD. Moores, S., *'Neutral on Our Side':...*, pp. 128–129, "were warned...", p. 128. See also: Letter from Boheman to Lundberg, dated April, 23, 1954, pp. 1–2, in: HP 24 VI/U.S.A; "Tillstånd till in- och utförsel av vapen, ammunition och krigsmateriel"; 1953–juni 1955; Folder 22; Box No. 62; År: 1955–1960; RIA; Arninge.

slowly about to change as the United States, due to pressure from its European NATO allies, began to reassess its non-disclosure policy. This reevaluation was done in basic hegemonic terms and stressed the need for the United States to provide its allies with material benefits in exchange for their consent to U.S. preponderance in world affairs.

Sweden was treated just as the other Western European allies when the United States decided to keep withholding this technology from it, because the NATO countries had also been denied access to guided missile technology and associated know-how. Even the British, the closest ally of the U.S., were not granted full access to American guided missile technology and related electronics. This fact was made evident by the Joint Intelligence Committee (henceforth: JIC) in a report in July 1954 on the issue of closer technical cooperation with Britain in the guided missiles field. The report had been undertaken as a consequence of the ongoing high-level discussions between the United States and Britain, and the Secretary of Defense had requested that a study should be undertaken so that U.S. policy could be decided upon. The report revealed that similar negotiations had been conducted in early 1953, at which time the JCS had urged that there was a necessity "for adequate restrictions on exchange of classified guided missile information." On March 10 that year the Guided Missile Interdepartmental Operational Requirements Group (GMIORG) was directed to notify the JCS:

> [...] in those instances where it is considered desirable not to release a particular facet of classified guided missiles information to the United Kingdom or Commonwealth Nations, in order that the Joint Chiefs of Staff may take the necessary action to withhold such information.[38]

In a draft memorandum for the Secretary of Defense the JCS stated:

> The Joint Chiefs of Staff consider that in the interest of the over-all security and defense of the United States it is undesirable to release certain areas of guided missiles and associated electronics information, such as countermeasure vulnerability of specific U.S. missiles, application of atomic warheads or other information which would tend to compromise RESTRICTED DATA, and certain aspects of fuzing and guidance systems.[39]

38 J.C.S. 927/183, "U.S.-U.K. Discussions on Closer Technical Cooperation – Guided Missiles and Associated and Associated Electronics ", dated July 14, 1954, p. 2, in: RG 218; CDF 1954–56; 350.05 (3-16-44) Sec. 33–37; Box No. 74; (Folder: Exchange of Technical Information with Foreign Countries; CCS 350.05 (3-16-44) Sec. 37); NARA II, College Park, MD.

39 "Enclosure 'B', Draft Memorandum for the Secretary of Defense", p. 1, included in, ibid.

There were thus several areas which the Chiefs felt were very sensitive, and they clearly recommended caution when releasing information pertaining to guided missiles even to as close an ally as Britain. Referring to a memorandum that had been written following the discussion with the British, the JCS commented that:

> [...] the memorandum states that each country would keep the other informed about its program in the field of guided missiles and associated electronics; also that the field of guided missiles and associated electronics would be reviewed periodically to determine those particular projects or techniques for which closer cooperation would be desirable. It is considered that such statements could be interpreted by the United Kingdom as justification for requests for information without regard to a real need-to-know. [...] The areas of information which it is not desirable to release should be delineated by a technical group such as GMIORG [...].[40]

What the JCS feared was a situation where the United States would be more or less compelled to divulge classified information to the British. From the above discussion it was evident that they felt that even a program which would make it possible to exchange information about ongoing research might spell disaster for U.S. national security, and that the issue had to be thoroughly worked through to determine exactly what kind of information the United States could afford to release. Considering the attitude of the JCS in this matter it is not hard to imagine that there would be an even harsher tone taken with regard to the rest of the NATO allies and Sweden. The above confirms that the Americans had been honest with the Swedes in early 1954, when they had told Boheman and others that the refusal to discuss such matters with Sweden had nothing to do with Sweden *per se*.

The NATO countries had also been demanding technical assistance with regard to guided missiles for some time, and the way in which the United States handled this issue in 1954 is a good example of how the Americans used their hegemonic position in negotiations with the allies. In Europe, SACEUR Alfred M. Gruenther and SHAPE were engaged in a study of the air defense system in the NATO area, and there was a growing consensus that the Western European nations would need considerable and continual technical assistance from the United States and Britain. The allies, according to SACEUR, wished foremost to receive information on air defense guided missiles. "The data requested", said the Joint Strategic Plans Committee (henceforth: JSPC), was "within

40 "Enclosure 'C', Discussion", pp. 1–2, included in, ibid.

specific areas of information which the Joint Chiefs of Staff have previously determined not desirable to release on an international basis."[41]

Despite the fact that the U.S. representative at the annual NATO Ministerial Conference in Paris in December 1953 had gone on the record as saying that the United States was prepared to seek the necessary legislative enactments to allow it to share pertinent information regarding new weapons with its allies (although he did not specify exactly what he was referring to), and that this would have significant impact upon NATO planning; the JSPC stated:

> [i]n view of the recent rejection of the European Army concept by France, and the resultant uncertainty in Europe, it is considered inappropriate to relax U.S. security policies, at this time.

Its conclusion was that:

> [t]he current policy which precludes the release on a NATO-wide basis *of classified data* on U.S. surface-to-air guided missiles and supporting fire control equipment should not be modified at this time.[42] [Italics added.]

The JSPC thus chose to blame the indecisive Europeans (especially the intransigent French) for the continuation of the restrictive American policy regarding guided missile technology. Note also that the italicized phrase in the second quote was basically the same policy as had been guiding the American attitude toward Swedish feelers in this field.

The JCS chose to heed the recommendations made, and stated that:

> [t]he Joint Chiefs of Staff provided guidance [...] to the U.S. representative to the Standing Group, North Atlantic Military Committee, to the effect that the release to foreign governments of classified information relating to guided missiles designed for air defense, or relating to their supporting fire control system, is not favored, except as may be permitted through bilateral agreements. Specifically, the Joint Chiefs of Staff did not favor the release at that time of such information on a NATO-wide basis.[43]

41 J.C.S. 2073/906, "Air Defense Weapons and Systems Information", dated October 5, 1954, p. 2, in: RG 218; Geographic File, 1954–56; Western Europe (3-12-48) Sec. 294–302; Box No. 48; (Folder: CCS 092 Western Europe (3-12-48) Sec. 302); NARA II, College Park, MD.

42 Ibid., pp. 2–3.

43 "Enclosure 'B', Facts Bearing on the Problem", p. 1, included in, ibid.

Thus, the NATO countries would have to wait before they could count on getting access to such technical information from the United States. The decision by the JCS basically put these countries and Sweden on an equal footing.

A Breach In the Wall

On January 10, 1955, the Swedish Air Attaché in Washington Colonel Stig Wennerström wrote to Captain Rosenberg, the new Chief of the GMB, that the American Liaison Officer for Guided Missile Affairs, Colonel Parrot, had told Wennerström that it was of paramount importance that Sweden joined the Mutual Weapons Development Program (henceforth: MWDP) to further the Swedish-American cooperation on guided missile development. Wennerström had passed this information on to Ambassador Boheman, who in turn had queried Foggy Bottom on the issue.[44] Wennerström's letter seems to suggest that the Swedish request was made in response to American prodding, a small but important detail that has previously gone unnoticed by scholars in the field. Apparently there were American officers who were anxious to make sure that Sweden got on the guided missile bandwagon.

It is worth noticing that this fits perfectly well with the new force planning document, entitled MC 48 "The Most Effective Pattern of Military Strength for the Next Few Years", adopted by the NATO Ministerial Meeting in December 1954, which focused heavily on the strengthening of the air defense system of Western Europe.[45] It was probably seen as pertinent to include the Swedes in such a system, and if the Americans were preparing to release guided missile technology to the Europeans, then Sweden would have to show that she was interested in the program. I will come back to this issue below.

During the spring of 1955 the State Department repeatedly received informal queries from Boheman, concerning Swedish participation in the MWDP. In a letter from Assistant Secretary of State Livingston T. Merchant to Assistant Secretary of Defense Struve Hensel, the Swedish interest was explicitly related "to their desire to establish arrangements looking to United States-Swedish cooperation in the field of guided missiles development [...]." A Swedish delegation had visited the United States in April 1954 with the same purpose in mind. The State Department thought that, from a foreign policy point of view, Swedish participation would be desirable and stressed that it would be in accordance with the goals set out in NSC 121. Merchant said that

44 Letter from Stig Wennerström to S. O. Rosenberg, December 29, 1955, RHA 1021:1, in: KFF, RBN, Serie: F III B, "Inkomna och avgångna skrivelser ordnade efter klassifikationssystem", A 1005:26-A 373:7, År: 1956, VOL. 11, KrA.

45 Duffield, John S., *Power Rules: The Evolution of NATO's Conventional Force Posture* (Stanford, California: Stanford University Press, 1995), p. 75.

the issue had been discussed earlier by members of Quarles' staff with officers from Foggy Bottom.[46]

Boheman's first inquiry regarding Swedish adherence to MWDP went back at least to late December 1954 when he stated to Raynor at BNA that he had heard that such a program was in operation "which the Swedes perhaps could adhere to by means of some form of an exchange of notes." At the time, Raynor feigned innocence and told Boheman that he had not heard of such a program, but that there might exist one which he was unaware of, and he promised Boheman to look into the matter for him.[47] Boheman thus hinted that the Swedes would prefer an arrangement similar to the MDAA agreement in the summer of 1952; with a quiet exchange of notes on the governmental level instead of a more formal signing of an agreement. As we shall see, this was, perhaps coincidentally, sensed by the American Embassy in Stockholm which was to offer its recommendations on the matter later on.

Merchant emphasized to Hensel that even if the Swedish government was unwilling to accept a program that involved aid within the meaning of the Battle Act, under which economic assistance could be withheld if the recipient did not abide by the trade embargo regulations, "[s]ome degree of formalization might nevertheless be desirable in connection with such cooperation with Sweden." Merchant's letter said:

> In this regard the Chairman of the State-Defense Military Information Control Committee, after consultation with Defense representatives (including representatives of Mr. Quarles' Office), has informed me that that committee is prepared to consider such exceptions to the national disclosure policy in pursuance of such cooperation as would be in the national interest.[48]

This hints at a remarkably open attitude toward the Swedish requests, and begs the question of how the United States was treating similar requests from other nations at this time. While I cannot answer this in this thesis, it is safe to say that after Sweden's consent to American hegemony, its eagerness to support American objectives, coupled with the U.S. view of Sweden as strategically important, resulted in Swedish

46 Letter from Merchant to Hensel, dated March 23, 1955, p. 1, in: RG 59; MOFAS-SEA, 1943–1957; Box No. 27; (Folder: Sweden 1948–1955); NARA II, College Park, MD. See also: Moores, S., *'Neutral on Our Side':...*, p. 129.

47 Letter from Raynor to Hilliker, dated December 22, 1954, in: RG 59; 1955–1959 CDF; From: 758.11/11-1555 To: 758.55361/5-1556; Box No. 3466; (Folder: 758.5/1-755); NARA II, College Park, MD.

48 Letter from Merchant to Hensel, dated March 23, 1955, p. 2, in: RG 59; MOFAS-SEA, 1943–1957; Box No. 27; (Folder: Sweden 1948–1955); NARA II, College Park, MD.

requests being looked upon with great sympathy even though actual cooperation on guided missile issues was yet to materialize.

At the same time, at the end of March, 1955, Count Carl Douglas of the Swedish Embassy met with the Director of BNA Raynor. Douglas had previously intimated to Raynor that some Swedish officers were in Washington and that they desired to talk to people at the Pentagon and elsewhere about military communication issues. Raynor said that he had now found a corresponding interest on the part of the Pentagon in discussing "certain of these matters along the line of the 1953 discussions with the Swedes." However, the original request could not be entirely granted. Raynor said:

> [...] the requests made in the present instance had been so broad that they had encompassed certain matters on which it was not possible to talk. I said that I did not know what these matters were. I said we had been informed that if the Swedish Military Attaché would return to the Pentagon and discuss this matter we understood it should be possible to work out arrangements which would permit a considerable part of the discussions which the Swedes wanted to take place.[49]

The discussions in 1953 referred to by the American were most probably the conferences Brigadier General Bonde's party had had with General Eddleman and other American officers back then. The view that Swedish membership of the MWDP would further the aims of NSC 121 was shared by the new American Ambassador to Stockholm John Moors Cabot. Cabot explicitly thought that an excellent way of furthering the goals of NSC 121 was to assist the Swedes in their efforts to develop guided missiles.[50] With regard to the Swedish feelers toward a possible inclusion in the MWDP, the American Air Attaché Colonel MacDonald was of a slightly different opinion; stating that:

> [...] the Swedes have been defeating their own purposes by reason of their approach to this problem of cooperation in the guided missile field. They have been exploring every conceivable approach with the result that the separate agencies in Washington are not only confused but have become super cautious on this subject.[51]

49 758.5811/3-3055, Memorandum of Conversation between Raynor and Douglas, dated March 30, 1955, in: RG 59; 1955–1959 CDF; From 758.56/1-555 to 759.00-mayday/3-357; Box No. 3467; (Folder: 758.5811/1-455); NARA II, College Park, MD.

50 Moores, S., *'Neutral on Our Side':...*, p. 131.

51 Letter from MacDonald to Abbott, dated April 18, 1955, in: RG 59 1955–1959 CDF; From: 758.11/11-1555 To: 758.55361/5-1556; Box No. 3466; (Folder: 758.5/1-755); NARA II, College Park, MD.

There can be no doubt that a successful outcome to the Swedish guided missile development efforts was something that was seen as highly desirable in Washington, however. Referring to the discussions held between Quarles' staff and the group led by Oxenstierna in April 1954, the American Naval Attaché L. C. Mabley offered his advice in April 1955. Mabley underlined that perhaps there were ways of helping the Swedes even if the United States, due to its restriction policy, could not share information on guided missiles at this point. This could be done by aiding them in other areas so as to release much needed manpower and funding to be poured into the guided missile program—to the benefit of both Sweden and the United States. Wrote Mabley:

> Although emphasis during the discussion [...] seems to have been on guided missiles there are many other fields of development wherein the Swedes could use technical guidance with benefit to the U.S. as well as themselves. Therefore if the Swedish requests in the guided missile field involve information that the U.S. considers too sensitive, technical assistance in less sensitive fields would have the effect of releasing more Swedish funds and more Swedish technicians for use in their own guided missile development work.[52]

F. M. Sperry, the American Army Attaché, backed Mabley's views, and said that if the policies expressed in NSC 121 were to be followed, the best way "to assist Sweden to a position of strength and independence contributory to the national interest of the United States" would be by exchange of information, even negative information, with Sweden.[53] The term 'negative information' probably referred to information about developmental *cul-de-sacs*; i.e. if the United States because of classification issues could not tell the Swedes what to do, it could perhaps tell them what *not* to do.

While thinking that the Stockholm Embassy was probably not qualified to make judgments regarding what sort of arrangement would be best suitable for this purpose, Sperry wrote that:

> My experience in this country leads me to feel that the Swedes probably would prefer to expand an existing agreement such as the one providing for reciprocal aid on a reimbursable basis than to

52 Letter from Mabley to Abbott, dated April 19, 1955, in: RG 59; 1955–1959 CDF; From: 758.11/11-1555 To: 758.55361/5-1556; Box No. 3466; (Folder: 758.5/1-755); NARA II, College Park, MD.

53 Letter from Sperry to Abbott, dated April 26, 1955, in: RG 59; 1955–1959 CDF; From: 758.11/11-1555 To: 758.55361/5-1556; Box No. 3466; (Folder: 758.5/1-755); NARA II, College Park, MD.

> undertake a new agreement with its attendant danger of political criticism at home and abroad.[54]

Having received the views of the Service Attachés, the Counselor of the Embassy George Abbott wrote Grant Hilliker at BNA to tell him the Embassy's thoughts on the matter. Abbott repeated that the Embassy was not in a position to judge the sensitivity of specific military programs, but said that "we continue to favor the maximum possible cooperation with Sweden to achieve the objectives set forth in NSC-121."[55] Abbott also offered his, and the Embassy's view that the Pentagon was perhaps a bit too restrictive regarding Sweden. The standard of security in Sweden, wrote Abbott:

> [...] is high and compares favorably with many of our NATO partners, and we have the impression that the Pentagon, at least at certain levels, is unduly apprehensive on this point.[56]

Adding to this, again echoing Sperry's thoughts on the issue, Abbott wrote:

> As far as means of implementation of any program is concerned, we would like to again emphasize the Swedish liking for informal procedures and their allergy towards written agreements and any form of publicity in matters of this kind.[57]

I want the reader to note that this recommendation was more or less in sync with Boheman's suggestion in December 1954 cited above, that maybe Sweden could be made to adhere to the MWDP through an exchange of notes similar to the MDAA agreement. The Embassy thus also put in a good word for Sweden with regard to the security preparations in the country. Some months later, in August 1955, and after having pondered the advise offered by the Stockholm Embassy, Quarles wrote to the Assistant Secretary of State Livingston T. Merchant, to let him know that the issue had been discussed at the DOD and it was thought:

54 Ibid.

55 Letter from Abbott to Hilliker, dated April 27, 1955, in: RG 59; 1955–1959 CDF; From: 758.11/11-1555 To: 758.55361/5-1556; Box No. 3466; (Folder: 758.5/1-755); NARA II, College Park, MD. See also: Moores, S., *'Neutral on Our Side':...*, p. 129.

56 Letter from Abbott to Hilliker, dated April 27, 1955, in: RG 59; 1955–1959 CDF; From: 758.11/11-1555 To: 758.55361/5-1556; Box No. 3466; (Folder: 758.5/1-755); NARA II, College Park, MD.

57 Ibid.

> [...] that certain contributions which Sweden would be able to make in support of our own defense program would indicate that technical assistance to Sweden might be handled on a straight quid pro quo basis rather than through the Mutual Weapons Development Program.[58]

Quarles proposed that a more thorough evaluation should be done before a definite decision could be made.[59] It seems as if the view arrived at by Quarles staff had been heavily influenced by the advise solicited from the American Service Attachés in Stockholm. Essentially, the thought that the matter might best be handled on a *quid pro quo* basis was ideal to the Swedes since this implied that no new agreements had to be made; not even a new exchange of notes had to be undertaken. If the Swedes scratched the American back, then the Americans would scratch theirs.

But why did the Americans take this view? One probable reason is that the DOD knew that the Swedish government would never agree to participate formally in a NATO body and hence thought it futile and a waste of resources to try to convince the Swedes of the contrary.[60] Another contributory factor was that it would have been a very sensitive matter to handle this sort of cooperation through a NATO apparatus, such as the MWDP, when at the same time the allies themselves were not allowed access to such technology.[61] This was just one more example of the fact that the form of the Swedish consent to American hegemony was not what was important to the Americans—all that mattered was the content.

Due to its policy of neutrality the Swedish government remained constantly on the lookout for anything that threatened to give the secret

58 Letter from Quarles to Merchant, dated August 15, 1955, in: RG 59; 1955–1959 CDF; From: 758.11/11-1555 To: 758.55361/5-1556; Box No. 3466; (Folder: 758.5/1-755); NARA II, College Park, MD. See also: Nilsson, M., "'Gud vare tack och lov...", p. 41.

59 Letter from Quarles to Merchant, dated August 15, 1955, in: RG 59; 1955–1959 CDF; From: 758.11/11-1555 To: 758.55361/5-1556; Box No. 3466; (Folder: 758.5/1-755); NARA II, College Park, MD. See also: Nilsson, M., "'Gud vare tack och lov...", p. 41.

60 Simon Moores has reasoned along these lines in connection with the visit of the two guided missile specialists Galer and Schuldenfrei (see: Moores, S., *'Neutral on Our Side':...*, p. 130). For an account of their visit, see below in this thesis.

61 Support for this argument can be found in: Memorandum of Conversation between Crowley and Carl Douglas, dated November 9, 1956, in: RG 59; INR, 1945–1960; Lot 58D776; Box No. 13; (Folder: Sweden); NARA II, College Park, MD. Crowley told Douglas, the Counselor to the Swedish Embassy in Washington, that "it would be difficult to have a non-NATO member participate" in the MWDP (ibid.). Although Crowley did not become specific in his Memorandum I interpret this as supporting my conclusion.

collaboration with the United States away to the uninitiated Swedish public or the Russians. The fact that the Swedish fears focused so much on eliminating overt signs of cooperation had some almost comic consequences, e.g. the aversion to U.S. personnel wearing American uniform while in Sweden. Cabot drew Foggy Bottom's attention to this fact in December 1955 in a letter to the new Director of BNA Outerbridge Horsey. Cabot wrote that "the Swedes are very allergic to the use of foreign uniforms in Sweden", and he related an anecdote regarding the Swedish refusal to allow a visit by the USAF Symphony in March 1955 because the musicians would wear uniform. "It frankly did not occur to us that the Swedes would object to such a visit", said Cabot, "since it was of a purely cultural nature."[62]

Despite this episode bordering on parody, there was a more serious matter buried beneath the surface, said the Ambassador, and stated that it appeared as if the Swedes were concerned "lest the Russians be convinced that secret links exists between Sweden and NATO."[63] Once again, Sweden's policy of neutrality put very clear limits to the degree of cooperation that the Swedes could afford. The problematic factor for Sweden was of course that there *were* certain secret links between Sweden and NATO. If there had not been, there would hardly have been any reason for the Swedish government to fear that the Soviets would assume there were; surely it was more damaging to buy defense material from the United States than to allow an orchestra to wear uniform on their visit to Sweden. But this does not seem to have been the view of the Swedish authorities. The purchases of weapons from the Americans were never something that was hidden from the Russian, nor could it have been.

The Swedish government wished to portray this as nothing more than a commercial relationship, and this was possible as long as it seemed credible that these weapons were not pointed in only one direction. This is where the appearance of foreign uniforms on the streets of Stockholm and elsewhere was potentially devastating; how could the Swedes be expected to fire upon the same Americans that were 'crowding' the streets of the capital? In reality, the purchase of military hardware served to constrict the Swedish government's freedom of action much more than the occasional American uniform; but *symbolically* the uniform was much more powerful for it hinted at a common bond perhaps because of the fact that countries usually did not allow foreign military men to wear uniform in public unless they were close friends of the nation in question. Weapons could be sold to almost anybody, but you

62 Letter from Cabot to Horsey, dated December 7, 1955, p. 1, in: RG 59; Records of Ambassador John M. Cabot [henceforth: RAJOCa], 1945–1963; Folder 1 to Folder 10; Box No. 1; (Folder 2); NARA II, College Park, MD.

63 Ibid., p. 2.

did not walk around in uniform unless you were partners. The Swedish government could not afford to suggest such a thing, and therefore they viewed such instances with much perturbation.

Cabot wrote:

> As you of course know, there are many people in Sweden who believe that some such links should exist, that there must be joint planning to cover the possibility of a Soviet aggression in this area, and that Sweden must at the very least get our technical advice to help stiffen its defenses. In other words, they do want to be able to decide freely what to do in the event that Sweden itself is not attacked but her interests are affected by some aggression; all this sector of Swedish opinion does not want is to be irrevocably committed in the event of an aggression in some other part of the world.[64]

With his letter to Horsey, Cabot was arguing against numerous American visits including service personnel in uniform, because what was in the balance was the ongoing and expanding tacit military and security cooperation with Sweden. The Ambassador pointed this out to Horsey, saying that:

> I do not need to tell you that a fair amount is going on continuously beneath the table between the Swedes and ourselves. The Swedish military particularly wants a tie-up with Nato and is quite prepared to play ball with us in return for our playing ball with them on matters of mutual importance. This is valuable to us and I certainly do not wish to jeopardize it. So far, then, as the Swedes are concerned, the objection to numerous ostentatious military visits is that they give the impression that military links exist which in fact do not, *at least in connection with these visitors*.[65] [Italics added.]

It is necessary to pause here, and to reflect upon what Cabot was actually saying, i.e. the information between the lines, so to speak. In both instances above Cabot had only mentioned collaboration that was already going on (with the possible exception of the issue of joint planning, of which the degree actually undertaken is uncertain). He did not say that these people were wrong in assuming these things. Hence, one plausible way to interpret the Ambassador's words is that the Swedes were afraid that the secret collaboration that they were engaged in with the Americans would become known. This interpretation is further strengthened by the last italicized phrase, which made clear that it was not so much a worry that the Russians would draw the

64 Ibid., p. 2.

65 Ibid., pp. 2–3.

wrong conclusions, as that they would in fact draw the right ones. Uniformed visits risked drawing attention to the actual extent of current collaboration measures; the problem was that these visits had nothing to do with the real links to the United States and the Western Powers. This might force Undén to push in the Cabinet for breaking the covert cooperation off, something that would hurt American strategic interests in the area.

My conclusion above is directly supported by what the Ambassador said next. Cabot concluded, it was in the interest of the United States to refrain from such visits "unless something is to be gained by them." Cabot developed this thought when he wrote:

> We do not want to carry on a lot of overt military *activities in Sweden which, while themselves innocent, may lead to the uncovering of activities beneficial to us* which we don't want to have everyone know about. Moreover, we don't want to embarrass our friends in Sweden by forcing them to go to bat every time some general or congressman on a military affairs committee wants to come to Sweden on a vacation and feels he must give some appearance of an official reason for the visit. *We should save our ammunition for what is important for us.*[66]

It is thus indisputable that Cabot viewed the public's questions as warranted, and that they risked revealing the true nature of the Swedish-American relationship. Against this background it is most interesting to see what Cabot wrote next in his letter. He said that whenever there was a military visit to Sweden there was "presumably an adequate reason for it. Yet, he complained:

> [...] time and again I am asked about visits – for example, the proposed visit of a dozen high officers to discuss the training of pilots – for which the reasons given are obviously inadequate, or about other visits – for example, the several visits made and proposed by congressional groups to "study military installations" *when the reason is obviously phoney*. When such visits are made there are many people – Swedes as well as commie sympathizers – who must ask: "what's this all about? *Are the American military planning to operate on Swedish territory in the event of war? Are we secretly in Nato? Are Americans snooping and planning here with the consent of the Swedish government? Or the Swedish military?* Or without it?? Needless to say, the same questions would arise even more sharply in suspicious Russian minds

66 Ibid., p. 3.

> – and the Swedes know it. Do not forget that till the Soviet fleet visit in 1954 the Swedes would not even permit us to make naval visits to Stockholm.[67]

There are several intriguing things contained in this quote. One is why the reason for the congressional groups' visits was false. What was the real reason, if the purpose of studying military installations was only a cover? Could not a less controversial cover have been concocted? Hence, one wonders if the American Ambassador was once again listing things that could actually be exposed by these unrelated visits. But this poses the question whether the other possible reactions listed were also actually true? This would seem tenable, judging from what we know about Swedish-American collaboration during the Cold War, perhaps with the exception of the issue whether Sweden was a secret member of NATO. But one wonders why Cabot chose to highlight these particular potential reactions to American visits to Sweden? An assumption that Sweden is a hidden member of the Atlantic Pact does not seem to follow logically from the appearance of American military personnel and congressional groups in Sweden.

Cabot continued by taking another example of a visit, this time by a group from the National War College. The Ambassador questioned what the group would do in Sweden, and stated that if it was just a junket, i.e. a combination of business and pleasure, then it was not a good enough reason for spending the American tax payer's money. However, if "they are coming for any military reason", said Cabot:

> [...] it seems to me obvious that the visit of such a large group which is being trained in matters of military policy would in itself point the finger in a direction we do not want it pointed.[68]

He concluded the long letter by saying that:

> I trust that this rather lengthy letter will explain why I am allergic to any military activity of ours in this country which does not have a clear and important justification from the viewpoint of our national interests. Perhaps I should add we have had quite a lot of that kind, too, and I am all in favor of it; but I don't want to alert everyone by other visits having maximum reverberation and minimum reason.[69]

67 Ibid., p. 3. Note that Cabot apparently did not consider Communists to be Swedes.

68 Ibid., pp. 3–4.

69 Ibid., p. 4.

In some ways, then, Cabot apparently felt that U.S. relations with Sweden were becoming a bit too relaxed. Sweden was not just another Western European ally with whom the Americans could associate on whatever terms, or for whatever reasons. The United States would have to tread more carefully in order to preserve the important links to the Swedish military community, and if it wished to have continued consent to U.S. hegemony from the Swedish government.

There were also, at the time, certain operations undertaken by the Americans on Swedish territory, which the Swedish government did not know about. One such matter was revealed in a letter dated September 20, 1956, from Cabot to William P. Cochran, Jr., Esquire, Chargé d´Affaires ad interim at the American Embassy. It relates that the Americans were spying on the Soviets through the American Embassy, using an American agent by the name of Birdsall (this was most certainly not the operative's real name). Cabot wrote to Cochran:

> It certainly seems to me that it would be highly inappropriate to let it become public what Mr. Birdsall is really doing in the Embassy. I wish him to have any prerogative and facility we can prudently extend to him. But, I certainly do not wish it to be made of public record that we are conducting intelligence operations in the Embassy. As you suggest, this would not only be grist for the Commie mill but it would also be extremely difficult for the Swedes to accept in the light of their crackdowns on Communist espionage. After all, if it is known what Agency Birdsall represents, *the very least that can be assumed regarding him is the truth*, to wit, that he is conducting intelligence activities on Swedish soil directed at the Soviet bloc and *I don't see how the Swedes could reconcile this with their alliance-free policy.*[70] [Italics added.]

It goes without saying that American espionage activities in Stockholm conducted out of the Embassy would necessarily have rather great repercussions on Swedish-American relations, had they become publicly known. As is made evident by the quote above, Cabot did not have anything against the operation as such; he only wished to make sure that it remained secret. Thus, the Ambassador was prepared to gamble and play the odds since there was really no way the Embassy could control potential leaks within the Embassy, or prevent accidental exposure of Birdsall by Swedish counterintelligence or by the Swedish media.

70 Letter from Cabot to Cochran, Jr., dated September 20, 1956, p. 2, in: RG 59; RAJOCA, 1945–1963; Folder 11 to Folder 23; Box No. 2; (Folder 12); NARA II, College Park, MD.

I mentioned above that there were internal discussions in the United States about whether or not to provide guided missiles to Sweden, with some favoring such a strategy, and others opposing it. It seems, though, as if the Americans kept the inter-Departmental discussions secret from the Swedes at this point, because when the Counselor of the Swedish Embassy Count Carl Douglas inquired again a year later, in November 1956, about possible Swedish participation in the MWDP he got the answer from Crowley at BNA that no action had been taken on the American side due to the informal character of Boheman's feelers.[71]

On September 6, 1955, Wennerström informed Captain Rosenberg that two American guided missile specialists would arrive in Stockholm to visit the GMB, on September 21. The two specialists were Colonel Robert E. Galer (U.S. Marines Assistant, Guided Missiles Division, Bureau of Aeronautics of the Navy Department) and Mr. Marvin Schuldenfrei (a civilian, Engineering Assistant to Director of Guided Missiles Division, Bureau of Aeronautics of the Navy Department).[72]

The Americans preferred to see the visit as a result of a Swedish invitation, said Wennerström, but in reality it was, as in the case with Boheman's inquiry about possible Swedish membership in MWDP, an American initiative. Galer and Schuldenfrei would arrive dressed as civilians directly from Paris. The United States had established the Mutual Weapons Development Team inside the greater framework of the MWDP, and the Team was based in the French capital. A logical location, considering the connection between the MWDP and Article 3 in the NATO Treaty (the NATO headquarters were also located in Paris). The Mutual Weapons Development Team was supposed to assess what kind of technical and research and development assistance the different NATO member states were estimated to need, and at the same time evaluate if such assistance would yield a "positive result and be of considerable value for the defense of Western Europe."[73]

Hence, their activities were linked to NATO's defense planning. According to Moores, Galer and Schuldenfrei came to Stockholm with the approval of the Swedish Minister of Defense Torsten Nilsson, and "hoped to discuss matters using 'confidential' information, and, in the

71 Memorandum of Conversation between Crowley and Douglas, dated November 9, 1956, in: RG 59; INR, 1945–1960; Lot 58D776; Box No. 13; (Folder: Sweden); NARA II, College Park, MD.

72 Letter from Wennerström to the Chief of the GMB dated September 6 1955, RHA 1021:17, in KFF, Robotbyrån [henceforth: RBN], Serie: F III B, "Inkomna och avgångna skrivelser ordnade efter klassifikationssystem", A 01-A 7 , år: 1955, VOL. 5, KrA, p. 1. See also: Moores, S., *'Neutral on Our Side':...*, p. 130.

73 Letter from Wennerström to the Chief of the GMB dated September 6 1955, RHA 1021:17, in KFF, RBN, Serie: F III B, "Inkomna och avgångna skrivelser ordnade efter klassifikationssystem", A 01-A 7 , år: 1955, VOL. 5, KrA, p. 2. "positivt resultat och bli av påtagligt värde för Västeuropas försvar."

case of questions of a purely technical nature, *to furnish the Swedes with 'secret' data.*"[74] [Italics added.] Hence, there had clearly been a change in the year that had gone by since Oxenstierna's visit to the U.S. in April 1954 with regard to the kind of information that the Americans felt they could safely convey to the Swedes.

The result of these inquiries was communicated to the Department for Research and Development under the State Department. Wennerström wrote that "this team is constitutionally not tied only to NATO, but has so far only been acting in European NATO countries and, hence, has [at the moment] a pure NATO-character. Because of this, the common view is that it probably is not consistent with Swedish policy to establish contact."[75] Officially, the visit conducted by the two Americans had been labeled "Unclassified", since there was no official connection to the Team in Paris – after all, Galer and Schuldenfrei were representing the Navy Department. But Wennerström also accounted for the unofficial side of their visit. Before their departure to Stockholm, Galer and Schuldenfrei counseled the Mutual Weapons Development Team in Paris in order to "*avoid direct contact* [...]"[76] [italics added] between the Swedes and the NATO-Team. While in Paris they would "receive instructions pertaining to the visit to Sweden."[77] Afterwards they flew back to Paris and reported back to the Team, and received further instructions on how to report back to DOD.[78] Thus, Galer and Schuldenfrei were on an unofficial mission to find out what type of assistance Sweden needed on guided missile development.

What guided missile systems did they discuss, then? Well, there is yet to surface any written account of what was discussed during this visit, but Christer Ahlström has written that the mission concerned

74 Moores, S., *'Neutral on Our Side':...*, p. 130.

75 Letter from Wennerström to the Chief of the GMB dated September 6 1955, RHA 1021:17, in KFF, RBN, Serie: F III B, "Inkomna och avgångna skrivelser ordnade efter klassifikationssystem", A 01-A 7 , år: 1955, VOL. 5, KrA."[d]etta 'team' är författningsenligt icke knutet enbart till NATO, men har hitintills endast verkat inom europeiska NATO-länder och har därför f[ör] n[ärvarande] en ren NATO-karaktär. Man anser på grundval härav, att det troligen icke är förenligt med svensk politik att etablera kontakt."

76 Letter from Wennerström to the Chief of the GMB dated September 6 1955, RHA 1021:17, in KFF, RBN, Serie: F III B, "Inkomna och avgångna skrivelser ordnade efter klassifikationssystem", A 01-A 7 , år: 1955, VOL. 5, KrA, pp. 2–3. "*[f]ör att kringgå en direkt kontakt* [...]"

77 Letter from Wennerström to the Chief of the GMB dated September 6 1955, RHA 1021:17, in KFF, RBN, Serie: F III B, "Inkomna och avgångna skrivelser ordnade efter klassifikationssystem", A 01-A 7 , år: 1955, VOL. 5, KrA, p. 3. "[...] erhålla instruktioner före besöket i Sverige."

78 Letter from Wennerström to the Chief of the GMB dated September 6 1955, RHA 1021:17, in KFF, RBN, Serie: F III B, "Inkomna och avgångna skrivelser ordnade efter klassifikationssystem", A 01-A 7 , år: 1955, VOL. 5, KrA, p. 3.

ship-based guided missiles, which could seem probable since Galer and Schuldenfrei came from the Navy Department.[79] However, there is reason to believe that they discussed more than just ship-based missiles, as I will show below.

Of course, the visit of Galer and Schuldenfrei did little to remedy the acute need of assistance in the development of Swedish guided missiles. Something more had to be done. That was one of the reasons for the secret journey to the United States undertaken by the Swedish Supreme Commander General Swedlund in the fall of 1955. He took with him Air Attaché Colonel Stig Wennerström, Ambassador Boheman and a representative of the Wallenberg finance family, Karl-Arvid Norlin. Norlin was a board member of Enskilda, Wallenberg's private bank, and why he was a part of this traveling party has so far not been satisfactorily explained. The group headed by Swedlund met with the Vice Chief of Staff of the USAF, General Thomas White, and they discussed the guided missile issue.[80] General White told Swedlund and the others that the United States was currently conducting a thorough review of its weapons assistance policy towards foreign nations, and as a part of this review the special problems facing Sweden and the other neutrals would be examined.[81]

What White told his guests interests us since it is probably closely related to the discussion regarding guided missiles at the NATO meeting in December 1954. This shows that the Swedish guided missile community had influential friends in Washington that were prepared to really do something about the Swedish guided-missile-development situation. Certainly, it was not a matter of altruistic welfare work on the part of the Americans, but rather a realization that Sweden's ability to defend itself would benefit the United States too by fortifying NATO's northern flank. It was not just any collaboration, but a hegemonic collaboration, that the Americans wanted.

The meeting with General White was yet one more example of the more open attitude towards the Swedes, which had been noticeable since mid-1954. But the hegemonic collaboration with the Swedes still met with some resistance from different levels of the American bureaucracy, especially the State Department. Other examples of the amicable attitude could be seen in that same fall. In October 1955 the new Chief of the RSAF, Lieutenant General Axel Ljungdahl, was allowed to visit a series of American bases and companies to study various guided missile systems as well as methods utilized for measuring results of guided missile tests. Moreover, General Swedlund as well as Vice Admiral Stig H:son

79 Ahlström, C., *Stockholmsöverenskommelsen...*, p. 18, note 45. The Swedish term for ship-based guided missiles is: sjömålsrobotar.

80 Moores, S., *'Neutral on Our Side':...*, pp. 130–131.

81 Ibid., p. 131.

Ericson, Chief of the RSN, got to visit a Nike battery just outside of Washington D.C.[82]

A few days after General Swedlund had come back from the United States, in late November 1955, the Chief of the Swedish Defense Staff, Major General Richard Åkerman, had a conversation with Cabot and Major General Ridgely Gaither. Åkerman said that Swedlund thought that his visit to the U.S. had been extremely useful, and that they had been shown more than he had expected (referring specifically to the Nike launching site). Swedlund's visit had lasted for nearly four weeks and had been arranged as a clandestine one, but with the consent of the Swedish government. Cabot told the Swedish Major General that the United States had a great interest in seeing to it that Sweden's capability to defend itself was strong. The implicit meaning of this comment can hardly have escaped Åkerman, who certainly did not expect the Americans to bolster Sweden's defense so that it could defend its neutrality against Western violations as well. What was referred to was no doubt a very particular form of neutrality, one that discriminated heavily against, and would be entirely directed towards, the Soviet Union. After stating that Swedish neutrality policy was something that was the Swedes own business entirely, Cabot said that he:

> […] was not certain that in the event of general war Sweden's neutrality might not be more valuable to us than active Swedish participation on our side, provided Sweden was prepared to defend her neutrality strongly.[83]

It is obvious that the Ambassador was not hoping that Sweden would fend of American aircraft flying over Sweden, but that the comment was made with a reference to the Russians. After that Åkerman said that he was a bit perturbed over the fact that the Swedish plans, which called for the stockpiling of supplies needed for four months of fighting until help could be received from the outside, was draining the Swedish defense budget. Åkerman wondered if there was any chance that the Americans could step in and perhaps lend Sweden a hand after only one month, because this would allow the Swedes to divert funds to other much needed purposes. In reaction to this, Cabot said that he could not comment upon the military aspect of all of this, but that:

82 Ibid., p. 131, note 166.

83 Memorandum of Conversation between Cabot, Gaither, and Åkerman, dated November 27, 1955, pp. 1–2, quote on p. 2, in: RG 59; 1955–1959 CDF; From: 758.11/11-1555 To: 758.55361/5-1556; Box No. 3466; (Folder: 758.5/1-755); NARA II, College Park, MD.

> [...] the shorter time that Sweden was prepared to defend itself with its own resources, the better prepared we [the U.S.] had to be to get to Sweden the supplies she would need thereafter for defense and that we could not be so prepared without preliminary planning between their military and ours. It seemed to me that this would be contrary to Sweden's alliance-free policy.[84]

This supports the idea that planning talks on the staff level had not taken place between the United States and Sweden at this time, though secret cooperation was certainly going on in other areas, such as intelligence. At the same time it justifies the assumption that Cabot's prior comment regarding the hope that Sweden would defend its neutrality vigorously was not meant to apply to transgressions by the United States. It is hard to see that the Americans would be interested in supplying a power that was shooting at their aircraft.

General Gaither then interjected that it would not be possible for the United States to relieve Sweden after only one month even *with* prior preparations. It was just at the beginning of the war that everybody would be appealing for help, said Gaither, and the American production machine would not yet be up and running. Cabot tried to gloss over the Major General's frankness by commenting that in any case the time period needed would be shortened if properly prepared. The Ambassador then once more pointed out that the overall Swedish policy was of course a stumbling bloc for such preparations.[85] Åkerman, however, did not agree with Cabot.

> General Åkerman did not seem to think that this was an insurmountable obstacle. He indicated he thought that it might be possible for Sweden to enter into conversations provided they were kept very secret, and pointed to General Swedlund's secret visit as a case in point.[86]

While this confirms that such talks had not taken place at this time, it also suggests that they could be held. The reference to Swedlund's visit should probably not be taken to mean that such talks were held during the General's time in the United States, but it points to a possible venue for such discussions. The discussion then turned to the issue of technical assistance to Sweden and Cabot said that he had stressed to Washington the need for the United States to help the Swedes build up their defense, because this was in the best interest of the Americans. He hoped that some progress towards this end had been made on the other side of the

84 Ibid., p. 2.

85 Ibid., pp. 2–3.

86 Ibid., p. 3.

Atlantic, and urged Gaither to second them in Washington, to which the General warmly agreed and promised to do what he could. Cabot in this context mentioned particularly the Swedish interest in getting technical help regarding guided missiles, whereupon Åkerman confirmed the immense value that such help would have for the Swedish development effort. Said Åkerman:

> General Åkerman pointed out that even a few leads from the United States in regard to technical problems would be of immense benefit to the Swedes since it would enable them to avoid squandering their relatively slender resources on fruitless research.[87]

This statement suggests that Åkerman, and presumably other representatives of the Swedish military as well, had begun to consider their development program as a major *cul-de-sac*. The effort had yet to produce an operational missile and even the slightest help from the United States was perceived to be capable of mending the situation.

U.S. Hegemony, NATO Strategy & Planning, and the Guided Missile Issue

I have previously shown that the Americans had a long-standing aversion to releasing guided-missile-related technology to the Western European countries on a NATO-wide basis. The Western European allies had requested such technology already at the NATO meeting in Paris in December 1953, at which time it was not considered wise for the United States to adopt a more unrestrained policy. Now it is time to look even more closly still at this problem, and how it came increasingly to the fore in U.S. security policy evaluations during the fall of 1955.

In November that year George Perkins, who had now left Washington to become the U.S. Permanent Representative to the North Atlantic Council (henceforth: NAC), wrote a letter to the Director of RA B. E. Lane Timmons in which he expounded on a question that needed to be considered by Washington in preparation for the annual NATO meeting in December. The matter that according to Perkins needed urgent attention was that the United States had to take "significantly forward steps [...] with respect to the availability of new weapons and its aid policy."[88] The task at hand was to provide NATO forces with the strength required to implement the strategy in MC 48/1; NATO's new

87 Ibid.

88 Letter from Perkins to Timmons, dated November 10, 1955, p. 1, in: RG 59; RA; Records of the Director [henceforth: RecDir], 1955–1960; SF, 1956–60; Correspondence Files [henceforth: CorrF], 1955–59; Lot 61D252; Box No. 1; (Folder: RA Correspondence, 1955); NARA II, College Park, MD.

force planning document for the next few years, approved by the NAC on December 9, 1955. Perkins said in his letter:

> [...] there is a serious discussion of the problem of providing the strength required to implement MC48 strategy, or of adjusting military needs to get a maximum contribution toward MC48 needs out of available resources, the need to equip NATO forces with as modern weapons as NATO can produce will be exceedingly strong. The finger will point directly at U.S. Failure of U.S. to permit NATO and country replanning in light of MC48 to go forward on basis availability of modern weapon types will seriously affect possible success of enterprise as a whole and hence NATO.[89]

The planning document, which replaced MC 48 of December 1954, was one of a series of plans that appeared during 1954 which all cut substantially into NATO's conventional force requirements, and instead wished to provide for tactical nuclear weapons. Although the requirements for standing forces in MC 48/1 remained approximately at the same level as before, the reserves and back-up forces were greatly reduced.[90] In a report to the NAC on December 9, 1955, the NAMC stressed the poor shape of the air defense of the NATO area, and said that it was "greatly concerned over the slow progress being made in developing a sound system of Air Defense of Europe." It was acknowledged, however, that the issue was complicated by "many technical and political factors."[91]

This brought the question of guided missiles to the fore, and gave it a prominence it had not had before. Supplying the NATO countries with guided missiles was now one of the essential issues that had to be solved in order to achieve the goals of MC 48/1. Perkins wrote that:

> [...] it is absolutely essential that the United States be prepared to make available information on new weapons, but more important,

89 Ibid.

90 Jordan, Robert S., "Gruenther: Attempts to Retain NATO Solidarity" in Robert S. Jordan (ed.), *Generals in International Politics:...*, pp. 59–60. See also: Duffield, J. S., *Power Rules:...*, cf. pp. 75–77, 87–90.

91 M.C. 48/1 (Final), dated December 9, 1955, "North Atlantic Military Committee Decision on M.C. 48/1. A Report by the Military Committee on the Most Effective Pattern of NATO Military Strength for the Next Few Years", p. 6, accessed at: http://www.nato.int/docu/stratdoc/eng/a551209a.pdf (2007-02-12). These planning documents have also been dealt with by Dalsjö, and by the Swedish journalist Mikael Holmström, but the connection to the guided missile issue has, as far as I know, never been realized before (Dalsjö, R., *Life Line Lost:...*, pp. 60–63).

> be prepared to announce that it will, in due course, make the new weapons themselves available.[92]

This was necessary because without this, any attempt by the allied nations to implement MC 48/1 would be futile. But this issue was also important because of another factor, namely the risk of the United States losing its claim to hegemonic leadership and power. According to Perkins:

> [the] US can hardly continue to be looked upon as a leader in NATO partnership if it cannot rise to responsibilities of this kind.[93]

Perkins' evaluation of the current situation is absolutely crucial to an understanding of the importance of this issue to the United States, and it goes to show that the contemporary American policymakers were conscious of the hegemonic project, and the dangers facing it. In a way, the issue regarding the American monopoly on guided-missile-related technology and information had perhaps become a so-called reverse salient in the hegemonic project.[94] Earlier, the United States could afford to pursue this course, but new demands from the allies coupled with the new NATO requirements made the American position untenable. According to Perkins, if the U.S. wished to preserve its preponderant status, it would have to sacrifice a part of its privilege for the good of the alliance. The decision to release information regarding guided missiles (and other advanced weaponry) was thus explicitly advocated with a concern for U.S. hegemony, and the view of the United States among the allied countries of Western Europe.

The spirit of Perkins' message was incorporated into the Basic National Security Policy document NSC 5602/1 that was approved on March 15, 1956. In truly hegemonic language the document stated that:

92 Letter from Perkins to Timmons, dated November 10, 1955, p. 1, in: RG 59; RA; RecDir, 1955–1960; SF, 1956–60; CorrF, 1955–59; Lot 61D252; Box No. 1; (Folder: RA Correspondence, 1955); NARA II, College Park, MD.

93 Ibid., p. 1.

94 The term reverse salient has two somewhat different meanings. It is used in a military context, where it refers to a section of a frontline that is 'lagging' behind, i.e. where the enemy's resistance has not been subdued. In a history of technology context it has been used by Thomas P. Hughes when referring to a part of a technological system which is stopping the rest of the system from developing to a new 'higher' stage. Such reverse salients have to be overcome in order for the system not to stagnate and die (see: Hughes, Thomas P., *Networks of Power: Electrification in Western Society, 1880–1930* (Baltimore, MD: The Johns Hopkins University Press, 1983), pp. 14, 79–105).

> It is important for the United States to take the necessary steps to convince its allies, particularly its NATO allies, that U.S. strategy and policy serve their security as well as its own, and that the United States is committed to their defense and possesses the capability to fulfill that commitment. [...] The United States should provide new weapons (non-nuclear) [i.e. guided missiles] and advanced technology to allies capable of using them effectively, taking into account the protection of classified data, the essential requirements of U.S. forces, production capabilities and the likely available funds. [...] The United States should continue to provide military and other assistance, including where deemed appropriate new weapons and advanced technology, *to dependable allied nations where such assistance is necessary to enable them to make their appropriate contribution to collective military power*. Special attention in the technological field should be directed to assisting selected U.S. allies rapidly to develop their own advanced weapons systems [...].[95] [Italics added.]

Subsequent to the approval of MC 48/1 by the NAMC on November 14, 1955,[96] President Eisenhower announced during the NSC meeting at Camp David that the United States needed to share information on new advanced non-atomic weapons with its NATO allies, as well as the weapons themselves, in order to be able to make as firm a commitment as possible at the upcoming NATO meeting in December.[97] In a draft version of Secretary of Defense Charles E. Wilson's main statement to the NATO meeting it was said that:

> In order to further this modernization program, the United States is now prepared to release to NATO military and production authorities, and to the other NATO countries appropriate information pertaining to certain weapons which are now in various stages of production for United States forces. These weapons, not now included in military assistance programs, are surface-to-air and surface-to-surface guided missiles [...].[98]

95 FRUS: *Foreign Relations of the United States*; 1955–1957; Vol. XIX; *National Security Policy* (Washington, D.C.: United States Government Printing Office, 1990), pp. 247–248.

96 The attentive reader may now be confused, since I just said that the same document was approved by the NAC on December 9. The apparent confusion is due to the fact that all decisions made by the NAMC (which is a military agency) are subject to the approval of the NAC (which is a civilian agency).

97 Letter from Merchant to Dulles (see also attached letter to the Secretary of Defense Charles E. Wilson), dated November 29, 1955, p. 1, in: RG 59; RA; RecDir, 1955–1960; SF, 1956–60; CORRF, 1955–59; Lot 61D252; Box No. 1; (Folder: RA Correspondence, 1955); NARA II, College Park, MD.

98 "Proposed Remarks by the Secretary of Defense Before the North Atlantic

The fact that guided missiles were the first of these new types of weapons to be mentioned by Wilson points to the great significance the issue of guided missiles had for the Western European nations. What Wilson was offering was essentially the consolation to his fellow Ministers that help from the United States in this area would be forthcoming, even though the Secretary did not specify exactly when. Certain types of tactical nuclear missiles would be provided for U.S. forces in Europe during the next two years but with regard to the other countries within NATO, according to the draft, Wilson was to emphasize that "[b]efore further decision concerning other modern types of equipment for NATO can be made, a number of very difficult problems must be jointly solved."[99] Wilson was not to make any definite commitments; probably in an effort to buy time he was to take extra care to reassure the Europeans that the United States understood the importance of an effective air defense in Western Europe, and to say that the improvement of Western Europe's air defense and warning systems was of "first priority" for the United States.[100]

The impression that Wilson would mainly offer the European allies a pacifier, in the shape of promises of defense materiel deliveries to come, is enhanced by the fact that the budget for FY 1957 for the Mutual Security Program (henceforth: MSP) made no provisions for deliveries of these new weapons to Europe. This was commented upon by the Director of the MSP Hollister in a letter to Assistant Secretary of Defense, International Security Affairs (ISA), Gordon Gray in early January 1956. In a quite sharp tone, Hollister said that he felt it was his duty to comment that the budget did not make for anything other than replacement of old equipment; it did not provide enough for a significant build-up or modernization of existing forces. Hollister said:

> It thus does not deal with the apparently large equipment gap which is emerging in NATO as refinement of the concepts of MC 48 goes forward and as NATO countries face the task of replacing and modernizing existing conventional equipment. [...] The program makes no specific allowance for important new elements in the defense structure in the North Atlantic Treaty area, as, for example,

---

Council of Ministers Paris, France, December 16, 1955", p. 4, attached to: Letter from Elbrick to Dulles, dated December 12, 1955, in: RG 59; RA; RecDir, 1955–1960; SF, 1956–60; CorrF, 1955–59; Lot 61D252; Box No. 1; (Folder: RA Correspondence, 1955); NARA II, College Park, MD. Although this is only a draft it is unlikely that it was changed much prior to the meeting four days later. Comments regarding the draft offered by Elbrick further points to this conclusion.

99 Ibid., pp. 5–6.

100 Ibid., p. 5.

> a greatly improved air warning system, nor for many of the new or more modern types of weapons which our NATO allies have been requesting and which were discussed at the NATO ministerial meeting in Paris. On the contrary, the program largely assumes that our allies will be given the older types of weapons while those of a newer type will be retained in this country for our own armed forces. [...] I believe the program requires a thorough re-examination to determine whether, in terms of our broadest national interest, the equipment proposed for delivery is that which is most adapted to strengthening the forces of our allies.[101]

Hollister, too, stressed that the inability of the United States to deliver in this field was deeply problematic with concern to its relations with the allies. The MSP Director warned Gray that:

> [t]he failure to deliver in full and on time on a commitment as it may be interpreted by our allies is a source of serious, continuing friction.[102]

Thus was repeated the worry that the preponderant position of the United States was threatened by the failure to include provisions for guided missiles and other new types of weapons in the new budget proposal. Indeed, the consent given to U.S. hegemony by the European allies was not something to be taken for granted. The State Department seemed to agree with Hollister that the Pentagon's actions could jeopardize the effectiveness of the whole program. Lane Timmons at RA wrote Perkins in Paris saying that at Foggy Bottom it was feared that Congress might cut the proposed FY 1957 MDAP heavily "unless a high proportion of the newer long-lead time items is included."[103]

The possibility of Sweden benefiting from this announcement by Wilson was acknowledged immediately by military officials in Washington, and was even related in an article by United Press (henceforth: UP), carried by the Swedish local newspaper *Skånska Dagbladet*. The article was published by UP on January 3 and the Swedish translation was published the day after; it said that it was believed that the United States, and to some extent also Britain, would have to decide

101 Letter from Hollister to Gray, dated January 5, 1956, p. 1, attached to: Letter from Timmons to Perkins, dated January 24, 1956, in: RG 59; RA; RecDir, 1955–1960; SF, 1956–60; CorrF, 1955–59; Lot 61D252; Box No. 1; (Folder: RA Correspondence, 1956); NARA II, College Park, MD.

102 Ibid., p. 2.

103 Letter from Timmons to Perkins, dated January 24, 1956, in: RG 59; RA; RecDir, 1955–1960; SF, 1956–60; CorrF, 1955–59; Lot 61D252; Box No. 1; (Folder: RA Correspondence, 1956); NARA II, College Park, MD.

next year (i.e. 1957) "on the policy to furnish new weapons to the other NATO countries."[104] Cabot's translation of the article carried by *Skånska Dagbladet* read:

> The officials pointed out that while NATO is still uncertain about future NATO arms policies, there is one neutral European nation – Sweden – which is anxious to use United States know-how for some modern weapons such as guided missiles. [...] [T]he Swedish armed forces would like to have United States know-how in regard to certain aspects of the manufacture of guided missiles. They said the information here [in Washington] is that Swedish scientists have made good progress in designing guided missiles but if they had United States know-how on some vital aspects of manufacture Sweden could probably save millions of dollars brought about by trial and error experimenting on her own. [...] They said it would be unlikely for Sweden to receive know-how on modern weapons without the European NATO countries receiving this information first. However, they said that Sweden's desire to keep her armed forces modernized and ready for action is fully appreciated here, even if Sweden does not belong to NATO and is neutral, for that helps to contribute to the stability of northern Europe.[105]

It was thus acknowledged that a strengthening of Sweden's air defense system was beneficial to NATO. It is also interesting that the argument regarding the money-saving aspect of American support to Swedish guided missile research came very close to Major General Åkerman's words to Cabot and Gaither in late November 1955. So although it is impossible to identify who was the source for the UP article, it is not implausible to imagine that the source had some knowledge of this conversation.

Edwin E. Crowley at BNA did not think that the publication of this article would lead to any major harm being done, but it still considered it problematic that "information on highly classified discussions the Swedes and US leaked to the press." This was likely to reduce Swedish willingness to participate in any such discussions and arrangements, and, said Crowley, would hence be contrary to the United States' "interest in seeing that Sweden maintains as strong a defense establishment as possible." Even though it could not at this time be ruled out that the information came from a Swedish officer in Washington, he said, the fact that the Secretary General at the Swedish Foreign Ministry

104 785.5/1-956, Telegram from Cabot to Dulles, dated January 9, 1956, p. 1, in: RG 59; 1955–1959 CDF; From: 758.11/11-1555 To: 758.55361/5-1556; Box No. 3466; (Folder: 785.5/1-755); NARA II, College Park, MD.

105 Ibid., p. 2.

Arne Lundberg had brought the matter up with Ambassador Cabot suggested that the Americans could not ignore the incident completely.[106] Lundberg told Cabot that this was "a very touchy political subject now in Sweden", and continued by remarking that while there had been no serious repercussions so far, there might be.[107] Cabot had tried to calm Lundberg by stating that a search of Swedish newspapers for the article had turned up nothing at all, and it was actually Lundberg himself who provided the Ambassador with a copy of the article.[108] The Ambassador wrote to Secretary of State John Foster Dulles that:

> I must earnestly point out how unfortunate leaks of this kind are to any collaboration Sweden can afford to accept from us in military matters and I hope Department will emphasize undesirability of any repetition. If it is in our interest that Sweden should maintain as strong a defensive position as possible, such leaks are prejudicial to us from military as well as political standpoint.[109]

The leaking of this information shows that military circles in Washington were aware that, immediately after the NATO meeting in December 1955, there was a connection between the undertaking offered by the Secretary of Defense to furnish guided missiles and related technology to the allies, and Sweden's possibility of receiving the same. These officers apparently saw no real difference between Sweden and the NATO countries. They only observed that it would be difficult, but not unthinkable, to give such information and technology to a non-member before a decision had been made by the United States and Britain to release it to the NATO allies. Again it was not a matter of content but of form. Consent to U.S. hegemony was the guiding principle, and hence there was no reason why Sweden would be left out of the loop; a strong Swedish air defense was looked upon as beneficial to the defense of the northern NATO area, and Sweden still had the fourth

106 785.5/1-956, Telegram from Crowley to Parsons and Horsey, dated January 10, 1956, in: RG 59; 1955–1959 CDF; From: 758.11/11-1555 To: 758.55361/5-1556; Box No. 3466; (Folder: 785.5/1-755); NARA II, College Park, MD.

107 785.5/1-456, Telegram from Cabot to Dulles, dated January 4, 1956, in: RG 59; 1955–1959 CDF; From: 758.11/11-1555 To: 758.55361/5-1556; Box No. 3466; (Folder: 785.5/1-755); NARA II, College Park, MD.

108 785.5/1-1156, Telegram from Cabot to the State Department, dated January 11, 1956, and enclosed Memorandum of conversation between Cabot and Lundberg, dated January 9, 1956, in: RG 59; 1955–1959 CDF; From: 758.11/11-1555 To: 758.55361/5-1556; Box No. 3466; (Folder: 785.5/1-755); NARA II, College Park, MD.

109 785.5/1-456, Telegram from Cabot to Dulles, dated January 4, 1956, in: RG 59; 1955–1959 CDF; From: 758.11/11-1555 To: 758.55361/5-1556; Box No. 3466; (Folder: 785.5/1-755); NARA II, College Park, MD.

largest Air Force in the world. Its strength in battle, however, would depend on the quality of its armaments. Ambassador Boheman did not consider the information contained in the article to be very remarkable at all, and simply commented that it was a logical consequence of the realites confronting the U.S. when trying to defend Europe.[110]

The question regarding an American decision on releasing guided-missile-related technology and know-how dragged on, however, and not much had been done on this point by December 1956, when the annual NATO meeting recommended that the member states should be provided with air defense guided missiles. This finally resulted in an American decision in early 1957 that the United States should follow this recommendation and offer such guided missiles to its NATO allies. Denmark and Norway, for example, accepted and received Nike Ajax and Nike Hercules batteries.[111] In early November Count Carl Douglas of the Swedish Embassy approached Crowley at BNA regarding the possibility of Swedish-American cooperation on guided missile development.[112] Only a few days afterwards, the Assistant Secretary of Defense, ISA, Gordon Gray could tell the visiting Ambassador Cabot that "NATO member countries were beginning to receive such data *and accordingly this would facilitate matters for Sweden*."[113] [Italics added.]

Gray was, probably, already privy to information, already prior to the NATO meeting which allowed him to anticipate the U.S. government's decision in early 1957 to release such information. This further strengthens the hypothesis that the unwillingness to release information regarding guided missile development to Sweden has to be explained in terms of the United States' general policy in these matters, and not with reference to Sweden's non-alliance policy. Why else would the decision to release data to NATO allies benefit Sweden the least bit? The answer is that the Americans did not make a difference between apples and oranges; once again it was the content that mattered most—not the form. Prior to December 1956, NATO member or not, the American attitude was not an amicable one when it came to letting foreign nations study their

110 Letter from Boheman to Åström, dated May 10, 1957, p. 3, in: HP 24 VI/U.S.A; "Tillstånd till in- och utförsel av vapen, ammunition och krigsmateriel"; 1955 juli–1957; Folder 23; Box No. 62; År: 1955–1960; RIA; Arninge.

111 Dörfer, I., *System 37 Viggen...*, p. 90.

112 Memorandum of Conversation between Crowley and Douglas, dated November 9, 1956, in: RG 59; INR, 1945–1960; Lot 58D776; Box No. 13; (Folder: Sweden); NARA II, College Park, MD.

113 Moores, S., *'Neutral on Our Side':...*, p. 134. Moores has, mistakenly, consistently spelled Gray's name 'Grey'. More importantly, the importance of the connection between the NATO recommendation, the American decision in early 1957, and Gray's statement that the decision would benefit Sweden, is not commented upon by Moores.

guided missile endeavors. Gray's statement eliminates all other plausible explanations.

In early January 1957, the matter of possible assistance to the Swedish guided missile program was raised in a memorandum from Jacob Beam at EUR to Park Armstrong. Beam wrote to tell Armstrong that Ambassador Cabot had again called on Quarles and Gray to discuss the Swedish interest in getting U.S. technical support. Margrave, of the NAMC, had stated that the Pentagon had informally raised the issue, and thus EUR considered it important "to be able to give its views on the proposed help for Sweden's defense establishment." Beam now sought Armstrong's "concurrence in supporting exploration of the Swedish request with the other departments of the Government."[114] The issue obviously gained new importance after the NATO meeting in December 1956.

Also in early 1957, a memorandum spelling out the policy regarding Sweden and guided missiles was prepared within the State Department by an anonymous author. It was concluded that granting Swedish requests in this area was in line with the policy set forth in NSC 121, and that accordingly such requests should be treated sympathetically. The memorandum said:

> Under the terms of NSC 121, it is our policy to receive sympathetically Swedish requests for assistance in maintaining a strong military position without prejudicing the meeting of higher priority defense requirements. Now that dual-purpose weapons are to made available to NATO members, it is believed that, in application of this policy, *the Department should support any Swedish request received for assistance in this field and should seek to avoid tying this assistance to conditions which the Swedes could not accept*. The timing of any assistance extended should, of course, be arranged with due consideration for the higher priority requirements of our NATO partners.[115] [Italics added.]

The memorandum thus explicitly linked the decision to supply the NATO allies with guided missiles and the decision to grant Sweden access to these weapons as well; it could hardly have been formulated much more clearly than that. It was also stated that Sweden should not be expected to agree to conditions that would too obviously infringe upon

114 758.5/1-957, Memorandum from Beam to Armstrong, dated January 9, 1957, in: RG 59; 1955–1959 CDF; From: 758.11/11-1555 To: 758.55361/5-1556; Box No. 3466; (Folder: 758.5/1-755); NARA II, College Park, MD.

115 Untitled Memorandum by Unknown Author, undated early 1957, in: RG 59; EUR/SCAN; Records Relating to Swedish Affairs [henceforth: RRSWA], 1958–1964; POL 17 Political Affairs Sweden POL 17 Diplomatic Notes 1959–1963 to PS – Protective Service Sweden PS 7-1 Arrest (Swed. Seamen Visit New Orleans); Box No. 1; (Folder: Sweden Missile 1959–62); NARA II, College Park, MD.

the policy of neutrality. Once again the strategy chosen by the United States was colored by the official Swedish non-alignment policy, and this was not considered to be a problem as long the cooperation could be tailored around this policy and allow for the realization of U.S. strategic objectives in the area.

The connection between the NATO recommendation, the American decision in early 1957, and Gray's statement was not fully made by Moores, who related the information that Sweden would get access to missiles at the same time as the NATO countries as a fragment of historical knowledge. Moores did not explain why Gray made this implicit promise to Cabot, or why Sweden would at all benefit from the release of guided missile technology to NATO members, nor did he try to do so. But if one analyzes the events in the light of hegemonic theory, the pieces fall into place and the puzzle forms a coherent image of events. The explanation has to do with Sweden's strong consent to American hegemony, and unwavering support for U.S. policy objectives and the political and economic system promoted by the United States. It is important to understand this in the right way. Hegemonic theory does not help us to understand why the NATO countries had to have access to these weapons *before* Sweden; this seems pretty obvious. What demands an explanation is instead why Sweden, as a non-member of the alliance, would benefit from the fact that the United States released guided missile technology to formally allied countries.

The possibility for Sweden to purchase American missiles brought to the fore the question of the safeguarding of this technology, and how this cooperation should be handled. During the summer of 1957 Wennerström had been told by his American liaison officer that some kind of written agreement ought to be concluded between the United States and Sweden. In a briefing paper it was stated that the Americans might have worked out detailed specifications for how this sort of cooperation should be conducted with the NATO countries, and with other allied or associated powers. Furthermore, it was said that Sweden should try to review the text for agreements connected to the guided missile issue that the U.S. government had concluded with other nations, if there were such agreements.[116] In fact, the United States had no clear guidelines on how to handle this issue at that time. This was neither the first, or the last, time that the Americans raised the question regarding a new security agreement (we will see more of this in Chapter 6, p. 353), which could supplant or broaden the 1952 MDAA agreement. Wennerström had reported back to Stockholm by the middle of January that it would probably be necessary to conclude a new agreement along

116 "Allmänt om kontakterna med USA", dated July 18, 1957, p. 1–2, in: HP 24 VI/ U.S.A; "Tillstånd till in- och utförsel av vapen, ammunition och krigsmateriel"; 1955 juli–1957; Folder 23; Box No. 62; År: 1955–1960; RIA; Arninge.

the lines of the one from 1952, if Sweden was to continue to be eligible for purchases of advanced weapons systems from the United States.[117]

So why then, did not Sweden get instantaneous access to American guided missiles? How come that the purchase of Sidewinder was made almost two years later in early 1959? The reason for this, I will argue, was that bureaucratic wrangling over how this new policy should be realized slowed down the political process within the United States. Below I will provide further evidence that will show that the failure of the United States to act immediately upon its decision in early 1957 was the reason why Sweden, and the NATO countries, had to wait until early 1958 before negotiations regarding guided missile purchases could be initiated.

During the course of 1957 the Americans worked hard at making guided missiles available to their European allies. By the end of May the NAMC submitted a report on measures deemed necessary to implement the strategic concept entailed in MC 48/2, dated March 15 and approved by the NAC on May 9, 1957.[118] Tied to this document was MC 14/2, which was a document spelling out the overall strategic concept for the defense of the NATO area.[119] MC 14/2 remained in force until superseded by MC 14/3 in December 1967. The latter change was due to the perception, going back to the Kennedy administration, that Eisenhower's 'massive retaliation' was bankrupt as a strategic concept, and the development of a new approach called 'flexible response'.[120] With MC 48/2 the NAMC had derived a set of requirements that to its judgment had to be met within the next five years; it was basically a five-year plan for NATO designed to implement MC 14/2.[121]

With regard to Scandinavia, MC 14/2 was largely identical to its predecessor document MC 14/1 from December 1952. For example,

117 Letter from Åström to Boheman, dated February 12, 1957, in: HP 24 VI/U.S.A; "Tillstånd till in- och utförsel av vapen, ammunition och krigsmateriel"; 1955 juli–1957; Folder 23; Box No. 62; År: 1955–1960; RiA; Arninge.

118 MC 48/2 (Final Decision), dated May 23, 1957, "Final Decision on MC 48/2. A Report by the Military Committee on Measures to Implement the Strategic Concept", p. 1, accessed at: http://www.nato.int/docu/stratdoc/eng/a570523b.pdf (2007-02-12).

119 MC 14/2 (Revised) (Final Decision), dated May 23, 1957, "Final Decision on MC 14/2. A Report by the Military Committee on Overall Strategic Concept for the Defense of the North Atlantic Treaty Organization Area", accessed at: http://www.nato.int/docu/stratdoc/eng/a570523a.pdf (2007-02-14).

120 Sorley, Lewis, "Goodpaster: Maintaining Deterrence during Détente", in R. S. Jordan (ed.), *Generals in International Politics:...*, p. 130.

121 MC 48/2 (Final Decision), dated May 23, 1957, "Final Decision on MC 48/2. A Report by the Military Committee on Measures to Implement the Strategic Concept", p. 1, accessed at: http://www.nato.int/docu/stratdoc/eng/a570523b.pdf (2007-02-12).

as we remember from Chapter 4,[122] Sweden's economic, industrial and military strength (and also her strategic location) was reiterated almost *verbatim*; and the statement that the Soviets could not ignore the possibility that she might join the allies if the other Scandinavian countries were attacked was also repeated. However, there was one difference in formulation; instead of declaring that Sweden would be "most important in the defense of Scandinavia", as in MC 14/1,[123] it was now said that she "could contribute materially to NATO defense."[124] This clearly emphasized that Sweden could contribute directly to the NATO defense effort of the Scandinavian Peninsula.

The assumption that Scandinavia could only be defended as a whole was also retained, with the addition that this entity should be defended "as far forward as possible [...]."[125] Perhaps more significant in some respects was the comment regarding the area which was considered to be covered by NATO planning, and where the individual member countries had to step in and take responsibility. It was stated that:

> [w]hile NATO defense planning is limited to the defense of the Treaty area, it is necessary to take account of the dangers which may arise for NATO because of developments outside that area. [...] In this light, planning for the most efficient organization and the equipment of NATO forces must take account of the possible need for certain NATO countries to use some of their NATO forces to meet defense commitments elsewhere [...].[126]

While what was intended here was perhaps first and foremost French and British colonial areas, this phrasing could also, as Robert Dalsjö has remarked, apply to Sweden and U.S. intentions of assisting the Swedes if they were attacked.[127] Since Sweden was not a member of the alliance, there was no guarantee that the NATO countries would commit

122 See above, pp. 249–252.

123 MC 14/1 (Final), dated December 9, 1952, "Decision on M.C. 14/1: A Report by the Standing Group on Strategic Guidance", p. 18, accessed at: http://www.nato.int/docu/stratdoc/eng/a521209a.pdf (2007-03-06); see also: Weinberger, H., "A Small/Medium-sized Great Power in the North..." in K. L. Kleve (ed.), *The Cold War, Military Power...*, p. 55.

124 MC 14/2 (Revised) (Final Decision), dated May 23, 1957, "Final Decision on MC 14/2. A Report by the Military Committee on Overall Strategic Concept for the Defense of the North Atlantic Treaty Organization Area", p. 20, accessed at: http://www.nato.int/docu/stratdoc/eng/a570523a.pdf (2007-02-14).

125 Ibid., p. 21.

126 Ibid., p. 12.

127 Dalsjö, R., *Life Line Lost:...*, p. 61.

themselves to defend her territory if they themselves had not yet been drawn into a general war.

This was consistent with prior NATO planning, which demonstrated that NATO planners counted on the possibility that Sweden might join the West if she was attacked, and developments in Sweden after D-day had a quite large impact on NATO's campaign. This is clear from the "Capabilities Plan Allied Command Europe 1957", which outlined NATO's strategy in defending Europe from a Soviet attack, and showed that NATO's defense of Norway was, to a certain extent, heavily dependent upon Sweden's defenses. As mentioned in Chapter 4, one of the main assumptions for the study was "that Sweden and Switzerland would fight with the West if attacked [...]."[128]

The Western campaigns to defend Norway were based on two possible scenarios; namely, one where Sweden remained neutral due to a Soviet choice to respect Swedish neutrality, and another where Sweden was a co-belligerent due to a prior Soviet attack upon Sweden.[129] It was concluded that "Soviet actions with respect to Sweden, and the outcome thereof, are of the greatest importance to this campaign", and the document continued:

> Should the Soviet avoid Swedish soil, the forces available for use against Norway would be increased, but the means of access would be difficult. Should the Soviets attempt to drive into or across Sweden by force, it is estimated that Sweden would resist; her sizeable forces would then engage substantial Soviet forces, if the Soviets were to attempt offensive action.[130]

The link between Norway and Sweden was estimated to be of vital importance, and it was stated that "Trondheim must be held as

128 SGM 600-54, "Capabilities Plan Allied Command Europe 1957", dated September 10, 1954, p. 5, in: NATO; Standing Group Records Section [henceforth: staGRes]; NATO Archives, Brussels. I thank Hans Weinberger for making this document available to me. Perhaps it was a precursor to this document that the Norwegian Colonel Holterman was talking about to Swedlund in December 1953. Holterman had said to the Swedish Supreme Commander that there was a new defense plan for NATO available as of January 1, 1954, in which Sweden figured in a whole new way. Great benefits would accrue to Sweden due to this, Holterman had said: e.g. two American Marine Corps divisions could quickly be deployed in northern Norway, which would release Norwegian forces for use in southern Norway. Holterman explained that he was prepared to send at least one Norwegian division to Sweden to support operations in for instance the Skåne region (the southernmost part of Sweden) (Agrell, W., *Den stora lögnen...*, pp. 87–88).

129 SGM 600-54, "Capabilities Plan Allied Command Europe 1957", "Enclosure F", dated September 10, 1954, p. 2, in: NATO; staGRes; NATO Archives, Brussels.

130 Ibid., p. 6.

it supports the Oestlandet area, and is the main link with Sweden",[131] and it was further said that holding this area was "necessary [...] for the support of South Norway, and particularly the support of any Swedish participation which might develop."[132] This indicates that NATO plans took Sweden into account when planning for the aid of Norway. A key formulation was provided in the following paragraphs:

> Trondheim is estimated as being vulnerable initially only to sabotage and airborne raids. A threat from North Norway is not anticipated. *Unless and until a threat develops from the Swedish area, the home defense forces available in this region should be adequate.* [...] The threat to South Norway, except from airborne commando and sabotage operations, *is largely contingent upon the success of Soviet operations in* Denmark and *Southern Sweden.*[133] [Italics added.]

The document continued to state that:

> The conventional air forces available for, and allocated to, land/air operations in Norway are adequate to counter a threat of the magnitude which might develop in Northern Norway. *The composition and deployment of Norwegian air forces must take into account that of Swedish forces in view of the high probability of combined action in the event of war.* In this respect current programmes seem adequate.[134] [Italics added.]

Such cooperation was fully feasible due to the secret cooperation scheme that had been initiated between Sweden, Norway, and Denmark, after the end of the SDU negotiations in January 1949.[135] From this it is clear that NATO put enough faith in Swedish indirect or direct 'assistance' in the defense of Norway for the number of forces charged with the defense of Norway to depend on it. In other words, if Sweden failed to come through in this respect, NATO's plans would have to be considerably revised to fit the new strategic situation. It is also significant that there is no mention of a situation were Sweden had been occupied swiftly by the Soviet Union; Swedish resistance was expected to tie up a good number of Soviet divisions, and delay their advance considerably. Moreover, combined Norwegian-Swedish air operations

131 Ibid., p. 2.

132 Ibid., p. 5.

133 Ibid., p. 7.

134 Ibid., p. 7. See also: Weinberger, H., "A Small/Medium-sized Great Power in the North..." in K. L. Kleve (ed.), *The Cold War, Military Power...*, p. 56.

135 For the same, see: Weinberger, H., "A Small/Medium-sized Great Power in the North..." in K. L. Kleve (ed.), *The Cold War, Military Power...*, p. 56.

were deemed very probable. This no doubt also serves to explain why it was equally important to see to it that Sweden, as well as the NATO countries, was given access to guided missiles; the RSAF was expected to have direct bearing on the defense of Norway.

It is in this light one should view the NATO plans, drawn up in 1956, to send Norwegian fighter bombers and the SACEUR's nuclear strike force over Swedish territory. These operations were codenamed SNOWCAT (Support of Nuclear Operations With Conventional Attacks) and the idea was that the Norwegian fighter bombers would destroy enemy radar and saturate the Soviet command and control capacity so that the NATO Atomic Strike Force could get through to their targets. The flight paths to Eastern Germany, Poland, and the Baltic States went straight across Sweden. The operation was essentially a suicide missions for the Norwegian pilots, because the targets were at the limit of the aircraft operation radius. The planners counted upon the possibility that some of the aircraft which actually made it back could use Swedish airfields for emergency landings. The SNOWCAT operations were part of the SACEUR's planning until the mid-1970s.[136]

The plan assumed that returning aircraft could make emergency landings at Swedish airbases. Wilhelm Agrell has remarked that this is probably why a number of runways in middle and southern Sweden were extended and reinforced during the 1960s.[137] The importance of Sweden's airspace increased considerably in the mid-1950s, due to Eisenhower's new strategic concept focusing on all-out nuclear war, i.e. Massive Retaliation, outlined in NSC 162/2 approved in October 1953. The focus on the nuclear deterrent, and the so-called Polar Strategy, made Scandinavian airspace a crucial transit area for U.S. long-range bombers. Paradoxically, the same strategy downgraded the value of the flanks as staging areas for the defense of the central front.[138]

The SNOWCAT operation, while at first appearing to be adventurous to the extreme considering the possibility that Swedish air defenses might throw gravel in the machinery, becomes more understandable when one takes into account the planning that lay behind it, such as the Capabilities Plan of 1954. The plans of the British Bomber Command during the 1950s and 1960s are less well known, although we do know that the British in the late 1970s and early 1980s intended to send their bombers in huge numbers over Sweden. Upon reaching the Swedish west coast these aircraft, literally hundreds of British Vulcans and

136 Skogrand, Kjetil and Tamnes, R., *Fryktens likevekt. Atombomben, Norge og verden, 1945–1970* (Oslo: Tiden Norsk Forlag, 2001), pp. 197–206; see also: Weinberger, H., "A Small/Medium-sized Great Power in the North…" in K. L. Kleve (ed.), *The Cold War, Military Power…*, pp. 57–58.

137 Agrell, W., *Svenska förintelsevapen…*, p. 303.

138 Tamnes, R., *The United States and the…*, pp. 91–92, 99–100, 143–144.

Buccaneers and U.S. F111s and B52s, would have descended to low levels of 70 to 170 meters (200 to 500 feet), and then passed over Sweden on their way to their designated targets in the USSR and Eastern Europe. The Swedes were simply expected to turn a huge blind eye.[139] It is not known how much the Swedes knew about these plans, but at the same time it is obvious that this planning cannot have been based solely upon the hopes of the Western war planners. It is reasonable to assume that the Swedes were at least briefed about these plans (perhaps by the Norwegians), so that the Swedish military did not start shooting at the clutter of Allied aircraft over Swedish territory.[140] Johan Gribbe has shown that cooperation with NATO in wartime was practiced in the war games organized by the National Defense College (Försvarshögskolan) every year from 1952. Through these courses the top stratum of Sweden's military and civilian leadership was discretely made aware of the option of cooperation with the West. This was an important way for the military and the government to speak of the unspeakable, i.e. the failure and abandonment of Sweden's policy of neutrality.[141]

The strategic concept outlined in MC 14/2 was heavily influenced by the emergence of nuclear weapons on a massive scale on both sides in the Cold War. One remarkable aspect spelled out in the document was the aggressive posture of the NATO forces, and the willingness to initiate a massive nuclear exchange in case of hostilities in Europe, even if the Soviet Union did *not* employ nuclear weapons. Even though it was acknowledged that the Soviet Union might not resort to the use of nuclear weapons initially, and even if a large-scale war resulted from "military operations of a limited nature which the Soviets did not expect to result in general war," it was stated that:

> In either case there will almost certainly be Soviet land, sea and air campaigns with the purpose of isolating and seizing NATO Europe. Since NATO would be unable to prevent the rapid overrunning

139 Campbell, Duncan, "The 'deterrent' goes to war" in *New Statesman*, May 1, 1981, p. 8; see also: Weinberger, H., "A Small/Medium-sized Great Power in the North..." in K. L. Kleve (ed.), *The Cold War, Military Power...*, p. 58.

140 One could of course assume that the Allies would first plaster the Swedish defense installations just to make sure that the Swedes did not shoot at them. This would have been possible since the Americans and the British knew very well where the Swedish airbases and anti-aircraft installations were located. However, this was not a possibility for the Bomber Command, because these aircraft flew too low for them to be able to drop nuclear bombs on Sweden.

141 Gribbe, J., "En inre angelägenhet för fritt tankeutbyte. Synen på västsamverkan i Försvarshögskolans krigsspel 1952–1979" (Stockholm: TRITA-HST Working paper, 2003/1); Gribbe, J., "'...därest fienden lyckas taga svenskt territorium' Sverige och Väst i Försvarshögskolans krigsspel" in *Kungl Krigsvetenskapsakademiens handlingar och tidskrift*, No. 5, 2003, pp. 93–98.

> of Europe unless NATO immediately employed nuclear weapons both strategically and tactically, *we must be prepared to take the initiative in their use.* [...] In case of general war, therefore, NATO defense depends upon an immediate exploitation of our nuclear capability, *whether or not the Soviets employ nuclear weapons.*[142] [Italics added.]

The importance of a strong and capable air defense was emphasized in MC 48/2, where this factor was listed as the number one survival measure. It said:

> The capacity of our forces to survive and to retaliate, immediately and effectively, requires that we have an ability to oppose actively an initial Soviet nuclear attack. We must, therefore, develop and maintain an effective air defense system. This NATO system must provide for the coordination, and control as practicable, *of all aircraft and missiles associated with air defense.*[143] [Italics added.]

It was foreseen that the threat posed to the NATO area by the Soviet Union demanded a high degree of integration and increased interdependence of the NATO air defense system, and one of the necessary measures identified in the report was "the introduction of air defense missiles [...]."[144] These documents made clear that guided missiles were a vital part of the defense of the NATO area, and thus ushered in the need for the United States to actually release the new weapons which had been demanded by the allies, and Sweden, for a long time.

The reason why it had taken so long for the United States to come through on its promise was explained in early November 1957 in a high level conversation in preparation for the NATO Heads of Government meeting in Paris on December 16–18, 1957. The participants in the discussion were senior officers of the American government and state apparatus; including prominent characters such as the new Secretary of State John Foster Dulles, Director of the PPS Paul H. Nitze, Special Assistant to the President for National Security Affairs General Robert Cutler, George Perkins, former Special Assistant to the President Nelson A. Rockefeller, Special Assistant to the President Harold E.

142 MC 14/2 (Revised) (Final Decision), dated May 23, 1957, "Final Decision on MC 14/2. A Report by the Military Committee on Overall Strategic Concept for the Defense of the North Atlantic Treaty Organization Area", p. 9, accessed at: http://www.nato.int/docu/stratdoc/eng/a570523a.pdf (2007-02-14).

143 MC 48/2 (Final Decision), dated May 23, 1957, "Final Decision on MC 48/2. A Report by the Military Committee on Measures to Implement the Strategic Concept", p. 6, accessed at: http://www.nato.int/docu/stratdoc/eng/a570523b.pdf (2007-02-12).

144 Ibid., p. 12.

Stassen, Deputy Under Secretary of State Robert Murphy, and U.S. Special Representative in Europe 1952–53 William H. Draper, Jr..[145] The conversation centered on ways and means of strengthening NATO, and addressed indirectly the issue of guided missile deliveries to Western Europe.

Draper said:

> [...] the problem lies in developing a real policy on the disclosure of classified research and development information to our allies. He referred to the valuable work that had already been done in the NATO Mutual Weapons Development Program, sponsored by the US.[146]

Draper's statement shows that even immediately before the coming NATO meeting, in late 1957, the U.S. government did not have a formal and coherent policy for sharing information on new weapons, such as guided missiles, with the allies. Dulles offered the views of the U.S. government on guided missile deliveries to the Belgian Ambassador Robert Silvercruys and Belgian Minister Georges Carlier on November 18. The Belgian Ambassador wanted to sound out the Americans prior to the meeting of the Heads of Government in Paris, and queried Dulles as to the possible outline of "the blueprint" of U.S. thinking. Dulles was cautious, saying only that the U.S. government was still developing its thoughts regarding the meeting, but stated that the United States:

> [...] hope to share our thinking regarding development of new weapons. Supplying these weapons to our NATO allies, and deciding where new weapons, including guided missiles, should be positioned, are matters of mutual interest.[147]

Considering that Draper made his statement in an all-American forum, it must be assumed that this was more earnest and closer to the true state of things than Dulles' assurances to the Belgian Ambassador almost two weeks later. It seems plausible to assume that a coherent policy had not yet been worked out before the meeting a month later.

145 Memorandum of Conversation between Dulles et al., dated November 6, 1957, p. 1, in: RG 59; EXSEC; The Secretary's and the Undersecretary's Memoranda of Conversation [henceforth: SUMCON], 1953–1964; Lot 64D199; June 1957 to December 1957; Box No. 7; (Folder: Secy's M. of Con. November – 1957); NARA II, College Park, MD.

146 Ibid., p. 6.

147 Memorandum of Conversation between Dulles, Carlier, and Silvercruys, dated November 18, 1957, p. 1, in: RG 59; EXSEC; SUMCON, 1953–1964; Lot 64D199; June 1957 to December 1957; Box No. 7; (Folder: Secy's M. of Con. November – 1957); NARA II, College Park, MD.

That there was no such policy when Dulles spoke to the Belgians is confirmed in a draft position paper on November 15 prepared by the Office of the Assistant Secretary of Defense, ISA, on the subject of cooperation in development and production of non-nuclear weapons.

At this time the United States had agreements for exchange of information regarding research, development, and technical production data classified as 'Secret' with only two countries; Britain and Canada.

> No other exchange agreements are in effect at this time. Improvement in security arrangements with other countries would permit an extension of these agreements as desired. [...] The Sample Weapons Program announced by Secretary Wilson at last year's NATO Ministerial Meeting offers U.S. assistance in the form of samples of modern weapons and technical data to those countries who have a technological capability to utilize this information and satisfactory security arrangements to protect U.S. classified data.[148]

We can thus conclude that the NATO allies, in November/December 1957, had not been granted access to anything more than Sweden had with regard to guided missiles. Perhaps it is worth noticing the focus above on technological capability and satisfactory security arrangements. Maybe we ought to view the specialized Swedish development, and the subsequent cancellation of some projects in favor of license production of systems with the same characteristics (such as RB 340 and RB 341), in the light of this expressed American wish (I will deal with this in the next chapter).

So far, only Italy, France, the Netherlands, Norway, and Belgium had made inquiries about the Sample Weapons Program. Some security checks and technical surveys had been made, and some twenty Italian technicians had toured U.S. missile facilities for a month.[149] The Americans expected that the OSP could serve as a valve to relieve the discontent of the European allies with regard to the guided missile issue, hold the alliance together, and strengthen U.S. hegemony in the process.[150] It was also thought that the OSP would "permit more efficient use of the

148 "Cooperation in Weapons Development and Production Exclusive of Nuclear Components", pp. 9–10, quote on p. 10, attached to a Memorandum by the Acting Assistant Secretary of Defense (ISA) John H. Irwin, II, dated November 15, 1957, in: RG 59; RA; RecDir, 1955–1960; SF, 1956–60; CorrF, 1955–59; Lot 61D252; Box No. 1; (Folder: RA Correspondence, 1957); NARA II, College Park, MD.

149 Ibid., p. 10.

150 Untitled Memorandum, dated December 23, 1957, p. 1, in: RG 59; Bureau of European Affairs [henceforth: EUR]; RA; Political-Military Numeric Files [henceforth: PMNF] 1953–1962; 2.4 MWDP General, 1960–61 to 3.4 Civil Emergency Planning; Box No. 3; (Folder: OSP-General); NARA II, College Park, MD.

resources available to NATO for the common defense and to further strengthen the Alliance." The memorandum displayed fears that the slow progress concerning the sharing of technical information with the allies was damaging the Alliance. It stated that:

> [t]he U.S. needs its NATO allies more than ever before. Doubts have been generated in some Europeans concerning U.S. policies and purposes. Supporting European production, especially in the field of new weapons, would help reduce the Europeans sense of overdependence on the U.S. and their feeling of being merely pawns in a struggle between two colossi. It would also demonstrate in a tangible way, through their pocketbooks, our interest in the survival of our NATO partners.[151]

This paragraph underlines the importance for a hegemonic state of seeing to it that material benefits accrue to the consenting nations; otherwise the incentive for giving ones consent slowly, but steadily, evaporates. The document further stated that the United States had already "decided to provide certain fairly highly classified equipment to most of our NATO allies for their military forces [...]."[152] Another judgment to the same effect had been included in a memorandum for the Chairman of the JCS in March 1956, where it was said that:

> [i]f we do not now, in the FY 57 program, appropriate funds to provide our allies with the newer weapons in being in these years, our relative deterrent and fighting capabilities will be diminished to a serious degree. *If we should fail to make this provision, our allies would be keenly aware of this changed situation, and the impact thereof on their political attitudes could be drastic.* Also, the deterrent effect of our defenses on the Soviet forces would be weakened.[153] [Italics added.]

Clearly, the political effects that were spoken of above did not refer to any risk of the Western European countries going Communist. Rather it referred to the risk of the common front, i.e. the Alliance, being weakened; and a common front, in this context, meant a front led by the United States in the role of hegemonic power. What was feared was that the Europeans would sense the lack of initiative on the part of the United States, and subsequently suspend their consent to U.S. hegemony in Europe, perhaps turning to a regional solution to

151 Ibid., p. 1.

152 Ibid., p. 2.

153 Memorandum for the Chairman of the JCS, dated March 12, 1956, p. 6, in: RG 218; ChF; AR, 1953–57; 471.6 (1 Aug 56) to 500; Box No. 44; (Folder: 471.6 (1956)); NARA II, College Park, MD.

the guided missile issue, such as joint R&D efforts etc. This paragraph also highlights, perhaps better than any other, the perceived strategic importance of guided missiles for the defense of Western Europe. The memorandum perhaps overstressed the importance of this new weapon, but it is evident that supplying such dual-purpose weapons to Europe was high on the U.S. agenda.

But what about Sweden; was she left behind due to her non-participation in the Alliance? As we shall see, this was certainly not the case. In fact, Sweden had been given a privileged treatment. A State Department memorandum dated November 26, 1957, shows that Sweden had already been cleared for future procurement of guided missiles by the JCS and the State Department. The RSAF had secretly contacted the Americans, without the prior knowledge of the Swedish government, in order to have something concrete to show the Swedish authorities when requesting funding for the purchase of American missiles. Said the memorandum:

> The Swedish air force has been in contact with the U.S. Joint Chiefs of Staff, and has secured the approval of JCS and the Department of State to begin preliminary comparative studies of various missile weapons for purposes of possible later procurement through U.S. or NATO channels. *The weapons under discussion will be limited to the types being made available to our* NATO *allies*. The Swedish Government has not yet been apprised of this Swedish air force activity, it being the view of the Swedish air force that when an approach is made to the Swedish Government, specific recommendations as to choice of missiles should accompany the first approach.[154] [Italics added.]

Here is the best evidence of the fact that Sweden's purchases of guided missiles from the United States was intrinsically dovetailed with the policy on releasing such, and associated, technology to the European allies. This is also an important indication of how pertinent consent to U.S. hegemony was for the United States when it came down to an evaluation of which countries that would receive guided missiles. In terms of consent to American hegemony, level of industrial security, and level of technological capability, Sweden was now at least on a par with the NATO countries.

154 Memorandum entitled "Swedish Position on Use of Missiles", dated November 26, 1957, in: RG 59; Bureau of European Affairs Country Director for Scandinavia, Iceland and Finland (EUR/SCAN) [henceforth: EUR/SCAN]; RRSWA, 1958–1964; POL 17 Political Affairs Sweden POL 17 Diplomatic Notes 1959–1963 to PS – Protective Service Sweden PS 7-1 Arrest (Swed. Seamen Visit New Orleans); Box No. 1; (Folder: Sweden Missile 1959–62); NARA II, College Park, MD.

This conclusion is given additional weight by a memo written in August 1957 where the Swedish interest in guided missiles was recounted. The internal Swedish debate on whether or not to procure atomic weapons was related, and an interesting observation was made by Burke Elbrick at EUR, the author of the memo. According to Elbrick, due to the high costs involved in developing such warheads domestically:

> […] the Swedes at this time would probably be disposed to adopt dual purpose missiles without warheads in the belief that if needed the warheads could in any emergency be made available from the U.S.[155]

The requests made by Boheman for guided missiles from the United States, concluded Elbrick, had been for systems of this dual character; the most recent inquiries concerning Nike and Honest John missiles. So far the United States had not followed up on any of these requests, since the Americans had not yet adopted a definite position on the advisability of the Swedes getting their hands on such systems.[156] However, this position was now to be changed, and the reason was explicitly stated as being the release of such systems to the NATO allies. Wrote Elbrick:

> Now that dual-purpose weapons are to made available to NATO members, it is believed that the Department [of State] should support any Swedish request received for assistance in this field and should seek to avoid tying this assistance to conditions which the Swedes could not accept.[157]

There are two main points that need to be observed in this paragraph. Firstly, the link between the decision to supply these kinds of weapons to the NATO allies, and the decision to grant Sweden access to the same kinds of missiles is obvious. Secondly, Elbrick did not say that such sales should not come with conditions attached; he only said that these conditions should not be of such a kind that they would be impossible for the Swedes to accept. Elbrick recommended Deputy Under Secretary of State Murphy that he should:

> […] inform the Department of Defense of our views on this subject, say we expect to hear from the Swedes again, and seek agreement to

155 758.5611/8-2657, Memorandum from Elbrick to Murphy, dated August 26, 1957, p. 1, in: RG 59; 1955–1959 CDF; From 758.56/1-555 to 759.00-may-day/3-357; Box No. 3467; (Folder: 758.56/1-555); NARA II, College Park, MD. Interestingly, Elbrick did not say if the Swedes were wrong in assuming this.

156 Ibid., p. 1.

157 Ibid., p. 2.

our telling the Swedes, when asked, that Defense would be favorably disposed to discussing with them what could be done.[158]

Even though the Americans had not yet formulated a coherent plan regarding how the new weapons were to be distributed to the allies, the Swedes were included in the receiving group of countries. From late August 1957 Foggy Bottom recommended that Swedish guided missile requests ought to be favorably received by the United States, and the decision was directly linked to the much bigger issue of releasing this technology to the NATO countries. This means that the State Department had swung over to a favorable position even before the RSAF's secret negotiations with the Americans. We will have reason to come back to the question regarding the dual-purpose nature of some of the systems finally requested by the Swedes. Less than five months later, in January 1958, and after having secured the clearance of the JCS, the Deputy Chief of the RSAFB Torsten Rapp came to the United States at the head of a Swedish delegation to study guided missile options (I will return to this issue below).

Also in 1957, real steps were taken to assure the compatibility of Swedish and NATO defense materiel, most clearly with regard to the RSAF. In June 1957 the JIC submitted a report to the JCS containing the JIC's recommendations regarding the release of certain NATO Standardization Agreements (henceforth: STANAG) to Sweden. The report was produced in response to a request from the Acting Deputy U.S. Representative to NATO's Standing Group, who in turn had acted on behalf of a British request to release certain STANAG information to the Swedes. More specifically, the information considered for release concerned aircraft attachments and fittings, and the British was in favor of such release "as a measure to facilitate cooperation between Sweden and the NATO nations."[159]

It was concluded that the release of this information met the conditions of release as set forth in the U.S. Disclosure Policy MIC 206/29 of 1955, and was also in accordance with NATO security directives C-M(55)15 and SGM-300-56 from 1955 and 1956 respectively. The report

158 Ibid., p. 2.

159 JCS 927/323, "Report by the Joint Intelligence Committee to the Joint Chiefs of Staff on Release of Certain NATO Standardization Agreements (STANAGS) to Sweden", dated June 17, 1957, p. 2036, in: RG 218; CDF 1957; 350.05 (3-16-48) Sec. 8 to 350.05 (3-16-44) Sec. 55; Box No. 32; (Folder: Exchange of Technical Information with Foreign Countries); NARA II, College Park, MD. See also: JCS 927/320, "Note by the Secretaries to the Joint Chiefs of Staff on Release of Certain NATO Standardization Agreements (STANAGS) to Sweden", dated May 21, 1957, p. 2026, in: RG 218; CDF 1957; 350.05 (3-16-48) Sec. 8 to 350.05 (3-16-44) Sec. 55; Box No. 32; (Folder: Exchange of Technical Information with Foreign Countries); NARA II, College Park, MD.

stated that if Sweden could adopt the STANAGS under consideration, it would obviate the necessity for the Swedes to make modification to aircraft purchased from NATO countries, and such "cooperation between NATO and Sweden would appear to be in the interest of NATO."[160] The JIC recommended that the JCS should forward a memorandum to the U.S. Representative to the Standing Group, saying that this would benefit both NATO and Sweden, that the RSAF had a need-to-know, and that there was no objection to the release from a U.S. military security viewpoint.[161] One can only speculate as to why the Swedes had a need-to-know in this case, and how significant this is for the understanding of Sweden's role in the defense of NATO's northern flank in this period.

The standards to be released concerned: electric supplies (various receptacles and connectors), communication (telebriefing equipment and related procedures), oxygen and air (supply pressure and hoses, etc.), storage and handling of fuels and lubricants, suspension of external armaments (such as air/ground rockets, bomb and suspension systems – bomb lugs), and finally refueling pressure (pressure refueling couplings and associated equipment) and connections for refueling and refueling pressure (nozzle grounding plugs and sockets etc.).[162]

This was the prelude to the reconfiguration of the fueling pressure, nozzles, fuel etc. of the Swedish aircraft J32 Lansen around 1960 to NATO standard. In Sweden, however, this was referred to as a change to *international* rather than NATO standard, and was part of an unwritten, but officially sanctioned, order according to which collaboration with the West was to be facilitated through 'concrete actions'. When these standardization measures were decided upon, the unofficial policy prescribed that the actual motive behind them must not be stated. Consequently, the switch to NATO standard fuel was rationalized by commercial reasons, i.e. that this kind of fuel was more ubiquitous on the world market. All this was revealed in interviews held by the Neutrality Policy Commission with Åke Sundén, who was the Chief of the Aircraft Section at the RSAFB 1958–1967, but the Commission chose not to pay too much attention to the information provided by Sundén, and published only a watered down version of Sundén's statements.[163]

160 Ibid., pp. 2036–2037, quote on p. 2037.

161 Ibid., pp. 2037–2038.

162 JCS 927/320, "Note by the Secretaries to the Joint Chiefs of Staff on Release of Certain NATO Standardization Agreements (STANAGS) to Sweden", dated May 21, 1957, pp. 2028–2029, in: RG 218; CDF 1957; 350.05 (3-16-48) Sec. 8 to 350.05 (3-16-44) Sec. 55; Box No. 32; (Folder: Exchange of Technical Information with Foreign Countries); NARA II, College Park, MD.

163 For this and more information on sources, see: Nilsson, M., "'Gud vare tack och lov…", p. 50.

Towards the summer of 1956 Ambassador Cabot was able to see a possible solution to the Swedish guided missile dilemma. The USAF was planning to erect a long-range radar station in either Norway or Sweden, and the Pentagon let Cabot know that they would send someone to Stockholm to discuss the radar issue with both the Swedes and the American Embassy. It was not a small radar station that the Americans thought that Sweden might want to have in their front yard. It was going to be 150 x 400 ft, or approximately 45 x 120 meters, and its purpose was to monitor a Soviet guided missile site east of Stalingrad. The idea was that the USAF would help the Swedes with their guided missile development if the Swedes were willing to become involved in the American giant radar project—in fact, it seems as if the Americans were prepared to let the Swedes manage the project themselves. Just as the Pentagon had said to Cabot, in June two Americans arrived in Stockholm to discuss the issue. The Pentagon had sent Colonel Frank Harrold and Colonel Charles Richman, Head of the Electronics Intelligence Group of the USAF, Directorate of Intelligence.[164] This station would have been just one in a series of COMINT (Communications Intelligence) and detection radar stations that the U.S. built, and were planning to build, in areas adjacent to the Soviet Union.[165]

Colonel Harrold met privately with Supreme Commander General Swedlund and the two obviously talked the matter through. Harrold asked Swedlund if the Americans could be allowed to conduct a survey to identify possible radar sites, something which, apparently, Swedlund agreed to.[166] After the meeting the Chief of the Swedish Defense Staff Major General Richard Åkerman rendezvoused with Ambassador Cabot, to convey to him what in essence was the Swedish counterdemand in this barter trade. Sweden wanted, Åkerman said, to acquire an American prototype to serve as a model for its guided missile development. Cabot promised him that he would push this issue once he got back to Washington. Moores writes that Major General Westring extended the wish list before Cabot flew home to Washington by including the Falcon air-to-air missile as well as the Talos surface-to-air missile. The Pentagon was apparently well aware of the need to be more forthcoming with the Swedes if the radar station was ever to be built in Sweden, and was willing to donate $30 million towards the Swedish

164 Moores, S., *'Neutral on Our Side':...*, pp. 132–133.

165 Berdal, M. R., *The United States, Norway and...*, pp. 38–39.

166 The fact that the Supreme Commander had, or regarded himself as having, the authority to agree to such a proposal seems a bit odd. One would assume that such a thing would need clearance from the government. Perhaps, the line between a radar station and collaborating on missile development (not to mention everything else) was not all clear to Swedlund either.

guided missile program. Ambassador Cabot did not think that the Swedish government would agree to this generous proposal, and it does not seem as if the Swedes ever found out about the Pentagon's offer.[167] Soon thereafter the Americans apparently dropped the idea because there was no immediate need for such a station in Sweden at the time. Mats Berdal quotes a document saying that:

> '[...] the requirement for a Swedish site does not presently appear to exist. We have found, however, that the Swedish representatives appear eager to assist in many ways, and in the event a detection site is required in Sweden, negotiations could be opened.'[168]

This means that the Americans were really just sounding the Swedes out to see if the United States could count on any goodwill from Sweden. Having concluded that this was the case, the Americans were perfectly happy to let the matter slide for the time being. As we shall see below, the issue was to resurface once again before being scrapped once and for all.

Judging from Moores' account it seems as if the Sidewinder air-to-air missile was not one of the items on the wish list that Åkerman and Westring turned over to Cabot in June 1956. Perhaps this is not all that strange considering that Harrold and Richman were USAF men and Sidewinder was originally developed by, and intended for, the U.S. Navy.[169] The fact is that the USAF did not procure Sidewinder until 1957, and then only after two years of intense missile tests (the first tests were conducted June 12–16, 1955, at the Holloman Air Force Base in New Mexico).[170] Hence, in June 1956 the Sidewinder was not part of the USAF inventory and could not be offered. The Sidewinder issue was likely handled through another channel, directly with the U.S. Navy and Chief of Naval Operations (henceforth: CNO), i.e. the Chief of the U.S. Navy, the Swedish descendant Admiral Arleigh Burke.

It is not known whether or not Galer and Schuldenfrei, when they visited Stockholm two years before, in the fall of 1955 (see above in this chapter), also discussed the air-to-air missile Sidewinder with the Swedes, and if their visit was in fact a prelude to Sweden's decision to

167 Moores, S., *'Neutral on Our Side':...*, pp. 133, 185–186.

168 Berdal, M. R., *The United States, Norway and...*, p. 39.

169 Westrum, R., *Sidewinder...*, p. 133. In the United States the Navy have their own aircraft, separate from the Air Force.

170 Ibid. About the purchase, p. 161; about the prolonged test period, pp. 131–161; about the initial tests, p. 137. After the first Sidewinder version the U.S. Air Force insisted on producing their own Sidewinders. Ron Westrum calls this effort a shambles, and states that the Air Force did not get a really good version again until 1975 and the Sidewinder AIM-9L (ibid., p. 195).

procure this guided missile. But there certainly is a strong possibility that they did, since Sidewinder was manufactured and developed by the U.S. Navy.[171] Anyhow, once Cabot had returned to Washington in November 1956 he started to try to make clear to the recalcitrant officials at Foggy Bottom that "if the Eisenhower administration wanted to 'get a boot out of the door', it had to offer the Swedes a quid pro quo deal."[172] According to Moores, Cabot told Foggy Bottom that the Swedes had to be *enticed* into collaborating by offering something that the Swedes wanted—i.e. guided missiles. This was an absolute must if Washington wanted Sweden's compliance on the radar issue, he said.[173]

By arguing along these lines, writes Moores, Cabot was able to win Major General Millard Lewis over to his side. Lewis was the Director of USAF Intelligence, and was the one responsible for the radar idea. But still, he says, there were intransigent elements in the Pentagon that continued to slow down American help to Sweden's guided missile development. Quarles was one of these elements, since he allegedly had his doubts as to whether it would be wise to release highly classified information to Sweden. Moores also writes that "little, if any, softening of the US line could be discerned" by November 1956.[174] This is of course only to be expected since the question of the American attitude to Swedish missile purchases was intimately tied to the overall U.S. disclosure policy. As long as the NATO countries were not getting access to these systems, neither would the Swedes. But things were no doubt beginning to move, and one important incident occurred not even a month after Cabot's return to Washington.

The Soviet launch of Sputnik in October 1957, created great worries in the West and especially in the United States.[175] Strengthening the defenses of the West suddenly seemed even more acute than before, and this was probably an important cause of a relaxation in U.S. policy towards friendly nations. The fact that Sweden was included in this category does not necessarily mean that she was a secret ally of the United States; however, if the U.S. government unilaterally saw Sweden as its ally then that would no doubt have influenced their decision. There is plenty of evidence to this effect.[176] Moores writes that from

171 Babcock, E., *Sidewinder: Invention and Early Years...*, p. 6. Sidewinder was actually named after a desert rattlesnake. The name was considered to be appropriate because of the fact that both snake and missile were heat seeking, i.e. they homed in on the temperature differences between the prey and its surroundings.

172 Moores, S., *'Neutral on Our Side':...*, p. 133.

173 Ibid., p. 133.

174 Ibid., pp. 133–134.

175 Dörfer, I., *System 37 Viggen...*, p. 90.

176 By the end of 1959 the issue of Swedish non-alignment and the policy of

the American standpoint, giving Sweden access to advanced military technology "paved the way for Sweden to become more embedded within the West's peacetime defense structure." Moreover, by supplying war materiel to Sweden the Pentagon had a greater opportunity to influence the structure of Sweden's defense forces, and to increase the standardization "of both Swedish and US equipment and tactics."[177] Thus, the Swedish policy of neutrality and the need for a strong defense, paradoxically, offered Washington an ideal cover for realizing its intentions.[178]

On the Swedish side, the GMB had some persuading to do themselves. If American guided missiles were going to be procured then funding would be needed to cover the costs. At the same time that the USAF procured Sidewinder, i.e. 1957, the GMB applied for funding for procurement of guided missiles abroad during fiscal years 1957/58 and 1958/59. The GMB stressed that if they received the personnel reinforcements they had requested, the air-to-air missile RB 321 could be delivered in 1961 at the earliest; but if their wishes could not be granted now, deliveries would be delayed until at least 1963.

> With reference to stated times of deliveries *it is considered to be of utmost importance that guided missiles immediately be procured abroad* in order to arm existing fighter aircraft. For this reason negotiations have been conducted concerning purchases from abroad, which have resulted in good prospects of purchasing certain guided missiles already during fiscal year 1957/58.[179] [Italics added.]

The negotiations referred to by the GMB was most certainly the ones between the RSAF and the Americans prior to November 1957, which had

---

neutrality had in fact become a non-existent issue in the United States. When the NSC Planning Board met to discuss Scandinavia in December that year the discussion (which was to lead to the NSC 6006/1 policy paper on Scandinavia) centered instead on the half-hearted Norwegian and Danish support for NATO (for this, see: Moores, S., *'Neutral on Our Side':...*, p. 165). The former CNO Admiral Elmo Zumwalt even referred to the relationship between the United States and Sweden as an unofficial alliance (Zakheim, Dov. "The United States and the Nordic Countries During the Cold War" in *Cooperation and Conflict*, Vol. 33(2) 1998, p. 122).

177 Moores, S., *'Neutral on Our Side':...*, p. 207.

178 Ibid., p. 210.

179 "RCH A 6:12", in KFF, FRB, Serie: F IIIA, "Inkomna och avgångna skrivelser ordnade efter klassifikationssystem", A 40-M 3:30, År: 1957, VOL. 20, KrA. *"Med hänvisning till sagda leveranstider synes det vara av största vikt att robotar snarast anskaffas från utlandet* för beväpning av befintliga jakt. Med anledning härav har underhandlingar förts betr[äffande] inköp från utlandet, vilket givit vid handen att stora utsikter finnes att få inköpa vissa robotar redan under budgetåret 1957/58."

been kept secret from the government. According to the GMB, however, the foreign guided missiles were not as effective as the Swedish air-to-air missile RB 321, but the existing fighter aircraft would substantially enhance their fighting capability [öka avsevärt i stridsvärde] if these guided missiles were to be purchased. They suggested that 10 million SEK be put aside for this explicit purpose. Despite this fact, domestic development work ought to be continued at the current pace.[180]

Certainly it must be considered an understatement that the Swedish fighter aircraft fleet would enhance its fighting capability if it were to be equipped with foreign guided missiles. It is especially so since there was no viable alternative to this course of action. The optimistic statement that RB 321 was superior to its American counterparts, Sidewinder and Falcon, might have been true on paper. But the fact of the matter was that RB 321 had never even been successfully flown at this point, and had certainly not been tested in combat. These circumstances were precisely those that had led the Swedes out of the country in search for modern weaponry for their fighter aircraft fleet in the first place. As we have seen, the Americans were well aware of this.

The Swedish Air Attaché Wennerström visited the GMB in the spring of 1957. During his visit he received inquiries regarding procurement of American air-to-air guided missiles. Sweden was interested in buying three American guided missiles: Falcon, Sidewinder and Sparrow.[181] These guided missiles had been presented to the members of the GMC in a report on foreign guided missile systems dated February 28, 1957.[182] From the vantage point of the GMB, at that time, the possibility of obtaining any of these guided missiles seemed rather small. But it was expected that this situation would soon change for the better.[183] This forecast was perhaps helped along the way by the assuring message delivered by Gordon Gray to Ambassador Cabot in November 1956, which Cabot no doubt forwarded to the Swedes. It is not known whether Wennerström's efforts were part of the RSAF secret negotiations, which secured a positive response from the JCS and Foggy Bottom by late November 1957. However, it is possible that they were.

Spurred by the amicable U.S. attitude, Supreme Commander Swedlund told the government at a military briefing on November 21,

180 "RCH A 6:12", in KFF, FRB, Serie: F 111A, "Inkomna och avgångna skrivelser ordnade efter klassifikationssystem", A 40-M 3:30, År: 1957, VOL. 20, KrA.

181 "RHA 7:8", in KFF, FRB, Serie: F 111A, "Inkomna och avgångna skrivelser ordnade efter klassifikationssystem", A 40-M 3:30, År: 1957, VOL. 20, KrA.

182 Annex NO 1 to "NR 2/1957.", "Protokoll från FRR sammanträde den 4. september 1951.", entitled: "Robotutvecklingen utomlands", in KFF, FRR, Serie: A 1, "Protokoll från sammanträden 1948–1957", År: 1948–1957, VOL. 1, KrA, p. 2.

183 "RHA 7:8", in KFF, FRB, Serie: F 111A, "Inkomna och avgångna skrivelser ordnade efter klassifikationssystem", A 40-M 3:30, År: 1957, VOL. 20, KrA.

1957 that he wished to send a delegation to the United States to try to get more information concerning guided missiles. After discussing the issue for a while, and after Foreign Minister Undén had stressed that it was important that this delegation should not inquire about nuclear weapons, the government told Swedlund that they could not see any problem with it and consented to the request.[184] The American missile systems that the Swedes were interested in were air-to-air missiles Falcon and Sidewinder, and surface-to-air missiles Hawk and Tartar. The surface-to-air missiles Talos and Nike Hercules were thought to be too big and too expensive for Sweden. The British Firestreak missile, while perhaps acceptable from a technical point of view, was also considered to cost too much.[185] It seems that Swedlund did not inform the members of the government of the RSAF's secret negotiations with the United States at the briefing on November 21, even though he apparently informed them a week later that the Americans were now positive to Swedish inquiries. This conclusion is supported by Undén's personal notes (made sometime between November 26–30), where he wrote that some sort of decision had in fact been made by the U.S. government at the end of November 1957 to allow Sweden to procure certain types of missiles. Undén also knew that this was due to earlier Swedish probes on the subject.[186]

Moores has written that Swedlund did not know about the positive American decision when he briefed the government on November 21,[187] but this is most likely wrong. As we saw above (p.), the State Department memorandum dated November 26 said that the RSAF had secretly contacted the Americans, *without the prior* knowledge of the Swedish government. This could not have been the case if the RSAF had contacted the Americans *after* November 21, since Swedlund then informed the government that he wished to send a negotiating party to the United States. The government would then have had prior knowledge. The contacts which were kept secret from the government

184 *Östen Undén. Anteckningar 1952–1966*, Published by Karl Molin (Stockholm: Kungl. Samfundet för utgivande av handskrifter rörande Skandinaviens historia, Handlingar del 25, 2002), p. 550.

185 "PM ang inköp av RB från USA", dated November 24, 1957, pp. 1–3, in: HP 24 VI/U.S.A; "Tillstånd till in- och utförsel av vapen, ammunition och krigsmateriel"; 1955 juli–1957; Folder 23; Box No. 62; År: 1955–1960; RIA; Arninge.

186 *Östen Undén. Anteckningar 1952–1966*, p. 550.

187 Moores, S., *'Neutral on Our Side':...*, p. 188. Moores based his conclusion on the fact that Colonel Löfgren was informed by Taylor on November 22, i.e. the day after Swedlund's briefing, that the DOD was prepared to discuss the guided missile issue. However, Swedlund most likely already knew about the positive attitude of the State Department and the JCS. Hence what the Supreme Commander was unaware of at the briefing was only the DOD's position.

must thus have been made *prior* to November 21. Moreover, the memorandum also said that the RSAF had secured the approval of the JCS and the State Department, and it is not likely that Swedlund had not been informed of this prior to his briefing the government.

On the American side, the Department of the Air Force had been given the task of overseeing the discussions with the Swedes, and Colonel Charles Taylor, Chief of the Policy and Management Group of the USAF Collection and Dissemination Department, told the Swedish Military Attaché Stig Löfgren when they met the day after Swedlund's briefing in Stockholm that a series of conferences would probably be necessary, the first of which could probably be held in mid-December.[188] Despite this, however, estimates that came from the Planning Division of the Air Staff at this time still conveyed an uncertainty about the possibility to procure American missiles. Regarding air-to-air missiles for example, a document dated December 5, 1957, said that inquiries were to be done regarding the possibility of procuring such guided missiles.[189] But on the other hand, by then Captain Beskow of the RSN had already received three pages of confidential information pertaining to Sidewinder 1A, one handbook covering both Sidewinder 1 and 1A, as well as one booklet on the guided missile systems of the U.S. Department of the Navy from the CNO Admiral Arleigh Burke.[190]

Besides the NATO decision of December 1956 to give allies access to guided missile technology and related data, there were some domestic developments in Sweden that may have helped to swing the pendulum to some extent. In December 1957, it was decided to start the construction of Stril-60, the Swedish computerized and fully automatic control and reporting system, and at the same time the first preliminary operational specification for the system J37 Viggen interceptor aircraft was presented. These two systems, Stril-60 and J37 Viggen, were to influence and shape Sweden's air defense for the next 30 years.[191] This was clearly a boost for the effectiveness of the Swedish ability to defend

188 Letter from Löfgren to the Chief of Section II of the Defense Staff Holger Henning, dated November 22, 1957, in: HP 24 VI/U.S.A; "Tillstånd till in- och utförsel av vapen, ammunition och krigsmateriel"; 1955 juli–1957; Folder 23; Box No. 62; År: 1955–1960; RiA; Arninge. See also: Moores, S., *'Neutral on Our Side':...*, p. 188.

189 "NR H 38", "Inköp av amerikanska robotar. Underlag för diskussion.", in: Högkvarteret [henceforth: Hgkv], Flygstaben KH [henceforth: FS/KH, Planeringsavdelningen [henceforth: PA], "Inkomna och utgående skrivelser", Serie: F III, VOL. 1, År: 1955–59, KrA, p. 3.

190 Moores, S., *'Neutral on Our Side':...*, p. 173.

191 Dörfer, Ingemar, *System 37 Viggen...*, p. 91.

itself, and the whole Scandinavian Peninsula, an effort that would be made infinitely more effective with American guided missiles. This was something that the Americans were very aware of.[192]

In January, 1958, Torsten Rapp headed a delegation of Swedish officers on a visit to the U.S. to discuss guided missile procurement.[193] On his return to Stockholm, on March 20, Rapp briefed the government regarding the planned guided missile purchases,[194] and in mid-December 1958 the Americans notified the Swedes that the United States fully comprehended Sweden's foreign policy, and that they were willing to discuss the sale of a whole range of guided missiles.[195] Rapp's group represented all three Services, even though it was dominated by RSAF officers, and met with the Americans at the Pentagon on January 20–22, 1959. The first day was devoted to an exposition of Sweden's military capabilities, and the Swedes provided detailed characteristics of Swedish weapon systems, including aircraft. The RSAF had a definite weakness in its inability to defend against attacking enemy aircraft at altitudes over 50,000 feet, as well as at comparatively low levels. It was felt that American surface-to-air and air-to-air guided missiles would cure these weaknesses. On the following day the guided missile systems being made available to Sweden were described, and the last day was devoted to discussions on price and sales procedures.[196]

When one looks at how these negotiations were structured, the disingenuousness of the 'neutral technology thesis' becomes blatantly obvious. In order to gain access to American technology Rapp had to lay bare the capabilities of the entire Swedish Armed Forces. Strengths and weaknesses were openly put on the table for the Americans to see. Such disclosures (and let us not forget that the Swedes had made this into a habit by the late 1950s) was a necessary concomitant to American military technology, and were one of the more obvious consequences that followed from these purchases. It goes without saying that the Swedish military could never have been so open-hearted with the Russians, and hence purchases of military technology from the East were unthinkable. So when the Swedish government purchased guided missiles from the United States, it was also making a clear statement as to its allegiance in the Cold War.

192 For the awareness of American officials of the connection between guided missiles and Sweden's ability to defend itself, Norway, and Denmark, see: ibid., p. 135.

193 *Had There Been A War...* p. 123.

194 *Östen Undén. Anteckningar 1952–1966*, p. 557.

195 Moores, S., *'Neutral on Our Side':...*, p. 189.

196 758.5611/1-3158, Memorandum from Kerrigan to Parsons and Moline, dated January 31, 1958, p. 1, in: RG 59; 1955–1959 CDF; From 758.56/1-555 to 759.00-mayday/3-357; Box No. 3467; NARA II, College Park, MD.

Ingemar Dörfer has written that Rapp was at the time mostly interested in surface-to-air guided missiles Hawk and BOMARC. "Rapp was", he concludes, "promised almost anything."[197] If the American military officers came to the meeting with a reasonably positive image of the Swedish defense forces, then this perception seems to have been even more pronounced at the end of the three-day conference for the Americans were apparently impressed by Rapp and his companions. William B. Kerrigan at BNA noted that the Swedish representation "were all very obviously officers of ability far out of the ordinary", and said that the U.S. military representatives had told him that they had found the described capabilities of Swedish weapons and aircraft "very impressive."[198] During the discussions on the third day the following exchange took place, according to Kerrigan:

> In the course of a description of our sales procedure, General Rapp got the mistaken impression that the U.S. intended to exact rather substantial political quid pro quos for the sale of missiles. After discussing the matter with Colonel Taylor, I mentioned to General Rapp that speaking personally and unofficially, it did not appear to me that there were any serious political problems on the American side that would hinder an agreement on sales of missiles. This statement appeared to reassure him. Other than this single minor misunderstanding which was probably a result of language difficulties, the talks were on the wholly professional and technical level of an exchange of information designed to help the Swedes decide which weapons best met their needs.[199]

The episode downplayed by Kerrigan was not due to a misunderstanding at all. Colonel Taylor's comment probably arose from the resurrection of the American demand to barter U.S. guided missiles for the right to build a radar station on Swedish soil intended to gather electronic intelligence on Soviet missile tests.[200] As Moores has shown, it was Ambassador Cabot who first suggested this arrangement to the Swedes in the summer of 1956.[201] The matter had remained dormant until the demand was reiterated to Rapp one and a half years later, in early 1958. According to the Neutrality Policy Commission, the matter

197 Dörfer, Ingemar, *System 37 Viggen...*, p. 90.

198 758.5611/1-3158, Memorandum from Kerrigan to Parsons and Moline, dated January 31, 1958, p. 1, in: RG 59; 1955–1959 CDF; From 758.56/1-555 to 759.00-mayday/3-357; Box No. 3467; NARA II, College Park, MD.

199 Ibid., p. 2.

200 *Had There Been A War...*, p. 123; Moores, S., *'Neutral on Our Side':...*, p. 189; Berdal, M. R., *The United States, Norway and...*, p. 39.

201 Moores, S., *'Neutral on Our Side':...*, p. 132.

was promptly dropped when Sweden did not approve.[202] This suggests that the radar station was not really a matter of any significance for the Americans, otherwise they would probably not have dropped the issue so quickly. It seems as if the Americans were only testing their ground to see how far the Swedes would go in order to get access to guided missiles. This is contrary to Moores' conclusion that the radar issue played a decisive role in 1956–1957 in creating a more constructive USAF attitude towards Sweden's missile troubles.[203] However, the document referred to by Berdal above also seems to support my conclusion.

After having concluded the conference at the Pentagon successfully, the Swedish team toured various guided missile bases, where they were demonstrated the missiles discussed during the three-day meeting. Reasonably, the Swedish military representatives had to see the missiles in action before making their technical recommendations to the Swedish Ministry of Defense upon their return to the Swedish capital. The Americans expressed a wish to negotiate through the Swedish Air Attaché in Washington rather than via the Embassy in Stockholm, if Sweden decided to procure any of the guided missile systems.[204]

This corresponds fairly well to Simon Moores' findings. The American attitude was positive, provided the Swedes could guarantee that these high technology items were protected by suitable security arrangements. According to the Neutrality Policy Commission, the Swedish military counterintelligence establishment "was tasked to organize industry security, thereby guaranteeing security at the companies receiving U.S. technology."[205] What this echoed was the American wish, referred above, that such preparations had to be made if the United States was to release advanced technology to the Western Europeans. Erlander and the Swedish government immediately gave consent to a review of Swedish industrial security arrangements.[206]

On March 19, after coming back from the United States, General Rapp wrote to Deputy Assistant Secretary of Defense Charles H. Shuff to inform him about the Swedish thoughts regarding the missile systems

202 *Had There Been A War...*, p. 123.

203 Moores, S., *'Neutral on Our Side':...*, p. 186.

204 758.5611/1-3158, Memorandum from Kerrigan to Parsons and Moline, dated January 31, 1958, p. 2, in: RG 59; 1955–1959 CDF; From 758.56/1-555 to 759.00-mayday/3-357; Box No. 3467; NARA II, College Park, MD.

205 *Had There Been A War...*, p. 124. This was the starting point for the secret military counterintelligence organization IB, registering political dissenters, i.e. communists and the like, and the Social Democratic Shop Stewards (socialdemokratiska arbetsplatsombud, SAPO). These latter people oversaw the shop floor to control communist workers. In short, the illegal, political supervision of Swedish citizens studied by the Parliamentary Counterintelligence Committee (Säkerhetstjänstkommissionen).

206 Moores, S., *'Neutral on Our Side':...*, p. 191.

that Rapp's delegation had seen while in the U.S. In the letter Rapp outlined possible procurement scenarios regarding the air-to-air missiles Sidewinder and Falcon, and surface-to-air missiles Hawk, BOMARC, and Tartar.[207] Two days later Boheman wrote to Secretary of State John Foster Dulles informing him that the Swedish government wished to purchase guided missiles from the United States, and referred to the MDAA agreement of 1952 to justify the Swedish request.[208] While Tartar had a non-nuclear warhead and was intended for the RSN's destroyers, Hawk and BOMARC were dual-purpose missiles, meaning that they could be fitted with both a conventional and a nuclear warhead. As I have shown, the Americans had expected that Sweden would request permission to purchase such dual-purpose missiles, and apparently consented to the Swedes having this option even if they were to develop their own nuclear weapons, despite the fact that this was not desirable to the Americans. The American reply was positive while reserved,

207 Letter from Rapp to Shuff, dated March 19, 1958, pp. 1–7, in: HP 24 VI/U.S.A; "Tillstånd till in- och utförsel av vapen, ammunition och krigsmateriel"; 1958–febr. 1957; Folder 24; Box No. 62; År: 1955–1960; Ria; Arninge. See also: Moores, S., *'Neutral on Our Side':...*, pp. 190–191. Moores concluded from Rapp's letter that Sweden had decided to purchase the Sidewinder missile, as well as "a combination of the MG-10 [HAC's fire control system], Falcon GAR-10 and GAR-2 missiles." (Moores, S., *'Neutral on Our Side':...*, p. 190). There are, however, a few things that are incorrect in that statement. Firstly, the letter was just a tentative shopping list, not a definite procurement decision. Secondly, the name of the first Falcon missile was not GAR-10, but GAR-1D. True, there was one place in the document were the designation GAR-10 appeared, but that was a misprint. Thirdly, he stated in the text that Rapp requested an early-warning radar station that could detect IRBMs, but in the footnote he said that it was unknown whether this was the same ICBM radar that the USAF had been discussing with the Swedes since 1955 (Moores, S., *'Neutral on Our Side':...*, pp. 190–191). It is apparent, however, from the letter that it was not the same station. Rapp asked Shuff if he deemed it possible for Sweden to procure such a radar from the United States (see letter, p. 3). If this was the same radar that the Americans had tried to build at their own expense in Sweden since 1955, there would have been no reason for Rapp to ask if it was possible for Sweden to buy the radar.

208 758.5612/3-459, Memorandum from Willoughby to Timmons, dated March 4, 1959, p. 1, in: RG 59; 1955–1959 CDF; From 758.56/1-555 to 759.00-mayday/3-357; Box No. 3467; NARA II, College Park, MD. Letter from Boheman to Dulles, dated March 21, 1958, in: RG 59; EUR/SCAN; RRSWA, 1958–1964; POL 17 Political Affairs Sweden POL 17 Diplomatic Notes 1959–1963 to PS – Protective Service Sweden PS 7-1 Arrest (Swed. Seamen Visit New Orleans); Box No. 1; (Folder: Sweden Missile 1959–62); NARA II, College Park, MD. 758.5612/3-2158, Memorandum for the Director of the Office of Programming and Control Office of the Assistant Secretary of Defense (ISA) from Barnes, dated April 18, 1958, in: RG 59; 1955–1959 CDF; From 758.56/1-555 to 759.00-mayday/3-357; Box No. 3467; NARA II, College Park, MD. Letter from Boheman to Dulles, dated March 21, 1958, in: HP 24 VI/U.S.A; "Tillstånd till in- och utförsel av vapen, ammunition och krigsmateriel"; 1958–febr. 1957; Folder 24; Box No. 62; År: 1955–1960; RIA; Arninge. See also: Moores, S., *'Neutral on Our Side':...*, pp. 190–191.

and reminded the Swedes that before any purchases could be made the Americans wanted to send the State-Defense Military Information Control Committee (henceforth: S-DMICC) to Sweden to study security arrangements and preparedness.[209]

Boheman wrote back to Dulles a week later to let him know that the Swedish government concurred with the American proposal.[210] Things were now really beginning to move forward. Ambassador White met with Åström at the Foreign Ministry on April 22, and handed over a list covering all the security aspects which the S-DMICC would study while in Sweden. The list was quite extensive, reaching from basic security legislation, and governmental departmental security, physical and personnel security, to security in industry and classified matter control. According to Åström, the U.S. Ambassador and himself thought that the S-DMICC was perhaps a bit too ambitious, and both of them jokingly concluded that a study of all these issues would be virtually a doctoral thesis.[211]

In the meantime, the DOD would undertake cost and availability studies. The Swedes would be informed that this study was taking place, but no further talks should be held with them pending the outcome of the S-DMICC security survey.[212] The wish to keep the Swedes in suspense should not be taken as a sign of American unwillingness to conclude the deal; rather it should be construed as applied business strategy—one should never let the customer get the upper hand. The longer the U.S. could keep the Swedes guessing, the greater was the prospect of securing additional concessions on matters of security policy from the Swedish

209 Letter from Yardney to Boheman, dated April 3, 1958, in: HP 24 VI/U.S.A; "Tillstånd till in- och utförsel av vapen, ammunition och krigsmateriel"; 1958–febr. 1957; Folder 24; Box No. 62; År: 1955–1960; RiA; Arninge. See also: Moores, S., *'Neutral on Our Side':...*, p. 191.

210 Letter from Boheman to Dulles, dated April 11, 1958, in: HP 24 VI/U.S.A; "Tillstånd till in- och utförsel av vapen, ammunition och krigsmateriel"; 1958–febr. 1957; Folder 24; Box No. 62; År: 1955–1960; RiA; Arninge. See also: Moores, S., *'Neutral on Our Side':...*, p. 191.

211 Memorandum by Åström, dated April 22, 1958, in: HP 24 VI/U.S.A; "Tillstånd till in- och utförsel av vapen, ammunition och krigsmateriel"; 1958–febr. 1957; Folder 24; Box No. 62; År: 1955–1960; RiA; Arninge. See also: Moores, S., *'Neutral on Our Side':...*, p. 191. In Moores' version of the event, only Åström (White is not mentioned) thought that *the list* was virtually the size of a doctoral thesis. However, this is not an accurate account of what Åström's memorandum said. It is clear that both Åström and White had the same reaction to the list, and that it was the result of it that they commented on. The fact that both the Ambassador and Åström thought the S-DMICC mission overly ambitious is probably of some importance here, for it suggests that White, too, considered Swedish security to be adequate.

212 758.5612/3-2158, Memorandum for the Director of the Office of Programming and Control Office of the Assistant Secretary of Defense (ISA) from Barnes, dated April 18, 1958, in: RG 59; 1955–1959 CDF; From 758.56/1-555 to 759.00-mayday/3-357; Box No. 3467; NARA II, College Park, MD.

government. Between May 5 and 12, 1958, the seven men strong S-DMICC visited Sweden, for the purpose of evaluating the Swedish security measures in defense-related industries. The S-DMICC's personal guide was none other than Torsten Rapp, who took them to guided-missile-producing SAAB in Linköping, the electronics firm LM Ericsson in Mölndal, and to the engine producing Svenska Flygmotor AB in Trollhättan. The S-DMICC then returned to the United States and released its very positive report in August. The report stated that physical security at these companies was excellent and highly developed, and the personnel security arrangements were effectively applied. In conclusion the Commission thought that the overall industrial security arrangements in Sweden would be able to safeguard all classified American information. Simon Moores has concluded that "[t]he S-DMICC's visit to Stockholm proved to be the key that would later unlock the door for the Swedish authorities to acquire guided missiles."[213] While this conclusion is not entirely wrong, I believe it slightly exaggerates the importance of the S-DMICC's visit. Surely, deliveries could have been delayed if the S-DMICC had found great flaws in Swedish security arrangements, but as I have shown above, the State Department had already decided in August 1957 to recommend that Swedish inquiries regarding guided missiles should be considered favorably.

In his memo to Murphy in August, Elbrick had acknowledged that "Swedish security arrangements appear to be very good."[214] American officials had expressed opinions to the same effect before, and it seems highly unlikely that they really expected to find anything to the contrary. The visit to Sweden by the S-DMICC in the spring of 1958 was thus perhaps more of a formality. The key had been sitting in the door for some time; the only thing needed at this point was torque, which was what the S-DMICC offered. The report submitted by the S-DMICC was thus probably of less importance than Moores suggests. But what the episode does is that it further underlines that the United States made no difference between NATO and non-NATO states, as long as security was satisfactory, and as long as the country had the technological capability to utilize the technology granted them.

The irony was of course that all of this was done in vain, because the Swedish Air Attaché Wennerström, whom the Americans had wanted as a mediator in the negotiations, turned out to be a master spy and immediately made a mockery of the security program by providing the

213 Moores, S., *'Neutral on Our Side':...*, p. 181, pp. 191–194, quote on p. 192. For the names of the members of the S-DMICC and the Swedish delegation, see: ibid., pp. 191–192.

214 758.5611/8-2657, Memorandum from Elbrick to Murphy, dated August 26, 1957, p. 1, in: RG 59; 1955–1959 CDF; From 758.56/1-555 to 759.00-mayday/3-357; Box No. 3467; (Folder: 758.56/1-555); NARA II, College Park, MD.

Soviets with detailed information about the Sidewinder missile. The Soviets eventually copied the design of Sidewinder and the result—nicknamed Atoll by the West—was seen at an air show on June 9, 1961.[215] It did not take the Soviets very long to produce their own version of Sidewinder once they got their hands on the technical specifications.

According to Moores, the Americans viewed the Swedish refusal to permit the placement of the radar station in Sweden as an act of defiance, and it prompted immediate action in the form of a temporary reversal of the disclosure policy.[216] However, it seems as if the temporary reversal was brought on not by Rapp's refusal in January, but rather by another negative response from the Swedes in August. For on August 14, the State Department notified Frederick E. Nolting Jr., the Deputy Chief of Mission of the USRO in Paris, that the Pentagon intended to offer Sweden 2,000 Sidewinder missiles and that the USAF was to send a representative to Stockholm to discuss the issue. Nolting replied on August 18 that he had no objection to this, and a handwritten notation on the telegram said that Colonel Taylor of the USAF was in Stockholm to talk with the Swedes about Sidewinder and the possible installation of long range ICBM radar detection system in Sweden.[217]

There are no traces in these telegrams of any adverse attitude towards the Swedes due to their earlier refusal to have the radar station erected in Sweden. It would seem then, that the reason for the temporary reversal was to be found after August 18. Since Taylor had been the one to broach the subject regarding the radar station to Rapp in Washington, it was perhaps no surprise that the issue came up again. The fact that Taylor was tied to the USAF's Collection and Dissemination Department suggests that the radar station was intended for intelligence gathering for the USAF.

The reversal in disclosure policy, although it is unclear exactly how it was manifested and what information was denied, was a bit worrying

215 Westrum, R., *Sidewinder…*, p. 206. Prior to his mission in Washington D.C. Wennerström had been based in Moscow, from where he reported on the Soviet missile development ("NR 1/1950", "Protokoll från FRR sammanträde den 26. januari 1950.", in KFF, FRR, Serie: A I, "Protokoll från sammanträden 1948–1957", År: 1948–1957, VOL. 1, KrA, p. 2). Apparently, the Swedes exaggerated the security measures introduced in their talks with the S-DMICC, allegedly because Sweden was so desperately in need of classified technical information (for this, see: Moores, S., *'Neutral on Our Side':…*, p. 192, note 91).

216 Moores, S., *'Neutral on Our Side':…*, p. 194.

217 Telegram from State Department to USRO in Paris, dated August 14, 1958; and handwritten notation on reply from Nolting, dated August 18, 1958, in: RG 59; EUR/SCAN; RRSWA, 1958–1964; POL 17 Political Affairs Sweden POL 17 Diplomatic Notes 1959–1963 to PS – Protective Service Sweden PS 7-1 Arrest (Swed. Seamen Visit New Orleans); Box No. 1; (Folder Sweden Missile 1959–62); NARA II, College Park, MD. See also: Moores, S., *'Neutral on Our Side':…*, p. 193.

to the Swedes, and Taylor informed Colonel Lindskog later that fall that a decision regarding the radar station was soon going to be reached in Washington, and pointed out that key decision-makers at the State and Defense Departments were still unsure about Sweden's political reliability. According to Taylor, this group included Deputy Secretary of Defense Quarles.[218] Whether or not this was actually the case, or if the Americans were just resorting to scare tactics in a last effort to get a positive response from the Swedes concerning the radar station, it seems clear that this was not an issue that the U.S. government wanted to permanently damage the cooperation with Sweden. Besides, the Americans had already promised to sell Sidewinder to Sweden, and the whole Swedish guided missile development program was rearranged as a result, with projects under way for a long time now being cancelled. Immediately after Rapp's visit in February 1958, the GMB stated that the funds for the air-to-air missile RB 321 were to be used for procuring a foreign guided missile (i.e. Sidewinder).[219] The GMB announced the future purchase of Sidewinder in July,[220] and in August 1958, paralleling the reversal of the U.S. disclosure policy, the Swedish Sidewinder Program was established at the Pentagon as "Case No. Sweden OPC-159" (later re-designated "20-ZAA-P2").[221] This supports the conclusion that Swedish intransigence over the radar station was not a major source of consternation for the Americans at this time. With such a program established at the Pentagon, the actual signing of the contract in early 1959 can be considered to have been a formality.

218 Moores, S., *'Neutral on Our Side':...*, p. 194.

219 "NR H 21", "PM ang mtrlanskaffningen 1958/65 för alt Adam.", in: Hgkv, FS/KH, PA, "Inkomna och utgående skrivelser", Serie: F III, VOL. 1, År: 1955–59, KrA, p. 2. Brief reference is made to this decision in: Moores, S., *'Neutral on Our Side':...*, p. 190. Also in February 1958, immediately after the visit of Torsten Rapp (now head of the RSAFB) to the United States, Rapp traveled to Britain accompanied by Colonel Ulf Cappelen-Smith, among others, to negotiate with the Ministry of Supply (MOS) to make Sweden eligible to purchase a surface-to-air missile system for the Swedish Air Force. As stated above, the systems that the Swedes were interested in were Bloodhound MK 1 och Thunderbird MK 2. The Bloodhound MK 2 system later replaced the projected Swedish RB 322 missile (for this, see: "RHA 373:3", in KFF, FRB, Serie: F IIIB, "Inkomna och avgångna skrivelser ordnade efter klassifikationssystem", A 1005-A 373:10, År: 1958, VOL. 25, KrA). These two systems were almost identical.

220 "NR H 117", "Orientering angående sjuårsplanen 1959/66 för anskaffning av flygmateriel.", in: Hgkv, FS/KH, PA, "Inkomna och utgående skrivelser", Serie: F III, VOL. 1, År: 1955–59, KrA, p. 2.

221 Letter from Gösta Lundström to the Director of Naval Intelligence, "RHM 44-24:12", in KFF, Robotavdelningen, Serie: F IIIC, "Inkomna och avgångna skrivelser ordnade efter klassifikationssystem", M 44-24:7-M 44-24:75, År: 1967, VOL. 188, KrA.

Consequently, in December 1958 Washington informed the Swedes that it had decided to shelve the plans for building the radar station on Swedish soil, and after that there were to be no further significant restrictions regarding the access to American guided missile systems. "Washington," writes Moores, "after having spent over three and a half years trying to ensure that its radar project came to fruition, was becoming tired of the Swedish position [...]."[222] At the same time he points out that it is not clear whether the plans were shelved due to the Swedish intransigence, or if the project had simply become uninteresting to Washington.[223] The Swedes had expressed their accommodating opposition by refusing the radar station to be built, and had gone victorious from the tug-of-war with the hegemon.

The Foundering of the Swedish Development Program

By the late 1950s most of the Swedish domestic development projects had been dropped, due to a combination of lack of funding, the concomitant lack of engineering resources, and the preceding American refusal to aid Swedish missile R&D. This becomes abundantly clear when one looks at the Swedish surface-to-air missile RB 322. The American technological monopoly position, and the unwillingness to release know-how to its allies, greatly contributed to this missile not being realized during the 1950s. But in the end it was the British Bloodhound missile that was finally procured by the RSAF instead, not an American system. Paradoxically then, the death knell of RB 322, and of most of the other missile projects as well, was delivered when foreign missiles became available to the Swedes. The Americans decision to share this technology with other nations complicated the matter for the RSAF, however. Suddenly, the RSA could procure a surface-to-air missile as well, worsening the rivalry between the RSAF and the RSA. Procurements allotted to the military by the Riksdag ended up being split up between the two Services, not making either of them very happy. The issue of Hawk and the inter-service rivalry over it will be dealt with in the next chapter.

The decision to develop RB 322 had been taken in April 1951,[224] and it was intended to be one of three ingredients in the future Swedish air defense system, the other two being fighter aircraft and anti-aircraft artillery.[225] However, matters remained dormant for some time and RB 322 was taken up for renewed discussion in September 1954, at which

222 Moores, S., *'Neutral on Our Side':...*, pp. 193–194.

223 Ibid., p. 194, note 99.

224 "NR 4/1951" ,"Protokoll från FRR sammanträde den 18. april 1951.", in KFF, FRR, Serie: A I, "Protokoll från sammanträden 1948–1957", År: 1948–1957, VOL. I, KrA, p. 1.

225 "NR 6/1951" ,"Protokoll från FRR sammanträde den 4. september 1951.", in KFF,

time it was concluded that the guidance system and jet engine presented such major problems that it was estimated that a guided missile could not be forthcoming until sometime in the 1960s.[226] This was symptomatic of the development situation in the 1950s and illustrative of the fact that the Swedish guided missile program was in dire straits without foreign assistance.

Another problem that was manifest by late 1953 and early 1954 was scarcity of personnel resources. By November 1953, the number of people employed by the GMB had increased to 212 from the modest number of 45–50 people in 1948, but it was longer allowed to hire any more people even though the GMB thought additional people absolutely necessary.[227] That situation, coupled with the non-existent foreign assistance, explains why so many projects were cancelled during the following five years. This was one of the last major guided-missile-related issues that Captain Oxenstierna had to deal with in his position as Chief of the GMB —on September 30, 1954, he resigned and Captain Rosenberg took over the burden of producing a Swedish missile system.[228]

Together with air-to-air missile RB 321, surface-to-air missile RB 322 was seen as being the most pertinent issue during the spring of 1955. The latter missile was the first (and in fact the only) air-to-air missile ever contemplated by the RSAF.[229] It was, for both tactical and strategic reasons, considered to be very important for Sweden's ability to clear the sky of enemy aircraft in a future war.[230] Large sums of money were now being thrown into the development process, but by 1957 it became clear that development was not going as planned. The official reason

---

FRR, Serie: A I, "Protokoll från sammanträden 1948–1957", År: 1948–1957, VOL. 1, KrA, p. 1.

226 "NR 6/1954. Protokoll från FRR sammanträde den 17 september 1954.", in KFF, FRR, Serie: A I, "Protokoll från sammanträden 1948–1957", År: 1948–1957, VOL. 1, KrA, p. 2.

227 Annex NO 2 to "NR 4/1954. Protokoll från FRR sammanträde den 5. mars 1954" entitled "PM angående konsekvenserna av inskränkningar beträffande personalanställning m m å försvarets robotvapenbyrå." in KFF, FRR, Serie: A I, "Protokoll från sammanträden 1948–1957", År: 1948–1957, VOL. 1, KrA, pp. 1–2. "en ytterligare utökning av personalen är ofrånkomlig".

228 "NR 6/1954. Protokoll från FRR sammanträde den 17 september 1954.", in KFF, FRR, Serie: A I, "Protokoll från sammanträden 1948–1957", År: 1948–1957, VOL. 1, KrA, p. 4.

229 "CLR INK 20/47", "Protokoll fört vid CLR sammanträde den 6/2 1947.", in: KFF, CLR, Serie: A II, "Inkomna och avgångna handlingar med diarium", År: 1945-1948, VOL. 1, KrA, p. 6.

230 " NR 2/1948", "Protokoll från FRR sammanträde den 11/6 1948 (militära ledamöterna).", in KFF, FRR, Serie: A I, "Protokoll från sammanträden 1948-1957", År: 1948–1957, VOL. 1, KrA, p. 2.

for this, at least according to the RSAFB, was the fact that increases in personnel had not been forthcoming to the amount deemed necessary. The RSAFB asked the Riksdag for more funds and the importance of providing a surface-to-air missile for the Defense made the Minister of Defense urge parliament to grant the request.[231]

But by then it had become increasingly clear to the military establishment that there was not going to be a Swedish surface-to-air missile. Moreover, procuring guided missiles abroad was now a realistic possibility, a situation that largely rendered a Swedish guided missile obsolete. The non-existent domestic option did, however, continue to play a role in the process of getting funds from the Riksdag. Hence, in 1958 the RSAFB asked the parliament for SEK 40 million, money that was to be used for continued domestic development, but also (in accordance with new directives from the Supreme Commander) for procurement "of a small number of foreign guided missiles."[232] This phrasing does not necessarily suggest that the RSAFB still expected that RB 322 would be realized. Rather, the part concerning foreign guided missiles should be interpreted as a crude and deliberate understatement by the RSAFB, which perhaps did not wish to appear completely incompetent in the eyes of the Riksdag. The funds were subsequently considerably revised upwards. The Minister of Defense Torsten Nilsson justified the revision by stating that surface-to-air missiles were considered to be of special importance by the studies made concerning the future organization of the Defense Forces.[233] In the end, appropriations increased fivefold between 1957 and 1959.

Likewise, the members of the GMC all waited in vain to see the testing of RB 321. In February 1957 they got to see film footage of the British Fireflash missile, a painful reminder of the inadequacy of the Swedish effort in this area thus far.[234] By late 1957 voices were being raised against further development of RB 321 in the Defense Staff, which stated that "it may be questioned whether the plans to develop RB 321 into an operational guided missile should not be shelved."[235] A few

231 "Statsverkspropositionen år 1957: BIL. 6: Fjärde huvudtiteln: Flygvapnet", in: *Bihang till Riksdagens protokoll år 1957*, första samlingen, andra bandet, Bilagor till Kungl. Maj:ts Proposition nr 1 angående statsverkets tillstånd och behov: fjärde–sjunde huvudtitlarna, p. 179, Riksdagsbiblioteket.

232 "Kungl. Maj:ts proposition nr 110 år 1958", in: *Bihang till Riksdagens protokoll år 1958*, första samlingen, elfte bandet, Kungl. Maj:ts Propositioner 110–119, p. 490, Riksdagsbiblioteket. "av ett mindre antal utländska robotar."

233 Ibid., pp. 491–492.

234 "NR 2/1957. Protokoll från FRR sammanträde den 28 febr 1957", in KFF, FRR, Serie: A I, "Protokoll från sammanträden 1948–1957", År: 1948–1957, VOL. 1, KrA, p. 2.

235 "NR H 38", "Inköp av amerikanska robotar. Underlag för diskussion.", Hgkv, Försvarsstabens planeringsavdelning, kvalificerat hemlig, År: 1955–59, VOL. 1,

months later, in February, 1958, the Air Staff had definitely given up all expectations that RB 321 would ever become operational. It would still remain as a post in the budgetary requests, however—the intention being to use these funds to procure a foreign air-to-air missile.[236] The two major domestic guided missile projects had thus been effectively killed after almost a decade of development.

Another project that never got off the ground was RB 330, a tactical nuclear surface-to-surface missile intended to strike Russian ground targets, an effort initiated by SAAB in cooperation with the RSAFB.[237] That RB 330 was intended to carry a nuclear warhead has been confirmed by the Chief of SAAB's Guided Missiles Division in Linköping in the 1950s Tore Gullstrand,[238] as well as by a document from the Planning Division of the Air Staff from late 1957 that stated that RB 330 was going to be "a tactical guided missile system for striking ground targets with nuclear weapons".[239] With this aim in mind, working groups on inertial guidance were established within the GMB in 1956, and they undertook a series of trips abroad for the purpose of gathering information on inertial guidance, visiting Canada and the United States in September 1957 (only one month after the Soviet Union launched its first inter-continental ballistic guided missile).[240] However, RB 330

KrA, p. 1. "det kan ifrågasättas, om planerna på att utveckla rb 321 till en strid-srb icke böra skrinläggas." I thank Johan Gribbe who drew my attention to this document.

236 "NR H 21", "PM ang mtrlanskaffningen 1958/65 för alt Adam.", in: Hgkv, FS/KH, PA, "Inkomna och utgående skrivelser", Serie: F III, VOL. 1, År: 1955–59, KrA, p. 2.

237 Cervin-Ellqvist, G. and König, B., *Robot 50 år.*, p. 18; see also: interview with Tore Gullstrand in *Östgöta Correspondenten*, 1994-02-11, p. A 5.

238 See interview with Tore Gullstrand in *Östgöta Correspondenten*, 1994-02-11, p. A 5.

239 "NR H 10", "Synpunkter på anpassning av flygmtrlanskaffningen med hänsyn till sannolika och minskade möjligheter till omsättning av befintlig och planerad mtrl.", in: Hgkv, FS/KH, PA, "Inkomna och utgående skrivelser", Serie: F III, VOL. 1, År: 1955–59, KrA, p. 1. I thank Johan Gribbe for pointing this document out to me.

240 "RHM 33:3", in KFF, FRB, Serie: F IIIB, "Inkomna och avgångna skrivelser ordnade efter klassifikationssystem", M 33-M 44:16, År: 1958, VOL. 30, KrA. "RHM 33:1", in KFF, FRB, Serie: F IIIB, "Inkomna och avgångna skrivelser ordnade efter klassifikationssystem", M 33-M 44:16, År: 1958, VOL. 30, KrA. Subsequent to the Soviet launch of this missile the imagined 'bomber gap', i.e. the false belief that the USSR had more intercontinental bombers than the U.S., was replaced by a fear of a major Soviet breakthrough in missile technology. When the Soviets sent the worlds first man-made satellite, *Sputnik*, into orbit a virtual campaign was launched in the United States against the specter of a 'missile gap' (Tamnes, R., *The United States and the…*, p. 99).

was cancelled in 1959 before a prototype could be tested.[241] The reason, according to the Chief of the RSN Stig H:son Ericson (who was given the information by Erlander and Undén), was that the Russians had stated to the Swedish government that they had reason to suspect that Sweden was about to develop a medium range ballistic missile that could reach Soviet territory. Due to this, Erlander and Undén had ordered that such development, if it was under way, should be cancelled.[242] Perhaps Wennerström had supplied this information to the Soviets.

The fact that the domestic development program was so slow to yield operational missiles naturally put the men in charge (for they were all men) in an awkward position when questioned about the poor results. Such inquiries forced these men to rationalize the Swedish development effort *ex post facto*. An example of such an endeavor could be found in the local Swedish newspaper *Vestmanlands Läns Tidning* on September 28, 1960, which featured an article entitled "Swedish Guided Missile Policy" ("Svensk robotpolitik"). The article was written by the Chief of the Air Staff's Press Relations Office Major Hans Neij,[243] who started out by saying:

> As an aviator one is often confronted with questions from the public regarding when the guided missiles will take over the tasks of manned aircraft, what Swedish guided missile research has accomplished during its more than a decade of activity, *why we purchase guided missiles from abroad instead of manufacturing them ourselves,* and many other questions concerning the guided-missile age and its problems.[244] [Italics added.]

241 Cervin-Ellqvist, G. and König, B., *Robot 50 år*, p. 18; see also: interview with Tore Gullstrand in *Östgöta Correspondenten*, 1994-02-11, p. A 5.

242 Ulving, Lars, "Svensk trygghet. Den svenska militärledningens överväganden kring kärnvapenanskaffning 1945–1961.", in K. Zetterberg (ed.), *Totalförsvar och atomvapen*, (Stockholm: Försvarshögskolan, 2001), p. 192. See also: Interview with Tore Gullstrand in *Östgöta Correspondenten*, 1994-02-11, p. A 5.

243 Neij was the Chief of the Air Staff's Press Relations Office (Pressdetaljen) 1955–1958. In October 1973 he became Chief of the Air Staff, a position he held for five years. After that he became Defense Attaché in Washington. All through his career he did a lot to further the Air Force's relations with the media. (For biographical information on Hans Neij, see: *Flygstaben – en minnesbok* (Stockholm: Högkvarteret, Försvarsmakten, 1994), p. 66.)

244 *Vestmanlands Läns Tidning*, "Svensk robotpolitik", September 28, 1960. "Som flygare möts man ofta av frågor från allmänheten om när robotarna skall överta de bemannade flygplanens uppgifter, vad den svenska robotforskningen uträttat under sin mer än ett decennium långa verksamhet, varför vi köper robotar från utlandet i stället för att tillverka dem själva och många andra frågor rörande robotåldern och dess problem."

Neij portrayed Sweden's missile policy as a very rational, and well considered, process. He explained that guided missiles were only procured once they were deemed able to carry out the tasks better than the current weapon platforms, or to complement them in different ways. In the article, he even explicitly stated that the Defense Staff and the RSAFB coordinated the needs of the different services to secure "a rational use of the country's collective technological ability".[245] As we have seen so far in this thesis, this was surely a euphemism. The RSAF had desperately wanted guided missiles for the last five or six years but been unable to secure them from the GMB, and this was part of the reason for, and to some extent a consequence of, the three services' retention of their own missile agencies and continued pursuit of independent procurement policies.

Having stressed that the domestic development had only resulted in one operational system (i.e. subsonic air-to-surface missile RB 304), Neij rhetorically asked if this was all that domestic missile research had come up with in all this time. The answer he provided was, maybe surprisingly, that the purpose of Swedish missile research had not been to facilitate domestic missile construction. On the contrary, said Neij:

> [the] number one priority has been, and still is, to acquire information and our own experience through monitoring of development abroad, basic research, limited design work, and test activities within the country. *Hereby we educate ourselves to become competent buyers of fully developed foreign missile systems*, which cannot be manufactured within the country due to our limited economic and technological resources. [Emphasis in original.]

And he continued:

> At the same time, we are also able to assess what is suitable for our own industry, e.g. the air-to-surface missile "304".

He rounded off by stating:

> Guided missiles are very expensive weapons. *Not least the development of missile systems demands high expenditure.* [Our knowledge and abilities are] to a large degree to our benefit in our dealings with foreign countries. *We do not run the risk of being*

245 *Vestmanlands Läns Tidning*, "Svensk robotpolitik", September 28, 1960. "ett rationellt utnyttjande av landets samlade tekniska förmåga"

fooled.[246] [Italics added; spaced characters instead of underlining in original.]

Thus Swedish missile research, according to Neij, was not primarily geared to providing domestic missile systems. This sheds a new and interesting light on the whole Swedish-American collaboration during this period. As we remember from Chapter 3 there were no explicit references to domestic development of operational guided missiles when the CD was established in 1945, nor when the GMB was created in 1948. However, it is noteworthy that Neij's phrasing corresponds very well to one of the top priorities formulated for the then newly created GMC in 1948, which was "to follow technological, tactical, and organizational development generally in Sweden as well as abroad".[247] This similarity can hardly have been an accident, and hence seems to imply that Neij was stating a deep and profound truth about the Swedish guided missile program in his article, just as he was with his statement about the limited economic and technological resources available to Sweden.

What does not fit the picture, though, is what he said about large development costs being a reason for not developing domestic missile systems. As we have seen, the historical evidence points in the opposite direction. In fact, a lot of money was spent on precisely that phase which was the most expensive, i.e. the development phase. It is reasonable to assume that Neij's article was probably intended to justify to the public why ten years of research had not yielded more results in the form of domestic missile systems. This was clearly the view of Swedish guided missile R&D that the Air Staff wanted the Swedish public to have. While the original intention of Sweden's guided missile R&D was to result in operational missiles, it is not unreasonable to assume that perhaps Neij's rationalization had in fact become an objective in the late 1950s. As we shall see in Chapter 6, the final empirical chapter, there is reason

246 *Vestmanlands Läns Tidning*, "Svensk robotpolitik", September 28, 1960. "[En] primäruppgift har varit och är alltjämt att genom uppföljning av utvecklingen i utlandet, grundforskning samt begränsat konstruktionsarbete och försöksverksamhet inom landet skaffa information och egna erfarenheter. Vi utbildar oss härigenom till KOMPETENTA köpare av färdigutvecklade utländska robotsystem, som inte kan tillverkas inom landet på grund av våra begränsade ekonomiska och tekniska resurser. Samtidigt kan vi också bedöma vad som ligger lämpligt till för vår egen industri, t.ex. attackroboten '304'. Robotar är mycket dyrbara vapen. Inte minst utvecklingen av robotvapensystem kräver dryga kostnader. [Vårt vetande och kunnande är] i hög grad till godo i våra affärer med utlandet. Vi riskerar inte att bli lurade."

247 Annex NO 1 to "NR 1/1948. Protokoll från FRR sammanträde den 19/5 1948" entitled "Instruktion för Försvarets R-vapenråd (FRR)", in KFF, FRR, Serie: A 1, "Protokoll från sammanträden 1948–1957", År: 1948–1957, VOL. 1, KrA, p. 1. "att följa den tekniska, taktiska och organisatoriska utvecklingen i stort såväl i Sverige som i utlandet".

to believe that the RSAFB, at least in one instance (the Falcon missile system), might have used domestic development as a way of preparing Swedish industry (i.e. SAAB) for license manufacturing of foreign missile systems. And if it happened once, then it is not unreasonable to assume that it could have happened several times.

Conclusion

In this chapter I have shown that the Swedish government actively sought to contribute materially to NATO's defense of Western Europe by urging the Americans to award OSP production contracts to Swedish companies. The government appointed the semi-official Swedish Export Association to act as an intermediary between NATO and Swedish companies. Prime contracts were, however, denied Sweden because of American sensitivities toward the NATO countries, which were expected to have been upset if lucrative production went to a non-member state. Nevertheless, Sweden manufactured military materiel for U.S. forces in Europe.

I have also traced the origins of the American decision to grant Sweden access to guided missile technology in the late 1950s, and shown that in order to understand this decision it is necessary to go beyond Sweden's relationship with the United States and look at the broader international picture, i.e. U.S. hegemonic policy towards the NATO countries. Sweden's fate in fact depended on the general American disclosure policy regarding this technology. The promise given by U.S. Secretary of Defense Wilson at the annual NATO meeting, in December 1956, and the subsequent decision taken in January 1957, also came to affect officially non-aligned Sweden. This was where the significance of Sweden's prior consent to U.S. hegemony, and the secret cooperation that it had supported, came into play.

It was not formal membership in NATO that was of key importance once the Americans had decided to release this technology to its allies. Rather, I argue, it had to do with Sweden's consent to American hegemony and the general hegemonic cooperation that had been an established fact since the middle of 1952. This seems to be the only plausible explanation of why a decision made by NATO to equip the member countries with guided missiles would facilitate Sweden's access to such weapons.

As we have seen, the Americans were even averse to letting the British have access to the most sensitive information regarding guided missiles, such as homing frequencies etc. This chapter has provided an important answer to the question why it took so long for Sweden to gain access to this coveted technology, something that has been wanting in other studies of the subject. The connection between the Swedish guided missile program and that of the NATO countries had been made visible already in September 1955 when Colonel Robert E. Galer and Marvin Schuldenfrei from the Guided Missiles Division, Bureau of Aeronautics of the Navy Department visited Sweden as unofficial representatives of NATO's Mutual Weapons Development Team in Paris. Galer and Schuldenfrei visited the GMB in order to find out in what areas Sweden needed assistance in their guided missile development. They then reported back to the Team's headquarters in Paris.

Hence, while Sweden's subsequent access to American guided missile technology and related data was preconditioned by Sweden's protracted and active consent to American hegemony, it was the prodding of the NATO allies that finally made the U.S. government change its attitude. Sweden did not (and indeed could not) by itself induce the American government to release this technology. In the end, what Sweden did mattered less than the considerations arising from the peculiar problems confronting a hegemonic, and utterly preponderant, nation that, due to the paradoxes of international relations, was in dire need of its allies' support.

The connection between the ability to procure guided missiles from abroad and the cancellation of several domestic development projects has also been demonstrated in this chapter. Paradoxically, it was when American assistance became readily available to Sweden, and when missiles could be purchased from the U.S., that the domestic option foundered. The lure of procuring American surface-to-air and air-to-air missiles proved too great for the Swedes to resist. As the next chapter will show, there was perhaps some truth to Neij's statements in the press where he intimated that Sweden's guided missile R&D was not primarily geared towards producing operational missiles, but towards making the Swedes able buyers and manufacturers of foreign missile systems.

We have also noted that NATO planning documents in the mid-1950s took Sweden's contribution into account when allocating forces to Norway. Swedish participation alongside of the Allies was taken more or less for granted, because the forces needed to defend Norway would have had to be increased if Swedish assistance would not be forthcoming. Moreover, the West's strategic bomber forces would fly over Sweden on their way toward their targets in the Soviet Union and Eastern Europe. Many hundreds of American, British, and Norwegian, aircraft would violate Swedish neutrality just prior to dropping their deadly load on the opposite side of the Baltic Sea. It does not seem plausible that the success of these vitally important operations could be left to chance, i.e. that it could simply be assumed that the Swedes would not start shooting Allied aircraft out of the sky. The Swedes must reasonably have been told about the Western plans to overfly her territory, and concurred to them in some way. While this conclusion may be controversial, it does seem to be inescapable unless one wants to assert that the Allied war planners were a bunch of wishful thinkers.

Chapter 6

Reaping the Deadly Fruit:

Purchases, and License Manufacturing, of American Missiles, 1959–1962

Account is taken of the possibility that [...] such missiles acquired from the U.S. might be used with considerable effect against Western [aircraft]. However, we believe this possibility to be so slight as to be negligible and outweighed by other considerations.

(Kohler to Magill, August 17, 1959.)[1]

Introduction

The last empirical chapter will address the fate of the domestic missile alternatives pursued by Sweden in the late 1950s, and will also analyze the purchases of American missile systems, and the license manufacturing of such systems by Sweden in the early 1960s. The guided missile deliveries should be seen as a material reward offered to Sweden for its prior, but also for its continued, support for, and consent to, American hegemony. This chapter offers the final piece of the puzzle of an understanding of the place of this technology in Swedish-American security relations, and is most closely tied to the fourth set of questions set out in Chapter 1.

Opting out of the Domestic Option

By the end of 1959 the RSAF had opted out of the domestic alternative for a surface-to-air missile altogether and instead decided on procuring Bloodhound MK. II from Britain, due for delivery during 1960. Large sums of money were directed towards missile procurement, and allotments continued to increase between 1960 and 1963.[2] This was

1 Memorandum from Kohler to Magill, dated August 17, 1959, p. 2, in: RG 59; EUR/SCAN; RRSWA, 1958–1964; POL 17 Political Affairs Sweden POL 17 Diplomatic Notes 1959–1963 to PS – Protective Service Sweden PS 7-1 Arrest (Swed. Seamen Visit New Orleans); Box No. 1; (Folder: Sweden Missile 1959–62); NARA II, College Park, MD.

2 "Statsliggare för budgetåret 1960/61", del 1, Statsverkets specialutgiftsstater under I–XIII huvudtitlarna jämte de affärsdrivande verken, p. 269, Riksdagsbiblioteket. "Statsliggare för budgetåret 1961/62", del 1, Statsverkets specialutgiftsstater under

partly because of the funds needed to pay for the American surface-to-air missile Hawk.

Allotments For A Surface-to-Air Missile, Fiscal Years (FY) 1955/56–1962/63:

	SEK Million:	% of Overall Defense Budget:
1955/56:	3.5	0.2
1956/57:	4.5	0.2
1957/58:	5.0	0.2
1958/59:	25.0	0.9
1959/60:	61.0	2.2
1960/61:	63.0	2.1
1961/62:	103.5	3.3
1962/63:	115.2	*

Sources: information about the defense budget for each year can be found in "Statsliggare för budgetåret ..." (p. 3 in every volume); for everything else, see footnotes above. *: No figures available. All figures have been rounded off to one decimal.

Large as these sums may seem, compared to what the United States spent on their guided missile development they were ridiculously small. For fiscal year 1958/59 the USAF spent more than SEK 10 billion, and for fiscal year 1959/60 this sum had increased to SEK 15 billion, i.e. half of the total procurement costs for the USAF.[3]

Hence, the RSAFB's SEK 115.2 million was not even one percent of what the USAF could spend on guided missile development. Of course, in the Swedish case, this was just for a surface-to-air missile. Were other types of guided missiles to be included the percentage of the defense budget spent on guided missiles would increase somewhat, but there would not be any drastic changes. It still goes without saying that the Americans could afford to make more mistakes in the development process than the Swedes could. From this perspective it is perhaps impressive that Sweden managed to produce an operational guided missile at all.

I–XIII huvudtitlarna jämte de affärsdrivande verken, p. 266, Riksdagsbiblioteket. "Statsliggare för budgetåret 1962/63", del 1:1, Statsverkets specialutgiftsstater under I–VII huvudtitlarna, p. 180, Riksdagsbiblioteket.

3 Annex NO 1 to "NR 2/1957. Protokoll från FRR sammanträde den 28 febr 1957.", entitled: "Robotutvecklingen utomlands", in KFF, FRR, Serie: A 1, "Protokoll från sammanträden 1948–1957", År: 1948–1957, VOL. 1, KrA, p. 9.

Why, then, was there such a big difference in appropriations between the period before 1958/59 and the period afterwards? The reason is to be found in the Defense Decision (försvarsbeslut) of 1958 (henceforth: FB 58), and the priority given in it to guided missiles in the defense of Sweden. This document, coupled with the American willingness to sell missiles to Sweden, paved the way for the increase in appropriations. The foundation for FB 58 had actually been laid as early as October 1954 in Supreme Commander Swedlund's long-term proposal for the future organization of the Swedish Armed Forces, called ÖB 54. This proposal was presented to the Parliamentary Defense Committee (försvarsberedningen; henceforth: PDC) in 1955, headed by the Minister of Defense Torsten Nilsson. Due to a series of negotiations concerning his proposal Swedlund presented another alternative, ÖB 57, two years later. Among other things, it was recommended that the Armed Forces should be organized so as to be able to hold the line against an attack until help could be gained from foreign friendly forces (read: the United States), but more importantly it recommended that the RSAF should be given a larger share of the defense appropriations. The ÖB 57 has been described as "a proposal for modernization and an important about turn in defense policy in the post-War period."[4]

The PDC presented its recommendations to the Standing Committee on Supply, which, in turn, expressed its view in a report presented to the Riksdag. The report was the immediate precursor to FB 58, and in it the Standing Committee quoted the PDC regarding the importance of modern guided missile technology. Said the report:

> Furthermore, the Committee would stress that technological development, especially atomic weapons and guided missiles, could be expected, within a few years, increasingly affect the organization of our Armed Forces. [...] The Committee considers it urgent, with regard to defensive capabilities, as well as defense economy, for technological development within our Armed Forces, in the guided missile field among others, *to be most forcefully promoted.* The organization of the work in the guided missile field ought to be designed so that the need for coordination between the different parts of the Armed Forces, and efficiency and expediency in the development work, is met. The possibility of buying missiles from abroad ought to be carefully monitored. [...] Research and testing in this field [nuclear weapons], and in the guided missile field, *ought*

4 Cars, Hans Christian, Skoglund, Claës, and Zetterberg, K., *Svensk försvarspolitik under efterkrigstiden* (Stockholm: Probus Förlag, 1986), pp. 28–30. Quote on p. 30. "ett moderniseringsförslag och en viktig omslagspunkt i försvarspolitiken under efterkrigstiden."

to be given the highest priority in appropriation of funding and personnel.[5] [Italics added.]

The Defense Minister restated the PDC's views (a committee that he himself chaired) in the Standing Committee's Report, but emphasized even more the importance of guided missiles, stating:

I consider it to be one of the crucial requirements for the effectiveness of our Armed Forces in a long-term perspective *that the work on development and procurement of guided missiles is intensified* so that such weapons can be provided to the Armed Forces as quickly as possible.[6] [Italics added.]

He then turned to the question of appropriations and formulated what he was prepared to do to see that this urgent and vital matter was taken care of in a fast and resilient manner.

I intend to suggest considerable increases in the allocations for guided missile activities during the coming fiscal year. Intensified research and development work in this area unavoidably demands increases in personnel.[7] [...] Funding for the development, etc., of

5 "Statsutskottets utlåtande NR B 53 år 1958", in: Bihang till B-riksdagens protokoll år 1958, sjätte samlingen, andra bandet, statsutskottets utlåtanden och memorial NR 43–107, p. 10, Riksdagsbiblioteket. "Beredningen vill vidare framhålla att den tekniska utvecklingen särskilt beträffande atomvapen och robotar torde komma att inom några år påverka vårt försvar och dess utformning i växande grad. [...] Beredningen bedömer som angeläget, med hänsyn till såväl försvarskraft som försvarsekonomi, att den tekniska utvecklingen inom försvaret, bland annat inom robotområdet, med all kraft främjas. Organisationen av arbetet på robotområdet bör vara så utformad att kraven på samordning mellan försvarets olika delar och på effektivitet och snabbhet i utvecklingsarbetet säkerställes. Möjligheterna till köp av robotar utomlands bör följas med största uppmärksamhet. [...] Forskning och försök inom detta område [atomvapen] och inom robotområdet bör ges högsta prioritet vid fördelning av anslag och personal."

6 "Statsutskottets utlåtande NR B 53 år 1958", in: Bihang till B-riksdagens protokoll år 1958, sjätte samlingen, andra bandet, statsutskottets utlåtanden och memorial NR 43–107, p. 18, Riksdagsbiblioteket. "Jag bedömer det vara en av de avgörande förutsättningarna för vårt försvars effektivitet på längre sikt, att arbetet med utveckling och anskaffning av robotar intensifieras så att dylika vapen så snabbt som möjligt kan tillföras krigsmakten."

7 "Statsutskottets utlåtande NR B 53 år 1958", in: Bihang till B-riksdagens protokoll år 1958, sjätte samlingen, andra bandet, statsutskottets utlåtanden och memorial NR 43–107, p. 19, Riksdagsbiblioteket. "Jag avser att föreslå väsentligt ökade anslag till verksamheten på robotområdet för nästa budgetår. Ett intensifierat forsknings- och utvecklingsarbete på detta område kräver ofrånkomligen personalförstärkningar."

> a domestic surface-to-air missile has so far been allocated under the appropriation for a *surface-to-air missile*. *Apart from that, funding is being asked for to undertake certain procurement of guided missiles from abroad.* The latter procurement is deemed appropriate with regard to educational and training activities within this part of our air defense. [...] The development of guided missiles should, in my opinion, be given special attention.[8] [First italics in original; second added.]

The PDC's recommendations thus echoed of the warning that the Americans had given Captain Oxenstierna and his group some four years ago, i.e. that not enough funding was being made available (see Chapter 5, p. 268). American hegemony was making itself felt. There can be no doubt that whatever prayers the guided missile developing community might have been mumbling in the four years preceding the report, they were all about to be answered in the following years. The old issue of personnel increases for the GMB was also addressed in this statement, something that quite a few had been crossing their fingers for. Attention should be given to the careful phrasing used when the issue of foreign guided missiles was brought up. It might seem odd that the Minister of Defense himself believed that the guided missile systems procured from abroad would just be used for educational and training purposes, but the fact is that internal discussions in the RSAFB suggest that they too looked at the procurement of foreign missiles as a learning experience, which would be very valuable in the development of a Swedish system.[9] However, the RFSAB probably did not think that Sweden would eventually become self-sufficient with regard to guided missiles, but simply thought that a lot could be learned from foreign missiles by taking them apart and then building Swedish missiles by reverse engineering. By 1958, the Air Staff had already excluded the possibility of RB 322 ever reaching production stage,[10] and it was now officially accepted that the

8 "Statsutskottets utlåtande NR B 53 år 1958", in: Bihang till B-riksdagens protokoll år 1958, sjätte samlingen, andra bandet, statsutskottets utlåtanden och memorial NR 43–107, p. 37, Riksdagsbiblioteket. "Under anslaget till *luftförsvarsrobot* har medel hitintills anvisats för utveckling m.m. av en inhemsk luftförsvarsrobot. Härutöver begäres medel för att genomföra viss anskaffning av robotar från utlandet. Sistnämnda anskaffning bedömes lämplig med hänsyn till utbildnings- och övningsverksamheten inom denna del av vårt luftförsvar. [...] Utvecklingen av robotar bör enligt min mening särskilt uppmärksammas."

9 "NR H 38", "Inköp av amerikanska robotar. Underlag för diskussion.", in: Hgkv, FS/KH, PA, "Inkomna och utgående skrivelser", Serie: F III, VOL. 1, År: 1955–59, KrA, pp. 2–3.

10 "NR H 21", "PM ang mtrlanskaffningen 1958/65 för alt Adam.", in: Hgkv, FS/KH, PA, "Inkomna och utgående skrivelser", Serie: F III, VOL. 1, År: 1955–59, KrA, p. 3.

"[...] delays bring to the fore the suggested procurement of foreign guided missiles."[11]

The Sidewinder Purchase

The contract for Sidewinder was finally signed by Torsten Rapp and CNO Admiral Arleigh on February 17, 1959. The signing meant that the RSAF bought 2,000 Sidewinder 1A missiles from the United States,[12] at a cost of $9,700,042, or SEK 50,246,217, to be paid over a two-year period.[13] This sum alone covered almost all of the United States' development costs for Sidewinder, which were about $10 million.[14] While 2,000 missiles may sound a lot, it must be remembered that in 1959 Sweden had 647 fighter aircraft.[15] This amounts to only 3 missiles per aircraft, and although the number of fighters decreased somewhat to 602 in 1961,[16] it was not nearly enough to make these aircraft effective in a prolonged combat situation. Even so, 3 missiles are better than none at all, and the affair was thus momentous in its own way. "The signing yesterday", Rapp wrote to his wife, "was splendidly ceremonial—without speeches—and it was so exciting that I could hardly write my name."[17] Rapp was obviously moved by the occasion and stated that Admiral Burke, who

11 "Kungl. Maj:ts proposition nr 110 år 1958", in: *Bihang till Riksdagens protokoll år 1958*, första samlingen, elfte bandet, Kungl. Maj:ts Propositioner 110–119, p. 490. "Dessa förseningar aktualiserar den föreslagna anskaffningen av utländska robotar."

12 *Had There Been A War...*, p. 123. For references and notes: Nilsson, M., "Gud vare tack och lov...", p. 41; "Kontrakt 123", "Memorandum of Understanding between the United States Navy and the Royal Swedish Air Force Relative to the Procurement of Sidewinder Rockets and Associated Equipment", in: KFF, Serie: F 111, "Hemliga avtal och kontrakt", År: 1959, VOL. 3, KrA. Apart from the 2,000 live missiles, Sweden also bought a total of 160 missile dummies from the Americans (see: Letter from Lt. Colonel Eric Malmberg to Rear Admiral W. J. Marshall on February 27, 1959, pp. 2–3, in: KFF, Serie: F 111, "Hemliga avtal och kontrakt", År: 1959, VOL. 3). The 2,000 guided missiles constituted almost exactly 2 percent of the total production of this version (Westrum, R., *Sidewinder...*, p. 173).

13 Letter from Rear Admiral W. J. Marshall to General Lindskog on 31 December 1958, "INKHM 44–324:1", pp. 1–2, in: KFF, Serie: F III, "Hemlig avtal och kontrakt", År: 1959, VOL. 3, KrA. The sum in SEK was written by hand on the document. See also: 758.56/1–1459, Memorandum for the Special Assistant for Mutual Security Coordination from Holcombe, Director of the Office of Programming and Control, dated January 14, 1959, p. 1, in: RG 59; 1955–1959 CDF; From 758.56/1–555 to 759.00-mayday/3–357; Box No. 3467; (Folder: 758.56/1–555); NARA II, College Park, MD.

14 Dörfer, I., *System 37 Viggen...*, p. 242.

15 Wennerholm, B., *Fjärde flygvapnet i världen?...*, p. 158.

16 For the number of aircraft, see: ibid., p. 158.

17 Letter from Torsten Rapp to his wife dated February 18, 1959, in Torsten Rapps arkiv, envelope XI in box KI, VOL. 11, KrA. "Underskrivandet i går var utmärkt

was a descendant of Swedish immigrants, was the best host he had ever met.[18] Sidewinder (or RB 24 as it was called in Sweden) took the place of the non-existent RB 321 and the work on this guided missile was abandoned.

At the same time that Rapp was in the United States to sign the contract for Sidewinder, Colonel Harald Jentzen, the Chief of the Weapons Section of the RSAB, was allowed to see several American guided missile systems—Nike Ajax, Nike Hercules, Hawk, Lacrosse, Honest John, BOMARC, and Falcon. The U.S. Army in particular displayed an increasing willingness to share classified data with the Swedes, and matters were made even easier when a group of officials from the U.S. Army, led by Colonel Charles Patterson, who was Deputy Director of Special Weapons at the Office of the Chief of Research and Development, was allowed to visit Sweden in October 1959.[19]

One can get an indication of just how privileged Sweden was in gaining access to Sidewinder at such an early stage, if one considers that besides Sweden only a few non-NATO countries were scheduled to receive Sidewinder at that time. Within the frameworks of the MAP, deliveries to Taiwan, the Philippines, Korea, and possibly Spain were projected for FY 1960.[20] Of these four countries, two (i.e. Taiwan and Korea) could be considered to be countries whose security was ranked very high by the United States. The Americans had gone to war over Korea; they had sworn to their willingness to do the same over Taiwan (the Second Formosa Crisis in 1958 was only a year in the past); and the Philippines had been granted formal independence from American colonial rule on July 4, 1946. Only Spain seems to stand out as a country which the United States did not have a clear commitment to defend. These were the kind of nations that were considered to receive Sidewinder in 1960; and then Sweden of course.

The Problem of Dual-Purpose Missiles

I have touched upon the subject of dual-purpose missiles before in this thesis, and the time has come to delve a little deeper into this issue and how American decision-makers viewed this problem in 1959. Even though the United States was prepared to sell BOMARC, Hawk, and

högtidligt – utan tal – och det var så spännande att jag knappt kunde skriva mitt namn."

18 Letter from Torsten Rapp to his wife dated February 18, 1959, in Torsten Rapps arkiv, envelope XI, box K1, VOL. 11, KrA. "den bästa värden jag någonsin träffat."

19 Moores, S., *'Neutral on Our Side':...*, p. 195.

20 Letter from Fessenden to Thurston, dated August 28, 1959, in: RG 59; RA; RecDir, 1955–1960; CorrF, 1955–59; Lot 61D252; Box No. 2; (Folder: RA Correspondence, 1959 Folder 2); NARA II, College Park, MD.

Tartar to Sweden, these systems posed a special problem since they either already were (as in the case of BOMARC), or were planned to become (as in the case of Hawk and Tartar), capable of carrying a nuclear warhead. In the spring of 1959 the State and Defense Departments were considering Swedish inquiries regarding the possibility of Sweden's purchasing the Falcon, Hawk, BOMARC, and Tartar missile systems. While Swedish nuclear capability was not considered an urgent problem, the new Secretary of State Christian Herter believed that the U.S. government had to include this possibility when considering the Swedish requests.[21]

Herter described the dilemma thus: while it was in the interest of the United States for Sweden to be able to defend itself against a Soviet attack and while the sale of the requested missile systems would certainly contribute to this end, it was also the U.S. government's policy to keep the amount of governments holding nuclear capability to an absolute minimum. A possible arrangement that the Swedes might have in mind, thought Herter, was "U.S. storage [of] nuclear warheads outside Sweden (presumably in some nearby area) and provision for making them available to Swedes in emergency."[22] Interestingly enough, this option does not seem to have posed a significant political problem for the Secretary; what did pose a problem, however, was if Sweden decided to procure nuclear warheads from another source, domestic or foreign, than the United States. This would have implications for NATO stockpiling arrangements which were most undesirable.[23] Herter

21 758.5612/4-1459, Telegram from Herter to U.S. Embassy in Stockholm and Paris, dated April 14, 1959, p. 1, in: RG 59; 1955–1959 CDF; From 758.56/1-555 to 759.00-mayday/3-357; Box No. 3467; (Folder: 758.56/1-555); NARA II, College Park, MD. The issue of U.S. policy towards Sweden on nuclear matters, and Sweden's own efforts to acquire atomic weapons, has been dealt with thoroughly by Thomas Jonter, see: Jonter, T., *Nuclear Research in Sweden: The Cooperation between Civilian and Military Research, 1947–1972* (Stockholm: SKI Report 02:18, 2002); Jonter, T., *Sweden and the Bomb: The Swedish Plans to Acquire Nuclear Weapons, 1947–1972* (Stockholm: SKI Report 01:33, 2001); Jonter, T., *Sverige,* USA *och kärnenergin. Framväxten av en svensk kärnämneskontroll 1945–1995* (Stockholm: SKI Report 99:21, 1999); Jonter, T. and van Dassen, Lars, "Making Historical Surveys of States' Nuclear Ambitions: Experiences from the Baltic Sea Region" in *The Nonproliferation Review*, Vol. 12, No. 1, March 2005, pp. 225–270. See also: Agrell, W., *Svenska förintelsevapen…*, pp. 17–86, 125–181, 235–321.

22 Ibid., pp. 1–2, quote of p. 1. Moores, S., *'Neutral on Our Side':…*, p. 196.

23 758.5612/4-1459, Telegram from Herter to U.S. Embassy in Stockholm and Paris, dated April 14, 1959, p. 1, in: RG 59; 1955–1959 CDF; From 758.56/1-555 to 759.00-mayday/3-357; Box No. 3467; (Folder: 758.56/1-555); NARA II, College Park, MD. See also: Moores, S., *'Neutral on Our Side':…*, p. 196. It is not clear what Herter was referring to here, and Moores does not provide a possible answer either. Maybe what he meant was that if the Swedes bought warheads from either Britain or France, the stockpiles in these countries would shrink. However, if new warheads were produced to satisfy the Swedish demand it is

wanted the views of other State Department officers regarding the question of supplying Sweden with these kinds of missiles, particularly with respect to the NATO control aspect, and wondered if perhaps a distinction made between tactical, i.e. surface-to-air and short range surface-to-surface missiles, and longer range strategic nuclear-capable weapons was reasonable. If this was considered feasible, wrote Herter, "this might provide rationale for U.S. aiding Sweden to acquire nuclear-capable weapons of type in which Sweden now interested."[24]

James C. H. Bonbright, the new Ambassador to Stockholm, replied almost two weeks later, and stated that of the two conflicting policy objectives addressed by the Secretary in his telegram the issue of strengthening Sweden's defense system ought to be given precedence. The distinction between non-nuclear and nuclear-capable weapons did not seem significant to the Embassy since the Swedes would probably soon be able to manufacture their own nuclear weapons. Regarding the possibility of distinguishing between tactical and strategic missiles, it was thought that such a distinction could and should be made; the Swedes, remarked Bonbright, had never expressed an interest in anything but defensive (i.e. tactical) atomic weapons.[25]

Herter, in his telegram, had stated that if Sweden opted for procuring non-U.S. nuclear warheads, this would leave the United States without any control of the missiles themselves.[26] The Ambassador replied that joint-use control of any missiles supplied by the United States to Sweden "would presumably fall within broad terms of Agreement of July 1, 1952".[27] What Herter and Bonbright seem to have been alluding to, when speaking of joint-use control, was an arrangement similar to NATO's 'two key' system, whereby the United States controlled the warheads (through the SACEUR, which was always an American), while

hard to see why this would affect stockpiling arrangements. Perhaps this was thought to diminish British and French production capacity while the Swedish warheads were being produced, which in turn would affect NATO stockpiling arrangements.

24 Ibid., p. 2.

25 758.5612/4-2559, Telegram from Bonbright to Herter, dated April 25, 1959, p 1, in: RG 59; 1955–1959 CDF; From 758.56/1-555 to 759.00 mayday/3-357; Box No. 3467; (Folder: 758.56/1-555); NARA II, College Park, MD. Regarding the imprtance of a strong Swedish defense, see also: Moores, S., *'Neutral on Our Side':...*, pp. 196–197.

26 758.5612/4-1459, Telegram from Herter to U.S. Embassy in Stockholm and Paris, dated April 14, 1959, p. 1, in: RG 59; 1955–1959 CDF; From 758.56/1-555 to 759.00-mayday/3-357; Box No. 3467; (Folder: 758.56/1-555); NARA II, College Park, MD. See also: Moores, S., *'Neutral on Our Side':...*, p. 196.

27 758.5612/4-2559, Telegram from Bonbright to Herter, dated April 25, 1959, p. 1, in: RG 59; 1955–1959 CDF; From 758.56/1-555 to 759.00 mayday/3-357; Box No. 3467; (Folder: 758.56/1-555); NARA II, College Park, MD.

granting the host country veto over their use.[28] This interpretation of the MDAA agreement would reasonably have come as a complete surprise to the Swedish government, and it is hard to see what the Ambassador was referring to. Wrote Ambassador Bonbright:

> Since Sweden would be purchasing these missiles it would naturally prefer to have complete freedom in their use. Embassy believes, however, that Swedish desire to obtain missiles would be so strong that they would accept agreement providing some degree of US control over their use. Extent of such control, and whether it would prove effective if put to test in emergency or war, is another question and would depend to some degree on precise missile system purchased.[29]

At the same time it was stated that the Swedes might turn to Britain or France for help if the United States was too rigid in its demands to control the missiles. However, Bonbright thought that this was more a negotiating lever than a serious option for Sweden.[30] But the State Department was more adamant than the Service Attachés in Stockholm, and Under Secretary of State Clarence Douglas Dillon wrote that:

28 Gaddis, J. L., *Strategies of Containment:...*, pp. 221–222; Calleo, D. P., *Beyond American Hegemony:...*, p. 48. This was a way for the U.S. to counteract the European demands for independent nuclear forces, something that the Americans had been strenuously against. Despite U.S. resistance both the British and the French had established independent nuclear capabilities by 1960, but U.S. Secretary of Defense Robert McNamara insisted that such forces would never be large enough to act as a deterrent. Moreover, without a unified command structure this upset the Kennedy administration's strategy of 'flexible response', and, said the Americans, the risk of accidents or miscalculations would thus be great (Gaddis, J. L., *Strategies of Containment:...*, p. 221). As a hegemonic power the United States saw its preponderance, and its national security, threatened by these nuclear forces (i.e. accommodating opposition), since the U.S. risked being drawn into a devastating war due to the use of nuclear forces outside its control. On the other hand, the European allies would never have accepted complete U.S. control over such weapons within their borders. The 'two key' system was thus a compromise to satisfy the demands for control of all parties. However, the hegemon held the upper hand with regard to nations without nuclear capability, since it controlled access to the warheads.

29 758.5612/4-2559, Telegram from Bonbright to Herter, dated April 25, 1959, pp. 1–2., in: RG 59; 1955–1959 CDF; From 758.56/1-555 to 759.00 mayday/3-357; Box No. 3467; (Folder: 758.56/1-555); NARA II, College Park, MD. Regarding the Swedish will to accept U.S. control, see also: Moores, S., *'Neutral on Our Side':...*, p. 197.

30 758.5612/4-2559, Telegram from Bonbright to Herter, dated April 25, 1959, p. 2., in: RG 59; 1955–1959 CDF; From 758.56/1-555 to 759.00 mayday/3-357; Box No. 3467; (Folder: 758.56/1-555); NARA II, College Park, MD. See also: Moores, S., *'Neutral on Our Side':...*, p. 197.

> [...] it would be unwise enter negotiations for sale such weapons without reasonable prospect of satisfying agreements either (a) re provision US nuclear warheads in event of emergency, or (b) re nuclear-capable weapons in event Swedes wished to leave question of nuclear warheads open for time being.[31]

The State Department stated that the United States, as a matter of general policy, was making available to non-U.S. NATO forces nuclear-capable weapons furnished directly to NATO governments "usually without explicit provision for US control over use of weapons as such."[32] In the case of IRBMs, however, there was a provision requiring the consent of the receiving government to allow the United States the last word in the decision to launch such missiles.[33] According to Dillon, the only way for the United States to sell these weapon systems to Sweden was if a bilateral agreement was concluded providing for either joint control of these missiles, or Swedish consent to use them only with American nuclear warheads.[34] The State Department questioned whether Sweden would be able to agree to any such arrangement which would entail a degree of U.S. control over the missile systems furnished, and whether there was any prospect of Sweden being prepared to make the necessary changes to its foreign policy. The Department concluded that:

> [l]eaving aside the question of explicit formal provisions for US control over use of nuclear-capable weapons per se, US de facto control over weapons delivery system through custody and control of nuclear warheads, even though stored outside Sweden, would obviously represent important policy departure for Swedes. Arrangements for storage of nuclear warheads in nearby country (possibly Germany or UK) could involve difficulties, although possibly of a subsidiary character.[35]

Interestingly, the choice of storing nuclear warheads outside Sweden, for delivery in case of emergency, was actually considered to

31 758.5612/4-2559, Telegram from Dillon to Embassy in Stockholm, dated May 21, 1959, p 1, in: RG 59; 1955–1959 CDF; From 758.56/1-555 to 759.00 mayday/3-357; Box No. 3467; (Folder: 758.56/1-555); NARA II, College Park, MD.

32 Ibid., p. 1.

33 Ibid., p. 1.

34 Moores, S., *'Neutral on Our Side':...*, p. 197.

35 758.5612/4-2559, Telegram from Dillon to Embassy in Stockholm, dated May 21, 1959, p 2, in: RG 59; 1955–1959 CDF; From 758.56/1-555 to 759.00 mayday/3-357; Box No. 3467; (Folder: 758.56/1-555); NARA II, College Park, MD. See also: Moores, S., *'Neutral on Our Side':...*, p. 197.

be an alternative by the State Department. The telegram pointed, in an almost comic way, to the fact that actual delivery of these warheads could involve difficulties, something which must be considered to be an understatement if the attack on Sweden involved a preceding or simultaneous attack on continental Europe. The State Department stressed to the Embassy that "analysis of possible arrangements with Sweden in this field does not represent US policy and should be considered on strictly NOFORN [i.e. Not Releasable to Foreign Nationals] basis unless instructed otherwise."[36]

This information was of course considered extremely sensitive, not the least because of the fact that the Swedes might try to take advantage of it in their negotiations with the Americans. While the different Departments could not really agree on how to handle the issue with Swedes, there was not really any serious opposition to the idea of selling dual-purpose missiles. Consequently, in the early summer of 1959 the State and Defense Departments agreed to offer Sweden the Tartar, Hawk, Falcon, and BOMARC missiles with conventional warheads. The Swedes were to be informed that it would be unwise of them to make any assumptions regarding the possibility of the United States furnishing nuclear warheads in the future.[37] The Americans realized that to offer these missiles to Sweden was the only sensible thing to do, considering that the Swedes might eventually develop their own nuclear weapons anyhow. These weapons would be very expensive, and technically inferior, and would draw valuable funding away from other areas which were equally or even more important.[38]

A Sweden armed with American missiles could of course cause some trouble for the West, should the Swedish government decide to shoot at 'friendly' aircraft. A memorandum dated August 17 contained a rare trace of an American official explicitly mentioning the possibility of these missile systems being used against the United States. The memorandum written by the Assistant Secretary for European Affairs Foy Kohler at EUR repeated that the State and Defense Departments agreed on making these missile systems available to Sweden, and stated that "[a]lthough Sweden is not an ally of the United States and cannot be expected to become an ally under present international circumstances,

36 Ibid., p. 3.

37 758.5612/6-259, Telegram from State Department to the U.S. Embassy in Stockholm, dated June 2, 1959, p 1, in: RG 59; 1955–1959 CDF; From 758.56/1-555 to 759.00 mayday/3-357; Box No. 3467; (Folder: 758.56/1-555); NARA II, College Park, MD. See also: Moores, S., *'Neutral on Our Side':...*, p. 197. According to Moores, this issue was not resolved until the fall of 1959 (Moores, S., *'Neutral on Our Side':...*, pp. 195, 197–198). However, it seems as if what happened in the fall was only that the Americans agreed to inform the Swedes about the decision, which had already been made that summer.

38 Moores, S., *'Neutral on Our Side':...*, p. 197.

it identifies itself closely with the West […].” It also revealed that the Americans were about to sell the Hawk system to Switzerland, another neutral. Then Kohler turned to the issue of American overflights of Sweden in wartime, and the possible reaction that these might provoke. He said:

> Account is taken of the possibility that, in the event of a major war in which Sweden […] was seeking to maintain its neutrality, such missiles acquired from the U.S. might be used with considerable effect against Western attempts to overfly either country. However, we believe this possibility to be so slight as to be negligible and outweighed by other considerations.[39]

This shows that the issue had in fact been discussed between officers at the State Department and the Pentagon, but it is still not known how they had come to the conclusion that the risk of Sweden turning the American missile systems against American aircraft was so minute. It is also unclear what other considerations Kohler referred to, but one can assume that the defensive power that such missiles would lend Sweden in its fight against Russian aircraft was conceived to be very valuable. But was this valuable enough for the Americans to think that it would be worth relying on nothing but the hope that Sweden would not defend its territory against intrusions from the West as well?

Either the United States was prepared to gamble that the Swedes would not choose to defend their neutrality against *all* intrusions, but only against those perpetrated by the Russians, or they perceived that they had good reason not to worry about this scenario. A gambling attitude does not match the meticulous attention given the security arrangements that the United States wished Sweden to make in order to receive the missile systems in the first place. Why make such an effort to make sure that the Russians did not get their hands on the technology, if the weapons risked being used against American forces anyway due to Swedish domestic security considerations?

One could argue that the American policymakers understood the limits of hegemonic power; i.e. it was not possible to control the security policy of another nation state. This would of course be especially true in the chaotic first few weeks of another global war—a nuclear war guided by the logic of the Eisenhower administration's massive retaliation

39 Memorandum from Kohler to Magill, dated August 17, 1959, p. 2, in: RG 59; EUR/SCAN; RRSWA, 1958–1964; POL 17 Political Affairs Sweden POL 17 Diplomatic Notes 1959–1963 to PS – Protective Service Sweden PS 7-1 Arrest (Swed. Seamen Visit New Orleans); Box No. 1; (Folder: Sweden Missile 1959–62); NARA II, College Park, MD. See also: Moores, S., *'Neutral on Our Side':…*, p. 198.

doctrine. But it is perhaps more plausible to assume that the Americans had received some tacit guarantee from Swedish military officials that the missiles would under no circumstances be turned against the West. After all, if war came the government would evacuated to Britain, but the military would still be in the country. There are no documents that outright support such a conclusion, except some statements by the Chief of the RSAF Nordenskiöld to that effect,[40] but considering the close cooperation that had been going on between the two countries for nearly a decade, and considering the well documented tendency not to put such matters down on paper, it is not totally implausible. Moreover, it is a fact that NIE 49 expected the RSAF not to carry out the Swedish government's orders to deal with Western intrusion more harshly in wartime.[41] Future scholarly work might perhaps discover what motivated the Americans when reaching this conclusion.

On October 20, 1959, General Lindskog, who had replaced Stig Wennerström as Air Attaché in Washington, met with a Mr. Nelson from the BNA and Hammond and Margrave from the NAMC, at the invitation of Hammond. The latter informed Lindskog that now would be an appropriate time for the Swedish government to submit a detailed and comprehensive formal request regarding U.S. guided missile systems in which the Swedes were interested. The information concerning Sweden's needs and desires in the guided missile field that the NAMC had at the moment was scattered and incomplete, said Hammond. This was so because word of what the Swedes were interested in had been received through three channels—State, Defense, the three Services, and in some instances even through the individual manufacturers. The United States, said Hammond, needed a government-to-government request before any transactions or commitments could be approved.[42]

Hammond had recently been to Stockholm, where he had pointed out to his hosts that the Swedes needed to provide the American government with such a comprehensive statement of their desires in this field. According to him, such a note ought to distinguish clearly

40 For example, Nordenskiöld had told the Chief of the British Air Staff, Lord Tedder, in the spring of 1949 that he would disobey the Swedish government, if necessary, to aid the Western air offensive (for this, see: Telegram from Colonel Hardy to Washington, dated April 13, 1949, in: RG 59; BNA, 1941–1953; SF, 1941–1953; South Africa, Union of, Policy – Sweden, External Political Affairs: U.S.; Box No. 18; NARA II, College Park, MD).

41 Moores, S. and Widén, J., "Sverige under det kalla kriget – den amerikanska underrättelsetjänstens perspektiv, mars 1952" in *Kungl Krigsvetenskapsakademi ens handlingar och tidskrift*, 2003:5, p. 89.

42 758.5612/10-2059, Memorandum of Conversation between Lindskog, Hammond, Margrave, and Nelson, dated October 10, 1959, p. 1, in: RG 59; 1955–1959 CDF; From 758.56/1-555 to 759.00-mayday/3-357; Box No. 3467; (Folder: 758.56/1-555); NARA II, College Park, MD.

between on the one hand which missiles and components Sweden wished to purchase, and on the other which they wished to produce under license. The Swedes had been asked to be very specific; any modification in the model desired was to be included. If the Swedes did not have this detailed information about the American missile systems, then they should indicate the objectives which they wanted to achieve with the missile so that American technicians could evaluate and determine the model best suited to Swedish requirements. According to the memorandum, General Lindskog was very attentive, took extensive notes during the conversation and agreed to the proposed procedure.[43]

In late October 1959 the Chief of the Swedish Defense Staff Major General Curt Göransson visited Ambassador Bonbright (the Ambassador referred to him as a member of the Swedish General Staff), and told him in confidence about his planned visit to the United States. The Ambassador wrote Deputy Assistant Secretary of State Ivan White to inform him of the visit, even though the Swedish Major General would not meet with anybody from the State Department during his stay in the U.S. Göransson, who was particularly interested in guided missiles, was to meet with Pentagon officials and hoped to get a chance to talk to some American military leaders, as well as seeing some U.S. bases.[44]

Bonbright expressed the same kind of bewilderment at the multitude of Swedish visitors interested in guided missiles that Hammond had stated to Lindskog only a week earlier, stating to White that it seemed to the Embassy "that just about everybody in the Swedish armed forces who is interested in missiles is headed for the United States these days, and it is not clear to us that all of the visits have been coordinated." Göransson's visit had been cleared by the Swedish government, however, and Göransson had explained to Bonbright that it was of a very sensitive character. Bonbright reported:

> He stressed the fact that his position is a particularly delicate one [as Chief of the Swedish Defense Staff] and that, for this reason, he is anxious to avoid any publicity whatever, either before, during or after the trip. He told me that very few Swedes know of it and he surrounded his whole account with an aura of secrecy. He also indicated that in view of his delicate position he would not go

43 Ibid., pp. 1–2.

44 Letter from Bonbright to White, dated October 27, 1959, p. 1, in: RG 59; Bureau of European Affairs Office of Northern European Affairs [henceforth: EURONOE]; Records Relating to Sweden [henceforth: RRS], 1957–1975; Box No. 2; DEF 2-Gen. Repports [sic] and Statistics, 1966–1969 to POL 33-8 – Law of the Sea, 1960–1969; Lot File: 72D339; (Folder: DEF-7 Visits and Missions, 1959–1969); NARA II, College Park, MD.

> without governmental approval (by which he meant at least the Minister of Defense), which I believe he now has. Here any question about his whereabouts will be answered with the statement that he is "out of the country on leave of absence". [45]

Göransson's visit raised anew, said the Ambassador, the question of where the United States stood with regard to providing dual-purpose weapons systems to Sweden. He was obviously aware of the fact that a favorable decision had been reached in the United States, but also noted that he had yet to hear whether this decision had been communicated to the Swedes. Bonbright hoped that this did not mean that any new difficulties had arisen, which would keep the Swedes from procuring such missiles from the United States, and he said that he would appreciate it if the Deputy Assistant Secretary could "give it a jog."[46]

As it turned out, Göransson would have quite an extensive itinerary. He was planning to arrive in New York on November 8, and although the purpose of the visit was unofficial the Swedes had requested that he meet with officials from the DOD with corresponding duties. It was understood that he wished to address the guided missile procurement issue on behalf of the Swedish Defense Staff. Göransson also wanted to visit quite a few defense installations in the U.S., including the Detroit Arsenal, White Sands, the Nike installation at Elgin Air Force Base, a SAC base, and Hughes Aircraft Company (henceforth: HAC). Göransson also wished to talk to the appropriate people to thank them in person for making urgently needed materiel available to Sweden, and "to discuss further possibilities in the same field […]."[47]

The Pentagon was prepared to handle the visit on an unofficial basis and would provide unclassified tours of the defense installations mentioned by the Swedes. Certain classified discussions would also be permitted, on condition that the Swedish Embassy in Washington presented an official request to the Americans. Secretary of State Herter noted that no commitment had been made to the Swedes at this moment because it was still unsure what exactly the Swedes were after. It now appeared as if Sweden wished to produce at least one entire missile [no doubt the Falcon missile], and they had recently requested data for production purposes which the United States had so far been unable to disclose. More information regarding the Swedish requirements was needed, said Herter, as well as a formal governmental request, in order for the U.S. government to be able to make an unambiguous decision on

45 Ibid., p. 1.

46 Ibid., pp. 1–2.

47 758.5811/10-3159, Telegram from Herter to Embassy in Stockholm, dated October 31, 1959, in: RG 59; 1955–1959 CDF; From 758.56/1-555 to 759.00-may-day/3-357; Box No. 3467; (Folder: 758.5811/1-455); NARA II, College Park, MD.

the matter. Apparently, Lindskog had informed him that the Swedish government was now preparing such a formal request.[48]

Göransson's requests were met; he visited the JCS and was given a briefing on its organization, as well as a briefing regarding the relationship between the JCS and the DOD. He also met with the Chief of Staff of the USAF.[49] On December 2, 1959, the new Swedish Ambassador to Washington Gunnar Jarring submitted an eight-page comprehensive and formal survey of Sweden's requirements in the guided missile field to Secretary of State Herter.[50] The survey included Hawk, BOMARC, Falcon, and Sidewinder; no other missiles were contemplated at the moment, although Jarring wrote that he would perhaps submit a new request for other weapons not listed in this one.[51]

However, before any deliveries could be made a new security agreement had to be signed between Sweden and the United States.[52] As we have seen in Chapter 5, the issue was not at all new. Both the Swedes and the Americans had been aware of the necessity for such an agreement at least since January 1957, and probably earlier, and a draft for a security agreement had been prepared, and informally presented to the Swedes, by the Americans already in late September 1959.[53] The issue regarding the U.S. attitude towards Swedish nuclear capability was resolved once and for all in April 1960 with the approval of NSC 6006/1, the new policy document on Scandinavia. It stated that the United States would discourage Sweden from acquiring atomic weapons, but at the same time, states Moores, it included Sweden under the U.S. nuclear umbrella.[54] It should be noted that the inclusion of Sweden under the U.S. atomic strike force was not mentioned explicitly in NSC 6006/1; however, it seems likely that American assistance to Sweden would come in the form of strategic air support, since ground troops were not

48 Ibid., pp. 1–2.

49 Memorandum for J–5, dated November 4, 1959, in: RG 218; CDF, 1959; 9167/3440 Italy (1959) to 9170/9105 Europe (East) (15 July 1959); Box No. 145; (Folder: 9168.4/5420 Sweden); NARA II, College Park, MD.

50 Letter from Jarring to Herter, dated December 2, 1959, p. 1, in: RG 59; CDF, 1960–63; from 758.56/1-1360 to 759.00 (W)/5-261; Box No. 1843; (Folder: 758.5622/12-660); NARA II, College Park, MD.

51 Ibid., p. 8.

52 Moores, S., *'Neutral on Our Side':...*, p. 199.

53 Letter from Jarring to Åström, dated September 25, 1959; and draft, dated September 21, 1959, pp. 1–3, in: HP 24 VI/U.S.A; "Tillstånd till in- och utförsel av vapen, ammunition och krigsmateriel"; 1958–febr. 1957; Folder 24; Box No. 62; År: 1955–1960; RIA; Arninge.

54 Moores, S., *'Neutral on Our Side':...*, pp. 198–203. NSC 6006/1 can be found in: FRUS: *Foreign Relations of the United States*; 1958–1960; Vol. VI, Part 2; *Western Europe* (Washington, D.C.: United States Government Printing Office, 1993), pp. 672–688.

a probable alternative. The Swedish Defense staff, in a strategic study made in the spring of 1960 (SSP/60), assumed that the United States would strike the Soviet Union with nuclear weapons *two* days after a Soviet atomic attack on Sweden.[55] Bengt Wallerfelt, who was the first researcher to analyze this study, asked whether the Swedes made this curiously precise assumption on their own or if they had recieved an American guarantee. I will return to these issues below.[56]

Licensed to Produce

Torsten Rapp returned to the United States shortly after the Swedish purchase of Sidewinder. He took with him yet another delegation of officers in order to discuss more guided missile purchases.[57] What was discussed during his visit in January 1960 was probably, besides Hawk and BOMARC,[58] the air-to-air missile Falcon. Sidewinder, which had been the main armament on the A, B and D-versions of the J35 Draken aircraft, was to be replaced by Falcon on the F-version.[59] Sweden had been informed by the State Department in the fall of 1959 that Washington intended to let Sweden purchase Falcon, and was now making its move to acquire the missile.[60]

The RSAF's procurement of Falcon resulted in the discontinuation of two domestic missile systems, RB 340 and RB 341.[61] These missiles were originally intended for the Swedish J35D Draken and J37 Viggen aircraft; the RSAFB had ordered preliminary project specifications from SAAB on November 1 1957, and by November 22 a proposal had been produced. This was the end of a series of project investigations which had been carried out all through 1957 as part of the typifying work for the aircraft. It is of some importance for the argument that I will provide below that one of the missiles (RB 340) would have had a semi-active

55 Wallerfelt, Bengt, *Si vis pacem – para bellum. Svensk säkerhetspolitik och krigsplanläggning 1945–1975* (Stockholm: Probus Förlag, 1999), pp. 106–131. See also: Agrell, W., *Svenska förintelsevapen...*, p. 303.

56 Wallerfelt, B., *Si vis pacem – para bellum...*, p. 113.

57 *Had There Been A War...*, p. 123.

58 Dörfer, I., *System 37 Viggen...*, p. 92.

59 Ibid., p. 92.

60 Moores, S., *'Neutral on Our Side':...*, p. 198.

61 The digit 3 was kind of a prefix to the specific missile designation. Sometimes the RB 304E missile was called RB 04E, and RB 24 (or Sidewinder 1A) was sometimes called RB 324. Likewise, the Hawk missile was called RB 67 just as often as it was called RB 367. The RB 340 and RB 341 guided missiles could hence have been called RB 40 or RB 41. In the RSAFB's archive, all guided missiles were registered using the prefix 3.

radar homing device,[62] while the other (RB 341) would have had an infrared (henceforth: IR) homing device.[63]

In January 1959 (i.e. prior to the signing of the Sidewinder contract) the RSAFB wrote a letter to SAAB asking if the company was interested in manufacturing air-to-air missiles under license. This was not a question pertaining to just any missiles, but concerned the export versions of the Falcon missile (called HM-55 and HM-58) from the American manufacturer HAC.[64] The RSAFB immediately received a positive answer from SAAB,[65] and by December 1959 the RSAFB had decided to procure the Falcon missile. The State Department had given its blessings to a technical assistance agreement between the RSAFB and HAC, according to which HAC would provide Sweden with necessary basic data and specifications as well as furnish information regarding what type of modifications had to be made. [66]

When SAAB received their first order in March 1960,[67] representatives from SAAB, LM Ericsson, and Svenska Radio AB had already visited HAC and studied missiles and production processes on location in the United States. Moreover, the RSAFB had then signed a contract with HAC.[68] This did not mean, however, that, the deal was sealed. There were still some matters that had to be resolved before Sweden could be allowed to acquire the Falcon missile. The preconditions were spelled out in a memorandum for the Munitions Control Division at the State Department at the end of March 1960.

The memorandum stated that is was in the overall interest of the United States to furnish Sweden with operational weapons developed

62 This means that a radar in the aircraft is pointed at the target and that the reflected signal is received by a receiver in the missile's radome. *Active* radar homing is when both the transmitter and receiver are placed inside the missile.

63 Cerwin-Ellqvist, G and König, B., *Robot 50 år*, p. 20; see also: *SAABs robot- och elektronikhistoria, del 1*, "Jaktrobotsystem 340/341", 1:5:1. At low altitudes reflections from ground terrain could confuse the missile's homing device, that was why an IR-version was needed.

64 Cerwin-Ellqvist, G and König, B., *Robot 50 år*, p. 20; Taylor, John W. R. and Taylor, Michael J. H., *Guided missiles of the World* (London: Ian Allan, 1972), p. 34. It just so happened that the semi-active radar homing version of Falcon (designated AIM 4-A) was the first guided weapon of any kind that the USAF procured (Taylor, J. W. R. and Taylor, M. J. H., *Guided missiles of the World*, p. 86). Hughes Aircraft Company had begun work on it as early as 1947, and the first 4,000 production models appeared in 1954 (ibid., p. 33).

65 *SAABs robot- och elektronikhistoria, del 1*, "Jaktrobotsystem 27/28", 1:8:1.

66 Letter from Jarring to Herter, dated December 2, 1959, pp. 4–5, in: RG 59; CDF, 1960–63; from 758.56/1-1360 to 759.00 (W)/5-261; Box No. 1843; (Folder: 758.5622/12-660); NARA II, College Park, MD.

67 Cerwin-Ellqvist, G and König, B., *Robot 50 år*, p. 27.

68 *SAABs robot- och elektronikhistoria, del 1*, "Jaktrobotsystem 27/28", 1:8:1.

in the U.S. "to the greatest degree practicable." The Swedes had already been granted access to a considerable amount of information regarding such weapon systems; but with regard to guided missiles it was stated that the American inventory was not entirely satisfactory to Sweden. A memorandum stated that:

> Sweden does not consider that any of the U.S. developed missiles would be compatible with their aircraft and therefore are anxious to have the U.S. develop an all-weather radar guided and infra red homing air-to-air missile which would be suitable to their military needs and for which there is presently no U.S. military requirement.[69]

The Americans wished to meet the Swedish requirement, and the DOD suggested firstly, that a new security agreement would have to be negotiated with Sweden. Secondly, that the Department of the Air Force would be responsible for security and control with regard to the information and technology released from the United States to Sweden. Thirdly, that a technical assistance agreement should be created (the one already cleared with the State Department) which would provide Sweden with U.S. government-owned rights in components, assemblies, or other missile assistance as well as providing for reciprocity of Swedish-owned information and improvements made on the American designs. Any contract concluded between the Swedes and HAC would be subject to this security agreement, which would be designed to prohibit commercial firms from selling U.S. government secrets abroad. The memorandum made clear that the U.S. government would more or less dictate the terms of the contract signed by HAC and the RSAFB.[70]

A paragraph also spelled out what the consequences would be with regard to the NATO countries. It was concluded that any information that would be released to Sweden, as part of this agreement, would have to be offered to NATO first.[71] The United States had to tread carefully and respect the sentiments of the NATO allies if she was to be able to keep claiming to be the leader of the Western Bloc. Sweden could not

69 Draft Memorandum for the Munitions Control Division, dated March 29, 1960, p. 1, in: RG 59; EURONOE; RRE, 1957–1975; DEF 2-Gen. Repports [sic] and Statistics, 1966–1969 to POL 33-8 – Law of the Sea, 1960–1969; Lot File: 72D339; Box No. 2; (Folder: Folder: Defense Affairs Sweden; DEF 4 Collective Defense and Alliances, 1960–1966); NARA II, College Park, MD. For the security agreement, see also: Moores, S., *'Neutral on Our Side':...*, p. 199.

70 Draft Memorandum for the Munitions Control Division, dated March 29, 1960, pp. 1–3, in: RG 59; EURONOE; RRE, 1957–1975; DEF 2-Gen. Repports [sic] and Statistics, 1966–1969 to POL 33-8 – Law of the Sea, 1960–1969; Lot File: 72D339; Box No. 2; (Folder: Folder: Defense Affairs Sweden; DEF 4 Collective Defense and Alliances, 1960–1966); NARA II, College Park, MD.

71 Ibid., p. 3.

be given too preferential a treatment, it was thought, because this might cause the United States considerable embarrassment when facing their NATO allies. Where the line actually went seems to have been a matter of dispute, however.

The arrangement visualized in the memorandum quoted above became American policy in relation to Sweden on this issue. The Pentagon's decision dragged on due to a conflict between the Services that caused internal problems. The contentious issue seems to have been the U.S. policy that such information and hardware should be made available to NATO before it was given to Sweden, and that this had apparently not been done by late May, at which time the United States was in the midst of negotiating the new security agreement with Sweden.[72] This question needed to be addressed since the Swedes were the first ones to possibly be granted this technological know-how and hardware, and it had to be sorted out before the U.S. government could give its approval. The manner in which it was resolved provides an excellent example of the plasticity of U.S. hegemonic policy.

A conference regarding the impending Falcon affair was held in the Pentagon on March 30, 1960. The issues that were up for discussion were: whether the participation of the U.S. government in this affair would constitute grant aid under the Mutual Security Act of 1954, as amended; what governmental agreements would be necessary to implement the proposed transaction; what benefits should accrue to the United States for granting the contract between HAC and the RSAFB; and, finally, which steps had to be taken "to carry out the United States policy that defense information should not be made available to countries outside NATO unless NATO allies also had it available to them."[73]

The flexibility of American policy becomes very clear when we look at the handling of the first question, i.e. whether the involvement of the U.S. government constituted grant aid. It was concluded that it did not, since it would be a transaction of money for hardware and information, which formally belonged to HAC. However, the memorandum from the conference revealed that the affair would involve the transfer of technology and know-how with a classification category that at the time was still to be denied to Sweden. This was why HAC was seeking the consent of the American government in the first place. The transaction, it was said, would:

72 Memorandum to Willoughby for Paris Ambassadors' Meeting May 23–26, 1960, dated May 19, 1960, p. 2, in: RG 59; EURONOE; RRE, 1957–1975; DEF 2-Gen. Repports [sic] and Statistics, 1966–1969 to POL 33-8 – Law of the Sea, 1960–1969; Lot File: 72D339; Box No. 2; (Folder: POL-1 General Policy and Background, 1963–1969); NARA II, College Park, MD.

73 Memorandum for the File, dated March 31, 1960, p. 1, in: RG 59; CDF, 1960–63; From: 758.00(w)/6-862 To: 758.5/5-660; Box No. 1842; (Folder: 758.5/5-660); NARA II, College Park, MD.

> […] require a defense-economic benefit in exchange for *a limited exception from our security control* of the rights and information.[74] [Italics added.]

The Americans were contemplating releasing not only technology that had not yet been made available to the NATO allies, but at the same time, as implied by the quote above, materiel that should really be denied Sweden due to its classification categorization. With regard to what benefits the American government should be granted in order to approve of the affair it was decided that an annex would be added to the basic contract, which, in brief, would provide *inter alia*:

> […] that Hughes and KFF [RSAFB] would deliver to the United States, upon request, a copy of all documents and other forms of evidence of data and information for its governmental purposes, *including delivery to other governments for mutual defense*; that Hughes and KFF grants royalty-free license to practice any invention developed by Hughes in implementing the contract or by KFF in improving the devices covered by the contract […].[75] [Italics added.]

The italicized phrase above corresponds almost exactly to the part in the MDAA agreement of 1952 that prescribed that material furnished to Sweden should be made available to the other signatories as well. The final technical assistance agreement did not contain these exact words, but the spirit of the paragraph quoted above was included (see below) The U.S. government was seeking to maximize the profits accruing from the Falcon affair. But the episode also shows that the Americans were prepared to sidestep their own regulations if the U.S. government would benefit from it.

The problem of how to make this information available to NATO remained, however. The difficulty was to identify the information which was to be offered to NATO, as well as the manner in which it could be accomplished. It was agreed that the NATO countries did not have to be granted access to the missile system itself, and it was proposed that the United States would change its classification category with regard to NATO "so that the necessary information could be released to them if they wished to make arrangements with the private owner or controller of such rights for use in production." The offer to NATO was to be made by describing the items involved.[76]

The DOD had presented the Swedes with a proposal for a security agreement in March 1960, at which time it was made clear that such

74 Ibid., p. 1.

75 Ibid., p. 2.

76 Ibid., p. 2.

an agreement was an absolute precondition for the U.S. government to transfer classified data to Sweden. The Swedish Ambassador Gunnar Jarring included the American draft in his letter to Foreign Minister Undén.[77] In late June 1960 Jarring asked Assistant Secretary for European Affairs Kohler about the present status of the Falcon missiles, and whether the new security agreement currently being negotiated between the U.S. and Swedish governments represented an enlargement of the exchange of notes agreement of 1952. Kohler, however, replied that the negotiations were moving forward and that results that could lead to the production of the Falcon missiles in Sweden might be expected in a few weeks.[78]

The draft texts of the security agreement, the technical assistance agreement, and an agreement for the adequate protection of patents, were included in a memorandum from William C. Burdett at BNA to Foy Kohler in early October 1960, preparing him for a meeting with Swedish Ambassador Jarring the same day. Jarring was coming to discuss the agreements and Kohler had been authorized to negotiate the agreements for the United States by Under Secretary of State Hare. The draft for the technical cooperation assistance agreement, was rather out of the ordinary, and Article II in particular deserves an extensive quote:

> The Government of Sweden shall, in accordance with detailed arrangements to be entered into by authorized representatives of the respective Governments, cause to be promptly furnished to the Government of the United States of America such information and data as may be a consequence of production, maintenance, repair, test, research, or development carried out under any contract authorized under Article I, and subcontracts thereunder. The scope of information and data required by the preceding sentence shall be limited to designs, changes, developments, modifications, and improvements in weapons or weapon systems, their components, assemblies, or end items or derivatives thereof, or in the production, maintenance, and repair thereof, developed by the Government of Sweden, prime contractors and any subcontractors, or their employees, whether such information and data are reported in writings, sound recordings, drawings, photographs, models, or any other form of recording or graphic reproduction. *Such information and data shall be furnished*

77 Letter (with attachment) from Jarring to Undén, dated March 31, 1960, in: HP 24 VI/U.S.A; "Tillstånd till in- och utförsel av vapen, ammunition och krigsmateriel"; 1960 mars–1960 oktober; Folder 25; Box No. 63; År: 1960–1962; RIA; Arninge.

78 758.5612/6-2960, Memorandum of Conversation between Jarring, Kohler, and London, dated June 29, 1960, p. 1, in: RG 59; CDF, 1960–63; from 758.56/1-1360 to 759.00 (W)/5-261; Box No. 1843; (Folder: 758.5622/12-660); NARA II, College Park, MD.

without cost to the Government of the United States of America except for reasonable charges for reproduction, handling, mailing, and other similar administrative expense. [...] The Government of Sweden shall to the extent it may now or hereafter lawfully do so, grant to the Government of the United States of America an irrevocable, nonexclusive, nontransferable, and royalty-free license to: a. use, and cause to be used, such information and data by or for the Government of the United States of America *for its defense purposes throughout the world as determined by the Government of the United States of America*; and b. *practice and cause to be practiced by or for the Government of the United States of America for its defense purposes throughout the world* as determined by the Government of the United States of America, all inventions, improvements, and discoveries made in the course of production, maintenance, repair, test, research, or development carried out under any contract authorized under Article I and subcontracts thereunder.[79] [Italics added.]

With this agreement, the United States wanted to make sure that *all* improvements that might follow from the transaction would be made available to the U.S. government free of charge. Any kind of information, in any form available, was to be turned over to the United States, seemingly without prior request. This information, hardware, or data could be used freely by the United States for its defense purposes throughout the world. The Swedish government would have no say whatsoever with regard to this issue, because this was an exclusive hegemonic prerogative. This effectively meant that the Swedish government could become a supplier of American defense materiel to be used anywhere in the world.

The United States would be able to choose for itself how and where it wished to put the weapons produced as a consequence of the technical cooperation assistance agreement to use. Moreover, what was referred to was not the array of results derived from the often ambiguous and nebulous process called *basic research*. What was talked about here was improvements on already existing weapons systems such as guided missiles, which could be more or less immediately and successfully utilized by the U.S. Armed Forces. This fact was made even more evident in Article III, which said that:

At the request of the Government of the United States of America, the Government of Sweden shall furnish, or cause to be furnished, to

79 "Agreement Between the Government of the United States of America and the Government of Sweden for Technical Cooperation Assistance", pp. 1–2, attached to Memorandum From Burdett to Kohler, dated October 7, 1960, in: RG 59; EURONOE; RRE, 1957–1975; DEF 2-Gen. Repports [sic] and Statistics, 1966–1969 to POL 33-8 – Law of the Sea, 1960–1969; Lot File: 72D339; Box No. 2; (Folder: DEF-7 Visits and Missions, 1959–1969); NARA II, College Park, MD.

> the Government of the United States of America, at charges no greater than those made to the Government of Sweden for its own purposes, prototypes, and related spare parts, of any weapons or weapons system, their components, assemblies, or and end items or derivatives coming within the scope of Article II. Utilization, operation and maintenance manuals, detailed manufacturing drawings, specifications, bills of material, photographs, and or other similar documentation relating to such prototypes and spare parts shall be furnished without cost, except for reasonable charges for reproduction, handling, mailing, and other similar administrative expenses.[80]

According to Thomas Jonter, Sverker Åström at the Foreign Ministry wrote to Ambassador Jarring in late September and told him that the American way of handling the matter was not acceptable to Sweden. The American wish that the new security agreement should replace the old MDAA agreement of 1952 was not palatable to the Swedish government, said Åström. The new agreement should instead be complementary to the one from 1952.[81] The agreement that Åström was most upset about was the one concerning technical cooperation quoted above. While acknowledging that Sweden had agreed to the conclusion of a general security agreement, Åström said that an addition to this agreement regarding technical cooperation *on a governmental level* was unacceptable for Sweden for political reasons. He also stated that the technical cooperation agreement seemed to be of the type which was common between allied states, somthing which added to the unacceptability of the proposal. But the Swedish government's protestations were only directed towards the proposal that the agreement should be signed by the two governments. Therefore Åström added that Jarring should point out to the Americans that Sweden would not object if the content of the American draft was to be part of individual agreements between the Swedish services (such as the RSAFB) and their American business partners.[82] Åström wrote:

> On the other hand—and you should stress this point—we cannot for obvious reasons have any objections in principle to the stipulations in the draft, where applicable and when the parties have agreed on them,

80 Ibid., p. 3.

81 Jonter, T., "Det amerikanska spåret...", p. 174. Jonter wrote that Åström had written to Jarring in early August, but that is incorrect. The mistake was due to another date typed on the document.

82 Letter from Åström to Jarring, dated September 28, 1960, pp. 1–2, and 4, in: HP 24 VI/U.S.A; "Tillstånd till in- och utförsel av vapen, ammunition och krigsmateriel"; 1960 mars–1960 oktober; Folder 25; Box No. 63; År: 1960–1962; RIA; Arninge.

being a part of individual business agreements between our service authorities and their American contractual partners.[83]

From Åström's comments on the draft for the technical cooperation agreement it is evident that he did not, at least not at this point, object to the stipulations having to do with the right of the United States to use such information and data for its defense purposes.[84] Åström thus left the door open for another solution to the issue, and this would eventually result in an agrement being made (see below). As long as the government could claim that it had not signed something which might compromise its policy of neutrality, everything was in order. It mattered little that the result of the agreement was the same either way. Political and geostrategic realities demanded that the government should play this game. The Americans did not object, since they got what they wanted from the arrangement. To the U.S. government it did not matter whether the results generated by the technical cooperation came into its possession formally via the hands of the Swedish government or the Swedish military. Once again, form mattered more than content for the Swedes.

It also appears as if the Swedish military did not entirely understand the importance of a new security agreement to the Americans. An example of this was provided in November 1960 when Major General Thunberg, Vice Chief of the RSAFB, sent a new priority list covering the information and hardware that the Swedes considered it necessary to have in order for the Swedish manufacturers to meet their deadlines without considerable delays to his American counterpart Major General W. T. Thurman. Thunberg acknowledged the fact that the new security agreements had not yet been signed by the two governments, but he also expressed his confidence that they soon would be, and hoped that the Swedes could be allowed to obtain the information and hardware sought after anyway.[85] However, this posed a problem to the Americans in their handling of the case.

83 Ibid., p. 4. "Däremot – och detta bör Du framhålla – kan vi uppenbarligen icke ha några principiella erinringar mot att bestämmelserna i avtalsutkastet, i tillämpliga delar och när parterna därom enas, ingår i enskilda affärsuppgörelser mellan våra förvaltningsmyndigheter och deras amerikanska avtalspartners."

84 "Agreement Between the Government of the United States of America and the Government of Sweden for Technical Cooperation Assistance", dated August 5, 1960, pp. 1–2, in: HP 24 VI/U.S.A; "Tillstånd till in- och utförsel av vapen, ammunition och krigsmateriel"; 1960 mars–1960 oktober; Folder 25; Box No. 63; År: 1960–1962; RIA; Arninge.

85 758.5622/12-860, Letter from Thunberg to Thurman, dated November 27, 1960, pp. 1–2, in: RG 59; CDF, 1960–63; from 758.56/1-1360 to 759.00 (W)/5-261; Box No. 1843; (Folder: 758.5622/12-660); NARA II, College Park, MD.

The Assistant Chief of the DOD's Special Projects & Sales Division George A. Chadwick wrote to William B. Cobb, Jr. at the State Department in December, saying that:

> [i]t is becoming increasingly difficult to administer the United States portion of this project because of the type and classification of the information being requested [...]. Much of the material and data requested [...] will have to be withheld until the agreements with the Swedish Government are signed. It is therefore requested that expeditious action be taken to complete the agreements between the United States and some authorized agency of the Swedish Government.[86]

The security agreement was deemed the most urgent to conclude, since the transfer of the Falcon missile was dependent upon this. There has, however, been some confusion among scholars as to what exactly happened next. It has been said that the Swedish government eventually retreated, and that it accepted the American proposal in December 1960, and that the security agreement was signed on January 30, 1961.[87] From this, it seemed that the exchange covered all three agreements, and that it superseded the one from 1952. However, this exchange only concerned one of the three different agreements that the Americans wanted to conclude with Sweden, i.e. the security agreement. Contrary to what Jonter seemed to suggest, the security agreement did not replace the MDAA agreement, but took the form that Åström and the Swedish government had wished for all along. It was an addition to the old agreement, stating that it had been found desirable to widen the scope of the latter to cover security measures regarding all information, equipment, materiel, or services which had been given a security classification by either government.[88] The Americans, moreover, publicly referred to the new agreement as being a part of a mutual defense assistance effort,[89] something which must have irritated the Swedes somewhat.

86 758.5622/12-860, Memorandum from Chadwick to Cobb, Jr., dated December 8, 1960, pp. 1–2, in: RG 59; CDF, 1960–63; from 758.56/1-1360 to 759.00 (W)/5-261; Box No. 1843; (Folder: 758.5622/12-660); NARA II, College Park, MD.

87 Jonter, T., "Det amerikanska spåret...", pp. 174–177. See also: Moores, S., *'Neutral on Our Side':...*, p. 199. Moores based his account on Jonter's article. It could be mentioned that these agreements were not the focus of Jonter's article.

88 SÖ 1961:51; *Had There Been A War...*, p. 124. Jonter confirmed to the author in a telephone conversation on July 30, 2007 that this was what he was aiming to say in his article "Det amerikanska spåret...", but that this point got lost.

89 "Mutual Defense Assistance: Security of Information, Equipment, Materials, or Services. Agreement Between the UNITED STATES OF AMERICA and SWEDEN", in: HP 24 VI/U.S.A; "Tillstånd till in- och utförsel av vapen, ammunition och

The Memorandum of Understanding of 1961

The Americans had not dropped the idea of a technical cooperation agreement with Sweden just because the security agreement had been signed, and negotiations resumed soon thereafter. The Americans picked up on the Swedish government's wish to conclude the technical cooperation assistance agreement on a non-governmental level. The technical cooperation agreement had now been turned into a Memorandum of Understanding (henceforth: MOU), and the Swedish Ministry of Defense had been replaced as a party to the agreement by the RSAFB.[90] An American delegation headed by a Mr. Hero visited Sweden in the summer of 1961 to discuss the MOU, an agreement on security procedures implementing the newly concluded security agreement, as well as a patent agreement.[91] The result of the conference, which was held between June 8 and June 12, was that an agreement was reached on the formulation and signing of a Memorandum of Understanding as well as implementing procedures for the general security agreement.[92]

On August 7, 1961, the Vice Chief of the RSAFB Major General Lage Thunberg signed the MOU (henceforth: MOU 1961), and assured the Americans that the security procedures agreed upon would be dutifully implemented by Sweden. The content of the MOU 1961 was essentially the same as the draft for the technical cooperation assistance agreement. It still gave the United States the right of any information, improvement etc., that was the result of any contract concluded between the RSAFB and an American company or government agency. The paragraphs stating that the United States could freely use these results for its own defense purposes were still included, although the part 'throughout the world' had been taken out.[93] These sections now read:

---

krigsmateriel"; 1960 nov–maj 1961; Folder 26; Box No. 63; År: 1960–1962; RiA; Arninge.

90 Draft for "Memorandum of Understanding Between the United States Department of Defense and the Royal Swedish Air Force Board, an Agency of the Government of Sweden…", dated April 5, 1961, pp. 1–5, in: HP 24 VI/U.S.A; "Tillstånd till in- och utförsel av vapen, ammunition och krigsmateriel"; 1960 nov-maj 1961; Folder 26; Box No. 63; År: 1960–1962; RiA; Arninge.

91 Letter from Åström to Jarring, dated June 14, 1961, pp. 1–4, in: HP 24 VI/U.S.A; "Tillstånd till in- och utförsel av vapen, ammunition och krigsmateriel"; 1961 juni–juni 1962; Folder 27; Box No. 63; År: 1960–1962; RiA; Arninge. For the implementation of the security agreement, see also: Jonter, T., "Det amerikanska spåret…", p. 175.

92 Letter from Roswell L. Gilpatric to Lage Thunberg, dated July 25, 1961, in: HP 24 VI/U.S.A; "Tillstånd till in- och utförsel av vapen, ammunition och krigsmateriel"; 1961 juni–juni 1962; Folder 27; Box No. 63; År: 1960–1962; RiA; Arninge.

93 Letter from Thunberg to Gilpatric, dated August 7, 1961; "Memorandum of Understanding Between the United States Department of Defense and the Royal Swedish Air Force Board, an Agency of the Government of Sweden", dated August 7, 1961, pp. 1–5, in: HP 24 VI/U.S.A; "Tillstånd till in- och utförsel

> [...] a. use, and cause to be used, such information and data by or for the United States for defense purposes as determined by the United States; and b. practice and cause to be practiced by or for the United States for defense purposes as determined by the United States, all inventions, improvements, and discoveries made in the course of production, maintenance, repair, test, research, or development carried out under any authorized contracts [...].[94]

This was acceptable to the Swedes since the Swedish government was not technically a signatory, even though the effect was the same, i.e. that Sweden (through the RSAFB) became a part of the U.S. supply chain. The agreement was not limited to the Falcon affair, but was limited to matters handled by the RSAFB. This meant that all contracts concluded with between the RSAFB and an American counterpart were subject to the stipulations of the MOU 1961. The existence of this memorandum has previously gone unnoticed by scholars. The way in which this matter was handled was quite remarkable, because it so clearly revealed the rationale behind the Swedish government's policy of neutrality. Åström had urged Jarring to tell the Americans that even though the MOU 1961 could not be concluded on the governmental level, it could be concluded on a lower level. This is something different from not wanting to put anything in writing, due to a will not to be bound by formal agreements. The MOU 1961 *was* a formal agreement, and the Swedish government did not have a problem with that, as long as a government representative did not put his signature on it. This was repeated almost exactly a year later with the signing of the "Memorandum of Understanding Concerning Technical Information" (henceforth: MOU 1962).

The MOU 1961 differed considerably both in form and content from the security and (as we shall see below) patent agreements. While the latter two were turned into exchanges of notes, the MOU 1961 took the form of a treaty, signed by both parties. Too, the final versions of the security and patent agreements were heavily edited by the Swedish government, and differed to a considerable extent from the draft versions. That was not the case with MOU 1961. It remained true to the draft version, with only barely noticable modifications. The reason for these discrepancies was that the MOU 1961 was concluded at the Service level, while the

---

av vapen, ammunition och krigsmateriel"; 1961 juni–juni 1962; Folder 27; Box No. 63; År: 1960–1962; RIA; Arninge.

94 "Memorandum of Understanding Between the United States Department of Defense and the Royal Swedish Air Force Board, an Agency of the Government of Sweden", dated August 7, 1961, pp. 2–3, in: HP 24 VI/U.S.A; "Tillstånd till in- och utförsel av vapen, ammunition och krigsmateriel"; 1961 juni–juni 1962; Folder 27; Box No. 63; År: 1960–1962; RIA; Arninge.

other two were concluded at the government level. Governmental agreements had to be made public, and it was hence very important that the language in them did not openly contradict official policy or in any way limit the government's freedom of action. This never became an issue with regard to the MOU 1961 which remained secret. How come the MOU 1961 was not also modified, and turned into an exchange of notes between the two governments? The answer probably lies with the Americans as much as with the Swedes. Although we do not know for sure, an educated guess would be that the U.S. government would not compromise on its right to make use of any improvements, and since this prerogative could not be masked in an exchange of notes on the governmental level, the Swedish government prefered to play the neutrality card and sit this one out.

The precise consequences of the MOU 1961 are not known, since no research has been done on the matter. Did the U.S. ever demand to get acces to equipment or information regarding Swedish improvements to American designs? Were such improved weapon systems, designs etc., ever used by the U.S. or its allies in combat (in South East Asia for instance)? Did the Russians know about the MOU 1961? We just do not know at this time. The content of the MOU 1961 has never been interpreted in this way prior to this study. More correctly, it has never been interpreted at all, perhaps because no one has actually read what was in it. For example, Jonter, who wrote about the three agreements in the report of the Intelligence Committee (Säkerhetstjänstkommissionen), did not refer to the original documents.[95]

With the conclusion of the MOU 1961 the Swedish government confirmed its commitment to U.S. hegemonic policy, without having to take public responsibility for the consequences. Officially, Sweden was non-aligned in peacetime to be able to declare herself neutral in time of war. Off the record, however, the government was treading deeper and deeper into a quasi-alliance quagmire.

Production Begins

Once the security agreement had been agreed upon Sweden could start its license production of the Falcon missiles. The Swedish names for Falcon HM-55 and HM-58 became RB 27 and RB 28 respectively, and as I mentioned above, the homing devices in RB 340 and RB 341 were of some importance to this story. The reason for this is that RB 27 had a semi-active radar homing device (just like RB 340), and that RB 28 had an IR homing device (just like RB 341).[96] The Swedish missiles were hence essentially replicas of the two American Falcon versions, and considering what has been said above it is not unreasonable to assume

95 Jonter, T., "Det amerikanska spåret…", pp. 173–175.

96 Cerwin-Ellqvist, G and König, B., *Robot 50 år*, pp. 27–28.

that the RSAFB never intended RB 340 and RB 341 to become operational. Instead, it seems as if these systems were dreamed up in order to prepare the engineers at SAAB for the license production of Falcon.

SAAB's work on RB 340 and RB 341 was subsequently cancelled in November 1960.[97] It seems, however, as if SAAB had not been kept in the loop about the RSAFB's intentions in ordering RB 340 and RB 341, for in October 1960, i.e. one month before the project was cancelled, SAAB produced a short presentation of the RB 340/341 missile system for the RSAFB. The reason was to make the higher echelons of the RSAF aware of the system's capabilities, allegedly because "SAAB had gained knowledge of the Air Force's interest in the Falcon missiles."[98] The presentation was nothing less than a sales pitch for SAAB's system; among other things were mentioned possible future developments such as a surface-to-air and/or an air-launched anti-ship missile.[99] Among the potential offspring of the RB 340/341 project was a surface-to-air missile called RB 345, and a plan for such a missile was actually drawn up. It would essentially have been an RB 340 with an additional boost engine,[100] but nothing ever became of it.[101] A study was also made for basing two surface-to-air missiles, which were to be entirely funded by SAAB and designated RB 361 and RB 362, on RB 340 and RB 341. A number of possible solutions were presented, the aim being a system which did not require either high development costs or a long development time. It was to be a light missile system for use against low-altitude targets that could be easily incorporated into the current anti-aircraft forces in the RSA, RSAF and RSN. Not surprisingly, nothing ever became of RB 361 or RB 362 either.[102]

Considering that the license order had been placed in March 1960, and that SAAB had been asked about its interest in license manufacturing Falcon back in January 1959, it seems a bit odd that SAAB did not realize that the RSAF wanted the Falcon missiles instead of SAAB's missiles. Nevertheless, it remains the only plausible explanations of why SAAB produced this report. Apparently, SAAB thought that there was still a chance to salvage RB 340 and RB 341. There was certainly cause for the company to be concerned over a cancellation. Revenues would be lost so it could be imagined that avoiding this prospect would have seemed

97 Ibid., p. 20.

98 *saabs robot- och elektronikhistoria, del 1*, "Jaktrobotsystem 340/341", 1:5:1. "Saab fått kännedom om Flygvapnets intresse för Falcon-robotarna.

99 Ibid, 1:5:2.

100 The RB 345 was also called Aurora (see: "NR H 117", "Orientering angående sjuårsplanen 1959/66 för anskaffning av flygmateriel.", in: Hgkv, FS/KH, PA, "Inkomna och utgående skrivelser", Serie: F III, VOL. 1, År: 1955–59, KrA, p. 3).

101 *saabs robot- och elektronikhistoria, del 1*, "Jaktrobotsystem 340/341", 1:5:14.

102 *saabs robot- och elektronikhistoria, del 1*, "Jaktrobotsystem 27/28", 1:8:11.

desirable to them. Why would the RSAFB keep its contractor in the dark like that? It does not seem likely that the RSAFB did not know what they were going to do until November 1960. Guided missiles were expensive, and the RSAFB had already spent more than SEK 30 million on RB 340/341,[103] so it must be taken for granted that the people calling the shots in the RSAFB knew that they could not afford to purchase the production rights to the Falcon missile *and* pay for the production of RB 340/341 at the same time.

The story becomes even more intriguing if we consider the fact that HAC, the producer of Falcon, also took part in the production of subsystems for the Swedish guided missiles RB 340/41. In the GMB's archive there is evidence of this taking place already in early 1959, although it seems as if this collaboration, at least initially, was intended to be paralleling the possible manufacturing of Falcon at SAAB. Here we find out that the RSAFB was interested in commencing cooperation with HAC, which was supposed to act as a consultant on the development of RB 340.[104]

Shortly afterwards, SAAB sent an offer to four U.S. firms querying if they would be interested in manufacturing a power-supply system for this missile.[105] Apparently, it was SAAB who had found the design specifications for the original power-supply system unsatisfactory. Through their inquiries in the United States they settled for a design for a so-called auxiliary power unit proposed by the Airesearch Manufacturing Co., Division of the Garrett Corporation. Torsten Rapp and the RSAFB were anxious to support SAAB on this matter.[106] It seems as if SAAB, at this time, did not have any information about the forthcoming cancellation of their RB 340 and RB 341 program. As late as in May, 1959, they said in a letter to the RSAFB that the collaboration with HAC would not affect "prior agreements regarding responsibility and work distribution".[107]

103 As a comparison it could be noted that the cost of the development of the two Falcon guided missiles in the United States had been between $80 and $130 million, see: Dörfer, I., *System 37 Viggen…*, p. 242.

104 RHM 44-340:32, in KFF, RBN, Serie: F III B, "Inkomna och avgångna skrivelser ordnade efter klassifikationssystem", M 44-340:17-M 44-365:20 , år: 1959, VOL. 45.

105 RHM 44-340:68, in KFF, RBN, Serie: F III B, "Inkomna och avgångna skrivelser ordnade efter klassifikationssystem", M 44-340:17-M 44-365:20 , år: 1959, VOL. 45.

106 "RHM 44-340:57", in: KFF, RBN, Serie: F III B, "Inkomna och avgångna skrivelser ordnade efter klassifikationssystem", M 44-340:17-M 44-365:20, År: 1959, VOL. 45. KrA.

107 "RHM 44-340:55", in: KFF, RBN, Serie: F III B, "Inkomna och avgångna skrivelser ordnade efter klassifikationssystem", M 44-340:17-M 44-365:20, År: 1959, VOL. 45. KrA. "tidigare överenskommelser beträffande ansvar och arbetsfördelning".

There were already other American companies involved in the early development of the Swedish guided missiles too. In order to be able to construct an IR homing device for RB 341 SAAB needed to gain knowledge about such devices and about infrared technology in general. There were four areas where such knowledge was extremely pertinent: IR radiation from possible targets and backgrounds, IR detectors and their cooling, IR materials and their optics, utility-adapted systemic work on detection and signal processing. In addition there was more traditional technology such as mechanics, electronics, and servo technology where they needed more knowledge. For the purpose of studying infrared radiation SAAB purchased measuring equipment from Barnes Engineering in the United States, a company which remained a frequent distributor of specialized equipment for SAAB during the late 1950s and early 1960s.[108]

Finally, on September 23, 1959, the head of the RSAFB Torsten Rapp formally approved the cooperation with HAC. At the same time the cooperation was also approved by the USAF, which made an official statement to the effect that contract proposals were to be sent to the State Department.[109] However, a Technical Assistance Agreement between the RSAFB, its subcontractors SAAB and LM Ericsson, and HAC had already been made a couple of months before this. In this agreement the Swedes expressed their wish to obtain technical assistance from the Americans in the design and development of the B-3 weapon system for the F-version of J35 Draken.[110] The interesting thing about this document is that it specifically stated that the Falcon missiles would be the main armaments on J35 Draken, and it also said that it was a continuation of a preceding agreement dated May 18, 1959.[111]

This means that the RSAFB no later than September 23, 1959, and maybe as early as May 18 that year, had sealed the fate of the RB 340 and RB 341 missiles. From this point on, the cancellation of the domestic systems in November 1960 (including the work done by SAAB) was a mere formality. This gives us ample reason to question the stated reason for the RSAFB's interest in collaborating with HAC on the development of RB 340 and RB 341 in early 1959. Had the RSAFB already made the decision to cancel the domestic development when it had SAAB query

108 *SAABs robot- och elektronikhistoria, del 6*, "IR-verksamheten vid SAAB", 6:3:1.

109 Letter from Torsten Rapp to Robotbyrån dated September 23, 1959. RHM 44-340:97, in KFF, RBN, Serie: F III B, "Inkomna och avgångna skrivelser ordnade efter klassifikations-system", M 44-340:17-M 44-365:20 , år: 1959, VOL. 45.

110 Kontrakt NR 129 (INK H 94802) "Extension Technical Assistance Agreement", p. 1, in KFF, 1959–1960, Serie: F III, "Hemliga avtal och kontrakt", VOL. 4, KrA.

111 Ibid., p. 1; see also: Kontrakt NR 129 (INK H 94802) "Statement of Work", p. 1, in KFF, 1959–1960, Serie: F III, "Hemliga avtal och kontrakt", VOL. 4, KrA. The Agreement from May 18 is unfortunately nowhere to be found in the GMB's archive.

the American firms about their interest in manufacturing a power-supply system for RB 340? I have not been able answer this question in this thesis, but there is ample reason to assume that it had. As things turned out, however, SAAB's indignation was probably appeased by the fact that it got to produce Falcon under license, and hence got a piece of the pie any way. But as I stated above, the company forfeited future royalty revenues entitled to them if the Swedish systems would have been exported.[112]

Under the initial Falcon agreement HAC would furnish the RSAFB with information, based on which negotiations concerning definitive, detailed contracts and licenses could be conducted. The fact is that HAC explicitly stated that, under normal circumstances they would not, for only $26,215, furnish *any* buyer with the type of technical information and services now supplied to SAAB. The company was only willing to do this because of the fact that the RSAFB was acting on behalf of the Swedish government, which reassured HAC that the information would be used only as a basis for more specified contracts and licenses. Hence it seems as if the Swedish government was the only part in the cooperation that distinguished between the government and the RSAFB. The agreement explicitly prohibited the RSAFB (and its subcontractors) from spreading and/or utilizing this information for the purpose of production or assembly prior to that time when contracts had been entered into with the American firm.[113] The consequence of this was of course that the Swedes were prohibited from using this knowledge in the production of RB 340 and RB 341, or any other domestic missile system. An amendment spelled out what the final contracts would entail:

112 The Swedish guided missiles would probably have been exported to Denmark, Finland and Austria, since these were the countries that purchased SAAB 35 Draken later on. Finland bought Falcon, too, at this time. SAAB had to buy and restore RB 27s from the United States, the RB 28s were, however, new, manufactured by SAAB (*saabs robot- och elektronikhistoria, del 1*, "Jaktrobotsystem 27/28", 1:8:11). Marcus Wallenberg, the Director of SEB, and whose family controlled SAAB, had personally negotiated the deal with Finland's President Kekkonen, Prime Minister Sukselainen, and Defense Minister in the spring of 1961. The Swedish government was willing to approve export licenses on a commercial basis. This was a decision that probably made Wallenberg very happy considering that the affair was worth SEK 150,000,000 (*Östen Undén. Anteckningar 1952–1966*, pp. 632–633). For more on the Draken system, see: Widfeldt, Bo, *Saab 35 Draken. En historisk återblick på projektering, utveckling, tillverkning och tjänsten inom de svenska, finska, danska och österrikiska flygvapnen* (Nässjö: Air Historic Research, 1995).

113 Kontrakt NR 129 (INK H 94802) "Extension Technical Assistance Agreement", pp. 1–2, in KFF, 1959–1960, Serie: F III, "Hemliga avtal och kontrakt", VOL. 4, KrA.

"These definitive contracts will be for the purchase of equipment from HAC, licensed production of HAC equipment [the Falcon missiles], engineering support and technical assistance [...]."[114]

In this document, which had been signed a month before Torsten Rapp formally approved the collaboration between SAAB and HAC, SAAB's participation was predetermined. The final contracts were entered into on February 27, 1961,[115] i.e. not even month after the security agreement had been concluded. Even though SAAB was not the only manufacturer involved in the license production of Falcon it was put in charge of the whole manufacturing process, being responsible for organizational, technical, and economic issues and co-planning together with the other manufacturers. The company was thus responsible for the finished product and was the only company that the RSAFB dealt with directly.[116]

On January 1, 1962, Major Åke Sandbladh at RFN, the guided missile testing site in the north of Sweden, filed a report regarding a visit to the United States made between October 17 and November 7, 1961. The aim of this journey had been to study target- and testing ranges, prior to the testing of RB 327 and RB 328 in Sweden.[117] Sandbladh visited both Holloman Air Force Base, Alamogordo, New Mexico, and the White Sands Missile Range, El Paso. He studied various target drones, such as the Australian Jindivik, the American Ryan Q2-C (also called Firebee), and the French CT 20. The properties of these drones made them interchangeable, but it seems as if the American Firebee drew the longest straw with the Swede. Sandbladh was so impressed by the Firebee that it was given a separate chapter in his report. It was all a matter of identifying the most suitable target for the testing of Falcon in Sweden, and the Swedish Major probed the drone inside out. He studied the fuselage, engine, and all other equipment inside it, as well as the equipment on the ground. He looked at launching and landing methodology, including the equipment and facilities needed for this purpose, and for reparations and pre-flight checks. He also studied the organization of the testing ranges, and security aspects at the sites.[118]

114 Kontrakt NR 129 (INK H 94802) "Statement of Work", p. 1, in KFF, 1959–1960, Serie: F III, "Hemliga avtal och kontrakt", VOL. 4, KrA.

115 Kontrakt NR 141 (INK H 14868), 142 (INK H 14869), 143 (INK H 14870) and 144 (INK H 14871) in KFF, 1960–1961, Serie: F III, "Hemliga avtal och kontrakt", VOL. 5, KrA.

116 *saabs robot- och elektronikhistoria, del 1*, "Jaktrobotsystem 27/28", 1:8:1–2.

117 "FCHA 373:3", in KFF, Robotavdelningens försöksplats i Vidsel [henceforth: Ravdvid], Serie: F I, "Avgående och inkommande handlingar HA 1-HA 8", 1962, VOL. 1, KrA.

118 "FCHA 373:3", "Bilaga 0", pp. 1–2, in KFF, Ravdvid, Serie: F I, "Avgående och

The first tests of the license-produced Falcon actually took place in the United States at the Holloman Air Force Base. Nine RB 27s and six RB 28s were fired from an American Convair F-102 equipped with the M6-10 Aircraft and Weapon Control System, slightly modified to be more like the B-3 weapon system intended for the Swedish J35 F Draken aircraft. The results from these tests were considered to be "remarkably good", and they would later serve as a comparison for the upcoming tests in Sweden.[119] The protracted domestic tests were carried out at RFN between June 1963 and July 1965, and this time the missiles were launched from a J35 F Draken aircraft equipped with the new B-3 weapon system.[120] A total of twenty-five RB 27s were fired, of which only two had live warheads.[121] Twenty-two of these passed the target close enough to destroy it, while three missed their target, and the tests were considered to be very successful. Twenty RB 28s were also fired, of which only one had a live warhead. There were some initial problems, however, and the first three missiles failed. But once the problem had been identified the following seventeen missiles performed satisfactorily.[122]

However, while the tests at higher altitudes went well, the performance of RB 28 at low altitudes was not satisfactory to the RSAFB. Discussions between the RSAFB and SAAB ensued, and the RSAFB suggested that if the underperformance could not be mended a new guided missile based on the discontinued RB 340 with the addition of a doppler type radar homing device, ought to be contemplated. This missile, which was to be called RB 342, was intended for the 37 Viggen aircraft that was under development, but the idea was soon dropped and not much work was ever performed on RB 342.[123]

---

inkommande handlingar HA 1-HA 8", 1962, VOL. 1, KrA.

119 *saabs robot- och elektronikhistoria, del 1*, "Jaktrobotsystem 27/28", 1:8:5. "anmärkningsvärt gott".

120 Ibid., 1:8:5.

121 Ibid., 1:8:6. The warhead in RB 27 was of an unconventional type, a so-called *continuous-rod*. The proximity fuze triggers an explosion of 6.5 kilos of TNT (trinitrotoluene) in the warhead, which then releases an expanding ring of 204 long steel rods. These rods are welded together at alternate ends forming a zigzag pattern, and they are wrapped lengthwise around the explosives. This ring becomes approximately 18 meters in diameter, until it eventually breaks up into individual fragments. The continuous-rod warhead focused all of its energy in a tight belt at a right angle to the axis of the missile. The effect was very deadly indeed, slicing an aircraft fuselage in half as if it were made of butter. The Sidewinder had the same kind of warhead (Ibid.,1:8:9; see also: Westrum, R., *Sidewinder…*, p. 159–160).

122 Ibid., 1:8:6; For information about the warhead and proximity fuze, see: ibid., 1:8:10–11.

123 *saabs robot- och elektronikhistoria, del 1*, "Jaktrobotsystem 340/341", 1:5:14.

In June 1961 SAAB offered a cost estimate for the production of 4,200 missiles to the RSAFB, stating that it would cost the RSAFB SEK 80 million (SEK 720 million at today's values [2007]). At the same time the company established a Licensing Office responsible for the considerable amount of paperwork generated from the communication between HAC and SAAB (all in all about 35,000 documents).[124] However, it soon became obvious that this initial estimate was too optimistic, and in July 1962 SAAB delivered a new cost estimate based on recently discovered new needs. In the estimate, SAAB stated that the development costs had increased by SEK 7 million to a total of SEK 87 million, and the cost of production of the test series had increased by SEK 10.8 million to almost SEK 35 million. This created a stir within the RSAFB which responded to SAAB, in October the same year, that due to the cost increases the scale and the delivery schedule had to be revised. According to the RSAFB the increases prior to this could be regarded as balanced by the value of license production to the Swedish guided missile industry, but the latest increase was of a totally different kind. The economic constraints had been broken, and if prices were not reduced the only viable option for the RSAFB might be to cancel the whole license-production effort. However, the RSAFB did not resort to this desperate measure. Instead, in April 1963 it cut down the amount of missiles to be produced considerably, to the not very impressive number of 600.[125] The first RB 28 from the test series was delivered by SAAB on December 23, 1962, and the first RB 27 from the corresponding series was delivered on July 1, 1963.[126] Today, the only thing left to remind the afterworld about the domestic systems is one RB 340 displayed on the town square in Vidsel in northern Sweden.[127]

The Purchase of Hawk

Even though the RB 345 never came into being it is interesting because it had striking similarities with the Hawk system. It was to be a mobile surface-to-air missile mounted, just like Hawk (RB 367), on each battery in groups of three.[128] Just as in the case of RB 340 and RB 341, the similarity between the Swedish and the American missile system was hardly a coincidence. Hawk was different from the other missiles that I write about in this thesis, because it was purchased for the RSA by the Royal Swedish Army Board (henceforth: RSAB).[129] Hawk was an Army

124 Cerwin-Ellqvist, G. and König, B., *Robot 50 år*, p. 28.

125 *saabs robot- och elektronikhistoria, del 1*, "Jaktrobotsystem 27/28", 1:8:3–4.

126 Cerwin-Ellqvist, G and König, B., *Robot 50 år*, p. 28.

127 *saabs robot- och elektronikhistoria, del 1*, "Jaktrobotsystem 340/341", 1:5:14.

128 Ibid., 1:5:14–15.

129 Armétygförvaltningen, KATF.

missile in the United States too, and because of this the GMB did not handle the case. This task was instead given to the RSA's Guided Missile Office (Robotkontoret).

Prior to the purchase of Hawk, the RSA had only had manually (and to some extent mechanically) aligned anti-aircraft guns at their disposal. Hawk, on the other hand, was a mobile, fully-automatic anti-aircraft system. During the post-War period the speed and performance of aircraft improved substantially, e.g. through the introduction of the supersonic engine, and this, in turn, demanded more of the anti-aircraft forces regarding coverage both in the horizontal and on the vertical.[130]

In November 1957 the Swedish Air Attaché to Washington wrote to the Chief of the Defense Staff to inform him of the existence of Hawk. According to the Air Attaché, approximately one month earlier the Swedish Army Attaché had attended an exhibition featuring Army ordnance materiel at the Association of the Army's Annual Meeting at the Sheraton Park Hotel in Washington. The main contractor of Hawk, Raytheon Manufacturing Company, and its subcontractor Northrop Aircraft Company arranged the exhibition. This was the first time that Hawk was shown in public, and Sweden had a representative in the audience.[131] In the fall of 1959 Washington decided to sell the Hawk missile to other nations on a government-to-government basis, and Sweden was informed of this decision. The Americans thought that if Sweden had procured Hawk, it might coax its Norwegian and Danish neighbors into expediting their plans to procure such weapons within the NATO framework.[132]

As part of the effort to build up NATO's defences laid out in MC 70 from 1958, thirteen Hawk battalions had been proposed for the FY 59 MAP, and one of these had been recommended for deployment in Norway. However, the American offer was turned down because of Norwegian apprehensions that the Hawk missiles would be equipped with atomic warheads.[133] Thus the connection between Hawk and NATO's infrastructure program, and the decision to release missile data

130 "775 001:544", in Arméthygförvaltningen [henceforth: KATF], Vapenavdelningen [henceforth: vavd], Robotkontoret [henceforth: RBK], Serie: F 1, " Avgångna och inkomna skrivelser ordnade efter klassifikationssystem", NR 775 001:449-775 002:89, År: 1965, VOL. 23, KrA. The information regarding the Swedish designations of these radar stations, CW , pulse and Doppler principles is taken from: Nordbeck, Sven, "Luftvärnsrobotsystem 67 och 77" in C. Herlitz (red.), *Luftvärnets historia*, pp. 177–179.

131 Letter dated Nov. 14, NR "775 001:1", in KATF, vavd, RBK, Serie: F 1, " Avgångna och inkomna skrivelser ordnade efter klassifikationssystem", NR 397 006:1-776 001:30, År: 1955–59, VOL. 3, KrA. Hawk was actually an acronym meaning: Homing All the Way Killer.

132 Moores, S., *'Neutral on Our Side':...*, pp. 198–199.

133 Tamnes, R., *The United States and the...*, pp. 149, 152, 161.

taken at the NATO meeting in December 1956, becomes evident once again.

The United States wished to see a Sweden that was more integrated into the Western European defense structure, and technology became a tool with which the Americans tried to accomplish this goal. This approach later became American policy in NSC 6006/1, where it was stated that it would contribute to the defense preparedness of the Western Powers if Sweden modernized its defense posture and established "early warning, air control and advanced weapons systems [i.e. guided missiles] (without nuclear warheads) which are *compatible with and complementary to* those planned for installation in the territory of neighboring U.S. allies."[134] [Italics added.] The Americans were laying the material foundations for Sweden's participation in the upcoming war. Technology was embodying U.S. policy objectives in Sweden. To ensure this, the Pentagon announced its willingness to assist Sweden in its missile development efforts.[135] The same policy paper also stated that the United States would assist Sweden if she was attacked by the Soviet Union,[136] a decision which, in effect, erased NATO membership as a prerequisite for assistance from the United States. Indeed, one of the great differences between members and non-members was now virtually non-existent, at least in relation to Sweden.

In Ambassador Jarring's report to Herter on December 2, 1959, it was stated that the Hawk weapon system was of interest to the Swedes, but that more information was needed in order for them to make an adequate evaluation of the equipment and its performance, and "the consequences of an integration [...] into the Swedish Air Defense."[137] Sweden wished to receive detailed information regarding target acquisitioning, tracking and illuminating radars, computing and control functions, as well as data on launchers and the missile itself. It was stressed that the required number of units, although relatively limited, could not be stated until Sweden had been furnished with this information. Other information, such as required manpower, integration problems, performance check-outs, and cost figures, was also asked for. "Should it be established that Sweden's requirements would be adequately served by the Hawk

134 FRUS: *Foreign Relations of the United States*; 1958–1960; Vol. VI, Part 2; *Western Europe*, p. 676. See also: Moores, S., *'Neutral on Our Side':...*, p. 200.

135 Moores, S., *'Neutral on Our Side':...*, p. 204.

136 FRUS: *Foreign Relations of the United States*; 1958–1960; Vol. VI, Part 2; *Western Europe*, p. 679.

137 Letter from Jarring to Herter, dated December 2, 1959, p. 1, in: RG 59; CDF, 1960–63; from 758.56/1-1360 to 759.00 (W)/5-261; Box No. 1843; (Folder: 758.5622/12-660); NARA II, College Park, MD.

Weapon System," wrote Jarring, "the Swedish Government is interested in procuring the system in the United States."[138]

In January, 1960, Colonel Jentzen reported the experiences he had had during a recent six-week visit to the United States. He had visited a number of the U.S. Army's bases and studied weapons development there, and he thought that the hospitality shown had been greater than he had initially expected. In fact, "[v]irtually all the information requested had been made available" to Jentzen.[139] The U.S. Army Attaché in Stockholm reported back to Washington in February that the Swedes were about to make a decision between Hawk or BOMARC, and that this decision was contingent upon a definite American decision regarding Sweden's possibility of purchasing Hawk. In fact, it was stated that the Swedes were unable to make a decision between the two missiles until the American position on this matter had become known.[140]

In early March Jarring met with Under Secretary for Political Affairs Livingston T. Merchant to discuss the Hawk purchase. In a memorandum written two days prior to this meeting it was said that the American government was unable at that time to give the Swedes a definite answer due to the fact that sufficient internal coordination on the American side had not yet been achieved. It was expected, however, that the DOD would suggest that Sweden should approach the NATO Hawk Board in order to establish if their needs could be met by NATO's coordinated Hawk production program. Only if such an approach should fail for some reason, should the Swedes request to purchase the missile from the United States. It was expected that such a request would be sympathetically received, with the possible exception of certain parts e.g. fuze requirements.[141]

When Merchant met with Jarring the Swedish Ambassador informed him that the Hawk missile system had been theoretically evaluated in Sweden, and that it appeared suitable for Sweden's defense requirements. Jarring said that Sweden wished to purchase one Hawk battalion from the United States. Merchant gave Jarring the answer which had been drafted in the memo two days earlier, namely that Sweden might approach the NATO Hawk Board and purchase the system from the European production source. The Swedish Ambassador replied that he

138 Ibid., p. 2.

139 Moores, S., *'Neutral on Our Side':...*, p. 195.

140 Telegram from U.S. Army Attaché in Stockholm to Washington, dated February 3, 1960, in: RG 59; EUR; RA; PMNF 1953–1962; 2.4 MDWP General, 1960–61 to 3.4 Civil Emergency Planning; Box No. 3; (Folder: Swedish Defense Planning and Capabilities); NARA II, College Park. MD.

141 758.5611/3-760, Memorandum from Margrave to Cobb, Jr., dated March 7, 1960, in: RG 59; CDF, 1960–63; from 758.56/1-1360 to 759.00 (w)/5-261; Box No. 1843; (Folder: 758.5622/12-660); NARA II, College Park, MD.

would forward Merchant's suggestion to the Swedish government, but added that he personally could see that this would be difficult from a public relations standpoint; "there might be reservations regarding an approach to NATO in Europe that did not extend to the United States." Merchant added that the American study of the overall Swedish missile requirements had not yet been completed, but that it was in progress and that a reply would be forthcoming as soon as the result of the study had been received.[142] Jarring thus assumed that it would be more difficult for Sweden to purchase Hawk from a European production source than from the United States, due to the possible detrimental effects on the neutrality policy of this overt connection with NATO. I will return to this below.

What the Americans were referring the Swedes to was the European Hawk Program. The Hawk missile was a novelty in more than one way: not only was it the first surface-to-air missile designed to strike supersonic targets at low altitude; it also embodied the success of the new policy established by the NATO Heads of Government in December 1957. The policy stated that the United States promised to aid the member states in producing technically advanced weaponry on a large scale in Europe. To supervise this project NATO established a new entity called the Hawk Production Organization. Five NATO countries were represented in this organization: Belgium, France, Germany, the Netherlands and Italy. One company in each country produced selected components and sub-assemblies for the missile. The companies were: Ateliers de Constructions Electriques de Charleroi in Belgium; Thomson-Houston in France; Telefunken in Germany; N. V. Philips in the Netherlands and Finmeccanica in Italy. These firms pooled their resources in a French-registered consortium called SETEL (Société Européenne de Téléguidage), a company which sub-contracted a plethora of technical and industrial tasks to the constituent companies, and which also assembled the final product.[143]

The arrangement had striking similarities with the license-manufacturing deal made between Sweden and the United States regarding the air-to-air Missiles RB 27 och RB 28 – although on a much

142 758.56/3-960, Memorandum of Conversation between Jarring, Merchant, Long, and Cobb, Jr., dated March 9, 1960, pp. 1–2, quote on p. 2, in: RG 59; CDF, 1960–63; from 758.56/1-1360 to 759.00 (W)/5-261; Box No. 1843; (Folder: 758.5622/12-660); NARA II, College Park, MD. See also: Promemoria regarding this matter, dated March 25, 1960; as well as Letter from Jarring to Åström, dated April 14, 1960, in: HP 24 VI/U.S.A; "Tillstånd till in- och utförsel av vapen, ammunition och krigsmateriel"; 1960 mars–1960 oktober; Folder 25; Box No. 63; År: 1960–1962; RIA; Arninge.

143 "The Hawk Production Organization", "775 001:2", in KATF, vavd, RBK, Serie: F I, " Avgångna och inkomna skrivelser ordnade efter klassifikationssystem", NR 397 006:1-776 001:30, År: 1955–59, VOL. 3, KrA.

larger scale. The American companies Raytheon and Northrop worked directly with their European counterparts on transferring American technological know-how pertaining to the production of Hawk.[144] For example, the Dutch company Philips produced the Battle Command Centers; the American company Thomson-Houston's subsidiary in France produced the Target Illuminating Radar and parts of, or the entire, missile was produced by the German firm Telefunken.[145]

Some of these firms, such as Telefunken and Thomson-Houston, were already a part of NATO's, at least to that date, most important infrastructure project the NATO Air Defense Ground Environment (henceforth: NADGE). The NADGE project was an effort initiated to modernize NATO's Early Warning radar system[146] in Europe, a computerized system intended to direct manned interceptor aircraft and surface-to-air missiles towards intruding aircraft from the Eastern Bloc. This system was to disregard geographical borders, and was thus a materialization of Article 5 in the North Atlantic Treaty, which stated that an attack upon one alliance member was an attack upon all members. Article 5 was being built into material objects.[147] NATO's joint infrastructure program concerned all fixed installations, such as airfields, communications systems, missile sites, ports, petroleum pipelines, storage facilities and air defense radar systems.[148]

In connection with this, it is interesting to note that Hawk, Nike Hercules and Mace batteries were part of NATO's infrastructure program, and that these missiles were dispersed all over Europe.[149] This fact also stresses the significance of Sweden's purchase of Hawk. As a consequence, a central ingredient of NATO's infrastructure program could now be found a few 100 kilometers further east on the Scandinavian Peninsula than would otherwise have been the case. On March 31, 1960, the State Department wrote to Ambassador Jarring and formally recommended Sweden to buy Hawk from the European production

144 Ibid.

145 "775 001:115", p. 2, in KATF, vavd, RBK, Serie: F I, " Avgångna och inkomna skrivelser ordnade efter klassifikationssystem", NR 397 000:41-775 001:122, År: 1963, VOL. 18, KrA.

146 The purpose of this system was to alert, immediately, a defending nation to an imminent threat. In this particular case, the imminent threat was Soviet bombers or missiles.

147 Huston, James A., *One for All: NATO Strategy and Logistics through the Formative Period, (1949–1969)* (Newark: University of Delaware Press, 1984), pp. 168–171. A consortium, called NADGECO, was established for the NADGE project, and this included British Marconi and HAC. Marconi went on to become the main contractor for the Swedish digital Combat Operations System Stril 60.

148 Huston, J. A., *One for All...*, pp. 157–158.

149 Huston, James A., *One for All...*, p. 166.

line, if possible, since that would be a particular advantage for Sweden. Showing its responsibility as a hegemonic power, it was added that should this option not be possible, the Americans were willing to provide the system to the Swedes "subject to delivery priorities accorded to the United States forces and to forces of other NATO countries." The Americans also stated that a response to the request for information on the BOMARC and Falcon systems made by the Swedes on December 2, 1959, was still in preparation and would be transmitted as soon as it had been completed.[150] Sweden's earlier uneasiness about procuring Hawk from a European production source was expressed again by Jarring in mid-April, when he called on Deputy Under Secretary of State Hare and Cobb, Jr. of BNA. The Swedish government, said Jarring:

> [...] wished to continue to follow the practice of buying missiles in the United States. [...] Sweden, for political reasons, continuing *its policy of non-alliance with Europe*, preferred to deal directly with the United States Government. For these reasons the Swedish Government feels it is unable to use the European production source.[151] [Italics added.]

General Bonde, Chief of the RSA, repeated Jarring's words in early May in a meeting with U.S. Army Attaché Colonel Schmidt at the American Embassy in Stockholm. Bonde, whose forces were the ones to be equipped with Hawk, said that the suggestion that Sweden should go through the NATO Coordination Board was "embarrassing for Sweden since she is not a member of NATO." Bonde spoke frankly with Schmidt and stated that having to go through NATO was as if Sweden had no diplomatic relations with the United States, but had to go through a third party in order to do business with the Americans. More importantly, "the Swedish Government," said Bonde, "cannot officially approach NATO for the purchase of weapons."[152]

A strange double-sidedness is revealed here; on the one hand, the Swedish government wished to portray purchases of military technology

150 Letter from State Department to Jarring, dated March 31, 1960, in: RG 59; EUR/SCAN; RRSWA, 1958–1964; POL 17 Political Affairs Sweden POL 17 Diplomatic Notes 1959–1963 to PS – Protective Service Sweden PS 7-1 Arrest (Swed. Seamen Visit New Orleans); Box No. 1; (Folder: Lot 67D-75); NARA II, College Park, MD.

151 758.5612/4-1260, Memorandum of Conversation between Jarring, Hare, and Cobb, Jr., dated April 12, 1960, in: RG 59; CDF, 1960–63; from 758.56/1-1360 to 759.00 (W)/5-261; Box No. 1843; (Folder: 758.5622/12-660); NARA II, College Park, MD.

152 Letter from Timmons to Brandin, dated May 12, 1960, p. 1, in: RG 59; EURONOE; RRS, 1957–1975; DEF 2-Gen. Repports [sic] and Statistics, 1966–1969 to POL 33-8 – Law of the Sea, 1960–1969; Lot File: 72D339; Box No. 2; (Folder: DEF 12-5C Missiles – Redeye 1959–1969); NARA II, College Park, MD.

as apolitical. Such affairs did not entail any further commitments on the part of the Swedish government, it was said, and hence they were rather uncontroversial (the purchases of these missiles were made public in the press). On the other hand, the government itself acknowledged the political implications involved when it worried that procuring Hawk from the NATO production line could be frowned upon by potential spectators. Compared with the terms agreed to when signing the technical assistance agreement in connection with the Falcon affair, the purchase of Hawk from a NATO source seems relatively unproblematic, to say the least. Still the Swedish government strongly resisted this option. What lay behind this seeming inconsistency? The answer, it seems, was perception (but perception based on geopolitical realities). Buying Hawk from a European NATO country could be *perceived* by outside observers as Sweden somehow being tied to NATO. Deliveries of military technology from NATO smacked more of 'alliance' than did any direct relation with the United States. As Jarring indicated above, the NATO option was more sensitive because it risked upsetting the delicate balance of power in Europe.

In contrast, the Falcon agreements were secret, and would thus not be perceived by anyone at all. The negative Swedish government attitude towards the European option was a direct display of accommodating opposition, and yet another example of how the materiality of political and geostrategic realities affected the material manifestations of Sweden's neutrality policy, which in turn shaped Swedish-American cooperation during the Cold War. The Swedish government was as prepared as ever to give its consent to American hegemony, but such consent was not unconditional and it could not be given in just any way. The Americans surely recognized the difficulties under which the Swedes labored, and were clearly prepared to compromise where and when it was considered possible to do so without infringing upon U.S. policy objectives.

The NSC gave its approval to the decision to sell guided missiles to Sweden on a case-by-case basis in April 1960.[153] In a note from the State Department on July 13, 1960, to the Swedish Chargé d'Affaires ad interim Nils Montan, the Americans formally acknowledged the problem for Sweden in buying Hawk from Europe, while informing Sweden about the American decision. Foggy Bottom stated that since the Swedish government felt unable to acquire Hawk from European sources the U.S. government was now prepared to receive purchase orders from Sweden.[154] The American note was immediately followed by a response from the Swedes, mediated by Montan, on July 26. In this

153 Moores, S., *'Neutral on Our Side':...*, p. 199.

154 Note from State Department to Montan, dated July 13, 1960, in: RG 59; EUR/SCAN; RRSWA, 1958–1964; POL 17 Political Affairs Sweden POL 17 Diplomatic Notes 1959–1963 to PS – Protective Service Sweden PS 7-1 Arrest (Swed. Seamen

letter Sweden requested from the Americans an offer for the purchase of "[o]ne HAWK battalion, consisting of 88 missiles, maintenance and spare parts included, equipment not peculiar to the HAWK system excluded, direct support equipment for one HAWK battalion, a quantity of dummy missiles for training purposes, if available."[155]

Even though the Swedish interest in the Hawk system was great, it took some months before the formal decision to purchase the American missile was actually taken; by the turn of the year 1960/61 the RSAFB finally decided to purchase Hawk from U.S. Army.[156] In the letter sent to the Procurement Division [Inköpsavdelningen] at the RSAB in early January 1961, the Swedish Army Attaché, Colonel Stig Löfgren, reported that the American Department of the Army estimated the total cost at $18.1 million.[157] This sum did not, however, include the cost for the Swedish personnel receiving training on the Hawk system in the United States.[158] The first training period stretched from October to December, 1961.[159]

Hence the Americans understood the Swedish position and were prepared to work around it; the United States, it was reasoned, would "have to make the best of it." The reason why the United States wished Sweden to purchase Hawk, as well as Sidewinder, from Europe was that the Americans in this way avoided, when the deal was made public, any charges from the NATO allies that U.S. commercial interests had something to do with the fact that the Swedes purchased the missiles from the United States. This was sensitive since, as a consequence, European commercial interests would be hurt by it. Likewise, the wish to see to it that NATO requirements had been fulfilled before tending to the Swedish needs was warranted because it would have been "most embarrassing if

Visit New Orleans); Box No. 1; (Folder: Sweden Missile 1959–62); NARA II, College Park, MD.

155 Letter from Montan to State Department, dated July 26, 1960, in: RG 59; EUR/SCAN; RRSWA, 1958–1964; POL 17 Political Affairs Sweden POL 17 Diplomatic Notes 1959–1963 to PS – Protective Service Sweden PS 7-1 Arrest (Swed. Seamen Visit New Orleans); Box No. 1; (Folder: Sweden Missile 1959–62); NARA II, College Park, MD.

156 Nordbeck, S., "Luftvärnsrobotsystem 67 och 77" in C. Herlitz (red.), *Luftvärnets historia*, p. 175.

157 "775 001:6", in KATF, vavd, RBK, Serie: F I, " Avgångna och inkomna skrivelser ordnade efter klassifikationssystem", NR 397 006:28-775 007:75, År: 1961, VOL. 13, KrA.

158 "775 001:24", in KATF, vavd, RBK, Serie: F I, " Avgångna och inkomna skrivelser ordnade efter klassifikationssystem", NR 397 006:28-775 007:75, År: 1961, VOL. 13, KrA.

159 "775 001:26", p. 2, in KATF, vavd, RBK, Serie: F I, " Avgångna och inkomna skrivelser ordnade efter klassifikationssystem", NR 397 006:28-775 007:75, År: 1961, VOL. 13, KrA.

for some reason the U.S. delayed furnishing the Hawk kits and samples which we have promised the NATO organization but was able to fill the Swedish commercial order."[160] To make sure that this situation could be rapidly defused, if it were to arise, it was recommended that USRO in Paris would be authorized to respond that the Swedes had been offered the chance to obtain their missiles from a European source, but that they had declined due to political considerations.[161]

This episode shows that the matter of giving Sweden a secondary rating, i.e. after the NATO countries, when procuring military material from the United States seems to have been mainly an issue of saving face in U.S. relations with the NATO Allies. Were it not for them, there seems to have been nothing intrinsic to Sweden's position which made it rank lower on the priority scale. As a hegemonic power, the United States had to see to it that its most devoted clients were kept happy: a way of doing this was for the United States try and make sure that the material benefits accruing to the consenting nations were allocated to the NATO countries first. But as the above has shown, this was more a matter of letter than of spirit, since Sweden was *de facto* given a very preferential treatment. Its access to guided missiles would not be circumscribed; instead the NATO countries would be offered the same information if they requested it.

It is instructive, concerning how to comprehend the Swedish position and the purchase of this missile in the larger context of the Cold War, that not only was Sweden obtaining access to the Hawk missile before most of the NATO countries, but Hawk had not, at this time, been fully deployed in the U.S. either. In fact, by mid-1960 it was expected that Hawk would become operational in the first half of fiscal year 1961.[162] Thus, the deployment of Hawk in the RSA paralleled the same process in the United States. The Swedish purchase of Hawk, and the American approval of the same, can be seen in relation to the putting into operation of similar battalions all over Europe and the United States under way at the same time. All this is a clear illustration of the ubiquitous character

160 Letter from Wolf to Fessenden, dated April 29, 1960, in: RG 59; EURONOE; RRS, 1957–1975; DEF 2-Gen. Repports [sic] and Statistics, 1966–1969 to POL 33-8 – Law of the Sea, 1960–1969; Lot File: 72D339; Box No. 2; (Folder: DEF 12-5C Missiles – Redeye 1959–1969); NARA II, College Park, MD.

161 Letter from Axelrod to Brandin, dated May 17, 1960, in: RG 59; EURONOE; RRS, 1957–1975; DEF 2-Gen. Repports [sic] and Statistics, 1966–1969 to POL 33-8 – Law of the Sea, 1960–1969; Lot File: 72D339; Box No. 2; (Folder: DEF 12-5C Missiles – Redeye 1959–1969); NARA II, College Park, MD.

162 "History of the U.S. Army Rocket & Guided Missile Agency" covering the period from January 1 to June 30, 1960, in: RG 156; EOHB; AGMP, 1940–62; Box No. 51; ARGMA Historical Summary; NARA II, College Park, MD.

of American hegemony at the time, and how technology was used as a tool to forge a cohesive Western defensive bloc.

The favored position of the Swedish Armed Forces becomes even clearer from a State Department office memorandum written in September 1960. Here it was confirmed that Sweden would receive their Hawk battalion before the NATO countries, as well as before any other country, with the exception of South Korea. However, the State Department seems to have taken this with considerable ease, not being too worried about the possible reactions of the NATO camp, because they had after all tried to convince the Swedes to buy Hawk from the Europeans and the European organization had been dutifully informed about the reason for the failure of this approach.[163]

On August 22, 1960, the Swedes sent a letter to the State Department announcing their intention of sending a group of representatives to Washington to conclude the purchase of the Hawk battalion. The visit was to be a very long one (it was planned to take place between September 14 to October 16), and would include the Chief of RSA Ordnance Major General Malm, the Chief of the Weapons Division Brigadier Jentzen, the Chief of the Robot Branch Colonel Grenander, as well as technical and legal personnel.[164] At the end of September a request from the Department of the Army to release secret Hawk information to Sweden (to act as a basis for a Swedish decision to purchase the system) was approved by the DOD, and, as a result, Sweden was offered to purchase Hawk on October 13, 1960. The Swedes promptly accepted the offer on November 9, and a year later, in October 1961, the Department of the Army stated that it considered the increase in the military potential of Sweden, by the acquisition of Hawk, to be advantageous to the United States. Deliveries were scheduled to begin during the second half of 1962.[165] In late October 1961, Army member

163 Memorandum from Smith to Fessenden, dated September 1, 1960, in: RG 59; EUR; Office of Atlantic Political and Military Affairs [henceforth: OAPOMA]; Military Assistance Policy/Mutual Security Program Files [henceforth: MAP/MSPROF] 1961–1963; Economic, Berlin to Yugoslavia; Box No. 3; (Folder: Sweden); NARA II, College Park, MD.

164 Telegram from U.S. Air Attaché in Stockholm to Department of the Army, dated August 29, 1960, in: RG 59; EUR; OAPOMA; MAP/MSPROF 1961–1963; Economic, Berlin to Yugoslavia; Box No. 3; (Folder: Sweden); NARA II, College Park, MD.

165 Memorandum from Conmy, Jr., of the State-Defense Military Information Control Committee (SDMICC) to the Chairman of the SDMICC, dated October 25, 1961, in: RG 59; EUR; OAPOMA; MAP/MSProf 1961–1963; Economic, Berlin to Yugoslavia; Box No. 3; (Folder: Sweden); NARA II, College Park, MD. The cooperation regarding guided missiles also spread into the civilian sector. Such an example was the launching of the first 'Swedish' space rocket, the Plutnik (alluding to the Russian Sputniks), in August, 1961, at RFN. The reason for the quotation marks around 'Swedish' is that both the rockets and the launching pad were provided by NASA. More American rockets were launched during the

of the S-DMICC Colonel Joseph B. Conmy, Jr. wrote a memo for the S-DMICC Chairman requesting:

> [...] that an exception to the National Disclosure Policy [...] be granted to permit the release to Sweden of the HAWK system and such SECRET information in the categories of organizational, tactical doctrine, operational deployment, technical data, production data, research and development, as is necessary that country to receive, operate and maintain the HAWK system.[166]

This, once again, just as in the case of the Falcon affair, points to the flexible nature of U.S. policy towards Sweden at this time; Sweden was not really supposed to have access to this information, but due to the perceived reliability of the Swedes, which in turn was due to Sweden's consent to U.S. hegemony and good track record with regard to earlier cooperation endeavors, exceptions to otherwise rigorous security standards could be granted.

The RSA's purchase of Hawk actually caused some trouble for the RSAF, then in the midst of trying to convince the Swedish Ministry of Defense to let it procure surface-to-air missile Bloodhound MK. II from Britain. Here, once again, the inter-service rivalries became explicit. The people back in Washington were kept abreast of the events by the American Army Attaché in Stockholm as they unfolded. In September 1960 he noted in a telegram to the Department of the Army that the RSAF "strongly opposed Hawk purchase", and similarly that the RSA felt that the RSAFB's purchase of Bloodhound II was an unnecessary expense for the Swedish Government.[167]

The RSAF spent SEK 94 million, or $18.2 million, on military equipment in the United States in 1960; in 1961 the figure was SEK 73 million ($14 million); and in mid-1962 it was estimated that the Swedes would spend over SEK 140.5 million ($27 million). The RSA was going to pay SEK 93.6 million ($18 million) for Hawk over a period of three years, and it was believed that the total amount, for all three Swedish Services, would

---

summers of 1962–1964 (this time with Nike-Cajun and Nike-Apache rockets), and people from NASA and the USAF participated directly in some of these experiments (for this, see: Wormbs, Nina, *Vem älskade Tele-*X? *Konflikter om satelliter i Norden 1974–1989* (Stockholm: Gidlunds Förlag, 2003), pp. 9–12).

166 Memorandum from Conmy, Jr., of the State-Defense Military Information Control Committee (SDMICC) to the Chairman of the SDMICC, dated October 25, 1961, in: RG 59; EUR; OAPOMA; MAP/MSPROF 1961–1963; Economic, Berlin to Yugoslavia; Box No. 3; (Folder: Sweden); NARA II, College Park, MD.

167 Telegram from U.S. Air Attaché in Stockholm to the Department of the Army, dated September 14, 1960, in: RG 59; EUR; OAPOMA; MAP/MSPROF 1961–1963; Economic, Berlin to Yugoslavia; Box No. 3; (Folder: Sweden); NARA II, College Park, MD.

come to about SEK 197.6 million ($38 million) by the end of 1962.[168] In the fall of 1962 the RSA planned on procuring yet another Hawk battalion from the United States, and while there were some problems with regard to coordination of the Swedish order and the American production schedules, there was otherwise "no current problem with regard to the procurement of additional missiles from the U.S." In connection with this the U.S. Army Attaché in Stockholm Colonel Thomas C. Rohan commented that the Swedish procurement plans:

> [...] will involve the expenditure of many millions of dollars in the U.S. and will tend to force their Army into further reliance on the U.S. for replacement parts, modifications, training, etc.[169]

The heavy Swedish dependence on American materiel was thus considered to be a good thing, both economically and strategically, for the United States. Rohan mentioned particularly the increasing level of dependency on the part of the Swedes upon the United States due to these procurements. This should make us wonder what effect this dependence would have had upon the possibility that the RSA would turn these battalions on American aircraft overflying Sweden in time of war.

In two articles in two local Swedish newspapers (*Örnsköldsviks Allehanda* and *Borås Tidning*) in the fall of 1962, RSA Captain Nils A. Löfström boasted about the RSA's newly acquired Hawk missile. The American missile was "an ace of trumps"[170] for the air defense, it was said, and would complement the already existing Bloodhound IIs. These articles, or one could use the singular 'article' for they were identical, are examples of the shaping of a mythology around the strong defenses of non-aligned Sweden—a true process of domestication of glory, to borrow a phrase from Ingemar Dörfer. An analysis of the symbolic values expressed in these (and other) articles is not the purpose

168 Memorandum from U.S. Air Attaché Harrold to Bonbright, dated June 8, 1962, in: RG 59; EURONOE; RRS, 1957–1975; DEF 2-Gen. Repports [sic] and Statistics, 1966–1969 to POL 33-8 – Law of the Sea, 1960–1969; Lot File: 72D339; Box No. 2; (Folder: Defense Affairs; DEF 12-5a Procurement. Sale. Sweden 1960–1969); NARA II, College Park, MD.

169 Memorandum from U.S. Army Attaché Rohan to Bonbright, dated August 7, 1962, pp. 1–2, quote on p. 2, in: RG 59; EURONOE; RRS, 1957–1975; DEF 2-Gen. Repports [sic] and Statistics, 1966–1969 to POL 33-8 – Law of the Sea, 1960–1969; Lot File: 72D339; Box No. 2; (Folder: Defense Affairs; DEF 12-5a Procurement. Sale. Sweden 1960–1969); NARA II, College Park, MD.

170 See: *Örnsköldsviks Allehanda*, "Fem robotsystem försvarar vårt luftrum. Försvaret får ett trumfess i amerikanska Hawk-roboten", August 17, 1962; and *Borås Tidning*, "Försvaret får trumfess i amerikanska Hawk-roboten", September 1, 1962.

of this study, however, but it is worth noting that the RSA, in a truly propagandistic fashion, wished to tell the world about the procurement of this and other systems.

Yet another article, an official pressrelease, was published in January 1963 after the system had been delivered to Sweden. This time, however, it was published in one of the largest national newspapers *Svenska Dagbladet*. The message was clear—the Swedish skies were now very unsafe for enemy aircraft to operate in; even attacks by enemy guided missiles would be made much more difficult. "It is valuable—not least psychologically" it was said "that an enemy's aircraft and guided missiles can be combated with success wherever they might appear."[171] Once again, semantics may be important when analyzing the meaning of these contemporary articles. Interestingly, all of them referred to the targets of these missiles as 'enemies'. This means that in none of them was it assumed that Sweden was only fending off intrusions into its airspace to protect its neutrality, but that Sweden was caught up in a general or limited war. As we have seen, Sweden considered itself to have only one possible enemy, i.e. the USSR. One could hence view these articles as having three simultaneous purposes: a) reassuring the Swedish public that Sweden's defenses were good and capable of handling any threat, b) deterring a Soviet attack by letting the Russians know that Sweden had advanced weapons at its disposal, and c) reassuring the United States that the missiles on Swedish soil were not aimed against U.S. aircraft or missiles.

Deepening the Hegemonic Collaboration, 1960–62

At the end of 1959 plans were being hatched within the American administration to produce a new policy paper, NSC 6006/1, dealing with Scandinavia, and in mid-March 1960 a final draft version of the paper became available for discussion in the NSC.[172] The document caused some trouble within the administration, which has subsequently lead to a scholarly debate as to its significance and meaning. What caused the stir was the inclusion of a paragraph (Paragraph 28 in the final version) spelling out that the United States should undertake unilaterally to come to the assistance of Sweden if she was attacked by the Soviet Union. At the NSC meeting on April 1 Secretary of State Herter objected to this paragraph, not so much to its principle, as to its practicality, and questioned how such an operation could be successful if the Swedes did not know about it.[173] He also brought up the problem of the defense

171 *Svenska Dagbladet*, "Arméns luftförsvar skall förnyas", January 17, 1963. "Det är värdefullt – inte minst psykologiskt – att en fiendes flyg och robotar kan bekämpas med framgång var de än uppträder."

172 *Had There Been A War...*, pp. 103–104.

173 For another interpretation, see: Bruzelius, Nils, *"near friendly or neutral shores":*

of other neutral nations, should this secret security guarantee become known, and moreover thought that it seemed to confer great benefits upon Sweden with little in return for the United States. Due to the Secretary's protestations the paragraph was temporarily cut out of the document, in order to be scrutinized by the State Department, while President Eisenhower approved the rest of the document on April 10, 1960.[174]

The State Department's evaluation was discussed in the NSC in October 1960, and the paragraph now said that in the event of a general war "'it was assumed that the United States would encourage and assist Swedish resistance by all means, military means included, consistent with U.S. prior obligations to NATO."[175] The Chairman of the NSC Planning Board opined against this formulation, especially when the State Department had stated that war planning with respect to Sweden was only *assumed*, and questioned whether the people at State had any idea what they were talking about. Said the Chairman:

> Doesn't the State Department know what the facts are? In my opinion, the last sentence of the first paragraph should have read: 'U.S. and NATO war plans take into account the possibility of Swedish involvement in a general war.'[176]

The Neutrality Policy Commission did not reflect further upon this comment, but I argue that it explains a lot. It seems as if the State Department, including Secretary Herter, was in fact badly informed with regard to current planning. As we have seen below, Swedish participation in a general war was seen as probable; so probable in fact, that NATO's Capabilities Plan was based upon this assumption, as was the size of the domestic Norwegian forces.

The contested Paragraph 28 was eventually approved on November 10, 1960, in an amended version of NSC 6006/1, and it then stated that within the frameworks of a general war the United States should "encourage and assist" Sweden, and in the event of an attack upon Sweden alone the U.S. was to "be prepared to come to the assistance of Sweden as part of NATO or UN response to the aggression."[177] Indeed, the meaning of the words 'encourage and assist' was far from obvious, since

---

The Deployment of the Fleet Ballistic Missile Submarines and US *Policy Towards Scandinavia, 1957–1963* (Stockholm: Stockholm Papers in the History and Philosophy of Technology TRITA-HOT 2053, 2007), pp. 63–75.

174 *Had There Been A War…*, pp. 106–107.

175 Ibid., p. 107.

176 Ibid., p. 108. See also: Dalsjö, R., *Life-Line Lost:…*, p. 62.

177 FRUS: *Foreign Relations of the United States*; 1958–1960; Vol. VI, Part 2; *Western Europe*, p. 679.

the document did not specify what this would entail. It could very well just have been a question of supplying Sweden, as far as possible, with essential raw materials, foodstuffs, etc.; nowhere was it mentioned that the United States would support Swedish defense with American troops. With regard to an isolated attack on Sweden, however, it becomes a bit clearer, but it was still only a matter of assisting Sweden as part of a UN or NATO response. This contingency does, though, seem to have envisioned the use of U.S. forces to repel a Soviet onslaught.[178]

The issue of U.S. military assistance to Sweden came up again when, in late 1960, the State Department began to draw up an annex to its Operations Plan for Sweden. The Operations Plan was closely tied to NSC 6006/1, and the first version seems to have been produced on October 7, 1960. In fact, the draft supplement, called Annex E, was:

> [a] result of a change, by November 7, 1960 NSC Action 2332, in the U.S. Policy Toward Scandinavia [...]; this change adopted a new paragraph in the policy paper, inserted between paragraphs 27 and 28, with particular regard to U.S. policy towards Sweden.[179]

Annex E dealt with U.S. assistance to Sweden, and was thus directly related to the controversial Paragraph 28 of NSC 6006/1. The document concluded that:

> [t]he security of Sweden is a matter of both political and military interest to the United States. Any Soviet aggression, whether direct or indirect, against the sovereignty of Sweden, will challenge the free world, primarily in NATO.[180]

It was furthermore stated:

> In the event of general war, Sweden's position cannot be predetermined; if Sweden does not become a co-beligerent [sic], it is important that it remain a friendly neutral. In any event, in view of the probable sensitivity of some NATO powers to any preferential *assistance guarantee* to a

178 Simon Moores has interpreted this to mean that the American nuclear umbrella was extended to cover Sweden (Moores, S., *'Neutral on Our Side':...*, p. 200).

179 Memorandum for the Board Assistants, dated January 17, 1961, p. 1, in: RG 59; EUR; RA; PMNF 1953–1962; 2.4 MWDP General, 1960–61 to 3.4 Civil Emergency Planning; Box No. 3; (Folder: Swedish Defense Planning and Capabilities); NARA II, College Park, MD.

180 "Annex E: U.S. Assistance for the Defense of Sweden", attached to: "Memorandum for the Board Assistants", dated January 17, 1961, p. 1, in: RG 59; EUR; RA; PMNF 1953–1962; 2.4 MWDP General, 1960–61 to 3.4 Civil Emergency Planning; Box No. 3; (Folder: Swedish Defense Planning and Capabilities); NARA II, College Park, MD.

> neutral non-member and the Swedish Government's careful avoidance of identification with NATO, any U.S. contingency preparations to assist Sweden to resist a Soviet Bloc aggression should be conducted with utmost discretion, with related contingency planning being conducted only on a unilateral basis.[181] [Italics added.]

This is perhaps one of the few documents were the Americans themselves applied the term 'assistance guarantee' to the projected help to Sweden. The statement that Sweden's position could not be predetermined, it should be noted, did not refer to all situations, but only to a general war where Swedish neutrality has been respected by the Soviet Union. This indicates that the Americans still expected Sweden to become a co-belligerent in case of an isolated attack upon her, or in the event of an isolated attack upon Norway.

Then was repeated the guidance given in Paragraph 28 in NSC 6006/1 concerning recommended United States action in the event of general and limited war, and it was said that the discreet liaison between Sweden's military establishment and "selected NATO powers" [i.e. Norway, Denmark] should be encouraged and maintained "as the basis for possible future active military cooperation."[182] Prior research has concluded that since the contingency planning was to be strictly unilateral, Sweden did not know about the American plans.[183] In this case, however, the unilateral character of the American planning should be taken to mean that there would be no planning in concert with other NATO countries or with Sweden. This does not per definition exclude the possibility that certain Swedish officials were made aware of U.S. thinking and contingency planning.

A memorandum was attached to the document relating the comments of BNA to the draft, and stated BNA's views to be:

> [...] that it is acceptable to the Pentagon, in that it permits the continuation of informal US-Swedish contact in service channels. Although the guidance may seem to be too stringent by restricting US military planning to an [sic] unilateral basis, BNA says that they consider it would be counter-productive to go further. In their view, it should not be US government policy to encourage bilateral or multilateral planning since to do so might compromise the Swedish military with their own Government and thus weaken the hand of

181 Ibid., p. 1.

182 Ibid., pp. 1–2, quote on p. 2.

183 *Peace and Security:...*, p. 201. In the concluding statements of the report it is said that "[...] the American policy laid down in the remarkable Guidelines for Policy and Operations from 1962 meant that Sweden had unknowingly secured powerful military support in the event of a Soviet attack." (ibid.)

> those Swedish elements which are disposed to build up the Swedish military establishment. The paper does not envisage NATO planning as such for emergencies involving Sweden, but simply recognizes [...] the fact that Norway and Denmark are now discreetly exchanging military information with Sweden. We and BNA think that this liaison which the NATO Scandinavian countries are carrying on is more effective than anything we, or other NATO Countries, might do.[184]

It was probably around this time that a telex link was established between the Swedish Defense Staff headquarters in Stockholm and the USAF headquarters in Wiesbaden, Germany. Operators in Stockholm and Wiesbaden tested the connection weekly, and a crypto device was added to the telex machine at the Defense Staff sometime in 1963–1965.[185] Furthermore, every year one to three highly placed officers within the Defense Staff, including the Chief of the Defense Staff, traveled down to Wiesbaden to discuss military and intelligence matters with the Americans there. These trips were rarely if ever openly talked about at the Defense Staff, according to the Chief of the Defense Staff's Attaché Department Gustaf Hamilton, and they were definitely not put down on paper.[186]

The discussion above resulted in the State Department's "Guidelines for Policy and Operations: Sweden" in June 1962, which was approved as official policy in November (and thereby replaced NSC 6006/1).[187] Much fuss has been made concerning this document, too, and the Inquiry on Security Policy even used the word "remarkable" when summarizing the document.[188] But what was it about these guidelines that was so remarkable? Not much in it was very new at all; in fact, the major difference between the Guidelines of 1962 (which had been under way since October 1960) and NSC 6006/1 seemed to be merely a question of semantics. The Guidelines stated that the United States should *undertake* to assist Sweden, while Paragraph 28 in NSC 6006/1 stated that

184 Office Memorandum from Millar to Magill, dated January 24, 1961, attached to: "Memorandum for the Board Assistants", dated January 17, 1961, in: RG 59; EUR; RA; PMNF 1953–1962; 2.4 MWDP General, 1960–61 to 3.4 Civil Emergency Planning; Box No. 3; (Folder: Swedish Defense Planning and Capabilities); NARA II, College Park, MD.

185 Nilsson, M., "'Gud vare tack och lov...", pp. 15–17; *Had There Been A War...*, p. 191; Berdal, M. R., *The United States, Norway and...*, p. 139.

186 Nilsson, M., "'Gud vare tack och lov...", p. 17.

187 "Guidelines for Policy and Operations: Sweden", dated June 1962, in: RG 59; EXSEC; Policy Guidelines [henceforth: PG], 1961–66; Spain to Uganda; Lot 67D396; Box No. 13; (Folder: Policy Guidelines Sweden); NARA II, College Park, MD. See also: *Peace and Security: ...*, p. 87.

188 *Peace and Security: ...*, p. 201.

the U.S. should *be prepared* to do the same.[189] Granted, semantics can be extremely important in the world of diplomacy, but, even so, I argue that there was essentially nothing new contained in "Guidelines for Policy and Operations" with regard to American military assistance to Sweden. In fact, the State Department had already more or less anticipated this in August 1949, when it said in its Policy Statement:

> Our interest in Sweden in present circumstances is enhanced by its strategic geographic location and its position of influence among the northern countries. It occupies a vital position on the northwestern flank of the USSR and has a commanding position in respect to both the Baltic Sea and the Danish Straits. In the light of our military commitments to Norway and Denmark under the North Atlantic Treaty, an attack on Sweden could not fail to create the most serious effects for us. We would be subjected to extreme pressure from Norway and Denmark to intervene, and we would be faced with the fact that hostile occupation of Sweden would render infinitely more difficult any future defense of those countries.[190]

Moreover, NSC 5515/1 from April 1, 1955, which was entitled "Study of Possible Hostile Soviet Actions", had stated that a Soviet attack upon Sweden should be considered a threat to the continental United States.[191] This document has, as far as I know, never been referred to by Swedish Cold War historians, but is probably worthy of some scholarly attention. Perhaps even more so than the Guidelines from 1962. Prior to 1955 it had often been said that Sweden was strategically important to the United States, but it had always been said within the context of defending Scandinavia and Western Europe. This being the case, an attack upon Sweden could have been said to constitute a threat to U.S. interests in Europe. What had not been said, however, was that an attack upon Sweden was a direct threat to the continental United States. That implied that the Americans viewed such an attack as part of a general war, and not just as a part of a regionally confined conflict in Scandinavia or Northern Europe. With this in mind it is perhaps easier

189 "Guidelines for Policy and Operations: Sweden", dated June 1962, p. 18, in: RG 59; EXSEC; PG, 1961–66; Spain to Uganda; Lot 67D396; Box No. 13; (Folder: Policy Guidelines Sweden); NARA II, College Park, MD. For the formulation in NSC 6006/1, see: FRUS: *Foreign Relations of the United States*; 1958–1960; Vol. VI, Part 2; *Western Europe*, p. 679. For a comparison between the two formulations, see: *Peace and Security: …*, pp. 89–90.

190 FRUS: *Foreign Relations of the United States*; 1949; Vol. IV; *Western Europe*, p. 773.

191 FRUS: *Foreign Relations of the United States*; 1955–1957; Vol. XIX; *National Security Policy*, p. 74. Attacks upon Iran, Yugoslavia, and Soviet occupation of Finland were also deemed to pose a serious threat to continental United States.

to understand why the U.S. government would undertake unilaterally to assist Sweden.

However, the best indication that there was essentially nothing new in the Guidelines of 1962 is the fact that the American contingency plan was referred to as an assistance guarantee by the Americans themselves, as early as January 1961. Moreover, this remark was made with regard to the *old* formulation of Paragraph 28 in NSC 6006/1, i.e. that the United States should *be prepared* to come to the assistance of Sweden. Thus one could be justified in asking: if the formulation of Paragraph 28 was seen as a guarantee by American officials, then what was so novel about the alternative writing in the Guidelines? The answer is probably that the Inquiry needed some public attention, and thus chose to stress the difference in wording just a bit too much. Furthermore, American assistance was still envisaged as arriving as a part of a NATO or U.N. response to the aggression.[192] It does, however, seem as if the matter had been more thought through by this time than it had when it was introduced in NSC 6006/1 two years earlier, for it stated that it would be "highly important to inform our NATO allies in advance of any such action we proposed to take."[193]

What *was* interesting about the Guidelines, however, was the broadening of the meaning of Soviet aggression to include both direct and *indirect* aggression, and the emphasis put on "[Sweden's] strategic position, and the contribution its defensive strength makes to the NATO northern flank."[194] It is perhaps difficult to assess what should count as indirect aggression; would it, for example, be an election victory by the Communist Party, or would a very successful election leaving the Communists in a position to heavily influence government policy do? The implication was clear: wherever aggression, direct or indirect, threatened the peace, the security of the United States was involved.

As I showed in Chapter 2,[195] this was the essence of the Truman Doctrine. The emphasis put on Sweden's strategic position and its importance for the security of NATO's northern flank runs like a straight line from Truman's explanation of the underlying rationale behind the

192 "Guidelines for Policy and Operations: Sweden", dated June 1962, p. 18, in: RG 59; EXSEC; PG, 1961–66; Spain to Uganda; Lot 67D396; Box No. 13; (Folder: Policy Guidelines Sweden); NARA II, College Park, MD.

193 Ibid. p. 18. See also: *Peace and Security: ...*, p. 90. This was perhaps an understatement; if U.S. assistance was to come as a joint NATO operation, then informing its allies was the least the United States would have to do. This may indicate that the United States was also planning on unilateral action, outside any UN or NATO response.

194 "Guidelines for Policy and Operations: Sweden", dated June 1962, p. 18, in: RG 59; EXSEC; PG, 1961–66; Spain to Uganda; Lot 67D396; Box No. 13; (Folder: Policy Guidelines Sweden); NARA II, College Park, MD.

195 See above, pp. 76–81.

Truman Doctrine. The Americans had tried to get a paragraph like this into the Atlantic Treaty in 1948, but failed due to objections on the part of its European allies.[196] Now, however, it was a matter of a unilateral U.S. undertaking, and since there was no one to object the formulation could be left standing. Another notable detail was the wish to continue "the military program for sending Swedish officers to train with U.S. officers in Germany."[197] That such a program existed had previously been unknown, yet it was not even mentioned in the Inquiry's report.[198]

The document also confirmed the importance of the gradual standardization of military equipment in Sweden with that of the NATO countries, as well as the overall military cooperation with the West undertaken since the early 1950s, when it stated that:

> [n]otwithstanding the Swedish non-participation in any alliance, the military tasks and missions of the Swedish Armed Forces are, to a great extent, in consonance with U.S. and NATO objectives, enhance the security of the United States and complement U.S. efforts to deter aggression on the NATO northern flank.[199]

Prior research has shown that on August 31, 1962, the MOU 1962 was signed by Sweden and the United States. Furthermore, it has been stated that it was a framework agreement, and that it was connected to the security agreement from January 30, 1961. The agreement itself was, it has been said, not classified, but tied to it were several classified so-called annexes, which specified the projects on which information was going to be exchanged. The contacts between the two parties were handled by specially appointed Project Officers.[200] While most of this is correct, there has been some confusion in prior research as to what the MOU 1962 actually was.

Jonter has written that the agreement was the exchange of notes published in SÖ 1962:38.[201] This is incorrect. The agreement in SÖ 1962:38 concerned only patent rights, and was concluded on October

196 See above, p. 89, note 216.

197 "Guidelines for Policy and Operations: Sweden", dated June 1962, pp. 15–16, in: RG 59; EXSEC; PG, 1961–66; Spain to Uganda; Lot 67D396; Box No. 13; (Folder: Policy Guidelines Sweden); NARA II, College Park, MD.

198 For this, see: *Peace and Security: ...*, pp. 87–93.

199 "Guidelines for Policy and Operations: Sweden", dated June 1962, p. 4, in: RG 59; EXSEC; PG, 1961–66; Spain to Uganda; Lot 67D396; Box No. 13; (Folder: Policy Guidelines Sweden); NARA II, College Park, MD. See also: *Peace and Security: ...*, p. 91.

200 Jonter, T., "Det amerikanska spåret...", p. 176; *Had There Been A War...*, p. 124; *Peace and Security: ...*, p. 102.

201 Jonter, T., "Det amerikanska spåret...", p. 176, note 117.

4, i.e. over a month later than the MOU 1962.[202] It had been modeled on similar U.S. agreements with the NATO countries.[203] Discussions soon ensued regarding procedures for reciprocal filing of classified patent applications. A conference was held at the Swedish Embassy in Paris on July 1–2, 1963, between Swedish and American representatives, and on November 17, 1964, the new procedures went into force.[204] Since Jonter thought that the patent agreement was the same as the MOU 1962, this explains the erroneous claim that the latter was unclassified.[205] In fact, the MOU 1962 was never made public, but the patent agreement was. Furthermore, the MOU 1962 was not a framework agreement, but a separate agreement intended to facilitate the exchange of classified information *before* actual contracts were signed, and spelled out the procedures for how this exchange of information (as specified in the Annexes) was to be administrated.[206]

The American draft for the MOU 1962, entitled "Memorandum of Understanding Concerning Technical Information Program", referred to

202 See: SÖ 1962:38. See also, e.g.: letter from Per Lind to Count Lewenhaupt (and included annex), dated July 9, 1962, in: HP 24 VI/U.S.A; "Tillstånd till in- och utförsel av vapen, ammunition och krigsmateriel"; 1962, juli–1964; Folder 28; Box No. 64; År: 1962–1970; RIA; Arninge.

203 "Promemoria ang. förhandlingar med USA om patentavtal", undated, p. 1, in: HP 24 VI/U.S.A; "Tillstånd till in- och utförsel av vapen, ammunition och krigsmateriel"; 1961 juni–juni 1962; Folder 27; Box No. 63; År: 1960–1962; RIA; Arninge.

204 See: protocol from the Paris conference, undated, pp. 1–9; see also: "ang. skriftväxling med Amerikas Förenta Stater avseende tillämpningsavtal till 1962 års skriftväxling rörande skyddet av patenträttigheter m.m. I samband med försvarsleveranser", dated November 27, 1964; transcript of "Procedures for Reciprocal Filing of Classified Patent Applications in the United States of American and Sweden", pp. 1–6, dated October 20, 1964, in: HP 24 VI/U.S.A; "Tillstånd till in- och utförsel av vapen, ammunition och krigsmateriel"; 1962, juli–1964; Folder 28; Box No. 64; År: 1962–1970; RIA; Arninge.

205 Jonter, T., "Det amerikanska spåret…", p. 176. Perhaps it deserves to be mentioned that the Neutrality Policy Commission may have contributed to the confusion. In its report it did not specify the purpose of the MOU 1962, but simply stated that "the parties negotiated classified appendices on technology exchange" within the framework of the agreement (*Had There Been A War…*, p. 124). However, a look at the original document would have brought order to the matter.

206 See untitled memorandum from Jarring to Åström dated may 21, 1962, p. 1, in: HP 24 VI/U.S.A; "Tillstånd till in- och utförsel av vapen, ammunition och krigsmateriel"; 1961 juni–juni 1962; Folder 27; Box No. 63; År: 1960–1962; RIA; Arninge. See also: "Memorandum of Understanding Concerning Technical Information", dated August 31, 1962, pp. 1–3, in: HP 24 VI/U.S.A; "Tillstånd till in- och utförsel av vapen, ammunition och krigsmateriel"; 1962, juli–1964; Folder 28; Box No. 64; År: 1962–1970; RIA; Arninge; *Peace and Security: …*, p. 102.

the MDAA agreement of 1952 as a "Mutual Security Agreement".[207] The Swedes felt that this was undesirable and consequently suggested that it should be removed. They also wished for the word 'program' to be eliminated from the title of the agreement.[208] The Americans granted the Swedish government's wishes. In the final version the word 'program' had been removed, and the MDAA agreement was simply referred to as an exchange of notes.[209] The Swedish government was well aware of the fact that Sweden was a party to the MDAA, but it preferred this fact to remain unspoken, at least in print, since this carried with it unwanted connotations.

The Swedes thought that the Americans understood the intricacies of the Swedish foreign policy, and that they would thus be prepared to accommodate Sweden with regard to the form for an agreement. It was even thought that because of this it was perhaps less important who signed and held the agreement together on the Swedish side.[210] This echoed the way that the conclusion of the MOU 1961 had been handled. The MOU 1961 definitely acted as a precedent for the conclusion of the MOU 1962, in the sense that the latter was also signed not by a representative of the Swedish government, but by the Vice Chiefs of the three services.[211] With the exchange of notes on October 4, 1962, regarding the patent agreement all three arrangements that the Americans had wished to make with Sweden in late 1960 (the other two being the general security agreement and the MOU 1961) had been

207 "Memorandum of understanding concerning technical information program" (Draft), undated, p. 1, in: HP 24 VI/U.S.A; "Tillstånd till in- och utförsel av vapen, ammunition och krigsmateriel"; 1961 juni–juni 1962; Folder 27; Box No. 63; År: 1960–1962; RIA; Arninge. When the U.S. government publicized the mdaa agreement in 1954 it had been designated Mutual Defense Assistance (see: "Mutual Defense Assistance: Reciprocal Procurement Rights for Military Supplies or Services. Agreement Between the UNITED STATES OF AMERICA and SWEDEN", undated, in: HP 24 VI/U.S.A; "Tillstånd till in- och utförsel av vapen, ammunition och krigsmateriel"; 1953–juni 1955; Folder 22; Box No. 62; År: 1955–1960; RIA; Arninge).

208 Letter from Åström to Jarring, dated June 20, 1962, in: HP 24 VI/U.S.A; "Tillstånd till in- och utförsel av vapen, ammunition och krigsmateriel"; 1961 juni–juni 1962; Folder 27; Box No. 63; År: 1960–1962; RIA; Arninge.

209 "Memorandum of Understanding Concerning Technical Information", dated August 31, 1962, pp. 1–3, in: HP 24 VI/U.S.A; "Tillstånd till in- och utförsel av vapen, ammunition och krigsmateriel"; 1962, juli–1964; Folder 28; Box No. 64; År: 1962–1970; RIA; Arninge.

210 "PM angående amerikanskt förslag till överenskommelse rörande informationsavtal", dated January 10, 1962, pp. 2–6, in: HP 24 VI/U.S.A; "Tillstånd till in- och utförsel av vapen, ammunition och krigsmateriel"; 1961 juni–juni 1962; Folder 27; Box No. 63; År: 1960–1962; RIA; Arninge.

211 Ibid., p. 3.

concluded. The hegemonic cooperation between Sweden and the United States had entered yet a higher stage.

Then, in late 1962, came an event that perhaps many an American official had been waiting for; Swedish Foreign Minister Östen Undén stepped down, after 17 years in this position, in favor of his younger Social Democratic colleague Torsten Nilsson. The shift was expected by the Americans to bring about a shift also in the conduct of Swedish foreign policy, as well as in the secret military collaboration scheme. In October that year, the new American Ambassador to Sweden, Graham Parsons, paid Nilsson a visit in his new office. The call was, Parsons wrote afterwards, "notable for warmth of reception", and the new Foreign Minister "reiterated desire for 'close connection' with US representative."[212] Nilsson made sure to imprint in the Ambassador's mind that his policies would be decidedly different from Undén's by stating that he was yet "another less leftward and more westward leaning of Social Democratic leaders in high office."[213] Parsons concluded his telegram by offering the Secretary immediate expectations with regard to future cooperation with Sweden. The Ambassador wrote:

> As comment, while it would be premature assert we likely to have more influence than previously on Swedish policy, we can be sure of better hearing and normal opportunity for contact through a friendly receptive Foreign Minister. Atmosphere strikingly different from that which my predecessors and I have experienced during seventeen years tenure of remote and withdrawn Unden [sic: Undén].[214]

Torsten Nilsson had thus struck a new chord with the American Ambassador, one that promised closer and more confidential relations between the two. As former Defense Minister, the secret military collaboration was well known to Nilsson, and this fact, in combination with him being more westward-leaning, and the series of recently concluded agreements, seemed to bode well for the future of Swedish-American hegemonic cooperation in the military field.

Conclusion

The period 1959–1961 was a busy one with regard to missile purchases; Falcon, Sidewinder, Hawk, as well as the British Bloodhound system were procured during this time. By the end of the 1950s it had become clear that the Swedish development effort would not result in any operational surface-to-air or air-to-air missiles. A shortage of personnel

212 758.13/10-362, Telegram from Parsons to Secretary of State, dated October 1, 1962, p. 1, in: RG 59; CDF, 1960–63; From: 758.00(w)/6-862 To: 758.5/5-660; Box No. 1842; (Folder: 758.13/1-1860); NARA II, College Park, MD.

213 Ibid., p. 1.

214 Ibid., p. 2.

and funding contributed to this outcome. The domestic projects RB 321 and RB 322 were cancelled, and do not seem to have been mourned by anyone, except perhaps by SAAB. The objective difficulties in engineering workable air-to-air and surface-to-air missiles seem to have been too great for the limited Swedish budget.

Air-to-air missiles were small and relatively light, and they all operated at supersonic speed. It would seem that the problems pertaining to guidance and steering of such missiles were too complex to be solved with the limited funding available to SAAB and the RSAFB. However, the experience gained through research on these aborted systems yielded other valuable results in the form of knowledge that could be brought to bear when purchasing and license-manufacturing foreign missiles. The RB 340/341 and RB 27/28 Falcon affair was a good example of this. Indeed, the development of RB 340/341 could be viewed as a program qualifying SAAB's engineers for license-production of the Falcon missiles. Whether that was really the case will perhaps never be known, but the available evidence point in that direction.

Sweden was among the first countries in the world to gain access to the American missile systems in question and must therefore be considered a very privileged buyer of American high-technology armaments. The risk that these systems might be turned against American planes overflying Swedish territory in wartime was considered to be so small as to be negligible: a remarkable attitude from U.S. (and British) planners if they in fact had not received some kind of reassurance from the Swedish government. This was, however, a reasonable conclusion considering the Swedish dependence on U.S. technology. The Americans seem to have reasoned that Sweden would not bite the hand that was feeding it.

By the end of the period dealt with in this chapter four agreements had been established providing for the exchange of classified information between the two countries: the security agreement in January 1961, the MOU 1961 signed in August that year, the MOU 1962 signed a year later, and finally the patent agreement concluded in October 1961. These would have great benefits for the development of the aircraft Viggen in the late 1960s and early 1970s, as well as for future Swedish missile development. The access to classified data and hardware was what characterized the period and beyond.

One of these agreements, the MOU 1961, has previously been unknown to scholars studying Swedish-American relations during the Cold War. The MOU 1961 actually made Sweden, represented by the RSAF, a part of the supply line for the U.S. military, since it stipulated that all modifications, improvements etc. of equipment originating in the United States, should be handed over to the DOD to be used by the U.S. at its discretion. The Swedish government was no longer free to decide for itself whether or not to share such information or technological systems with the United States. The MOU 1961 was probably also an important precedent for how to handle similar matters in the future, and the MOU 1962 was concluded according to the same pattern.

Chapter 7

Conclusion

Results

It is now time to go back to the questions posed in Chapter 1 and to summarize what answers this study has provided. The first set of questions was:

> *How did the United States deal with Sweden's neutrality policy, and how did Sweden handle* U.S. *preponderance during the period covered by this study. What was required for Sweden to gain access to advanced American military technology, such as guided missiles? Furthermore, what role did technology, i.e. the material objects, play in this process?*

The United States evidently used technology as a tool to get its way with Sweden during the period studied in this thesis. Technological pressure tactics were used together with more traditional diplomatic pressure to make the Swedish government understand the importance of offering its consent to U.S. hegemony in Western Europe. The Swedes tried to convince the Americans, foremost through the efforts of the Swedish Ambassador Boheman in Washington, that Sweden surely could be entrusted with American military technology, and that the neutrality rhetoric was merely politically expedient and not to be taken seriously by the United States. However, the American government wanted tangible evidence of were Sweden's allegiances lay. Prior research has not fully appreciated technology's role in Swedish-American relations.

Two factors were more significant than anything else in showing the policymakers of the United States that Sweden consented to American hegemony during the late 1940s and early 1950s; namely, Sweden's participation in the Marshall Plan and the COCOM trade embargo. Participation in these two hegemonic apparatuses had two simultaneous consequences; on the one hand it was a tangible proof of Sweden's adherence, and on the other it provided the United States with ready-made punitive tools if the Swedes strayed too far out of line.

Through its participation in the ERP Sweden contributed to the political, economic, and military formation of Western Europe in 1948–1951. This showed the American government that Sweden wished to belong to the Western camp in the years to come. The Swedish government tried to mask the political and military implications of their participation by portraying it in purely economic terms. Through her participation in the Marshall Plan, moreover, Sweden consented to a curtailment of its autonomy in foreign trade, and thereby affirmed a close ideological affinity to the West, and 'the politics of productivity'.

Sweden agreed to actively support its Western European neighbors in their efforts to rebuild their war-torn countries, and since the Marshall Plan had been cleverly engineered to discriminate the Soviet Union and Eastern Europe this meant that no equal hospitality was ever granted to the countries of the Eastern Bloc. Even more important than the ERP, however, was the Swedish government's tacit agreement to abide by the COCOM trade embargo. This question of prohibiting re-export of strategic materials to the East was a chief concern to the U.S. government, which feared that the European neutrals would act as transit countries for advanced technology to the East. Sweden could not be allowed to procure American military technology until the Americans could be sure that such equipment would not be exported eastward.

Sweden got the point, and (along with Switzerland) she soon became the most disciplined implementer of the COCOM policies out of all the Western European nations. No formal agreements were signed, except for a promise that Sweden would take into account the lists provided to them by the COCOM Secretariat in Paris. Although a cumulative process, the crucial document in this process, which finally satisfied the American government, was the Stockholm Agreement from the summer of 1951. Thus Sweden was never a member of COCOM; the government only extended the term 'contraband' to cover goods on the organization's lists.

So how did the Americans pressure Sweden into compliance? Well, in the case of the trade embargo issue the Americans used the twenty Bendix radar stations as a pawn. This protracted conflict showed that the room for accommodating opposition was much smaller before Sweden had finally acceded to American hegemony. Strategic considerations of course provided the backdrop for this episode since Sweden was not regarded as dependable due to its unwillingness to come down unambiguously on the Western side, i.e. to consent fully to U.S. hegemony. The decision to finally grant the export license for the stations in May 1951 was explicitly linked to Sweden's compliance with the trade embargo. This fact was not revealed to the Swedes, however, even though the U.S. Embassy in Stockholm recommended to Acheson that he should do so. The Swedish government was instead led to believe that the domestic supply situation in the U.S. had drastically changed so that the export license could now be granted. It is, however, hard to imagine that at least some Swedish officials did not make the connection themselves, even without any explicit U.S. statements.

From the outset it was made abundantly clear to Ambassador Boheman that the United States expected Sweden to toe the line if it wished to receive deliveries from the U.S. The Swedes, especially the military, did their utmost to appease the Americans, but despite the fact that some measures, such as Colonel Möller's presentation at the Pentagon in March 1949, impressed some American officers, nothing really changed. In the end, only Swedish consent to American hegemony could reverse the dire situation.

Having had less luck with the Swedish government, the Americans applied considerable pressure upon Swedish private companies, especially after the outbreak of full-blown war in Korea, most notably towards the ball and roller bearing producer SKF. The company eventually went so far as promising the Americans to cancel all its remaining orders to Eastern Europe in an attempt to gain U.S. acceptance. However, once the Swedish government and SKF had showed their willingness to comply with the trade embargo, the U.S. government was not unreasonable and proved very flexible when it came to the implementation of the embargo by the Swedish government. The American officials showed considerable skill in their management of the hegemonic cooperation with Sweden. Theirs was a true 'stick' and 'carrot' approach, and the U.S. government followed basically the same line when dealing with the Swedish government as it did when handling its relations with the other Western European governments. Although there were certainly idiosyncrasies in the U.S.-Swedish relation, as there are bound to be in all hegemonic relations, Sweden was generally not treated as an exception; what was true for the rest of Western Europe was also true for Sweden.

In mid-1951 the U.S. Government received the confirmation it needed, through the signing of the Stockholm Agreement, in order to extend Sweden basically the same courtesies enjoyed by the other Western European countries, including the NATO member states. This was reflected in the American attitude towards Swedish requests for purchases of military equipment. The American confirmation of Sweden's consent to its hegemony came with the announcement by Truman in early 1952 that Sweden was now eligible to buy armaments in the United States. In the summer that year came the formal acknowledgement of this fact: the exchange of notes in June/July that concluded the MDAA agreement between the two countries.

Here, too, the Americans proved very flexible. The very fact that there was only an exchange of notes and not a formal signing of an agreement is ample proof of that. Latitude was also granted Sweden with regard to the content of the agreement. For example, the Swedes were allowed to remove all explicit references that referred to the exchange as an agreement, all references to the MDAA (i.e. U.S. law) were excluded, and they were also granted exception to the clause stating that one of the ways in which the assistance received could be used, was to permit participation in the defense of the area of which the country was a part; in this case Scandinavia. All this was done with reference to the policy of neutrality, and thus shows us very concretely the material effects of this policy on the construction of the secret collaboration with the United States during this period. However, the Swedes had to promise not to use American technology for aggressive purposes, even though the government was granted some latitude on this issue as well. Instead of officially complying with U.S. legislation, Sweden referred to the U.N. Charter.

While much of this has been dealt with by other scholars this thesis sheds new light on these events by interpreting them within the context of hegemonic theory. By doing this, these events take on a new meaning and significance, and this study thus provides important new knowledge for our understanding of Swedish-American relations during the Cold War.

The second set of questions in Chapter 1 was:

> *Why did Sweden not get access to American guided missile technology before the late 1950s? How can we explain the apparent delay in this field compared with other types of military technology?*

I have shown that the reason why Sweden did not get guided missile technology from the United States until the end of the 1950s had nothing to do with the idiosyncrasies of Swedish foreign policy. Paradoxically, then, this area was perhaps one of the few that was not affected by this policy. The limited access was instead a reflection of the worries and considerations of the American Congress and government, where it was seen as harmful to U.S. strategic interests to release this technology. It is important to note, however, that we have to separate the issues of sharing operational guided missiles and sharing know-how related to such technologies. For during many years, up until the early 1950s the Americans did not really have any operational missile systems themselves; this is especially true of air-to-air missiles. Thus, there were not always any missiles available to the U.S. government even if they had wanted to supply such weapons to Sweden at an earlier stage.

Moreover, it was not only the Swedes who were unable to get access to guided-missile-related data from the United States. This experience was shared by the NATO countries as well. A call from the European allies to provide them with guided-missile-related technology at the annual NATO meeting in Paris in 1953 was refused by the United States, who regarded it as unsafe to release such information on a NATO-wide basis. Even the closest ally of the United States, Britain, was not granted complete access to classified information. In this sense, and in relation to this technology, there was no real difference in the way the United States treated Sweden, and the way in which it treated its NATO allies. The calls for the United States to grant the Europeans access to these weapons were repeated over the course of the three following years, and at the meeting in December 1956 Secretary of Defense Wilson finally committed the United States to providing modern weapons, such as guided missiles, to the NATO countries. Wilson's promise was then subsequently made into official policy in January the following year.

This decision in fact stemmed from American worries that its hegemonic position might be threatened if it failed to accommodate its European allies. Such concerns had in turn been caused by the change in general NATO war planning which after the approval of MC 48/1 in late 1955 gave priority to a strategy based on modern (including nuclear) weaponry and emphasizing the airborne threat to Western Europe. It was felt that the United States could not continue to aspire to leadership in the Atlantic Pact if it did not face up to the responsibilities that accrued to a truly hegemonic nation and to help the Europeans fulfill the force goals of MC 48/1. The decision to equip the allied nations with guided missiles also came to affect non-aligned Sweden. Apparently, it opened the door to the Swedes just as much as it did to the members of the Alliance. The explanation for this is that what mattered was not what official foreign and security policy a country professed to follow, but if it consented to American hegemony or not. No doubt, Sweden also benefited from its geostrategic position, which made it pertinent to American security interests, but this position alone would not have made Sweden eligible to purchase American missiles. This is the best explanation of why an American decision, made at the insistence of the NATO countries, to equip the member countries with guided missiles, would facilitate Sweden's access to such weapons.

Prior research has not paid any attention to this factor and to its importance in explaining why Sweden did not gain access to American missiles before the late 1950s. Instead, the focus has been on Sweden's policy and on the American view of it. I have shown that we have to elevate our perspective and take into account U.S. policy in general, as well as the intricacies of the interplay between the United States and the European NATO allies, if we are to make sense of this part of Sweden's Cold War history.

After the crucial decision had been made it was only one and a half years before Sweden announced in mid-1958 that it would procure Sidewinder. Due to the Swedish refusal to let the Americans build a large detection radar station in Sweden the amiable atmosphere was briefly interrupted, but from December 1958 the Americans raised no more objections. From then on Sweden got access to almost any missile system that it wanted, as long as it did not wish to equip these missiles with atomic warheads. Sweden then, in its turn, went to great lengths to satisfy the American demands for increased industrial security, and Sweden initiated a thorough secret security program to register and control suspected communists. It was truly a hegemonic relationship, filled with *quid pro quo*s. Despite not being a member of NATO Sweden was a privileged buyer of modern American weaponry, and it was among the first countries in the world to receive the American missile systems in question. It is noteworthy that of the countries that were scheduled to receive Sidewinder before Sweden, none of them were members of NATO, albeit they were certainly American allies in the Cold War.

The international factor is extremely important for a correct understanding of Sweden's security policy during the Cold War. There seem to be two strands of interpretations of Swedish-American relations during the Cold War: one that sees the United States' attitude towards Sweden as unique, and one that views it as part of a general pattern in U.S. foreign relations. The former view has so far dominated Swedish Cold War research. This is probably because of a limited knowledge of how the United States treated other Western European nations. However, more comparative histories are needed if we are to better comprehend what was truly unique to the Swedish experience, and, more importantly, what Sweden had in common with the NATO countries.

The third set of questions was:

> *Why did so many of the early Swedish guided missile development projects not result in operational missile systems? What purpose did the domestic R&D efforts have? Which were the priorities behind this ambitious R&D effort?*

Swedish foreign and security policy demanded, for reasons of credibility, that Swedish defenses should be strong enough to repel an attack or a violation of its neutral territory in wartime. To do this, Sweden opted for supplying its armed forces to a great extent from domestic sources. This was yet another material expression of the policy of neutrality. Therefore, it seemed only natural to the Swedes that guided missile development should also be undertaken domestically; everyone agreed that missiles would catapult the Swedish defense forces straight into the future. The Swedish domestic development, however, pretty soon turned out to be a *cul-de-sac*.

Swedish guided missile development hit a wall by the early 1950s and no help was forthcoming from the Americans. The difficulties encountered by the GMB in its efforts to produce an operational Swedish missile were directly connected to the American technological monopoly. However, the connection was not a straightforward one, and Sweden did not gain access to American guided missile know-how even though the Swedish consent to American hegemony was recognized by the U.S. government by 1952. As stated above, the reason why the Americans did not grant Sweden access to missile related data lay in overall U.S. policy, and the Swedish experience was shared by most Western European countries with the exception of Britain (and not even the British were given non-discriminatory access to American systems and know-how).

The Americans kept a close eye on the Swedish missile development efforts through the reporting of the Military Attachés, and the Swedes tried widely differing approaches in their efforts to get information from the Americans, reaching from suggestions of a barter of one technology for another, to more or less desperate calls to expose the state of Swedish guided-missile-related R&D to the Americans without any demands for reciprocity, with the hope that this would inspire reciprocal measures sometime in the future. None of this was to any avail. The lack of access to classified American information that characterized the Swedish development process went through drastic changes during the period investigated in this thesis. This did not have anything to do with anything that the Swedish government had done (apart from offering its continuing consent to American hegemony), but was dependent on much larger issues concerning domestic U.S. security policy regarding guided missile technology.

All of this made necessary a concentration of scarce resources on a few very important projects. But even though this was understood from very early on, and explicitly stated by people within the Swedish guided-missile-developing community, such concentration of development was nonetheless hard to attain due to administrative inertia, and armed service rivalry. Thus, the latter half of the 1940s and the early 1950s were characterized by a disparate development effort where valuable resources were scattered out on many different projects, and the Swedish development program was pushed further and further into the *cul-de-sac*. In fact, it was not until about the mid-1950s that the development was narrowed down to only three projects—aircraft based anti-ship missile RB 304, air-to-air missile RB 321, and surface-to-air missile RB 322. Of these, only RB 304 would eventually become operational (I will return to the probable reason for this below). There was thus a connection between the many guided missile projects initiated and the small results, in terms of operational guided missiles.

The personnel and finance-related problems experienced by the GMB were an important, limiting, factor from the very beginning. Only under circumstances where the Swedes did not have access to American and

British know-how, and where the GMB was heavily dependent upon the success of the domestic development program, could this become such a disturbance. The inability to purchase guided missiles and know-how from the United States necessitated a diverse and intensive domestic research effort that was then, paradoxically, frustrated for that very reason. There were some tiny exchanges with the surrounding world during this period, among them Britain and Norway, with some kind of (more or less) regular barter trade program being established between Norway and Sweden in early 1947. This did not do much to help the Swedish development forward, however, because Norway was not in a better position than Sweden, and by the mid-1950s it was obvious to the Swedes that the domestic development effort was not going to produce any operational surface-to-air or air-to-air missiles; at least not before the mid-1960s (at the earliest). The reasons for this were multiple, but the most important were the interrelated lack of engineers and funding.

This study is the first to examine Sweden's domestic guided missile development and the results are interesting for an understanding of the challenges that confronts a small nation trying to acquire advanced weapons technology. It shows that there are definitely limits to what can be achieved without access to huge amounts of funding and personnel. This is not only of interest to historians of technology. On the contrary, one of the main points of this thesis is that these issues deserve more attention by diplomatic historians as well. It was precisely because of Sweden's limited resources that it was necessary to seek assistance from abroad, and eventually to purchase missile systems from the United States. The consequences for Swedish security policy were immense.

The Swedes tried desperately to convey their situation to the Americans, asking them for assistance as well as opportunities to purchase missile systems, but to no avail. Instead it was the constant prodding by the NATO allies, and the awareness of its hegemonic responsibilities awakened by it, that finally made the United States decide to release guided-missile technology to Sweden. The Swedes acquired access to this technology because of the paradoxical problems that confronted the U.S. government in the mid and late 1950s. The United States was in great need of the support of its allies, and it was explicitly acknowledged by the Americans themselves that they could not continue to claim to be the leader of the Western democracies if they failed to give the allies what they wanted. To be sure, the NATO countries did not beg the U.S. government to release these weapons to the Swedish Armed Forces, but Sweden was included among those countries that got access to guided missiles due to her record of prior consent to American hegemony. The hegemonic cooperation between Sweden and the United States that had been established almost a decade earlier provided the vehicle which the issue of guided missiles was attached to. Without this prior collaboration, it is hard to imagine that Sweden would have received American missiles in the early 1960s.

In the short period between 1959 and 1961 the American Falcon, Sidewinder, Hawk missiles (as well as the British Bloodhound system) were procured, and the domestic projects RB 321 and RB 322 were cancelled. Sweden never managed to produce a workable surface-to-air or air-to-air missile during the period covered by my dissertation, but two other kinds of missiles were in operation by 1962—air-to-surface missile RB 304 and ship-based anti-ship missile RB 315. A tentative explanation of why the Swedes were able to engineer the latter kind, but not the former, is that the latter were relatively big and bulky sub-sonic missiles, while the former were smaller and lighter super-sonic missiles. It appears as if the technical challenges of building the complex guidance and steering mechanisms for super-sonic missiles were too great for the Swedes, at least with the limited amount of funding available.

However, the aborted missile systems did give the Swedish engineers valuable experience, which could then be put to good use when Sweden purchased and manufactured foreign missiles on license; the case of RB 340/341 and RB 27/28 Falcon was a perfect illustration of this. There are ample reasons to believe that the RSAFB never intended RB 340/341 to become operational, but that development was only begun in order to qualify the engineers at SAAB for license production of the American Falcon missiles.

Once the U.S. government had made the decision to release guided-missile technology to the European allies in early 1957, the exchange of information between Sweden and the United States changed both qualitatively and quantitatively, and the Swedes were given both classified information and material rewards in the form of guided missiles. By the end of the period investigated in this thesis, a framework for the exchange of classified information, a hegemonic apparatus, had been established, something that would be of great importance for the development of Swedens next military aircraft project system 37 Viggen in the late 1960s and early 1970s, not to mention future Swedish guided-missile developments.

Finally, the fourth set of questions was:

> *What role, if any, did Sweden's policy of neutrality, i.e. the material practices that constituted it, play in the overall secret military cooperation with the United States during this period? How did it affect the material actions undertaken, and how did it constrain and/or enable the historical subjects (i.e. actors) in the* U.S. *and in Sweden?*

As we have seen above, the material factors and interests that provided the rationale for the policy of neutrality, and that were reflected in this policy, had a very real effect on the way in which the secret Swedish-American collaboration was implemented. It must therefore be concluded that Gerard Aalders, as mentioned in Chapter 2, was wrong in concluding that the Cold War did not affect the Swedish policy in any notable way. Due to the great cost to the Swedish government if any of these efforts had become known, these interests suffused every aspect of the military technological relationship. Interestingly, though, the government never tried to hide the fact that it purchased guided missiles from the United States. These purchases were announced far and wide in Swedish newspapers, probably because of the propaganda value in letting it be known that Sweden had a modern defense. This fitted perfectly with the mythology of Swedish neutrality. The decision to start domestic development of guided missiles, and to continue these efforts all through the period in question, was a very concrete materialization of the Swedish policy of neutrality.

Once American systems could be purchased, the Swedish government does not seem to have sensed that there was any problem in making it common knowledge that Sweden was not self-sufficient in this area. In fact, it seems as if technology was viewed as something neutral in itself, as if there had been no concessions made in order to gain access to these weapons. The fact that the government always acknowledged that technology was indeed not neutral at all, through the refusal to contemplate purchases of Soviet missiles, was ignored and did not disturb this image. It is probable to assume that this, in reality, glaring inconsistency did not strike Swedish officials at all. This option was so unthinkable that it never even entered the minds of these actors. It was masked, unconsciously, by ideology and could thus it be overlooked without any effort. But there was probably also a conscious side to this problem. There seems to have been an assumption that Soviet technology would surely compromise Swedish neutrality in American eyes, and perhaps in the view of the Swedish public as well. But the retention of a large domestic defense-related production sector was important, because *all* armaments could not be allowed to come from abroad, since this would also lessen the credibility of the neutrality option. The Swedish position was thus more than a little Janus-faced.

Perhaps this is the reason why certain sectors, such as the aircraft industry, were 'blackboxed', at least partly, so that the full extent of foreign involvement would not become known. The government went about this blackboxing by propaganda measures speaking of the specific nature of Swedish armaments technology. According to this official account, Sweden's defense system could not be integrated with other nations due to technological discrepancies between Sweden's and, for instance, NATO's forces. But research has shown that Swedish technology was modified so as to enable cooperation with the West in the case of an emergency. Certainly, there were Swedish idiosyncrasies but these were most often out-weighed by the covert standardization effort.

All this clearly shows that the domestic development should not be interpreted as if this was the norm, and that consequently the policy of neutrality, and the material practices that it gave rise to (and which in turn acted to transform formulations of policy), were interrupted by short erratic bursts of concessions to the United States. This study argues that Swedish consent to U.S. hegemony was constant from the early 1950s up to 1962, and that the appearance of dissent should be interpreted in terms of accommodating opposition. Sweden consented to the hegemonic cooperation with the United States due to the material rewards, in the shape of sophisticated military technology and the increased sense of security that derived from it. But Sweden was at the same time an independent nation-state with its own peculiarities, and the government had to placate not only the general public, but also powerful business interests (of which the most influential was the omnipresent Wallenberg family) which gained enormously from the large defense contracts placed within its companies.

The Americans, after 1952, were almost always prepared to play along and accept these expressions of accommodating opposition as long as the Swedish government consented to its hegemony in general. Moreover, the United States was not all-powerful and could not control the actions of its allies. As a hegemonic power it could never be a puppet master. What it could do, however, due to its economic, political, and military preponderance, was to exert pressure to try and influence the Swedes to behave in ways that suited American policy objectives. While there were certainly issues on which the U.S. and the Swedish governments did not see eye to eye, military and strategic matters was not really one of those. The Soviet Union loomed large as the only possible enemy in the eyes of both Swedish and American policy-makers and military planners. On this arena they found common ground, and both governments were pragmatic in their interaction with each other. Each of them got basically what it wanted from the relationship. Sweden got military technology as well as enhanced national security, and the United States got a valuable ally in the Cold War as well as a strong bastion of defense in Scandinavia.

The interpretation in this study of the policy of neutrality's role in Swedish-American relations differs from prior studies in that it does not view this policy as antithetical to Swedish cooperation with the United States. The policy of neutrality served a political function in Sweden domestically, and a security function internationally. Once it became clear to U.S. policymakers that the Swedes were prepared to work around this policy, and that it did not pose a threat to American security interests in the region, the United States was prepared to compromise when dealing with the Swedish government. Swedish demands for secrecy and informality could be met, within limits of course. Thus, the policy of neutrality, or rather the material circumstances which made it necessary for the Social Democratic government in Stockholm, did affect the form and content of the hegemonic relationship, although not necessarily to the detriment of the United States. For example, the Swedish government tended to prefer not to sign formal agreements, and this was perfectly alright with the Americans, just as long as both sides understood that the result was still the same. The only real exception to this rule was the exchange of notes in the summer of 1952, which was signed by Östen Undén, whereby Sweden became eligible to buy military materiel in the United States, and this was basically only because the Americans insisted upon that such an exchange was necessary. But generally, the Swedish government preferred to handle matters through unilateral 'promises' or through semi-official or non-governmental proxies, as in the case of the MOU's of 1961 and 1962.

Technology's Consequences for U.S.-Swedish Relations and Sweden's Freedom of Action During the Cold War

In this thesis I have argued that the concept of hegemony is important for a correct understanding and interpretation of the secret military technological cooperation between Sweden and the United States during the Cold War. We have also seen that even before the onset of the Cold War, in the immediate postwar years, there was a certain amount of military cooperation between Sweden and the United States. However, the little that was soon ebbed away and was replaced by a harsh American attitude towards Sweden's neutrality policy and Swedish requests for export licenses for armaments.

The first real issue of conflict where Swedish government policy ran up against U.S. hegemonic objectives was the effort to form a Scandinavian Defense Union in 1948–49. The SDU proposal was a blunt and direct challenge to U.S. security interests in Scandinavia; indeed, it threatened the success of American hegemony in the region. Scandinavia as a whole, but especially Norway and Denmark (due to their geostrategic location as well as their island possessions Svalbard and Greenland), were too strategically important for the Americans to sit idly by while the Swedes led their neighbors astray.

More importantly, the SDU also challenged the United States' intention of getting Norway and Denmark into NATO, and hence the effectiveness of the entire Atlantic Alliance. The response chosen by the U.S. to counter this development was to use military technology as a an arbiter to achieve its policy objectives in the region. From the outset U.S. officials made clear to the Scandinavian countries that a SDU based on neutrality would not be given priority when trying to purchase arms from the United States. With the wartime experience fresh in mind, Norway and Denmark thus abandoned the SDU and signed up for membership in the NAT instead. The aftermath proved to be not so bad for Sweden after all, and a secret collaboration was initiated with the Western Powers, with Norway acting as proxy for NATO and the U.S. But even so, all through the period dealt with in this thesis, the Americans continued to press the Swedes to take a firm stand in the East-West conflict. The assurances of Boheman and other diplomats were not enough for the Americans; they wanted more tangible evidence of Sweden's position.

I have shown that the assumption that the United States gave up its ambition to get Sweden into NATO in early 1950 must be altered, and the problem put in another way. The goal of getting Sweden to join formally with an Atlantic Alliance was never part of the official U.S. policy objectives as set forth in NSC 28/1 approved in September 1948. Indeed, at that time, the Atlantic Pact did not yet exist, and NSC 28/1 spoke only of *integrating or associating* Sweden *in some way* with the Brussels Treaty. This can hardly be interpreted to mean that the U.S. was planning to include Sweden formally in a future NAT. What was sought in this document was merely to get Sweden to reject her subjective neutrality and align herself with the Western Powers in such a way as would have been collectively acceptable. Apart from certain individually expressed efforts to the contrary, e.g. those of Ambassador Matthews, it was never official U.S. policy to get Sweden to join an Atlantic Alliance after this point.

Moreover, I have argued that the explanation for the more amicable attitude towards Sweden on the part of the Americans, which was discernable after 1950, can best be understood if this process is analyzed with the help of the concept of hegemony. If that is done, it becomes clear that the improved relations were due not to any American reappraisal of Swedish foreign policy, but to Swedish consent to American hegemony through the promise to adhere to the COCOM embargo lists. The difference between these interpretations is an important one. As stated above, this became even more pronounced after the signing of the Stockholm Agreement in the summer of 1951.

In 1953 the Swedish government actively tried to support NATO's military build-up in Western Europe by prodding the Americans to place OSP production orders in Sweden. The Americans were very interested in this project and the Swedish Export Association acted as an intermediary in supplying its knowledge of Swedish industry. Certain contracts were awarded to Swedish companies that manufactured materiel for U.S. forces, but Sweden was not considered eligible for large prime contracts due to the fact that it would prove unpopular with the NATO countries if a non-NATO member received funds allotted for them.

However, Sweden did subcontract contracts for the production of explosives and propellants, mainly to Norway. This seems to have been the case at least as late as the mid-1950s.

Despite not being a member of NATO or any other regional defense alliance, Sweden was among the first countries in the world to gain access to the American missile systems in question and must therefore be considered very privileged as a buyer of American high technology armaments. The idea that these systems might be turned against American planes overflying Swedish territory in wartime was considered to be so small as to be negligible, a remarkable attitude from U.S. (and British) planners if they had not in fact received any kind of assurance from the Swedish government. This means that we will have to draw the conclusion that, with regard to Hakovirta's statement that the neutrality of a given state is more *credible* the more likely it is that it intends and is able to maintain its neutral position in a time of war,[1] Sweden's policy of neutrality was a huge failure, at least with regards to Britain and the United States.

The research that has been done in Russian archives thus far indicates that the Soviets did not think that Sweden's policy of neutrality was credible either.[2] The Soviet Union was not reassured by Sweden's high military expenditures, and viewed neutrality as nothing more than a camouflage. Sweden was generally deemed unreliable and, concludes Bo Petersson, "neutrality was not attributed an intrinsic value by Soviet analysts."[3] Indeed, this was even the conclusion drawn by the Neutrality

1 Hakovirta, Harto, *East-West Conflict and European Neutrality* (Oxford: Clarendon Press, 1988), pp. 26–38; quotes on p. 26 and p. 27 respectively.

2 This argument has also been advanced by Ingemar Dörfer, see: Dörfer, I., "Neutralitetsbygget... in *Kungl. Krigsvetenskapsakademiens handlingar och tidskrift*, 1995:1, p. 18.

3 Petersson, Bo, *The Soviet Union and Peacetime Neutrality in Europe. A Study of Soviet Political Language.* (Stockholm: Utrikespolitiska Institutet, 1990), pp. on military expenditures: 68, 75–77, on neutrality as camouflage: 137, quote on p. 138; Petersson, B., *Med Moskvas ögon. Bedömningar av svensk utrikespolitik under Stalin och Chrustjov.* (Smedjebacken: Bokförlaget Arena, 1994), pp. on neutrality as camouflage and its persistence over time: 141–142, on Sweden as unreliable: 60–62. See also: Berner, Örjan, *Soviet Policies Toward the Nordic Countries* (Lanham: University Press of America, 1986), pp. 41–104; Linder, Jan, *Svår neutralitet. Sverige under två sekel* (Stockholm: Infomanager Förlag, 2003), p. 166. An overview of the research regarding Swedish-Soviet relations can be found in: Petersson, M., "Svensk-sovjetiska säkerhetspolitiska relationer 1945–1960 – en översikt" in K. Zetterberg (ed.), *Att skåda Sovjetunionen i vitögat. Sex studier kring svenska relationer till Sovjetunionen under det kalla kriget* (Stockholm: Forskningsprojektet Försvaret och det kalla kriget (FOKK), Publikation nr. 2, 2004), pp. 17–60. See also: Molin K., "The Central Issues...", in M. Gehler and R. Steininger (eds.), The Neutrals and the European Integration, pp. 271–274.

Policy Commission in 1994.[4] There is also evidence that Polish planners saw Sweden as a threat to Polish security.[5] This was of course contrary to the intentions of the principal architect of Sweden's Soviet policy during 1945–1962, Foreign Minister Östen Undén,[6] for credibility was everything for Sweden if she was to be able to declare herself neutral and non-aligned after the outbreak of war. Thus, from this perspective, the Swedish neutrality policy could be deemed unsuccessful. Whether the Soviet attitude was due to their having knowledge of the Swedish-American cooperation, or if it sprang from a general mistrust of all capitalist nations, is not possible to say. However, the only way that one could know for sure if the policy had failed was if Sweden had been drawn into a war—i.e. after the fact. But even if there would have been a war, and even if Sweden would not have been attacked, we cannot say *a priori* that this was due exclusively, or even mainly, to the policy of neutrality.

We have seen in this dissertation that the Americans did not worry about their missiles being turned against U.S. aircraft overflying Sweden during wartime. While there is no documentary evidence that can tell us exactly why U.S. officials felt that they were justified in their assumption, I will allow myself to speculate regarding their possible reasoning. It is not implausible to assume that American officials and military officers reasoned that the probability of such actions decreased in direct proportion to the increase in Swedish dependence upon U.S. technology. The Americans were no doubt convinced that the Swedes knew that the United States would never supply spare parts, let alone deliver new missiles, to a nation which had just used up their limited missile supply by shooting at U.S. aircraft. This must have seemed like an infallible logic to them. Considering that U.S. officials knew full well that Swedish defense plans presupposed help from the outside, this was perhaps one of the main reasons why the Americans thought that the chance of a missile originating in the U.S. being turned against American aircrews was so small as to be negligible.

Surely, Sweden would have found itself in an extremely precarious situation if there had been a war. Both sides in the conflict would have sent their bombers over Swedish territory, and the members of the government would have had to decide whom they would fire upon. This would not have been an easy decision to make. Would they have discriminated against Soviet aircraft and risked Russian retaliation, i.e.

4 *Had There Been A War...*, p. 266.

5 For this, see: Agrell, W., "Silent allies and hostile neutrals: Nonaligned states in the Cold War" in S. G. Holtsmark, V. Mastny, and Andreas Wenger (eds.), *War Plans and Alliances in the Cold War: Threat Perception in the East and West* (London: Routledge, 2006), pp. 141–162.

6 Kronvall, O., *Östen Undéns sovjetsyn...*, p. 93.

war? Or would they have defended Swedish neutrality indiscriminately, and thereby alienated the Western Powers? As we have seen, NIE 49 expected the RSAF not to attack Western aircraft overflying Sweden in wartime, even if that meant going against direct orders from the Swedish government. It is not implausible to assume that Nordenskiöld's statements had made a lasting impression upon the Americans.

The limited supply of guided missiles, with the concomitant dependence upon further deliveries from the U.S., was a strong deterrent to Sweden against firing at American planes flying over Sweden. If defense against all intrusions had really been the aim of the Swedish preparations, then the government would have had to purchase many more missiles initially in order to be able to hold out longer. Furthermore, it is hard to understand how a country could base its defense upon the assumption that the U.S. would provide assistance, if just prior to that their aircraft had been blown out of the sky by the same government now asking for their help in defending its nation against a Soviet invasion. Of course, there is a remote chance that the Swedish government never thought twice about this glaring inconsistency, but to think that the military did not remark on the incompatibility of this doctrine with the dependence on U.S. technology is very hard indeed. Prior research has not made a point out of this, because it has not seen the restrictions put on governmental actions by technology. Here the history of technology, with its artifactual focus, can provide new insights into the nature of Swedish-American relations during the Cold War. This study thus contradicts the conclusion made by the Neutrality Policy Commission that the military's technological cooperation with the United States did not circumscribe Sweden's freedom of action.[7]

There was also another dimension to this issue, which was literally built into the hardware, and that was the question of missile homing frequencies. One of the major reasons for the United States not to release guided missile secrets to the Allies had been the fear that the frequencies on which the missiles' homing devices operated would become known to the Soviets. This would have rendered the American missiles ineffective since the Russians could target these frequencies and jam the receiver in the missiles. Maybe this was an option for U.S. aircraft entering Swedish airspace too. The Americans would of course have ready access to the frequencies of their own missiles, and could perhaps bring their losses down to a very acceptable minimum. One could then speculate that U.S. apprehensions about exporting guided missile technology to Sweden did not have so much to do with the risk of these missiles being turned against American planes, as with the risk of technical data regarding these missiles ending up in the hands of the

7 *Had There Been A War...*, pp. 265–266.

Russians. This would have been a reason to be fretful about industrial security for example.

On the other hand, when Kohler in August 1959 had talked about, and dismissed, the risk of the Swedes shooting at U.S. aircraft, he did in fact say that these missiles could be used with considerable effect. He would not have done that if he had known (which we probably have to assume that he would have) that the United States could get around this problem by resorting to technological gadgetry. Instead he said that other considerations outweighed this risk. It does not seem as if this would imply that the Americans did not have to worry because they could apply technological countermeasures. Thus, while we might never know the true reason, it seems most plausible to believe that what he had in mind were the combined effects of prior Swedish-American relations and Sweden's consent to U.S. hegemony, the Swedish dependence upon U.S. assistance for its defense, and the Swedish dependence on American military technology.

The secret cooperation with the United States and other Western powers was motivated by the Swedish need for assistance should the neutrality policy have failed and the country have been attacked. The underlying assumption behind these preparations was sound, Sweden did not have the power to withstand an onslaught for a longer period of time. There was a serious problem with this reasoning, however. The very existence of these ties to the United States and nato might have prejudiced the whole policy of neutrality in the eyes of the Russians. This scenario has not been much debated by historians, but it deserves more scholarly attention.[8] It is quite possible that the preparations and contacts with the West served to make the Swedish position less credible. The Soviets could not be sure that the Swedes were not planning to join, or aid, a Western offensive from day one. We know that this kind of misunderstanding drove much of the arms race during the Cold War, and indeed animated much of the Cold War conflict in general. What was intended as a defensive move by one party, was interpreted as an offensive move by the other. It is not unreasonable to assume that the same kind of logic could have affected Soviet planning with regard to Sweden. The paradoxal result would have been less security for the Swedish people.

The policy paper on Scandinavia NSC 6006/1 in a sense epitomized the American use of technology as a tool of hegemony. By stating as a main goal of U.S. policy in Scandinavia that Sweden's weapons systems should be compatible with, and complementary to, the ones being installed in neighboring NATO countries, and that such measures would contribute to the defense preparedness of the West, Washington was

8 For a noteworthy exception, see: Agrell, W., *Fred och fruktan. Sveriges säkerhetspolitiska historia 1918–2000* (Lund: Historiska Media, 2000), p. 157.

transforming an immaterial entity, i.e. strategic thinking, into material objects, e.g. guided missiles. Observe that this is not the same thing as saying that the United States would necessarily have been successful in doing so. After all, the Americans could not be sure, at least not as far as we know, of how these weapons would be utilized in the event of an armed conflict in Europe. However, American military technology was clearly seen as strengthening U.S. hegemony in Scandinavia. It is important to stress that technology does not contain politics in and by itself. It is the structure behind the artifacts, that put them in place and offers them support, which is political. The tools of hegemony can only serve their purpose as long as they are backed by the power and influence of the hegemon.

Bibliography

Primary Sources:

Archive Collections

The United States

The Dwight D. Eisenhower Library, Abilene, Kansas:

DWIGHT D. EISENHOWER; WHITE HOUSE OFFICE FILES:
National Security Council Staff Papers, 1948–61; Disaster File.

DWIGHT D. EISENHOWER; PAPERS AS PRESIDENT OF THE U.S.; ANN WHITMAN FILE:
Dulles-Herter Series.

National Archives and Records Administration, NARA II, College Park, Maryland:

RECORD GROUP 59; RECORDS OF THE U.S. DEPARTMENT OF STATE:
1955–1959 Central Decimal File; From: 758.11/11-1555 To: 758.55361/5-1556.
1955–1959 Central Decimal File; From 758.56/1-555 to 759.00 mayday/3-357.

Central Decimal File, 1960–63; from 758.56/1-1360 to 759.00 (W)/5-261.
Central Decimal File, 1960–63; From: 758.00(W)/6-862 To: 758.5/5-660.

Decimal File 1950–54; from 758.13/1-852 to 758.5MSP/12-1554.

Decimal File 1950–54; from 758.52/6-2551 to 758.713/10-3052.

Miscellaneous Lot Files; Lots 57 D 321, 323, & 324; Office Files of Leonard Unger, 1951–1956.

Miscellaneous Office Files of the Assistant Secretaries of State for European Affairs, 1943–1957.
Records of the Bureau of European Affairs; Subject Files Relating to European Defense Agreements, 1948–1954; Lot File 55D258.

Bureau of European Affairs Country Director for Scandinavia, Iceland and Finland (EUR/SCAN); Records Relating to Swedish Affairs, 1958–1964; POL

17 Political Affairs Sweden POL 17 Diplomatic Notes 1959–1963 to PS – Protective Service Sweden PS 7-1 Arrest (Swed. Seamen Visit New Orleans).

Bureau of European Affairs; Office of European Regional Affairs; Political-Military Numeric Files 1953–1962; 2.4 MWDP General, 1960–61 to 3.4 Civil Emergency Planning.

Bureau of European Affairs Office of Northern European Affairs; Records Relating to Sweden, 1957–1975; DEF 2-Gen. Repports [sic] and Statistics, 1966–1969 to POL 33-8 – Law of the Sea, 1960–1969; Lot File: 72D339.

Bureau of European Affairs; Office of Atlantic Political and Military Affairs; Military Assistance Policy/Mutual Security Program Files 1961–1963; Economic, Berlin to Yugoslavia.

Records of the Office of European Regional Affairs, 1946–1953; Subject Files, 1946–53; Files of Miriam Camp; Aviation France.

Records of the Office of European Regional Affairs; Records of the Director, 1955–1960; Subject Files, 1956–60; Correspondence Files, 1955–59; Lot 61D252.

Records of the Office of British Commonwealth and Northern European Affairs, 1941–1953; Subject Files, 1941–1953; South Africa, Union of, Policy – Sweden, External Political Affairs: U.S.

Records of the Office of British Commonwealth and Northern European Affairs, 1941–1953; Subject Files, 1941–1953; Country (Sweden, External Political Affairs: U.S.S.R..-Sweden, Wood Pulp).

Subject Files of the Bureau of Intelligence and Research (INR), 1945–1960; Lot 58D776.
Executive Secretariat; The Secretary's and the Undersecretary's Memoranda of Conversation, 1953–1964; Lot 64D199; June 1957 to December 1957.

Executive Secretariat; Policy Guidelines, 1961–66; Spain to Uganda; Lot 67D396.
Subject Files of the Bureau of Intelligence and Research (INR), 1945–1960; Lot 58D776.

Records of Ambassador John M. Cabot, 1945–1963; Folder 1 to Folder 10.

Records of Ambassador John M. Cabot, 1945–1963; Folder 11 to Folder 23.

RECORD GROUP 84; RECORDS OF THE FOREIGN SERVICE POSTS OF THE DEPARTMENT OF STATE:
Sweden; Stockholm Embassy; Top Secret General Records; 1943–1949.
Stockholm Legation and Embassy; Confidential File 1948.

Sweden; Stockholm Embassy; Classified General Records 1953–1963; 321.3 German Assets 1954 370.

Sweden; Stockholm Embassy; Classified General Records, 1950–1952; 050-350.

Sweden; Stockholm Embassy; Classified General Records, 1950–52; 511.2.

Stockholm Legation and Embassy; Confidential File 1945–1946; 861.7-879.6 (1945); 090-850 (1946).

RECORD GROUP 156; RECORDS OF THE OFFICE OF THE CHIEF OF ORDNANCE:

Executive Office Historical Branch; Records Relating to the Army Guided Missiles Program, 1940–62; GM 29/ Foreign Intelligence through GM 78 Consolidation Navy AAM Pro.

Executive Office Historical Branch; Records Relating to the Army Guided Missiles Program, 1940–62; GM 80/ Project Falcon MX-904 through GM 210/Snark.

RECORD GROUP 218; RECORDS OF THE JOINT CHIEFS OF STAFF:
Central Decimal File 1948–50; 381 (10-2-46) B.P. Pt. 2 to 381 (8-14-47) Sec. 3.

Central Decimal File 1954–56; 350.05 (3-16-44) Sec. 33–37; Box No. 74; (Folder:
Exchange of Technical Information with Foreign Countries; CCS 350.05 (3-16-44) Sec. 37).

Central Decimal File 1957; 350.05 (3-16-48) Sec. 8 to 350.05 (3-16-44) Sec. 55.

Central Decimal File, 1959; 9167/3440 Italy (1959) to 9170/9105 Europe (East) (15 July 1959).

Chairman's File; Admiral Radford, 1953–57; 471.6 (1 Aug 56) to 500.

Geographic File, 1954–56; Western Europe (3-12-48) Sec. 294–302; Box No. 48; (Folder: CCS 092 Western Europe (3-12-48) Sec. 302).

RECORD GROUP 319; RECORDS OF THE ARMY STAFF:
Army-Intelligence Project Decimal File 1951–1952; From 061 Swedish M/A to 823.1 Swedish M/A.

Army-Intelligence Project Decimal File 1951–1952; From 011 Sweden to 335.11 Sweden.

Army-Intelligence Project Decimal File 1951–1952; From 041.221 M/A Spain to 601.53 M/A Sweden.

Army-Intelligence Project Decimal File 1951–1952; From 335.11 Sweden to 676.3 Sweden, Horby.

Army-Intelligence Document File; From ID 662461 To ID 662710.

Army-Intelligence Document File; From ID 523271 To ID 523460.

Army-Intelligence Document File; From ID 394901 To ID 395100.

Army-Intelligence Document File; From ID 416521 To ID 416640.

Army-Intelligence Document File; From ID 382201 To ID 382390.

RECORD GROUP 330; RECORDS OF THE OFFICE OF THE SECRETARY OF DEFENSE:
International Security Affairs, Office of Military Assistance; Project Decimal File Apr 1949–May 1953; 333 Spain to 091.3-523 Thailand.

RECORD GROUP 341; RECORDS OF HEADQUARTERS UNITED STATES AIR FORCE (AIR STAFF):
Deputy Chief of Staff, Operations; Director of Intelligence; Dep Dir for Collection & Dissemination; Collection Operations Division; Air Attaché Branch; Station File, 1948–49; Spain to Sweden.

Deputy Chief of Staff, Operations; Director of Intelligence; Dep Dir for Collection & Dissemination; Collection Operations Division; Air Attaché Branch; Aircraft Operations Reports, 1948–51; Singapore to Union of South Africa.

RECORD GROUP 469; RECORDS OF THE U.S. FOREIGN ASSISTANCE AGENCIES:
1948–61; USRO; Executive Secretariat; Country Files, 1952–55; Spain-Turkey.

1948–61; Special Representative in Europe; Defense Production Division; Subject Files, 1950–52; Programs: Economic, UK-Military, UK.

1948–1961; Office of the Special Representative in Europe (1948–53); Office of the Special Assistant for Staff Planning; Subject Files of Henry J. Tosca [sic: Tasca, the name is misspelled on the box], Special Assistant, 1949–52.

1948–1961; Special Representative in Europe; Central Secretariat (1948–51); Country Files, 1948–52; S-T.
1948–1961; Office of the Special Representative in Europe (1948–53); Office of the Director of Administration Communications and Records Unit; Country Files, 1948–51; Sweden P-T.

Belgium

I. NATO Archives, Brussels:

STANDING GROUP RECORDS SECTION

Sweden

I. The Swedish Government Offices Records Center (Regeringskansliets arkiv- och dokumentcenter):

BOX: REGERINGSBESLUT 1995-06-29: "Neutralitetspolitikkommissionen – samtal med Sven Hasselrot" [The Neutrality Policy Commission – Conversation with Sven Hasselrot], 1993-04-02.

BOX NO. 16:
Public Record Office: FO 115/4364, 211/774, FO 188/669-188/673, FO 371/77394-371/166292.

II. The Swedish Military Archives (Krigsarkivet), Stockholm:

KUNGLIGA FLYGFÖRVALTNINGEN (KFF):
Robotbyrån, Serie: F II B, "Inkomna och avgångna handlingar ordnade efter klassifikationssystem", M 58-1300-M 93, År: 1954.

Robotbyrån, Serie: F III B, "Inkomna och avgångna skrivelser ordnade efter klassifikationssystem", A 01-A 7 , år: 1955.

Robotbyrån, Serie: F III B, "Inkomna och avgångna skrivelser ordnade efter klassifikationssystem", A 1005:26-A 373:7, År: 1956.

Robotbyrån, Serie: F 11IB, "Inkomna och avgångna skrivelser ordnade efter klassifikationssystem", A 40-M 3:50, År: 1957.

Robotbyrån, serie: F III B, "Inkomna och avgångna skrivelser ordnade efter klassifikationssystem", M 44-340:17-M 44-365:20 , år: 1959.

Robotbyrån, Serie: F 111 B, "Inkomna och avgångna skrivelser ordnade efter klassifikationssystem", M 44-340:71-M 44-365:120, År: 1960.

Försvarets robotvapenbyrå, Serie: F 111 A, "Inkomna och avgångna skrivelser ordnade efter klassifikationssystem", A 01-M 212, År: 1952.

Försvarets robotvapenbyrå, Serie: F III A, "Inkomna och avgångna skrivelser ordnade efter klassifikationssystem", A 40-M 3:30, År: 1957.

Försvarets robotvapenbyrå, Serie: F III B, "Inkomna och avgångna skrivelser ordnade efter klassifikationssystem", A 1005-A 373:10, År: 1958.

Robotvapenbyrån Försvarets robotvapenråd (FRR), Serie: A I, "Protokoll från sammanträden 1948–1957", År: 1948–1957.

Centrala ledningen för reaktions- och raketdrift (CLR), Serie: A II, "Protokoll från sammanträden", År: 1945–1948.

Centrala ledningen för reaktions- och raketdrift (CLR), Serie: A II, "Inkomna och avgångna handlingar med diarium", År: 1945–1948.

Centrala ledningen för reaktions- och raketdrift (CLR), Serie: F I, "Inkomna och avgångna handlingar med diarium", År: 1947.

Serie: F III, "Hemliga avtal och kontrakt", År: 1959.
1959–1960, Serie: F III, "Hemliga avtal och kontrakt".

1960–1961, Serie: F III, "Hemliga avtal och kontrakt".

Robotavdelningens försöksplats i Vidsel, Serie: F I, "Avgående och inkommande handlingar HA 1-HA 8", 1962.

ARMÉTYGFÖRVALTNINGEN (KATF): Vapenavdelningen, Robotkontoret (RBK), Serie: F I, "Avgångna och inkomna skrivelser ordnade efter klassifikationssystem", NR 397 006:1-776 001:30, År: 1955–59.

Vapenavdelningen, Robotkontoret (RBK), Serie: F I, " Avgångna och inkomna skrivelser ordnade efter klassifikationssystem", NR 397 006:28-775 007:75, År: 1961.

Vapenavdelningen, Robotkontoret (RBK), Serie: F I, " Avgångna och inkomna skrivelser ordnade efter klassifikationssystem", NR 397 000:41-775 001:122, År: 1963.

Vapenavdelningen, Robotkontoret (RBK), Serie: F I, " Avgångna och inkomna skrivelser ordnade efter klassifikationssystem", NR 775 001:449-775 002:89, År: 1965.

TORSTEN RAPPS ARKIV

III. The Archives of the Swedish Armed Forces Headquarters, Stockholm (Högkvarterets arkiv):

Flygstaben KH, Planeringsavdelningen, "Inkomna och utgående skrivelser", Serie: F III, År: 1955–59.

IV a. The National Archives of Sweden (Riksarkivet), Arninge:

UTRIKESDEPARTEMENTETS ARKIV: HP 24 VI/U.S.A; "Tillstånd till in- och utförsel av vapen, ammunition och krigsmateriel"; 1953–juni 1955; Folder 22; Box No. 62; År: 1955–1960.

HP 24 VI/U.S.A; "Tillstånd till in- och utförsel av vapen, ammunition och krigsmateriel"; 1960 mars–1960 oktober; Folder 25; Box No. 63; År: 1960–1962.

HP 24 VI/U.S.A; "Tillstånd till in- och utförsel av vapen, ammunition och krigsmateriel"; 1960 nov–maj 1961; Folder 26; Box No. 63; År: 1960–1962.

HP 24 VI/U.S.A; "Tillstånd till in- och utförsel av vapen, ammunition och krigsmateriel"; 1961 juni–juni 1962; Folder 27; Box No. 63; År: 1960–1962.

HP 24 VI/U.S.A; "Tillstånd till in- och utförsel av vapen, ammunition och krigsmateriel"; 1962, juli–1964; Folder 28; Box No. 64; År: 1962–1970.

SVERIGES ALLMÄNNA EXPORTFÖRENINGS ARKIV:
Sveriges Allmänna Exportförening; Serie: Korrespondens, samlingsserie; Signum: E I a; Box No. 1257; RiA, Arninge.

IV b. The National Archives of Sweden (Riksarkivet), Stockholm:

UTRIKESDEPARTEMENTETS ARKIV:
HP 24 VI/U.S.A; "Tillstånd till in- och utförsel av vapen, ammunition och krigsmateriel"; 1946, okt.–aug. 1950; Folder 20; Box No. 1353; År: 1946–1952.

V. SAAB Dynamics, Linköping:

SAABS *robot- och elektronikhistoria, del 1-6*, [SAAB's Missile and Electronics History, Parts 1–6] "SAAB och svensk robotverksamhet under 50-talet". Unpublished report which served as the basis for the 50th anniversary book *Robot 50 år*. (I got access to this report through former CEO of SAAB Dynamics Anders A. Andersson. This report can be accessed, on request, from the author.)

Governmental Publications:

The United States

THE FOREIGN RELATIONS OF THE UNITED STATES (FRUS):
1948; Vol. III, *Western Europe* (Washington, D.C.: United States Government Printing Office, 1974).

1949; Vol. IV; *Western Europe* (Washington, D.C.: United States Government Printing Office, 1975).

1950; Vol. I; *National Security Affairs; Foreign Economic Policy* (Washington, D.C.: United States Government Printing Office, 1977).

1950, Vol. III, *Western Europe* (Washington, D.C.: United States Government Printing Office, 1977).

1951; Vol. IV; *Europe: Political and Economic Developments*; Part 1 (Washington, D.C.: United States Government Printing Office, 1985).

1952–1954; Vol. VI, *Western Europe and Canada* (Washington, D.C.: United States Government Printing Office, 1986).

1955–1957; Vol. XIX; *National Security Policy* (Washington, D.C.: United States Government Printing Office, 1990).

1958–1960; Vol. VI, Part 2; *Western Europe* (Washington, D.C.: United States Government Printing Office, 1993).

Sweden

RIKSDAGSTRYCKET:
"Kungl. Maj:ts proposition nr 110 år 1958", in: *Bihang till Riksdagens protokoll år 1958*, första samlingen, elfte bandet, Kungl. Maj:ts Propositioner 110–119.

"Statsverkspropositionen år 1957: BIL. 6: Fjärde huvudtiteln: Flygvapnet", in: *Bihang till Riksdagens protokoll år 1957*, första samlingen, andra bandet, Bilagor till Kungl. Maj:

ts Proposition nr 1 angående statsverkets tillstånd och behov: fjärde–sjunde huvudtitlarna.

"Statsutskottets utlåtande NR B 53 år 1958", in: *Bihang till b-riksdagens protokoll år 1958*, sjätte samlingen, andra bandet, statsutskottets utlåtanden och memorial NR 43–107.

"Statsliggare för budgetåret 1960/61", del 1, Statsverkets specialutgiftsstater under I–XIII huvudtitlarna jämte de affärsdrivande verken.

"Statsliggare för budgetåret 1961/62", del 1, Statsverkets specialutgiftsstater under I–XIII huvudtitlarna jämte de affärsdrivande verken.

"Statsliggare för budgetåret 1962/63", del 1:1, Statsverkets specialutgiftsstater under I–VII huvudtitlarna.

STATENS OFFENTLIGA UTREDNINGAR (SOU):
SOU 1994:11 *Om kriget kommit… Förberedelser för mottagande av militärt bistånd 1949–1969* (Had There Been a War… Preparations for the reception of military assistance 1949–1969) (Stockholm: Fritzes, 1994).

SOU 1994:11 *Had There Been a War… Preparations for the reception of military assistance 1949–1969, Translation of SOU 1994:11* (Stockholm: Fritzes, 1994).

SOU 2002:108 *Fred och säkerhet. Svensk säkerhetspolitik 1969–89* [Peace and Security: Swedish Security Policy 1969–89.] (Stockholm: Fritzes, 2002).

SOU 2002:108 *Peace and Security: Swedish Security Policy 1969–89. Abridged Version and Translation of SOU 2002:108* (Stockholm: Swedish Government Official Reports, 2004).

SOU 2002:95 *Forskarrapporter till Säk erhetstjänstkommissionen* [Researcher Reports to the Intelligence Committee] (Stockholm: Fritzes offentliga publikationer, 2002).

SVERIGES ÖVERENSKOMMELSER MED FRÄMMANDE MAKTER (SÖ):
SÖ 1954:74.

SÖ 1961:51.

SÖ 1962:38

OTHER:
FOA *orienterar om Robotvapen* [A FOA Briefing Regarding Guided Missiles], No. 9, June, 1968.

Newspaper Articles:

Borås Tidning.
"Försvaret får trumfess i amerikanska Hawk-roboten" [The American Hawk Missile Is An Ace of Trumps For the Armed Forces], September 1, 1962.

Svenska Dagbladet.
"Arméns luftförsvar skall förnyas" [The

Army's Anti-Aircraft Defenses Will Be Modernized], January 17, 1963.

Vestmanlands Läns Tidning.
"Svensk robotpolitik" [Swedish Guided Missile Policy], September 28, 1960.

Östgöta Correspondenten.
Interview with Tore Gullstrand, 1994-02-11.

Örnsköldsviks Allehanda.
"Fem robotsystem försvarar vårt luftrum. Försvaret får ett trumfess i amerikanska Hawk-roboten" [Five Missile Systems Defend Our Airspace: The American Hawk Missile Is An Ace of Trumps For the Armed Forces], August 17, 1962.

Internet Sources:

Nationalencyklopedin, Swedish Online Encyclopedia:

http://www.ne.se
Biographical article on Henry Kjellson.

Smithsonian National Air and Space Museum, Washington D.C.:

http://www.nasm.si.edu
Article on German glide bomb Hs 293.

NATO Official website, Brussels, Belgium:

http://www.nato.int/docu/stratdoc/eng/

MC 14/1 (Final), dated December 9, 1952, "Decision on M.C. 14/1: A Report by the Standing Group on Strategy Guidance".

MC 14/2 (Revised) (Final Decision), dated May 23, 1957, "Final Decision on MC 14/2. A Report by the Military Committee on Overall Strategic Concept for the Defense of the North Atlantic Treaty Organization Area".

MC 48/1 (Final), dated December 9, 1955, "North Atlantic Military Committee Decision on M.C. 48/1. A Report by the Military Committee on the Most Effective Pattern of NATO Military Strength for the Next Few Years".

MC 48/2 (Final Decision), dated May 23, 1957, "Final Decision on MC 48/2. A Report by the Military Committee on Measures to Implement the Strategic Concept".

Literature:

Aalders, Gerard, *Swedish Neutrality and the Cold War, 1945–1949* (Amsterdam: Ph.D. diss., University of Nijmegen, 1989).

Aalders, G., "The Failure of the Scandinavian Defence Union, 1948–1949" in *Scandinavian Journal of History*, Vol. 15, No. 1, 1990, pp. 125–153.

Aalders, G. and Wiebes, Cees, *The Art of Cloaking Ownership: The Secret Collaboration and Protection of the German War Industry by the Neutrals: The Case of Sweden* (Amsterdam: Amsterdam University Press, 1996).

Abercrombie, Nicholas, et al., *The Dominant Ideology Thesis* (London: George Allen & Unwin, 1980).

Adler-Karlsson, Gunnar, *Western Economic Warfare 1947–1967: A Case Study in Foreign Economic Policy* (Stockholm: Ph.D thesis, Almqvist & Wiksell, 1968).

Agrell, Wilhelm, *Alliansfrihet och atombomber. Kontinuitet och förändring i den svenska försvarsdoktrinen*

1945–1982 [Non-alignment and Atomic Bombs: Continuity and Change in the Swedish Defense Doctrine, 1945–1982] (Stockholm: Liber Förlag, 1985).

Agrell, W., *Den stora lögnen. Ett säkerhetspolitiskt dubbelspel i alltför många akter.* [The Big Lie. A Security Policy Double Game in Too Many Acts] (Stockholm: Ordfront, 1991).

Agrell, W., *Svenska förintelsevapen. Utveckling av kemiska och nukleära stridsmedel 1928–70* [Swedish Weapons of Mass Destruction: The Development of Chemical and Nuclear Weapons, 1928–70] (Lund: Historiska Media, 2002).

Agrell, W., "Silent allies and hostile neutrals: Nonaligned states in the Cold War" in S. G. Holtsmark, V. Mastny, and A. Wenger (eds.), *War Plans and Alliances in the Cold War:...*, pp. 141–162.

Ahlin, Per, *Folkrätten i svensk säkerhetspolitik* [International Law in Swedish Security Policy] (Stockholm: Juristförlaget, 1993).

Ahlström, Christer, *Stockholmsöverenskommelsen rörande den svenska öst-väst-handeln – ett folkrättsligt bindande avtal?* [The Stockholm Agreement Concerning the Swedish East-West Trade] (Stockholm: Utrikespolitiska institutet, 1996).

Alenius, Mikael, "Informell allians? Den svenska militära underrättelsetjänstens samarbete med västmakterna 1950–1960" [Informal Alliance? The Swedish Military Intelligence Service's Cooperation With the Western Powers 1950–1960] in P. Iko and K. Zetterberg (ed.), *Kvalificerat hemlig...*, pp. 143–192.

Allen, Michael Thad and Hecht, Gabrielle (eds): *Technologies of Power: Essays in Honor of Thomas Parke Hughes and Agatha Chipley Hughes* (Cambridge, Mass. & London: MIT Press, 2001).

Ammann, Klaus, "Swiss Trade with the East in the Early Cold War" in J. Eloranta and J. Ojala (eds.), *East-West Trade and the Cold War*, pp. 113–131.

Andersson, Lennart and Hellström, Leif, *Bortom horisonten. Svensk flygspaning mot Sovjetunionen 1946–1952* [Beyond the Horizon: Swedish Aerial Surveillance of the Soviet Union, 1946–1952] (Stockholm: Stenbom, 2002).

Andrén, Nils, "Sweden: Neutrality, Defense and Disarmament", in H. Neuhold and H. Thalberg (eds.), *The European Neutrals in International Affairs*, pp. 39–58.

Andrén, N., "Swedish Defense: Traditions, Perceptions, and Policies" in M. H. Haltzel and J. Kruzel (eds.), *Between the Blocs:...*, pp. 177–181.

Andrén, N., *Maktbalans och alliansfrihet. Svensk utrikespolitik under 1900-talet* [Power Balance and Non-Alignment. Swedish Foreign Policy During the 20th Century.] (Stockholm: Norstedts juridik, 1996).

Appelqvist, Örjan, *Bruten Brygga. Gunnar Myrdal och Sveriges ekonomiska efterkrigspolitik 1943–1947* [Bridge Broken: Gunnar Myrdal and Sweden's Post-War Economic Policy, 1943–1947] (Stockholm: Santérus Förlag, 2000).

Artéus, Gunnar (ed.), *Svenska officersprofiler under 1900-talet* [Swedish Officer Profiles During the 20th Century] (Stockholm: Militärhögskolan & Axel Abrahamsons Förlag, 1996).

Artéus, G. and Leifland, Leif (eds.), *Svenska diplomatprofiler under 1900-talet* [Swedish Diplomat Profiles During the 20th Century] (Stockholm: Probus Förlag, 2001).

Artz, Lee and Ortega Murphy, Bren, *Cultural hegemony in the United States* (London: Sage Publications, 2000).

Askelin, Jan-Ivar, "Mardrömsscenariot var att piloterna blev mördade" [A Nightmare-Scenario Was that the Pilots Were Murdered] in *Framsyn* 1:2004 (Stockholm, 2004), pp. 28–31.

Aunesluoma, Juhana, *Britain, Sweden and the Cold War 1945–1954* (Oxford: St Anthony's College, 1998).

Aunesluoma, J., *Britain, Sweden and the Cold War 1945–1954: Understanding Neutrality* (Basingstoke: Palgrave Macmillan, 2003).

Babcock, Elizabeth, *Sidewinder: Invention and Early Years* (Annapolis, MD: Naval Institute Press, 1999).

Backer, John, H., *The Decision to Divide Germany: American Foreign Policy in Transition* (Durham, NC: Duke University Press, 1978).

Ball, Nicole and Leitenberg, Milton (eds.), *The Structure of the Defense Industry* (London: Croom Helm, 1983).

Barton J. Bernstein and Allen J. Matusow (eds.), *The Truman Administration: A Documentary History* (New York: Harper & Row Publishers, 1966).

Beaufre, André, (Translated from French by Joseph Green and R. H. Barry), NATO *and Europe* (London: Faber and Faber, 1967).

Berdal, Mats, R., *The United States, Norway and the Cold War, 1954–60* (Basingstoke: Macmillan Press Ltd., 1997).

Berner, Örjan, *Soviet Policies Toward the Nordic Countries* (Lanham: University Press of America, 1986).

Bjereld, Ulf and Demker, Marie, *Utrikespolitiken som slagfält. De svenska partierna och utrikesfrågorna* [Foreign Policy As Battle Field: The Swedish Parties and the Foreign Policy Issues] (Stockholm: Nerenius & Santérus, 1995).

Bjereld, U. and Ekengren, Ann-Marie, "Cold War Historiography in Sweden" in T. B. Olesen, *The Cold War – and the Nordic Countries:...*, pp. 143–175.

Björk, Kaj, "Anteckningar om neutralitetspolitikkommissionen" [Notes on the Neutrality Policy Commission] in *Tiden*, 1994, 86:7, pp. 31–34.

Björk, Ragnar and Johansson, Alf W. (eds.), *Samtidshistoria och politik. Vänbok till Karl Molin* [Contemporary History and Politics: A Tribute to Karl Molin] (Stockholm: Hjalmarsson & Högberg, 2004).

Björklund, Fredrika, *Samförstånd under oenighet. Svensk säkerhetspolitisk debatt under det kalla kriget* [Unity Under Discord: The Swedish Debate on National Security During the Cold War] (Stockholm: Almqvist & Wiksell International, 1992).

Bletz, Donald F., *The Role of the Military Professional in* U.S. *Foreign Policy* (New York: Praeger Publishers, 1972).

Blomberg, Mats, "Efter Neutralitetspolitikkommissionen: 50-talets verkliga säkerhetspolitik är bara delvis tecknad" [After the Neutrality Policy Commission: The Real Security Policy of the 1950s Is Only Partly Covered],

Kungl Krigsvetenskapsakademiens handlingar och tidskrift [The Royal Swedish Academy of War Sciences Proceedings and Journal], No. 1, 1995, pp. 97–11.

Blum, William, Killing Hope: U.S. Military and CIA Interventions since World War II (Monroe, ME: Common Courage Press, 1995).

Bocock, Robert, *Hegemony* (London: Tavistock Publications, 1986).

Bruzelius, Nils, *"near friendly or neutral shores": The Deployment of the Fleet Ballistic Missile Submarines and* US *Policy Towards Scandinavia, 1957–1963* (Stockholm: Stockholm Papers in the History and Philosophy of Technology TRITA-HOT 2053, 2007).

Buci-Glucksmann, Christine, *Gramsci and the State* (London: Lawrence and Wishart, 1980).

Buffet, Cyril and Heuser, Beatrice (eds.), *Haunted by History: Myths in International Relations* (Providence and Oxford: Berghahn Books, 1998).

Calleo, David P., *Beyond American Hegemony: The Future of the Western Alliance* (New York: Basic Books, Inc., Publishers, 1987).

Campbell, Duncan, "The 'deterrent' goes to war" in *New Statesman*, May 1, 1981, pp. 8–9 and 13.

Carlbom, Terry, *Rethinking Swedish Neutrality* (Lampeter: Saint David's University College, 1989).

Carlgren, Wilhelm M., "The Emergence of Sweden's Policy of Neutrality" in "Neutrality and Defence: The Swedish Experience" in *Revue Internationale d'Historie Militaire*, No. 57 (Stockholm: 1984), pp. 11–33.

Cars, Hans Christian; Skoglund, Claës; Zetterberg, Kent, *Svensk försvarspolitik under efterkrigstiden* [Swedish Defense Policy in the Postwar Era] (Stockholm: Probus Förlag, 1986).

Cervin-Ellqvist, Gunnar and König, Bertil, *Robot 50 år* [Guided Missile: 50 years] (Linköping: SAAB, 1997).

Churchill, Winston S., *The Second World War*, Volume VI, *Triumph and Tragedy* (London: Casell, 1954).

Cole, Paul Marion, *Neutralité du jour – The conduct of Swedish Security policy since 1945.* (Baltimore, MD: unpublished Ph.D. diss., Johns Hopkins School of Advanced International Studies, 1990).

Cole, P. M., "USA-samarbetet hemlighölls" [Cooperation with the USA Was Kept Secret] in *Dagens Nyheter*, October 21, 1990.

Cole, P. M., '

'Neutraliteten gavs upp" [Neutrality Was Abandoned] in *Dagens Nyheter*, October, 31, 1990.

Cook, Don, *Forging the Alliance:* NATO*, 1945–1950* (London: Secker & Warburg, 1989).

Costigliola, Frank, "Culture, Emotion, and the Creation of the Atlantic Identity, 1948–1952" in G. Lundestad (ed.), *No End to Alliance:...*, pp. 21–36.

Cox, Robert W., "Gramsci, Hegemony and International Relations: An Essay in Method" in S. Gill (ed.), *Gramsci, Historical Materialism and International Relations*, pp 49–66.

Cramér, Per, *Neutralitetsbegreppet* [The Concept of Neutrality] (Stockholm: Norstedts Förlag, 1989).

Dahl, Ann-Sofie, "The Myth of Swedish Neutrality" in C. Buffet and B. Heuser (eds.), *Haunted by History:...*, pp. 28–40.

Dallek, Robert, *Franklin D. Roosevelt and American Foreign Policy, 1932–1945* (Oxford: Oxford University Press paperback, 1981).

Dalsjö, Robert, "Tungt vägande kritik? En granskning av kritik mot teorin om nordisk balans" [Weighty Criticism? A Review of the Criticism Leveled Against the Theory of A Nordic Balance] in *Militärhistorisk tidskrift*, 1987, pp. 131–185.

Dalsjö, R., *Life-Line Lost: The Rise and Fall of 'Neutral' Sweden's Secret Reserve Option of Wartime Help from the West* (Stockholm: Santérus Academic Press, 2006).

Davidsson, Sune, in cooperation with Hans Bergendorff and Bo Ljung, *Sveriges utlandsberoende* [Sweden's Foreign Dependence] (Stockholm: Centralförbundet Folk & Försvar, 1976).

Day, Richard, B., *Cold War Capitalism: The View from Moscow, 1945–1975* (Armonk, NY: M. E. Sharpe, 1995).

Deighton, Anne, *The Impossible Peace: Britain, the Division of Germany, and the Origins of the Cold War* (Oxford: Clarendon Paperbacks, 1993).

Denham, Henry, *Inside the Nazi Ring: A Naval Attaché in Sweden, 1940–1945* (London: John Murray, 1984).

Deutscher, Isaac, *Stalin: A Political Biography* (London: Oxford University Press, 1949).

Dmytryshyn, Basil, USSR*: A Concise History* (New York: Charles Scribner's Sons, 1965).

Dobson, Alan P., "The USA, Britain, and the Question of Hegemony" in G. Lundestad (ed.), *No End to Alliance:...*, pp. 134–163.

Dobson, A. P. and Marsh, Steve, US *Foreign Policy Since 1945* (London: Routledge, 2001).

Dobson, A. P., US *Economic Statecraft for Survival 1933–1991: Of sanctions, embargoes and economic warfare* (London: Routledge, 2002).

Doel, Ronald E. and Söderqvist, Thomas (eds.), *The Historiography of Contemporary Science, Technology, and Medicine: Writing Science* (London: Routledge, 2006).

Dohlman, Ebba, *National Welfare and Economic Interdependence: The case of Sweden's Foreign Trade Policy* (Oxford: Clarendon Press, 1989).

Donovan, Robert J., *Conflict and Crisis: The Presidency of Harry S. Truman, 1945–1948* (New York: Norton & Company, Inc., 1977).

Duffield, John S., *Power Rules: The Evolution of* NATO*'s Conventional Force Posture* (Stanford, California: Stanford University Press, 1995).

Dörfer, Ingemar, "Soviet Policy in the North: Finland, Sweden, Norway, 1948–1949" (Paper prepared at Harvard University, 1964).

Dörfer, I, *System 37 Viggen. Arms, Technology and the Domestication of Glory* (Oslo: Universitetsforlag, 1973).

Dörfer, I., "System 37 Viggen – vapenteknisk forskning inom industrin" [System 37 Viggen: Weapons Technology Research Within Industry] in *Daedalus* (Stockholm: Sveriges Tekniska Museums Årsbok, 1986), pp. 147–156.

Dörfer, I., "Neutralitetsbygget – ett spökslott för inbördes beundran." [The Building of Neutrality: A Ghost House for Mutual Admiration], *Kungl Krigsvetenskaps akademiens handlingar och tidskrift*, No. 1, 1995, pp. 17–19.

Eisenberg, Carolyn, *Drawing the Line: The American Decision to Divide Germany, 1944–1949* (Cambridge: Cambridge University Press, 1996).

Eisenberg, C., "Rethinking the Division of Germany" in A. Hunter (ed.), *Rethinking the Cold War*, pp 47–61.

Ekecrantz, Stefan, *Hemlig utrikes-politik. Kalla kriget, utrikesnämnden och regeringen 1946–1959* [Secret Foreign Policy – The Cold War, the Advisory Council on Foreign Affairs and the Swedish Government 1946–1959] (Stockholm: Santérus Förlag, 2003).

Ekecrantz, S., "'Neutralitet', 'neutralitetspolitik' och 'alliansfrihet' – en begreppshistorisk propå till forskning och debatt" ["Neutrality", "Neutraliy Policy", and "Non-alignment": A Conceptual History and a Call for Research and Debate] in R. Björk and A. W. Johansson (eds.), *Samtidshistoria och politik…*, pp. 52–85.

Ekengren, Ann-Marie, *Sverige under kalla kriget 1945–1969. En forskningsöversikt.* [Sweden During the Cold War: A Historiography] (Stockholm: Forskningsprogrammet *Sverige under kalla kriget Arbetsrapport nr. 1*, 1997).

Ekman, Jan G., "Förhandlingarna om Skandinaviskt Försvarsförbund, 1948–49 [The Negotiations on a Scandinavian Defense Union, 1948–49] (Paper presented at the Department for Political Science at Stockholm University, spring 1967).

Eloranta, Jari and Ojala, Jari (eds.), *East-West Trade and the Cold War*, (Jyväskylä: University of Jyväskylä, 2005).

Engelbrekt, Kjell, "Den sjuttonde alliansmedlemmen?" [The Seventeenth Member of the Alliance?] in *Internationella Studier*, No. 4, 1999, pp. 61–72.

Etzhold, Thomas H. and Gaddis, John Lewis (eds.), *Containment: Documents on American Policy and Strategy, 1945–1950* (New York: Columbia University Press, 1978).

Femia, Joseph V., *Gramsci's Political Thought. Hegemony, Consciousness and Revolutionary Process* (Oxford: Clarendon, 1987).

Fleming, D. F., *The Cold War and its Origins 1917–1960*, Volume I, 1917–1950 (Garden City, NY: Doubleday & Company, Inc., 1961).

Flygstaben – en minnesbok [The Air Staff: A Memorial Book] (Stockholm: Högkvarteret, Försvarsmakten, 1994).

Francis L. Loewenheim, Harold D. Langley and Manfred Jonas (eds.), *Roosevelt and Churchill: Their Secret Wartime Correspondence* (London: Barrie & Jenkins, 1975).

Friis, Henning (ed.), *Scandinavia: Between East and West* (Westport, CT: Greenwood Press, Publishers, 1979 (originally published by Cornell University Press, 1950).

Gabriel, Jürg Martin, *The American Conception of Neutrality after 1941* (Hampshire: The MacMillan Press Ltd., 1988).

Gaddis, John Lewis, *Strategies of Containment: A Critical Appraisal of Postwar American National Security Policy* (New York: Oxford University Press, 1982).

Gaddis, J. L., *The Long Peace: Inquiries Into the History of the Cold War* (Oxford: Oxford University Press, 1987).

Gaddis, J. L., *We Now Know: Rethinking Cold War History* (Oxford: Clarendon Press, 1997).

Gaddis, J. L., *The Landscape of History: How Historians Map the Past* (Oxford: Oxford University Press, 2002).

Gaddis, J. L., *The Cold War: A New History* (New York: The Penguin Press, 2005).

Gardner, Lloyd C., *Spheres of Influence: The Great Powers Partition Europe, from Munich to Yalta* (Chicago: Elephant Paperbacks, 1993).

Gehler, Michael and Steininger, Rolf (eds.), *The Neutrals and the European Integration* (Wien, Köln, Weimar: Böhlau Verlag, 2000).

Gilbert, Martin, *Winston S. Churchill*, Volume VII, *Road to Victory, 1941–1945* (William Heinemann Ltd, 1986).

Gilbert, M., *Winston S. Churchill*, Volume VIII, *'Never Despair', 1945–1965* (William Heinemann Ltd, 1988).

Gill, Stephen (ed.), *Gramsci, Historical Materialism and International Relations* (Cambridge: Cambridge University Press, 1993).

Gill, S. and Law, David, "Global Hegemony and Structural Power of Capital" in S. Gill (ed.), *Gramsci, Historical Materialism and International Relations*, pp. 93–124.

Gimbel, John, *The Origins of the Marshall Plan* (Stanford, CA: Stanford University Press, 1976).

Goldmann, Kjell, "Neutralitetspolitiken som dubbelpolitik" [Neutrality Policy as Double Politics], *Internationella Studier*, No. 4, 1998, pp. 73–80.

Gribbe, Johan, "En inre angelägenhet för fritt tankeutbyte. Synen på västsamverkan i Försvarshögskolans krigsspel 1952–1979" [For the Free Exchange of Ideas: Cooperation with the West in Swedish War Games, 1952–1979] (Stockholm: TRITA-HST Working paper, 2003/1).

Gribbe, J., "'...därest fienden lyckas taga svenskt territorium' Sverige och Väst i Försvarshögskolans krigsspel" ["...If the Enemy Should Manage to Occupy Swedish Territory": Sweden and the West in Swedish War Games] in *Kungl Krigsvetenskapsa kademiens handlingar och tidskrift*, No. 5, 2003, pp. 93–98.

Gribbe, J. and Nilsson, Mikael, "The Foreign Domestic: Hard Artefacts and Soft Politics in Sweden During the First Half of the Cold War, 1945–1967" in ICON: *Journal of the International Committee for the History of Technology*, Vol. 11, 2005, pp. 51–62.

Gyldén, Nils, *Sweden's Security and Defence Policy—Through the Cold War and Towards the Turn of the Century* (Stockholm: Ministry of Defence, 1994).

Haglund, Magnus, "Jovisst var vi hemliga NATO-medlemmar!" [Certainly, We Were Secret NATO Members!] in *Kungl krigsvetenskapsakademiens handlingar och tidskrift*, No. 5, 2002, pp. 125–128.

Hahn, Peter L. and Heiss, Mary Ann, *Empire and Revolution: The United States and the Third World since 1945* (Columbus: Ohio State University Press, 2001).

Hakovirta, Harto, *East-West Conflict and European Neutrality* (Oxford: Clarendon Press, 1988).

Haltzel, Michael H. and Kruzel, Joseph (eds.), *Between the Blocs: Problems and Prospects for Europe's Neutral and Nonaligned States* (Cambridge, MA: Woodrow Wilson International Center for Scholars and Cambridge University Press, 1989).

Hanhimäki, Jussi M., *Scandinavia and the United States: An Insecure Friendship* (New York: Twayne Publishers, 1997).

Hanhimäki, J. M., "Security and Identity: the Nordic Countries and the United States since 1945" in G. Lundestad (ed.), *No End to Alliance:...*, pp. 81–91.

Haskel, Barbara G., *The Scandinavian Option: Opportunities and Opportunity Costs in Postwar Scandinavian Foreign Relations* (Oslo: Universitetsforlaget, 1976).

Haynes, Richard F., *The Awesome Power: Harry S. Truman as Commander in Chief* (Baton Rouge, Louisiana: Louisiana State University Press, 1973).

Hell, Johan Kristoffer, *Strategic Vulnerability: Understanding Sweden's National Security Policies During the Cold War* (Münster: Waxmann, 1996).

Herlitz, Carl (ed.), *Luftvärnets historia* [The History of the Anti-Aircraft Defense] (Stockholm: Militärhistoriska Förlaget, 1980).

Hill, C. N., *A Vertical Empire: The History of the* UK *Rocket and Space Programme, 1950–1971* (London: Imperial College Press, 2001).

Hoddeson, Lillian, "The conflict of memories and documents: Dilemmas and pragmatics of oral history" in R. E. Doel and T. Söderqvist (eds.), *The Historiography of Contemporary Science...*, pp. 187–198.

Hogan, Michael J., *The Marshall Plan: America, Britain, and the Reconstruction of Western Europe, 1947–1952* (Cambridge: Cambridge University Press, 1987).

Hogan, M. J. and Paterson, Thomas G. (eds.), *Explaining the History of American Foreign Relations* (Cambridge: Cambridge University Press, 1991).

Holmström, Barry, *Koreakriget i svensk debatt* [The Korean War in Swedish Debate] (Stockholm: Rabén & Sjögren, 1972).

Holmström, Mikael and von Sivers, Tom, *Tekniken som vapen.* [Technology as a Weapon] (Stockholm: Ingenjörsförlaget, 1985).

Holmström, M., "Svenskt dubbelspel under kalla kriget" [Swedish Double Game During the Cold War] in *Svenska Dagbladet*, October 11, 2003.

Holmström, Per and Olsson, Ulf, "Sweden" in Nicole Ball and Milton Leitenberg (eds.), *The Structure of the Defense Industry* (London: Croom Helm, 1983), pp. 140–180.

Holtsmark, Sven G., "Enemy springboard or benevolent buffer?: Soviet attitudes to Nordic co-operation, 1920–1955" in *Forsvarsstudier*, No. 6, 1992 (Oslo: Institutt for forsvarsstudier, 1992), pp. 59–65.

Holtsmark, S. G., Mastny, V., and Wenger, Andreas (eds.), *War Plans and Alliances in the Cold War: Threat Perception in the East and West* (London: Routledge, 2006).

Horowitz, David, *The Free World Colossus: A Critique of American Foreign Policy in the Cold War* (New York: Hill and Wang, 1965).

Howe Stephen, *Empire: A Very Short Introduction* (Oxford: Oxford University Press, 2002).

Hoxie, Gordon R., *Command Decision and the Presidency: A Study in National Security Policy and Organization* (New York: Thomas Y. Crowell Company, 1977).

Hughes, Thomas P., *Networks of Power: Electrification in Western Society, 1880–1930* (Baltimore, MD: The Johns Hopkins University Press, 1983).

Hunter, Allen (ed.), *Rethinking the Cold War* (Philadelphia: Temple University Press, 1998),

Huston, James A., *One for All: NATO Strategy and Logistics through the Formative Period, (1949–1969)* (Newark: University of Delaware Press, 1984).

Höll, Otmar (ed.), *Small States in Europe and Dependence* (Wien: Wilhelm Braumüller, Universitäts-Verlagsbuchhandlung, 1983).

Iko, Per and Zetterberg, Kent (ed.), *Kvalificerat hemlig. Fyra studier om svensk underrättelsetjänst och underrättelsetjänst i Sverige* [Top Secret: Four Studies About Swedish Intelligence Service and Intelligence Service in Sweden] (Stockholm: Försvarshögskolan, 2002).

Ireland, Timothy P., *Creating the Entangling Alliance: The Origins of the North Atlantic Treaty Organization* (Westport, Connecticut: Greenwood Press, 1981.

Ives, Peter, *Language and Hegemony in Gramsci* (London: Pluto Press, 2004).

Jenkins, Roy, *Truman* (London: William Collins Sons & Co. Ltd, 1986).

Johansson, Alf W. and Norman, Torbjörn, "The Swedish Policy of Neutrality in a Historical Perspective" in "Neutrality and Defence:..." in *Revue Internationale d'Historie Militaire*, No. 57, pp. 69–94.

Jonter, Thomas, *Socialiseringen som kom av sig. Sverige, oljan och USAs planer på en ny världsordning 1945–1949* [Socialization Derailed: Sweden, Oil, and the USA's Plans for a New World Order, 1945–1949] (Stockholm: Carlssons Bokförlag, 1995).

Jonter, T. and Silva, Charles (eds.), *Sverige inför en ny världsordning, 1945–1950. Formativa år för svensk utrikespolitik?* [Sweden Faced With A New World Order, 1945–1950: Formative Years in Swedish Foreign Policy?] (Stockholm: Utrikespolitiska institutet, 1995).

Jonter, T., "En tänkbar tolkningsram för svensk-amerikanska studier under kalla kriget" [A Possible Frame for Interpreting Swedish-American Relations During the Cold War] in T. Jonter and C. Silva, *Sverige inför en ny världsordning 1945–50...*, pp. 29–42.

Jonter, T., *Sverige, USA och kärnenergin. Framväxten av en svensk kärnämneskontroll 1945–1995* (Stockholm: SKI Report 99:21, 1999).

Jonter, T., *Sweden and the Bomb: The Swedish Plans to Acquire Nuclear Weapons, 1947–1972* (Stockholm: SKI Report 01:33, 2001).

Jonter, T., *Nuclear Research in Sweden: The Cooperation between Civilian and Military Research, 1947–1972* (Stockholm: SKI Report 02:18, 2002).

Jonter, T., "Det amerikanska spåret. En undersökning av IB:s bildande och eventuella kopplingar till USA." [The American Trail: An Investigation of the Establishment of IB and Possible Connections with the USA] in SOU 2002:95 *Forskarrapporter till Säkerhetstjänstkommissionen* [Researcher Reports to the Intelligence Committee] (Stockholm: Fritzes offentliga publikationer, 2002), pp. 135–191.

Jonter, T. and van Dassen, Lars, "Making Historical Surveys of States' Nuclear Ambitions: Experiences from the Baltic Sea Region" in *The Nonproliferation Review*, Vol. 12, No. 1, March 2005, pp. 225–270.

Jordan, Robert S., *Generals in International Politics: NATO's Supreme Allied Commander, Europe* (Lexington: The University Press of Kentucky, 1987).

Jordan, R. S., "Gruenther: Attempts to Retain NATO Solidarity" in R. S. Jordan (ed.), *Generals in International Politics:...*, pp. 53–72.

Joseph, Jonathan, *Hegemony: A Realist Analysis*, (London: Routledge, 2002).

Kaplan, Lawrence S., *A Community of Interests: NATO and the Military Assistance Program, 1948–1951* (Washington D.C.: Office of the Secretary of Defense, Historical Office, 1980).

Kaplan, L. S., *The Long Entanglement: NATO's First Fifty Years* (Westport, Connecticut: Praeger, 1999).

Karlsson, Birgit, *Handelspolitik eller politisk handling. Sveriges handel med öststaterna 1946–1952* [Trade Policy or Political Trade. Sweden's Trade with the Eastern Bloc 1946–1952] (Göteborg: Meddelanden från ekonomisk-historiska institutionen vid Göteborgs universitet 66, 1992).

Karlsson, B., "Sveriges relationer till Östeuropa 1945–1950" [Sweden's Relations to Eastern Europe 1945–1950] in C. Silva and T. Jonter (eds.) *Sverige inför en ny världsordning, 1945–1950...*, pp. 43–51.

Karsh, Efraim, *Neutrality and Small States* (London: Routledge, 1988).

Katzenstein, Peter J. (ed.), *Between Power and Plenty: The Foreign Economic Policies of Advanced Industrial States* (Madison, WI: University of Wisconsin Press, 1978).

Kennan, George F., *American Diplomacy, 1900–1950* (London: Secker & Warburg, 1952).

Kennan, G. F., *Memoirs, 1925–1950* (London: Hutchinson of London, 1967).

Keohane, Robert O., *After Hegemony: Cooperation and Discord in the World Political Economy* (Princeton: Princeton University Press, 1984).

Keohane, R.; Nye, Joseph S.; and Hoffman, Stanley, *After the Cold War: International Institutions and State Strategies in Europe, 1989–1991* (Cambridge, MA: Harvard University Press, 1993).

Keohane, R. O. and Nye, J. S., *Power and Interdependence* (third edition) (New York: Longman, (1977) 2001).

Kimball, Warren F. (ed.), *Churchill & Roosevelt: The Complete Correspondence*, Volume III, *Alliance Declining: February 1944–April 1945* (New Jersey: Princeton University Press, 1984).

Kissinger, Henry A., *The Troubled Partnership: A Re-appraisal of the Atlantic Alliance* (New York: McGraw-Hill, 1965).

Kleve, Karl L. (ed.), *The Cold War, Military Power and the Civilian Society. Report from the conference in Bodø, Sept 12–13, 2002* (Bodø: Norwegian Aviation Museum, 2003).

Klaus Knorr (ed.), NATO *and American Security* (Princeton, New Jersey: Princeton University Press, 1959).

Knorr, Klaus, "The Strained Alliance" in K. Knorr (ed.), NATO *and American Security*, pp. 3–10.

Knutsen, Torbjörn, "Amerikas fire ansikter" in *Internasjonal politikk*, Vol. 62, No. 3, 2004, pp. 303–325.

Kolko, Gabriel, *The Triumph of Conservatism: A Reinterpretation of American History* (New York: The Free Press, 1963).

Kolko, G., *The Politics of War: Allied Diplomacy and the World Crisis of 1943–1945* (London: Weidenfeld and Nicolson, 1968).

Kolko, G. and Kolko, Joyce, *The Limits of Power: The World and United States Foreign Policy, 1945–1954* (New York: Harper & Row, Publishers, 1972).

Krige, John, *American Hegemony and the Postwar Reconstruction of Science in Europe* (Cambridge, Massachusetts: The MIT Press, 2006).

Kronvall, Olof; Petersson, Magnus; Silva, Charles; and Skogrand, Kjetil, "Comments on Ola Tunander's article 'The Uneasy Imbrication of Nation-State and NATO: The Case of Sweden'", in *Cooperation and Conflict*, Vol. 35, No. 4, 2000, pp. 417–429.

Kronvall, O., "Rolf Sohlman" in G. Artéus and L. Leifland (eds.), *Svenska diplomatprofiler…*, pp. 274–316.

Kronvall, O., "Finland in Swedish Security Policy, 1948–1962", in K. L. Kleve (ed.), *The Cold War, Military Power …*, pp. 35–44.

Kronvall, O., *Den bräckliga barriären. Finland i svensk säkerhetspolitik 1948–1962.* [The Fragile Barrier: Finland in Swedish Security Policy, 1948–1962] (Stockholm: Försvarshögskolan, 2003).

Kronvall, O. and Petersson Magnus, *Svensk säkerhetspolitik i supermakternas skugga* 1945–1991 [Swedish Security Policy in the Shadow of the Super Powers 1945–1991.] (Stockholm: Santérus, 2005).

Kux, Stephan, "Europe's Neutral States: Partners or Profiteers in Western Security?" in *European Security Studies*, No. 4, 1986.

Laclau, Ernesto and Mouffe, Chantal, *Hegemony and Socialist Strategy. Towards a Radical Democratic Politics* (London: Verso Press, 1985).

LaFeber, Walter, *The Cambridge History of American Foreign Relations*, Volume II, *The American Search for Opportunity, 1865–1913* (Cambridge: Cambridge University Press, 1993).

LaFeber, W., *America, Russia, and the Cold War 1945–2002* (updated ninth edition) (New York: McGraw-Hill, 2004).

Lairson, Thomas D., "Revising Postrevisionism: Credibility and Hegemony in the Early Cold War" in A. Hunter (ed.), *Rethinking the Cold War*, pp. 63–90.

Leffler, Melvyn P., "National Security" in M. J. Hogan and T. G. Paterson (eds.), *Explaining the History of American Foreign Relations*, pp. 123–136.

Leffler, M., P., *A Preponderance of Power: National Security, the Truman Administration, and the Cold War* (Stanford, CA: Stanford University Press, 1992).

Leffler, M. P., *The Specter of Communism: The United States and the Origins of the Cold War, 1917–1953* (New York: Hill and Wang, 1994).

Leffler, M. P., "National Security and US Foreign Policy" in M. P. Leffler and D. S. Painter (eds.), *Origins of the Cold War:…*, pp. 15–52.

Leffler, M. P. and Painter, David. S. (eds.), *Origins of the Cold War: An International History* (London: Routledge, 1994).

Leifland, Leif, *Frostens år. Om USA:s diplomatiska utfrysning av Sverige* [The Years of the Deep-Freeze: The U.S. Diplomatic Deep-Freezing of its Relations with Sweden] (Stockholm: Nerenius & Santérus, 1997).

Linder, Jan, *Svår neutralitet. Sverige under två sekel* [Difficult Neutrality: Sweden Under Two Centuries] (Stockholm: Infomanager Förlag, 2003).

Linderborg, Åsa, *Socialdemokraterna skriver historia. Historieskrivning som ideologisk maktresurs, 1892–2000* [The Social Democrats Write History: History as Ideological Power Resource, 1892–2000] (Stockholm: Atlas, 2001).

Lindholm, Rolf H., *Sveriges neutralitet. Swedish Neutrality* (Lund: Juridiska Föreningen i Lund, 1987).

Lindholm, R. H., *Sverige under kalla kriget* [Sweden During the Cold War] (Stockholm: Santérus Förlag, 2000).

Lindström, Stefan, "The open society and its stewards", in K. L. Kleve (ed.), *The Cold War, Military Power...*, pp. 127–150.

Logue, John, "Sweden" in S. Victor Papacosma and Mark R. Rubin (eds.), *Europe's Neutral and Nonaligned States: Between NATO and the Warsaw Pact* (Wilmington, DE: SR Books, 1989).

Luders, Emma, *Sweden and Western Intelligence Collaboration: A Regime Study* (Stockholm: Stockholm International Studies 91:3, 1992).

Lundestad, Geir, *The American Non-Policy Towards Eastern Europe, 1943-1947: Universalism in an Area Not of Essential Interest to the United States* (Tromsø: Universitetsforlaget, 1978).

Lundestad, G., *America, Scandinavia, and the Cold War, 1945–1949* (New York: Columbia University Press, 1980).

Lundestad, G., "American-European Cooperation and Conflict: Past, Present, and Future" in G. Lundestad (ed.), *No End to Alliance:...*, pp. 245–262.

Lundestad, G., *The American "Empire" and Other Studies of US Foreign Policy in a Comparative Perspective* (Oxford: Oxford University Press, 1990).

Lundestad, G. (ed.), *No End to Alliance: The United States and Western Europe: Past, Present and Future* (New York: St. Martin's Press, 1998).

Lundestad, G., "How (Not) to Study the Origins of the Cold War" in O. A. Westad (ed.), *Reviewing the Cold War:...*, pp. 64–80.

Luntinen, Pertti, "Neutrality in Northern Europe before the First World War" in J. Nevakivi (ed.), *Neutrality in History...*, pp. 107–114.

Lödén, Hans, *"För säkerhets skull." Ideologi och säkerhet i svensk aktiv utrikespolitik 1950–1975* ["For the Sake of Security": Ideology and Security in Swedish Active Foreign Policy 1950–1975] (Stockholm: Nerenius & Santérus Förlag, 1999).

MacKenzie, Donald, *Inventing Accuracy: A Historical Sociology of Nuclear Guided missile Guidance* (Cambridge, Mass.: MIT Press, 1990).

Maddox, Robert, James, *From War to Cold War: The Education of Harry*

S. Truman (Boulder: Westview Press, 1988).

Maier, Charles S., "The Politics of Productivity: Foundations of American International Economic Policy after World War II" in P. J. Katzenstein (ed.), *Between Power and Plenty:...*, pp. 23–49.

Maier, C. S., "Hegemony and autonomy within the Western Alliance" in M. P. Leffler and D. S. Painter (eds.), *Origins of the Cold War:...*, pp. 154–174.

af Malmborg, Mikael, *Neutrality and State-building in Sweden* (Hampshire: Palgrave, 2001).

af Malmborg, M., "Sweden – NATO's Neutral 'Ally'? A Post-Revisionist Account" in Gustav Schmidt (ed.), *A History of* NATO..., pp. 295–314.

Mastny, Vojtech, *The Cold War and Soviet Insecurity: The Stalin Years* (Oxford: Oxford University Press, 1996).

McCormick, Thomas, J., *America's Half-Century: United States Foreign Policy in the Cold War and After* (second edition) (Baltimore, MD: The Johns Hopkins University Press, 1995).

McCullough, David, *Truman* (New York: Simon & Schuster, 1992).

McGlade, Jacqueline, "COCOM and the Containment of Western Trade and Relations" in J. Eloranta and J. Ojala (eds.), *East-West Trade and the Cold War*, pp. 47–61.

Mee, Jr., Charles L., *The Marshall Plan: The Launching of the Pax Americana* (New York: Simon and Schuster, 1984).

Millis, Walter, "The National Security Act" in R. G. O'Connor (ed.), *American Defense Policy in Perspective:...*, pp. 307–311.

Molin, Karl, *Omstridd neutralitet. Experternas kritik av svensk utrikespolitik 1948–1950* [Contested Neutrality: The Experts' Critique of Swedish Foreign Policy 1948–1950] (Stockholm: Tiden, 1991).

Molin, K., "The Central Issues of Swedish Neutrality Policy", in M. Gehler and R. Steininger (eds.), *The Neutrals and the European Integration*, pp. 261–275.

Molin, K. and Olesen, Thorsten B., "Security policy and Domestic Politics in Scandinavia 1948–49" in T. B. Olesen (ed.), *Interdependence Versus Integration:...*, pp. 62–81.

Moores, Simon, "'Neutral on our side': US Policy towards Sweden during the Eisenhower Administration", in *Cold War History*, Vol. 2, No. 3, April 01, 2002, pp. 29–62.

Moores, S. and Widén, Jerker, "Sverige under det kalla kriget – den amerikanska underrättelsetjänstens perspektiv, mars 1952" [Sweden's Position in the East-West Conflict as Mirrored by the CIA in March 1952] in *Kungl Krigsvetenskapsakademiens handlingar och tidskrift*, No. 5, 2003, pp. 78–92.

Moores, S., *'Neutral on Our Side': US-Swedish Military and Security Relations during the Eisenhower Administration* (London: unpublished Ph.D. diss., The London School of Economics and Political Science, September, 2004).

Mörth, Ulrika, *Sverige och Koreakriget: ett förkrigsskede* [The Korean War: A Pre-War Situation] (Stockholm: Försvarsdepartementet, 1988).

Mörth, U. and Sundelius, Bengt, *Interdependens, konflikt och säkerhetspolitik. Sverige och den amerikanska teknikexportkontrollen.* [Interdependence, Conflict, and Security Policy: Sweden and the American Export Controls] (Stockholm: Nerenius & Santérus Förlag, 1998).

NATO: *Facts About the North Atlantic Treaty Organization* (Paris: NATO Information Service, 1965).

Neuhold, Hanspeter, "Challanges To Neutrality in an Interdependent World" in M. H. Haltzel and J. Kruzel (eds.), *Between the Blocs:...*, pp. 89–93.

Neuhold, H. and Thalberg, Hans (eds.), *The European Neutrals in International Affairs* (Wien: Wilhelm Braumüller, Universitäts-Verlagsbuchhandlung, 1984).

Nevakivi, Jukka (ed.), *Neutrality in history: La neutralité dans l'histoire : proceedings of the Conference on the History of Neutrality organized in Helsinki 9–12 September 1992 under the auspices of the Commission of History of International Relations* (Helsinki: Suomen historiallinen seura, 1993).

Nilsson (now Dahl), Ann-Sofie, *Den moraliska stormakten. En studie av socialdemokratins internationella aktivism* [The Moral Super Power: A Study of Social Democracy's International Activism] (Stockholm: Timbro, 1991).

Nilsson, Mikael, "'Gud vare tack och lov så talade de inte om allt de gjorde': en källkritisk granskning av Neutralite tspolitikkommissionens betänkande SOU 1994:11." [Thank God They Did Not Speak About Everything They Did: A Source-Critical Examination of the Report from the Neutrality Policy Commission SOU 1994:11], (Stockholm: TRITA-HST Working paper, 2003/2).

Nilsson, M. and Stenlås, Niklas, "Cold War Neutrality and Technological Dependence: Sweden's Military Technology and the East-West Trade" in J. Eloranta and J. Ojala (eds.), *East-West Trade and the Cold War*, pp. 133–151.

Nordbeck, Sven, "Luftvärnsrobotsystem 67 och 77" [Surface-to-Air Missile System 67 and 77] in C. Herlitz (red.), *Luftvärnets historia.*, pp. 175–193.

Noreen, Erik, *Brobygge eller blockbildning? De norska och svenska utrikesledningarnas säkerhetspolitiska föreställningar 1945–1948* [Bridge-Building or Bloc-Building? The Norwegian and Swedish Foreign Ministries' Security Policy Perceptions 1945–1948] (Stockholm: Carlsson Bokförlag, 1994).

Noreen, E., "Från neutralitet till brobygge. Svensk utrikespolitik och säkerhetspolitik under de första efterkrigsåren" in T. Jonter and C. Silva (eds.), *Sverige inför en ny världsordning, 1945–50...*, pp. 53–68.

Norman, Torbjörn, "Stages in Swedish Neutrality" in J. Nevakivi (ed.), *Neutrality in History:...*, pp. .

Norman, T., "The Swedish Policy of Neutrality in a Historical Perspective" in "Neutrality and Defence: The Swedish Experience" in *Revue Internationale d'Historie Militaire*, No. 57, pp. 69–94.

O'Connor, Raymond G. (ed.), *American Defense Policy in Perspective: From Colonial Times to the Present* (New York: John Wiley & Sons, Inc., 1965).

Odom, Jr., Donald, *Swedish Foreign Policy Behavior: An Event Data Analysis* (Stockholm: Stockholm University, International Graduate School, 1990).

Olesen, Thorsten B. (ed.), *Interdependence Versus Integration:*

Denmark, Scandinavia and Western Europe, 1945–1960 (Odense: Odense University Press, 1995).

Olesen, T. B (ed.), *The Cold War – and the Nordic Countries: Historiography at a Crossroads* (Odense: University Press of Southern Denmark, 2004).

Olsson, Ulf, *The Creation of a Modern Arms Industry: Sweden 1939–1974* (Göteborg: Ekonomisk-historiska institutionen, Göteborgs universitet, 1977).

Olsson, U., *Furthering a fortune. Marcus Wallenberg. Swedish banker and industrialist.* (Stockholm: Ekerlid, 2001).

Oredsson, Sverker, "Erik Boheman" in G. Artéus and L. Leifland, *Svenska diplomatprofiler...*, pp. 169–211.

Osgood, Robert Endicott, NATO: *The Entangling Alliance* (Chicago: University of Chicago Press, 1962).

Ottosson, Sten, *Den (o)moraliska neutraliteten. Tre politikers och tre tidningars moraliska värdering av svensk utrikespolitik 1945–1952* [The (Im)moral Neutrality: Three Politicians' and Three Newpapers' Moral Judgments of Swedish Foreign Policy 1945–1952] (Stockholm: Santérus Förlag, 2000).

Ottosson, S., *Svensk självbild under det kalla kriget. En studie av stats- och utrikesministrarnas bild av Sverige 1950–1989* [The Swedish Self Image During the Cold War. A Study of the Prime and Foreign Ministers' view of Sweden 1950–1989] (Stockholm: Utrikespolitiska institutet, 2003).

Papacosma, S. Victor and Rubin, Mark R. (eds.), *Europe's Neutral and Nonaligned States: Between* NATO *and the Warsaw Pact* (Wilmington, DE: SR Books, 1989).

Pelletier, George Eugene, "Ridgeway: Trying to Make Good on the Promises", in R. S.Jordan, *Generals in International Politics:...*, pp. 31–52.

Pempel, T. J., "Restructuring Social Coalitions: State, Society, and Regime" in Rolf Torstendahl (ed.), *State Theory and State History* (London: Sage Publications, 1992), pp. 118–148.

Perkins, Bradford, *The Cambridge History of American Foreign Relations*, Volume 1, *The Creation of a Republican Empire, 1776–1865* (Cambridge: Cambridge University Press, 1993).

Petersen, Nikolaj, "'We Now Know'. The Nordic Countries and the Cold War: An Assessment", in T. B Olesen (ed.), *The Cold War – and the Nordic Countries:...*, pp. 177–191.

Petersson, Bo, *The Soviet Union and Peacetime Neutrality in Europe. A Study of Soviet Political Language.* (Stockholm: Utrikespolitiska Institutet, 1990).

Petersson, B., *Med Moskvas ögon. Bedömningar av svensk utrikes-politik under Stalin och Chrustjov.* [In Moscow's Eyes. Judgements of Swedish Foreign Policy Under Stalin and Chrustjov.] (Smedjebacken: Bokförlaget Arena, 1994).

Petersson, Magnus, "Sverige och Väst. Det säkerhetspolitiska samarbetet mellan Sverige, Norge och Västmakterna 1949–1969" [Sweden and the West: The Cooperation Between Sweden, Norway, and the Western Powers 1949–1969] in L. Wedin and G. Åselius (eds.), *Mellan byråkrati och krigskonst...*, pp. 165–234.

Petersson, M., "'She would not *fight* unless attacked'. Några kommentarer rörande NATO:s syn på Sverige 1949–1965." ["She Would Not *Fight* Unless Attacked": Some Comments on NATO's Perception of Sweden's Position in the Years 1949–1965.] in *Kungl. krigsvetenskapsakademiens handlingar och tidskrift*, No. 3, 2002, pp. 73–88.

Petersson, M., "The theory and practice of the 'Nordic Balance' during the 1950s and 60s, with special emphasis on Swedish-Norwegian relations", in K. L. Kleve (ed.), *The Cold War, Military Power...*, pp. 25–34.

Petersson, M., *"Brödrafolkens väl". Svensk-norska säkerhetsrelationer 1949–1969.* ["The Welfare of the Brother Folk": Swedish-Norwegian Security Relations 1949–1969] (Stockholm: Santérus Förlag, 2003).

Petersson, M., "Svensk-sovjetiska säkerhetspolitiska relationer 1945–1960 – en översikt" [Swedish-Soviet Security Policy Relations 1945–1960—An Overview] in K. Zetterberg (ed.), *Att skåda Sovjetunionen i vitögat...*, pp. 17–60.

Pleshakov, Constantine and Zubok, Vladislav, *Inside the Kremlin's Cold War: From Stalin to Khrushchev* (Cambridge, MA: Harvard University Press, 1996).

Porter, Andrew N., *European Imperialism, 1860–1914* (Basingstoke: Macmillan, 1994).

Prados, John, *Presidents' Secret Wars: CIA and Pentagon Covert Operations from World War II Through the Persian Gulf* (Chicago: Ivan R. Dee, Inc., 1996).

Reppy, Judith, "The United States" in N. Ball and M. Leitenberg (eds.), *The Structure of the Defense Industry*, pp. 21–49.

Riste, Olav, "'Janus Septentrionalis'? The Two Faces of Nordic Non-Alignment" in J. Nevakivi (ed.), *Neutrality in History...*, pp. 313–324.

Roberts, Geoffrey, *Stalin's Wars: From World War to Cold War, 1939–1953* (New Haven: Yale University Press, 2006).

Ross, Steven T., *American War Plans 1945–1950* (London: Frank Cass, 1996).

Rothstein, Bo, *The Social Democratic State: The Swedish Model and the Bureaucratic Problem of Social Reforms* (Pittsburgh: University of Pittsburgh Press, 1996).

Rubinstein, Alvin Z., *Soviet Foreign Policy Since World War II: Imperial and Global* (Boston, MA: Little, Brown and Company, 1981).

Ruggie John Gerard, *Constructing the World Polity: Essays on International Institutionalization* (London: Routledge, 1998).

Ruin, Olof, *I välfärdsstatens tjänst. Tage Erlander 1946–1969* [In the Service of the Welfare State: Tage Erlander, 1946–1969] (Stockholm: Tidens Förlag, 1986).

Sánchez, Peter M., *Panama Lost? U.S. Hegemony, Democracy, and the Canal* (Gainesville, FL: University Press of Florida, 2007).

Schlesinger, Jr., Arthur M., *The Imperial Presidency* (Boston, MA: Houghton Mifflin Company, 1973).

Schlesinger, Thomas O., *The United States and the European Neutrals* (Wien: Wilhelm Braumüller Universitäts-Verlagsbuchhandlung, 1991).

Schmidt, Gustav (ed.), *A History of NATO—The First Fifty Years*, Volume III (Hampshire: Palgrave, 2001).

Schollars, Todd J., "German wonder weapons: degraded production and effectiveness" in *Air Force Journal of Logistics*, Fall, 2003).

Scott, J. C., *Domination and the Arts of Resistance: Hidden Transcripts* (London: Yale University Press, 1990).

Service, Robert, *Stalin: A Biography* (London: Pan Books, 2005).

Sevón, Cay, *Visionen om Europa. Svensk neutralitet och europeisk återuppbyggnad 1945–1948* [The Vision of Europe: Swedish Neutrality and European Reconstruction, 1945–1948] (Helsinki: Suomen Historiallinen Seura, 1995).

Silva, Charles, "Att omvärdera svensk utrikespolitik 1945–1950" [To Reevaluate Swedish Foreign Policy 1945–1950] in T. Jonter and C. Silva (eds.), *Sverige inför en ny världsordning 1945–50...*, pp. 9–28.

Silva, C., *Keep Them Strong, Keep Them Friendly: Swedish-American Relations and the Pax Americana, 1948–1952*, (Stockholm: Ph.D. diss., University of Stockholm, 1999).

Skodvin, Magne, *Norden eller NATO? Utenriksdepartementet og alliansespørsmålet 1947–1949* [The Nordic Countries or NATO? The [Norwegian] Foreign Ministry and the Alliance Question, 1947–1949] (Oslo: Universitetsforlaget, 1971).

Skodvin, M., "Nordic or North Atlantic Alliance?: The Postwar Scandinavian Security Debate" in *Forsvarsstudier*, No. 3, 1990 (Oslo: Institutt for forsvarsstudier, 1990).

Skogmar, Gunnar, *De nya malmfälten. Det svenska uranet och inledningen till efterkrigstidens neutralitetspolitik* [The New Ore Fields: The Swedish Uranium and the Beginning of Sweden's Postwar Policy of Neutrality] (Stockholm: Forskningsprogrammet *Sverige under det kalla kriget Arbetsrapport nr 3*, 1997).

Skogrand, Kjetil and Tamnes, Rolf, *Fryktens likevekt. Atombomben, Norge og verden, 1945–1970* [The Balance of Fear: The Atomic Bomb, Norway and the World, 1945–1970] (Oslo: Tiden Norsk Forlag, 2001).

Skottsberg Åhman, Brita, "Scandinavian Foreign Policy, Past and Present" in H. Friis (ed.), *Scandinavia: Between East and West*, pp. 255–305.

Smith, Dan, *Pressure: How America Runs* NATO (London: Bloomsbury, 1989).

Sorley, Lewis, "Goodpaster: Maintaining Deterrence during Détente", in R. S. Jordan (ed.), *Generals in International Politics:...*, pp. 122–150.

Steen-Johansson, Cecilia, *Ett Folkbedrägeri.* DC *3:an och svensk säkerhetspolitik. Sverige och* NATO *1952–92.* [Public Deception. The DC 3 and Swedish Security Policy: Sweden and NATO 1952–92] (Stockholm: Carlssons Bokförlag, 1992).

Stivers, William, "The Incomplete Blockade: Soviet Zone Supply of West Berlin, 1948–49" in *Diplomatic History*, Vol. 21, No. 4, Fall 1997, pp. 569–602.

Stridsman, Jacob, "Erik Boheman – västvänlig svensk diplomat under det kalla kriget" [Erik Boheman—Western-Friendly Swedish Diplomat During the Cold War] in K. Zetterberg (ed.), *Hett och kallt...*, pp. 99–141.

Stråth, Bo, "The Swedish Image of Europe as the Other" in Bo Stråth (ed.), *Europe and the Other and Europe as the Other* (Brussels: P.I.E.-Peter Lang, 2000), pp. 359–383.

Stråth, B. (ed.), *Myth and Memory in the Construction of Community: Historical Patterns in Europe and Beyond* (Brussels: P.I.E.-Peter Lang, 2000).

Stråth, B., "Poverty, Neutrality and Welfare: Three Key Concepts in the Modern Foundation Myth of Sweden" in B. Stråth (ed.), *Myth and Memory…*, pp. 375–401.

Sundelius, Bengt, "National Security Dilemmas and Strategies" in M. H. Haltzel and J. Kruzel (eds.), *Between the Blocs:…*, pp. 108–110.

Sundelius, B., "Coping with Structural Security Threats" in O. Höll (ed.), *Small States in Europe and Dependence*, pp. 281–305.

Sundelius, B. (ed.), *The Committed Neutral: Sweden's Foreign Policy* (Boulder: Westview Press, 1989).

Suominen, Tapani (ed.), *Sverige i fred. Statsmannakonst eller opportunism? En antologi om 1812 års politik.* [Sweden During Peacetime: Statemanship Or Opportunism? An Anthology of the Policy of 1812] (Stockholm: Atlantis, 2002).

Talvitie, Heikki, "Sverige och 1812 års politik" [Sweden and the Policy of 1812] in T. Suominen (ed.), *Sverige i fred…*, pp. 15–74.

Tamnes, Rolf, *The United States and the Cold War in the High North* (Oslo: Ad Notam forlag AS, 1991).

Taylor, John W. R. and Taylor, Michael J. H., *Guided missiles of the World* (London: Ian Allan, 1972).

Theoharis, Athan G., *The Truman Presidency: The Origins of the Imperial Presidency and the National Security State* (Stanfordville, NY: Earl M. Coleman Enterprises, Inc., Publishers, 1979).

Therborn, Göran, *What Does the Ruling Class Do When It Rules?* (London: Verso, 1980).

Therborn, G., *The Ideology of Power and the Power of Ideology* (London: Verso, 1980).

Thorsell, Staffan, *Sverige i Vita Huset* [Sweden In the White House] (Stockholm: Bonnier Fakta, 2003).

Tunander, Ola, "The Uneasy Imbrication of Nation-State and NATO: The Case of Sweden" in *Cooperation and Conflict*, Vol. 34, No. 2, 1999, pp. 169–203.

Ulam, Adam B., *The Rivals: America and Russia Since World War II* (New York: The Viking Press, 1971).

Ulam, A. B., *Stalin: The Man and His Era* (New York: The Viking Press, 1973).

Ulam, A., B., *Expansion and Coexistence: Soviet Foreign Policy, 1917–73* (Second Edition) (New York: Holt, Reinhart and Winston, Inc., 1974).

Ulving, Lars, "Svensk trygghet. Den svenska militärledningens överväganden kring kärnvapenanskaffning 1945–1961." [Swedish Security: The Swedish Military's Considerations Regarding the Procurement of Nuclear Weapons, 1945–1961], in K. Zetterberg (ed.), *Totalförsvar och atomvapen*, pp. 159–217.

Vukadinovic, Radovan, "The Various Concepts of European Neutrality" in M. H. Haltzel and J. Kruzel (eds.), *Between the Blocs:…*, pp. 29–46.

Watson, Derek, *Molotov: A Biography* (Hampshire: Palgrave MacMillan, 2005).

Wahlbäck, Krister, "Norden och blockuppdelningen 1948–49" [The Nordic Countries and the Division into Blocs, 1948–1949] in *Internationella*

Studier, 1973:B (Stockholm: The Swedish Institute of International affairs, 1973).

Wahlbäck, K., "Sveriges naturliga ovilja att ta risker" [Sweden's Natural Inclination Not To Take Risks] in T. Suominen (ed.), *Sverige i fred...*, pp. 75–86.

Wedin, Lars and Åselius, Gunnar (eds.), *Mellan byråkrati och krigs-konst. Svenska strategier för det kalla kriget.* [Between Bureaucracy and the Art of War: Swedish Strategies for the Cold War] (Stockholm: Försvarshögskolan, 1999).

Weinberger, Hans, *Sievert: enhet och mångfald. En biografi över den svenska radiofysikens, radiobiologins och strålskyddets grundare Rolf Sievert* [Sievert: Unit and Disunity. A Biography of Rolf Sievert, the Founder of Swedish Radio Physics, Radio Biology, and Radiation Protection] (Stockholm: Stockholm Papers in History and Philosophy of Technology, TRITA-HOT 2021, 1990).

Weinberger, H., "'På sidan om de stora kraftlinjerna'? Teknik, vetenskap och det kalla kriget" [To the Side of the Great Power Lines? Technology, Science, and the Cold War] in S. Widmalm (ed.) *Vetenskapsbärarna...*, pp. 352–368.

Weinberger, H., "'Det rörde sig väl om ofarliga ting' Svensk neutralitetspolitik och synen på teknik" [It Probably Wasn't Anything Dangerous: Swedish Neutrality Policy and the View of Technology] in *Arbetarhistoria. Meddelande från Arbetarrörelsens Arkiv och Bibliotek*, No. 92, 4/1999, pp. 31–36.

Weinberger, H., "The Neutrality Flagpole: Swedish Neutrality Policy and Technological Alliances, 1945–1970" in M. T. Allen and G. Hecht (eds): *Technologies of Power:...*, pp. 295–331.

Weinberger, H., *Technology : and the de/construction of Swedish neutrality during the cold war* (Stockholm: Department of History of Science and Technology, Royal Institute of Technology TRITA-HST, 2002/2).

Weinberger H., "A Small/Medium-sized Great Power in the North: Swedish Neutrality During the Cold War Through the Geostrategic Lens" in K. L. Kleve (ed.), *The Cold War, Military Power...*, pp. 45–64.

Wennerholm, Bertil, *Fjärde flygvapnet i världen? Doktrinutveckling i det svenska flygvapnet i försvarsbesluten 1942–1958. Underlag, beslut och genomförande i nationellt och internationellt perspektiv.* [The Fourth Air Force In the World? Doctrine Development in the Swedish Air Force in the Long Term Parliamentary Defense Acts of 1942–1958. Foundations, Resolutions, and Implementation in a National and International Perspective] (Stockholm: Försvarshögskolan, 2006).

Westad, Odd Arne (ed.), *Reviewing the Cold War: Approaches, Interpretations, Theory* (London: Frank Cass, 2000).

Westrum, Ron, *Sidewinder: Creative Guided Missile Development at China Lake* (Annapolis, MD: Naval Institute Press, 1999).

Widfeldt, Bo, *Saab 35 Draken. En historisk återblick på projektering, utveckling, tillverkning och tjänsten inom de svenska, finska, danska och österrikiska flygvapnen* [Saab 35 Draken (Dragon): A Historical View on Projecting, Development, Manufacturing, and Service within the Swedish, Finnish, Danish, and

Austrian Air Forces] (Nässjö: Air Historic Research, 1995).

Widmalm, Sven (ed.), *Vetenskapsbärarna. Naturvetenskapen i det svenska samhället, 1889–1950.* [The Science Bearers: Natural Science in Swedish Society, 1889–1950] (Hedemora: Gidlunds Förlag, 1999).

Williams, Raymond, "Gramsci's relevance for the study of race and ethnicity", *Journal of Communication Inquiry, 10* (1986).

Wormbs, Nina, *Vem älskade Tele-x? Konflikter om satelliter i Norden 1974–1989* [Who Loved Tele-x? Conflicts Over Satellites in Norden, 1974–1989] (Stockholm: Gidlunds Förlag, 2003).

Yergin, Daniel, *Shattered Peace: The Origins of the Cold War and the National Security State* (Boston: Houghton Mifflin Company, 1977).

Zakheim, Dov. "The United States and the Nordic Countries During the Cold War" in *Cooperation and Conflict,* Vol. 33, No. 2, 1998, pp. 115–129.

Zetterberg, Kent, "Nils Swedlund" in G. Åselius (ed.), *Svenska officersprofiler under 1900-talet*, pp. 476-507.

Zetterberg, K., "Skandinavisk samverkan? 'N-D-S-1945–1949.' En studie i militärpolitiskt samarbete, strategisk analys och den säkerhetspolitiska prövningen av ett skandinaviskt försvarsförbund mellan krigsslutet och Atlantpaktens bildande." [Scandinavian Cooperation? "N-D-S-1945–1949." A Study of Political-Military Cooperation, Strategic Analysis, and the Security Political Testing of a Scandinavian Defense Union between the End of the War and the Formation of the Atlantic Pact] in K. Zetterberg (ed.), *Hotet från öster...*, pp. 17–74.

Zetterberg, K. (ed.), *Hotet från öster. Svensk säkerhetspolitik, krigsplanering och strategi, 1945–1958* [The Threat from the East: Swedish Security Policy, War Planning, and Strategy, 1945–1958] (Stockholm: Försvarshögskolan, 1997).

Zetterberg, K. (ed.), *Hett och kallt – Tre studier i svensk säkerhetspolitik och strategi gentemot västmakterna i början av det kalla kriget. Hot and Cold—Three Studies in Swedish Security Politics and Strategy towards the Western Powers in the beginning of the Cold War* (Stockholm: Försvarshögskolan, 1999).

Zetterberg, K. (ed.), *Totalförsvar och atomvapen* [Total Defense and Atomic Weapons], (Stockholm: Försvarshögskolan, 2001).

Zetterberg, K. (ed.), *Att skåda Sovjetunionen i vitögat. Sex studier kring svenska relationer till Sovjetunionen under det kalla kriget* [To Look the Soviet Union Straight in the Eye: Six Studies of Swedish Relations to the Soviet Union During the Cold War] (Stockholm: Forskningsprojektet Försvaret och det kalla kriget (FOKK), Publikation nr. 2, 2004).

Åselius, Gunnar (ed.), *Svenska officersprofiler under 1900-talet* [Swedish Officer Profiles During the 20th Century] (Stockholm: Militärhögskolan & Axel Abrahamsons Förlag, 1996).

Memoirs, Diaries, etc:

Acheson, Dean, *Present at the Creation: My Years at the State Department* (London: Hamish Hamilton, 1969).

Bohlen, Charles E., *Witness to History 1929–1969* (London: Weidenfeld and Nicolson, 1973).

Byrnes, James F., *Speaking Frankly* (New York: Harper & Brothers Publishers, 1947).

Sven Grafström. Anteckningar 1945–1954 [Sven Grafström: Notes, 1945–1954] Published by Stig Ekman (Stockholm: Kungl. Samfundet för utgivande av handskrifter rörande Skandinaviens historia, 1989).

Tage Erlanders dagböcker, 1945–1949 [The Diaries of Tage Erlander, 1945–1949], Published by Sven Erlander (ed.) (Hedemora: Gidlunds Förlag, 2001).

Tage Erlanders dagböcker, 1950–1951 [The Diaries of Tage Erlander, 1950–1951], Published by Sven Erlander (ed.) (Hedemora: Gidlunds Förlag, 2001).

Tage Erlanders dagböcker, 1952 [The Diaries of Tage Erlander, 1952], published by Sven Erlander (Hedemora: Gidluns Förlag, 2002).

Tage Erlanders dagböcker, 1953 [The Diaries of Tage Erlander, 1953] Published by Sven Erlander (ed.) (Hedemora: Gidlunds Förlag, 2001).

Truman, Harry S., *Memoirs*, Volume I, *Years of Decisions* (Garden City, NY: Double Day & Company, Inc., 1955).

Truman, H. S., *Memoirs*, Volume II, *Years of Trial and Hope* (Garden City, NY: Doubleday & Company, Inc., 1956).

Östen Undén. Anteckningar 1918–1952 [Östen Undén: Notes, 1918–1952], Published by Karl Molin (Stockholm: Kungl. Samfundet för utgivande av handskrifter rörande Skandinaviens historia, 2002).

Östen Undén. Anteckningar 1952–1966 [Östen Undén: Notes, 1952–1966], Published by Karl Molin (Stockholm: Kungl. Samfundet för utgivande av handskrifter rörande Skandinaviens historia, Handlingar del 25, 2002).

Index

E

F

G

J

K

L

N

Q

R

S

V

W

Y

Å

Ö